PHYSICS® | ACTIVE CHEMISTRY™ | ACTIVE BIOLOGY™ | EARTHCOMM®

**UNIT I ACTIVE PHYSICS ■ TEACHER'S EDITION**

# Coordinated Science for the 21st Century™

*An Integrated, Project-Based Approach*


*Arthur Eisenkraft, Ph.D.*
*Ruta Demery*
*Gary Freebury*
*Robert Ritter, Ph.D.*
*Michael Smith, Ph.D.*
*John B. Southard, Ph.D.*

84 Business Park Drive, Armonk, NY 10504 Phone (914) 273-2233
Fax (914) 273-2227 Toll Free (888) 698-TIME (8463) www.its-about-time.com

**It's About Time President**
Tom Laster

**Director of Product Development**
Barbara Zahm, Ph.D.

**Creative/Art Director**
John Nordland

**Design/Production**
Kathleen Bowen
Burmar Technical Corporation
Nancy Delmerico
Kadi Sarv
Jon Voss

**Illustrations**
Tomas Bunk
Dennis Falcon

**Project Editor**
Ruta Demery
*EarthComm, Active Physics, Active Chemistry, Active Biology, Coordinated Science for the **21st Century***

**Project Managers**
Ruta Demery
*EarthComm, Active Physics*
Barbara Zahm
*Active Physics Active Chemistry, Active Biology, Coordinated Science for the **21st Century***

**Project Coordinators**
Loretta Steeves
*Coordinated Science for the **21st Century***
Emily Crum
Matthew Smith
*EarthComm*

**Technical Art**
Stuart Armstrong
*EarthComm*
Burmar Technical Corporation
Kadi Sarv
*Active Physics, Active Chemistry Active Biology*

**Photo Research**
Caitlin Callahan
Kathleen Bowen
Jon Voss
Kadi Sarv
Jennifer Von Holstein

**Safety Reviewers**
Ed Robeck, Ph.D.
*EarthComm, Active Biology*
Gregory Puskar
*Active Physics*
Jack Breazale
*Active Chemistry*



Printed and bound in the United States of America

ISBN #1-58591-355-3 5 Volume Set ISBN #1-58591-354-5

2 3 4 5 VH 09 08 07 06

This project was supported, in part, by the
National Science Foundation
Opinions expressed are those of the authors and not necessarily those of the National Science Foundation or the donors of the American Geological Institute Foundation.

# UNIT 1 ACTIVE PHYSICS ▪ TEACHER'S EDITION

# Coordinated Science for the 21st Century™

**An Integrated, Project-Based Approach**

*Coordinated Science for the* ***21st Century*** is an innovative core curricula assembled from four proven inquiry-based programs. It is supported by the National Science Foundation and was developed by leading educators and scientists. Unit 1, *Active Physics,* was developed by the American Association of Physics Teachers and the American Institute of Physics.

Both *Active Physics* and *Active Chemistry,* Units 1 and 2, are projects directed by Arthur Eisenkraft, Ph.D., past president of the NSTA.

*Active Biology,* Unit 3, was developed to follow the same Active Learning Instructional Model as *Active Physics* and *Active Chemistry.*

*EarthComm,* Unit 4, was developed by the American Geological Institute, under the guidance of Michael Smith, Ph.D., former Director of Education and Outreach, and John Southard, Ph.D., of MIT.

Each unit of this course has been designed and built on the National Science Education Standards. Each utilizes the same instructional model and the same inquiry-based approach.

## Project Director, Active Physics and Active Chemistry

**Arthur Eisenkraft** has taught high school physics for over 28 years and is currently the Distinguished Professor of Science Education and a Senior Research Fellow at the University of Massachusetts, Boston. Dr. Eisenkraft is the author of numerous science and educational publications. He holds U.S. Patent #4447141 for a Laser Vision Testing System (which tests visual acuity for spatial frequency).

Dr. Eisenkraft has been recognized with numerous awards including: Presidential Award for Excellence in Science Teaching, 1986 from President Reagan; American Association of Physics Teachers (AAPT) Excellence in Pre-College Teaching Award, 1999; AAPT Distinguished Service Citation for "excellent contributions to the teaching of physics", 1989; Science Teacher of the Year, Disney American Teacher Awards in their American Teacher Awards program, 1991; Honorary Doctor of Science degree from Rensselaer Polytechnic Institute, 1993; Tandy Technology Scholar Award 2000.

In 1999 Dr. Eisenkraft was elected to a 3-year cycle as the President-Elect, President and Retiring President of the National Science Teachers Association (NSTA), the largest science teacher organization in the world. In 2003, he was elected a fellow of the American Association for the Advancement of Science (AAAS).

Dr. Eisenkraft has been involved with a number of projects and chaired many competition programs, including: the Toshiba/NSTA ExploraVisions Awards (1991 to the present); the Toyota TAPESTRY Grants (1990 to the present); the Duracell/NSTA Scholarship Competitions (1984 to 2000). He was a columnist and on the Advisory Board of *Quantum* (a science and math student magazine that was published by NSTA as a joint venture between the United States and Russia; 1989 to 2001). In 1993, he served as Executive Director for the XXIV International Physics Olympiad after being Academic Director for the United States Team for six years. He has served on a number of committees of the National Academy of Sciences including the content committee that helped write the National Science Education Standards.

Dr. Eisenkraft has appeared on *The Today Show*, *National Public Radio*, *Public Television*, *The Disney Channel* and numerous radio shows. He serves as an advisor to the ESPN Sports Figures Video Productions.

He is a frequent presenter and keynote speaker at National Conventions. He has published over 100 articles and presented over 200 papers and workshops. He has been featured in articles in *The New York Times*, *Education Week*, *Physics Today*, *Scientific American*, *The American Journal of Physics* and *The Physics Teacher*.

## Content Specialist, Active Chemistry

**Gary Freebury**, a noted chemistry teacher, educator, and writer worked as the Project Manager and Editor for the 3 Prototype chapters of *Active Chemistry* which are currently being field tested but are also in print, and he will continue to serve as the Project Manager and Editor to the project. He will be responsible for the writing of any introductory materials, producing the table of contents, indices, glossary, and reference materials. He will have a critical role in maintaining the integrity of safety standards across all units. He will also coordinate any modifications, changes, and additions to the materials based upon pilot and field-testing results.

Mr. Freebury has been teaching chemistry for more than 35 years. He has been the Safety Advisor for Montana Schools, past director of the Chemistry Olympiad, past chairman of the Montana Section of the American Chemical Society (ACS), member of the Executive Committee of the Montana Section of the ACS, and a past member of the Montana Science Advisory Council. Mr. Freebury has been the regional director and author of Scope, Sequence and Coordination (SS&C) – Integrated Science Curriculum and Co-director of the National Science Foundation supported Chemistry Concepts four-year program. He earned a B.S. degree at Eastern Montana College in mathematics and physical science, and an M.S. degree in chemistry at the University of Northern Iowa.

## Principal Investigator, EarthComm

**Michael Smith, Ph.D.,** is a former Director of Education at the American Geological Institute in Alexandria, Virginia. Dr. Smith worked as an exploration geologist and hydrogeologist. He began his Earth Science teaching career with Shady Side Academy in Pittsburgh, PA in 1988 and most recently taught Earth Science at the Charter School of Wilmington, DE. He earned a doctorate from the University of Pittsburgh's Cognitive Studies in Education Program and joined the faculty of the University of Delaware School of Education in 1995. Dr. Smith received the Outstanding Earth Science Teacher Award for Pennsylvania from the National Association of Geoscience Teachers in 1991, served as Secretary of the National Earth Science Teachers Association, and is a reviewer for Science Education and The Journal of Research in Science Teaching. He worked on the Delaware Teacher Standards, Delaware Science Assessment, National Board of Teacher Certification, and AAAS Project 2061 Curriculum Evaluation programs.

## Senior Writer, EarthComm

**John B. Southard, Ph.D.,** received his undergraduate degree from the Massachusetts Institute of Technology in 1960 and his doctorate in geology from Harvard University in 1966. After a National Science Foundation postdoctoral fellowship at the California Institute of Technology, he joined the faculty at the Massachusetts Institute of Technology, where he is currently Professor of Geology Emeritus. He was awarded the MIT School of Science teaching prize in 1989 and was one of the first cohorts of the MacVicar Fellows at MIT, in recognition of excellence in undergraduate teaching. He has taught numerous undergraduate courses in introductory geology, sedimentary geology, field geology, and environmental Earth Science both at MIT and in Harvard's adult education program. He was editor of the Journal of Sedimentary Petrology from 1992 to 1996, and he continues to do technical editing of scientific books and papers for SEPM, a professional society for sedimentary geology. Dr. Southard received the 2001 Neil Miner Award from the National Association of Geoscience Teachers.

## Primary Author, Active Biology Project Editor, EarthComm, Active Physics, Active Chemistry and Active Biology

**Ruta Demery** has helped bring to publication several National Science Foundation (NSF) projects. She was the project editor for *EarthComm*, *Active Physics*, *Active Chemistry*, and *Active Biology*. She was also a contributing writer for *Active Physics* and *Active Biology*, both students' and teachers' editions. Besides participating in the development and publishing of numerous innovative mathematics and science books for over 30 years, she has worked as a classroom science and mathematics teacher in both middle school and high school. She brings to her work a strong background in curriculum development and a keen interest in student assessment. When time permits, she also leads workshops to familiarize teachers with inquiry-based methods.

## Contributing Author, Active Biology and Active Physics

**Bob Ritter** is presently the principal of Holy Trinity High School in Edmonton, Alberta. Dr. Ritter began his teaching career in 1973, and since then he has had a variety of teaching assignments. He has worked as a classroom teacher, Science Consultant, and Department Head. He has also taught Biological Science to student teachers at the University of Alberta. He is presently involved with steering committees for "At Risk High School Students" and "High School Science." Dr. Ritter is frequently a presenter and speaker at national and regional conventions across Canada and the United States. He has initiated many creative projects, including establishing a science-mentor program in which students would have an opportunity to work with professional biologists. In 1993 Dr. Ritter received the Prime Minister's Award for Science and Technology Teaching. He has also been honored as Teacher of the Year and with an Award of Merit for contribution to science education.

# Acknowledgements

**Coordinated Science for the 21st Century** was developed by teams of leading science educators, university educators and classroom teachers with financial support from the National Science Foundation.

### NSF Program Officer

Gerhard Salinger
Instructional Materials Development (IMD)

## UNIT 1: ACTIVE PHYSICS

### Principal Investigators

Bernard V. Khoury
American Association of Physics Teachers

Dwight Edward Neuenschwander
American Institute of Physics

### Project Director

Arthur Eisenkraft
University of Massachusetts

### Primary and Contributing Authors

Richard Berg
University of Maryland
College Park, MD

Howard Brody
University of Pennsylvania
Philadelphia, PA

Chris Chiaverina
New Trier Township High School
Crystal Lake, IL

Ron DeFronzo
Eastbay Ed. Collaborative
Attleboro, MA

Ruta Demery
Blue Ink Editing
Stayner, ONCarl Duzen
Lower Merion High School
Havertown, PA

Jon L. Harkness
*Active Physics* Regional Coordinator
Wausau, WI

Ruth Howes
Ball State University
Muncie, IN

Douglas A. Johnson
Madison West High School
Madison, WI

Ernest Kuehl
Lawrence High School
Cedarhurst, NY

Robert L. Lehrman
Bayside, NY

Salvatore Levy
Roslyn High School
Roslyn, NY

Tom Liao
SUNY Stony Brook
Stony Brook, NY

Charles Payne
Ball State University
Muncie, IN

Mary Quinlan
Radnor High School
Radnor, PA

Harry Rheam
Eastern Senior High School
Atco, NJ

Bob Ritter
University of Alberta
Edmonton, AB

John Roeder
The Calhoun School
New York, NY

John J. Rusch
University of Wisconsin Superior
Superior, WI

Patty Rourke
Potomac School
McLean, VA

Ceanne Tzimopoulos
Omega Publishing
Medford, MA

Larry Weathers
The Bromfield School
Harvard, MA

David Wright
Tidewater Comm. College
Virginia Beach, VA

### Consultants

Peter Brancazio
Brooklyn College of CUNY
Brooklyn, NY

Robert Capen
Canyon del Oro High School
Tucson, AZ

Carole Escobar

Earl Graf
SUNY Stony Brook
Stony Brook, NY

Jack Hehn
American Association of Physics Teachers
College Park, MDDonald F. Kirwan
Louisiana State University
Baton Rouge, LA

Gayle Kirwan
Louisiana State University
Baton Rouge, LA

James La Porte
Virginia Tech
Blacksburg, VA

Charles Misner
University of Maryland
College Park, MD

Robert F. Neff
Suffern, NY

Ingrid Novodvorsky
Mountain View High School
Tucson, AZ

John Robson
University of Arizona
Tucson, AZ

Mark Sanders
Virginia Tech
Blacksburg, VA

Brian Schwartz
Brooklyn College of CUNY
New York, NY

Bruce Seiger
Wellesley High School
Newburyport, MA

Clifford Swartz
SUNY Stony Brook
Setauket, NY

Barbara Tinker
The Concord Consortium
Concord, MA

Robert E. Tinker
The Concord Consortium
Concord, MA

Joyce Weiskopf
Herndon, VA

Donna Willis
American Association of Physics Teachers
College Park, MD

### Safety Reviewer

Gregory Puskar
University of West Virginia
Morgantown, WV

### Equity Reviewer

Leo Edwards
Fayetteville State University
Fayetteville, NC

### Physics at Work

Alex Straus, writer
New York, NY
Mekea Hurwitz
photographer

### Physics InfoMall

Brian Adrian
Bethany College
Lindsborg, KS

### First Printing Reviewer

John L. Hubisz
North Carolina State University
Raleigh, NC

### Unit Reviewers

Robert Adams
Polytech High School
Woodside, DE

George A. Amann
F.D. Roosevelt High School
Rhinebeck, NY

Patrick Callahan
Catasaugua High School
Center Valley, PA

Beverly Cannon
Science and Engineering Magnet High School
Dallas, TX

Barbara Chauvin

Elizabeth Chesick
The Baldwin School
Haverford, PA

Chris Chiaverina
New Trier Township High School
Crystal Lake, IL

Andria Erzberger
Palo Alto Senior High School
Los Altos Hills, CA

Elizabeth Farrell Ramseyer
Niles West High School
Skokie, IL

Mary Gromko
President of Council of State Science Supervisors
Denver, CO

Thomas Guetzloff

Jon L. Harkness
*Active Physics* Regional Coordinator
Wausau, WI

Dawn Harman
Moon Valley High School
Phoenix, AZ

James Hill
Piner High School
Sonoma, CA

Bob Kearney

Claudia Khourey-Bowers
McKinley Senior High School

Steve Kliewer
Bullard High School
Fresno, CA

Ernest Kuehl
Roslyn High School
Cedarhurst, NY

Jane Nelson
University High School
Orlando, FL

Mary Quinlan
Radnor High School
Radnor, PA

John Roeder
The Calhoun School
New York, NY

Patty Rourke
Potomac School
McLean, VA

Gerhard Salinger
Fairfax, VA

Irene Slater
La Pietra School for Girls

### Pilot Test Teachers

John Agosta

Donald Campbell
Portage Central High School
Portage, MI

John Carlson
Norwalk Community Technical College
Norwalk, CT

Veanna Crawford
Alamo Heights High School
New Braunfels, TX

Janie Edmonds
West Milford High School
Randolph, NJ

Eddie Edwards
Amarillo Area Center for Advanced Learning
Amarillo, TX

Arthur Eisenkraft
University of Massachusetts

Tom Ford

Bill Franklin

Roger Goerke
St. Paul, MN

Tom Gordon
Greenwich High School
Greenwich, CT

Ariel Hepp

John Herrman
College of Steubenville
Steubenville, OH

Linda Hodges

Ernest Kuehl
Lawrence High School
Cedarhurst, NY

Fran Leary
Troy High School
Schenectady, NY

Harold Lefcourt

Cherie Lehman
West Lafayette High School
West Lafayette, IN

Kathy Malone
Shady Side Academy
Pittsburgh, PA

Bill Metzler
Westlake High School
Thornwood, NY

Elizabeth Farrell Ramseyer
Niles West High School
Skokie, IL

Daniel Repogle
Central Noble High School, Albion, IN

Evelyn Restivo
Maypearl High School
Maypearl, TX

Doug Rich
Fox Lane High School
Bedford, NY

John Roeder
The Calhoun School
New York, NY

Tom Senior
New Trier Township High School
Highland Park, IL

John Thayer
District of Columbia Public Schools
Silver Spring, MD

Carol-Ann Tripp
Providence Country Day
East Providence, RI

Yvette Van Hise
High Tech High School
Freehold, NJ

Jan Waarvick

Sandra Walton
Dubuque Senior High School, Dubuque, IA

Larry Wood
Fox Lane High School
Bedford, NY

## Field Test Coordinator

Marilyn Decker
Northeastern University
Acton, MA

## Field Test Workshop Staff

John Carlson

Marilyn Decker

Arthur Eisenkraft

Douglas Johnson

John Koser

Ernest Kuehl

Mary Quinlan

Elizabeth Farrell Ramseyer

John Roeder

## Field Test Evaluators

Susan Baker-Cohen

Susan Cloutier

George Hein

Judith Kelley

all from Lesley College, Cambridge, MA

## Field Test Teachers and Schools

Rob Adams
Polytech High School
Woodside, DE

Benjamin Allen
Falls Church High School,
Falls Church, VA

Robert Applebaum
New Trier High School
Winnetka, IL

Joe Arnett
Plano Sr. High School
Plano, TX

Bix Baker
GFW High School
Winthrop, MN

Debra Beightol
Fremont High School
Fremont, NE

Patrick Callahan
Catasaugua High School
Catasaugua, PA

George Coker
Bowling Green
High School
Bowling Green, KY

Janice Costabile
South Brunswick
High School
Monmouth Junction, NJ

Stanley Crum
Homestead High School
Fort Wayne, IN

Russel Davison
Brandon High School
Brandon, FL

Christine K. Deyo
Rochester Adams
High School
Rochester Hills, MI

Jim Doller
Fox Lane High School
Bedford, NY

Jessica Downing
Esparto High School
Esparto, CA

Douglas Fackelman
Brighton High School
Brighton, CO

Rick Forrest
Rochester High School
Rochester Hills, MI

Mark Freeman
Blacksburg High School
Blacksburg, VA

Jonathan Gillis
Enloe High School
Raleigh, NC

Karen Gruner
Holton Arms School
Bethesda, MD

Larry Harrison
DuPont Manual High
School, Louisville, KY

Alan Haught
Weaver High School
Hartford, CT

Steven Iona
Horizon High School
Thornton, CO

Phil Jowell
Oak Ridge High School
Conroe, TX

Deborah Knight
Windsor Forest High
School, Savannah, GA

Thomas Kobilarcik
Marist High School
Chicago, IL

Sheila Kolb
Plano Senior High School,
Plano, TX

Todd Lindsay
Park Hill High School
Kansas City, MO

Malinda Mann
South Putnam
High School
Greencastle, IN

Steve Martin
Maricopa High School
Maricopa, AZ

Nancy McGrory
North Quincy High
School, N. Quincy, MA

David Morton
Mountain Valley High
School, Rumford, ME

Charles Muller
Highland Park
High School
Highland Park, NJ

Fred Muller
Mercy High School
Burlingame, CA

Vivian O'Brien
Plymouth Regional
High School
Plymouth, NH

Robin Parkinson
Northridge High School
Layton, UT

Donald Perry
Newport High School
Bellevue, WA

Francis Poodry
Lincoln High School
Philadelphia, PA

John Potts
Custer County District
High School
Miles City, MT

Doug Rich
Fox Lane High School
Bedford, NY

John Roeder
The Calhoun School
New York, NY

Consuelo Rogers
Maryknoll Schools
Honolulu, HI

Lee Rossmaessler
Mott Middle College
High School
Flint, MI

John Rowe
Hughes Alternative
Center, Cincinnati, OH

Rebecca
Bonner Sanders
South Brunswick
High School
Monmouth Junction, NJ

David Schlipp
Narbonne High School
Harbor City, CA

Eric Shackelford
Notre Dame
High School
Sherman Oaks, CA

Robert Sorensen
Springville-Griffith
Institute
and Central School
Springville, NY

Teresa Stalions
Crittenden County High
School, Marion, KY

Roberta Tanner
Loveland High School
Loveland, CO

Anthony Umelo
Anacostia
Sr. High School
Washington, D.C.

Judy Vondruska
Mitchell High School
Mitchell, SD

Deborah Waldron
Yorktown High School
Arlington, VA

Ken Wester
The Mississippi
School for
Mathematics
and Science
Columbus, MS

Susan Willis
Conroe High School
Conroe, TX

# Your students can do science. Here are the reasons why. . .

When science learning is based on functional use and active involvement, students get turned on to science. The Active Learning Instructional Model outlined below gives **all** students the opportunity to succeed. Whether it's physics, chemistry, biology, or Earth Science, your students learn science, as they see how science works for them every day and everywhere.

Look for these features in each chapter of *Coordinated Science for the* ***21st Century***.

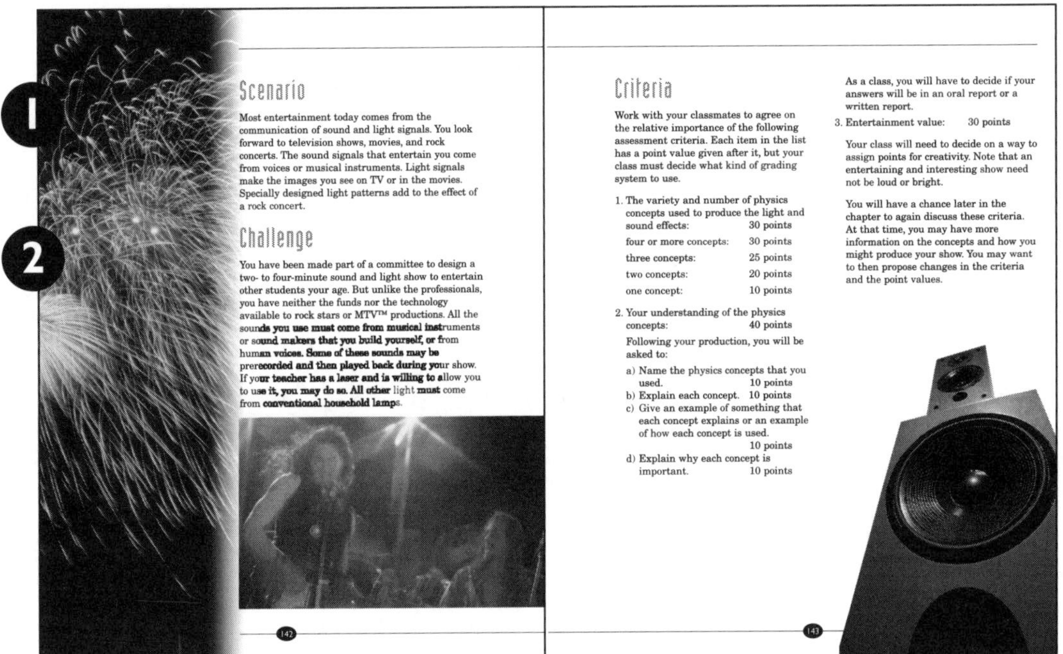

1

2

## Scenario

Most entertainment today comes from the communication of sound and light signals. You look forward to television shows, movies, and rock concerts. The sound signals that entertain you come from voices or musical instruments. Light signals make the images you see on TV or in the movies. Specially designed light patterns add to the effect of a rock concert.

## Challenge

You have been made part of a committee to design a two- to four-minute sound and light show to entertain other students your age. But unlike the professionals, you have neither the funds nor the technology available to rock stars or MTV™ productions. All the sounds you use must come from musical instruments or sound makers that you build yourself, or from human voices. Some of these sounds may be prerecorded and then played back during your show. If your teacher has a laser and is willing to allow you to use it, you may do so. All other light must come from conventional household lamps.

## Criteria

Work with your classmates to agree on the relative importance of the following assessment criteria. Each item in the list has a point value given after it, but your class must decide what kind of grading system to use.

1. The variety and number of physics concepts used to produce the light and sound effects: 30 points
   - four or more concepts: 30 points
   - three concepts: 25 points
   - two concepts: 20 points
   - one concept: 10 points
2. Your understanding of the physics concepts: 40 points

   Following your production, you will be asked to:

   a) Name the physics concepts that you used. 10 points
   b) Explain each concept. 10 points
   c) Give an example of something that each concept explains or an example of how each concept is used. 10 points
   d) Explain why each concept is important. 10 points

   As a class, you will have to decide if your answers will be in an oral report or a written report.
3. Entertainment value: 30 points

   Your class will need to decide on a way to assign points for creativity. Note that an entertaining and interesting show need not be loud or bright.

You will have a chance later in the chapter to again discuss these criteria. At that time, you may have more information on the concepts and how you might produce your show. You may want to then propose changes in the criteria and the point values.

142

143

## ❶ Scenario

Each chapter begins with a realistic event or situation. Students might actually have experienced the event or can imagine themselves participating in a similar situation at home, in school, or in your community. Chances are your students probably never thought about the science involved in each case, but now they will!

## ❷ Challenge

This feature presents students with a challenge that they are expected to complete by the end of the chapter. This challenge gives them the opportunity to learn science as they produce a realistic, science-based project. As they progress through the chapter they will accumulate all the scientific knowledge they need to successfully complete the challenge.

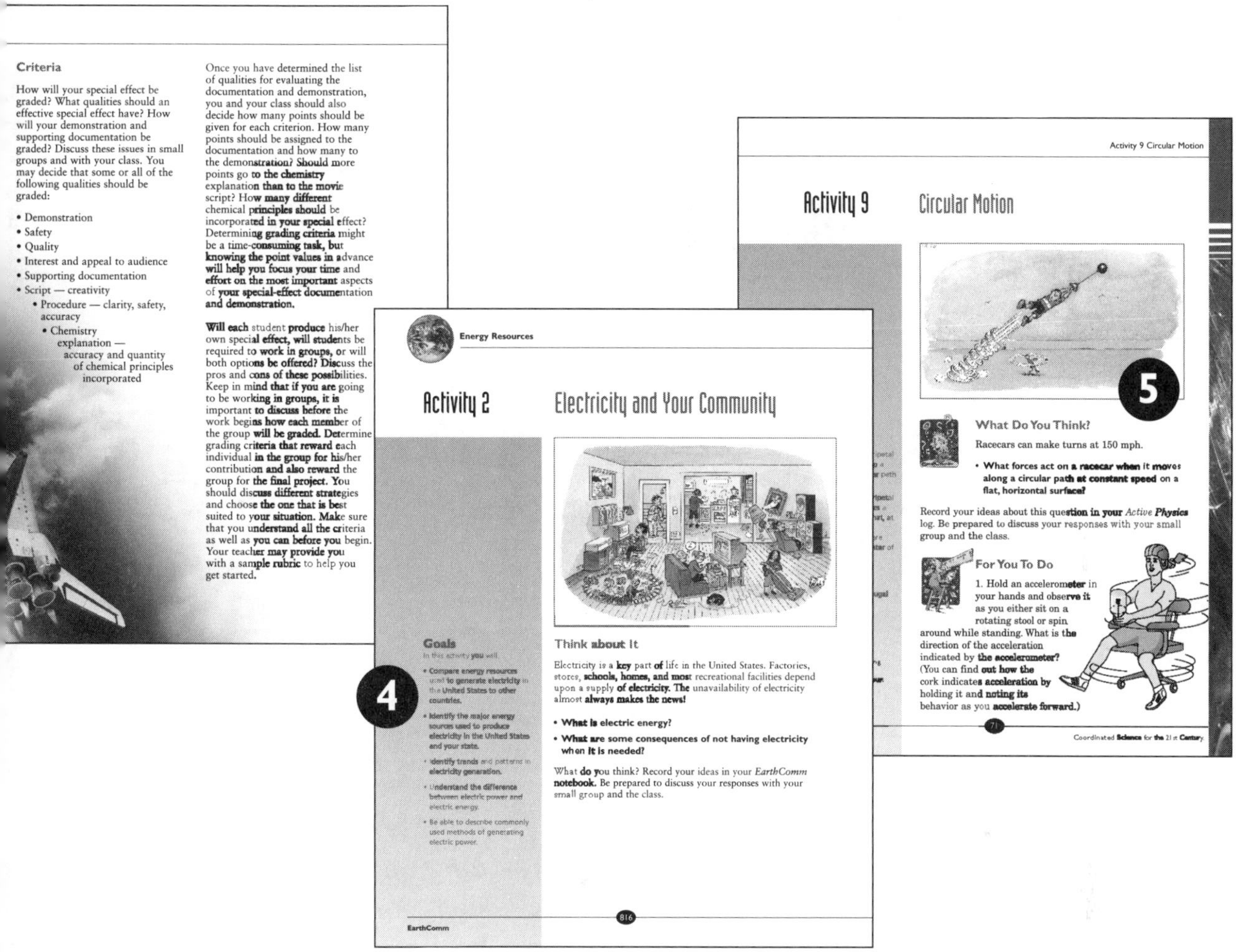

Criteria

How will your special effect be graded? What qualities should an effective special effect have? How will your demonstration and supporting documentation be graded? Discuss these issues in small groups and with your class. You may decide that some or all of the following qualities should be graded:

- Demonstration
- Safety
- Quality
- Interest and appeal to audience
- Supporting documentation
- Script — creativity
  - Procedure — clarity, safety, accuracy
    - Chemistry explanation — accuracy and quantity of chemical principles incorporated

Once you have determined the list of qualities for evaluating the documentation and demonstration, you and your class should also decide how many points should be given for each criterion. How many points should be assigned to the documentation and how many to the demonstration? Should more points go to the chemistry explanation than to the movie script? How many different chemical principles should be incorporated in your special effect? Determining grading criteria might be a time-consuming task, but knowing the point values in advance will help you focus your time and effort on the most important aspects of your special-effect documentation and demonstration.

Will each student produce his/her own special effect, will students be required to work in groups, or will both options be offered? Discuss the pros and cons of these possibilities. Keep in mind that if you are going to be working in groups, it is important to discuss before the work begins how each member of the group will be graded. Determine grading criteria that reward each individual in the group for his/her contribution and also reward the group for the final project. You should discuss different strategies and choose the one that is best suited to your situation. Make sure that you understand all the criteria as well as you can before you begin. Your teacher may provide you with a sample rubric to help you get started.

Activity 9 Circular Motion

Activity 9 Circular Motion

5

What Do You Think?

Racecars can make turns at 150 mph.

- What forces act on a racecar when it moves along a circular path at constant speed on a flat, horizontal surface?

Record your ideas about this question in your *Active Physics* log. Be prepared to discuss your responses with your small group and the class.

For You To Do

1. Hold an accelerometer in your hands and observe it as you either sit on a rotating stool or spin around while standing. What is the direction of the acceleration indicated by the accelerometer? (You can find out how the cork indicates acceleration by holding it and noting its behavior as you accelerate forward.)

71

Coordinated Science for the 21st Century

Energy Resources

Activity 2 Electricity and Your Community

4

Goals

In this activity you will:

- Compare energy resources used to generate electricity in the United States to other countries.
- Identify the major energy sources used to produce electricity in the United States and your state.
- Identify trends and patterns in electricity generation.
- Understand the difference between electric power and electric energy.
- Be able to describe commonly used methods of generating electric power.

Think about It

Electricity is a key part of life in the United States. Factories, stores, schools, homes, and most recreational facilities depend upon a supply of electricity. The unavailability of electricity almost always makes the news!

- What is electric energy?
- What are some consequences of not having electricity when it is needed?

What do you think? Record your ideas in your *EarthComm* notebook. Be prepared to discuss your responses with your small group and the class.

EarthComm 816

## 3 Criteria

Before your students begin the chapter and the challenge, they will explore together, with you and among themselves, exactly how they will be graded. Students thus become involved in evaluating their own learning process.

## Goals

A list of goals is provided at the beginning of each activity to let students know the learning outcomes expected from their active scientific inquiries.

## 5 What Do You Already Know?

Before starting each activity students' prior knowledge is tapped and shared as they discuss the introductory questions. Students are specifically told not to expect to come up with the "right" answers, but to share current understandings.

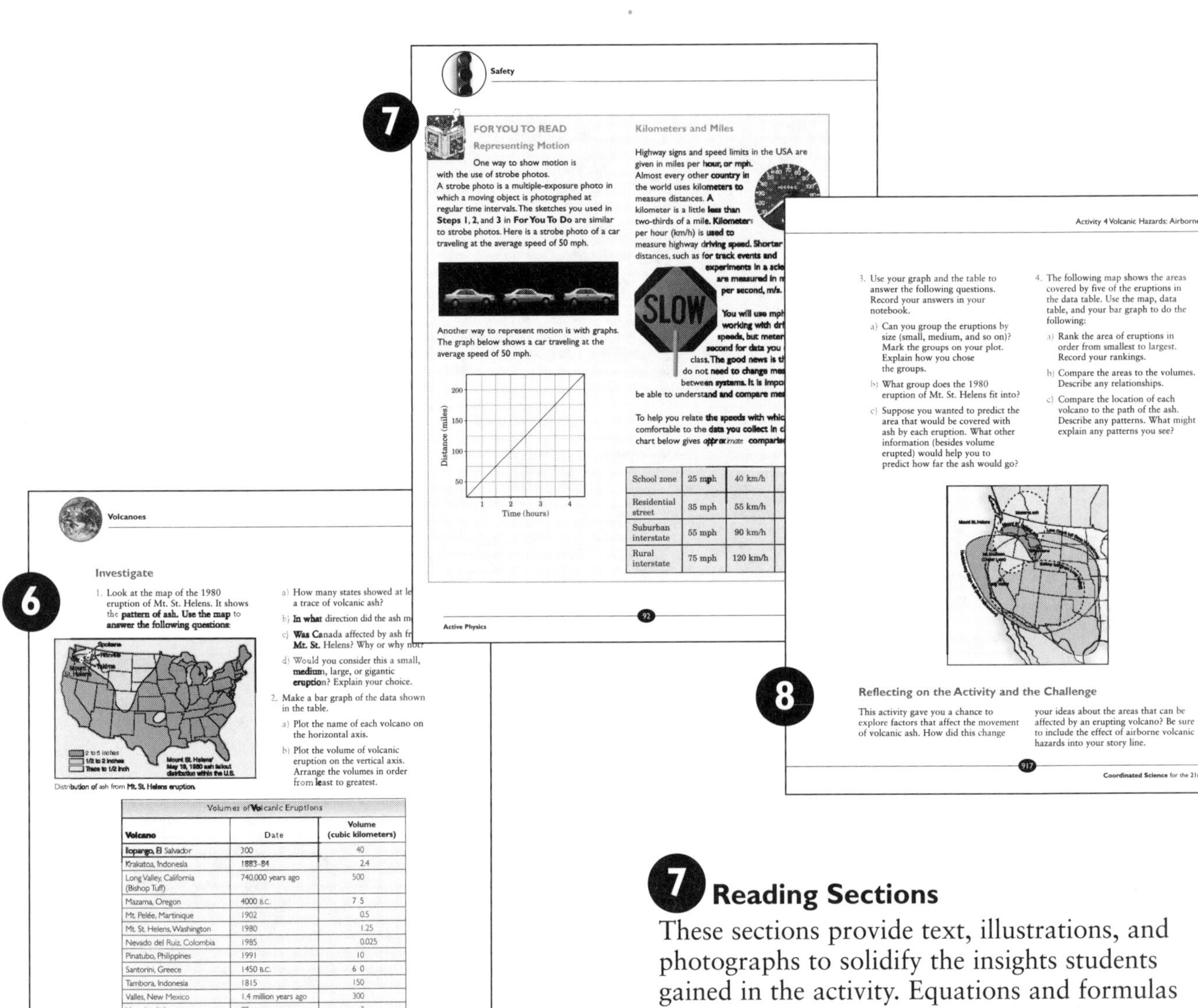

## 7 Reading Sections

These sections provide text, illustrations, and photographs to solidify the insights students gained in the activity. Equations and formulas are provided with easy-to-understand explanations. Science Words that may be new or unfamiliar are defined and explained. In some chapters, the Checking Up questions are included to guide the reading.

## 6 Investigate

In *Coordinated Science for the 21st Century*, students learn science by **doing** science. In small groups, or as a class, they will take part in scientific inquiry by doing hands-on experiments, participating in fieldwork, or searching for answers using the Internet or other reference materials.

## 8 Reflecting on the Activity and the Challenge

Each activity develops specific skills or concepts necessary for the challenge. This feature helps students see this big picture for themselves. It provides a brief summary of the activity and clarifies the purpose of the activity with respect to the challenge. Students thus see each piece of the chapter jigsaw puzzle.

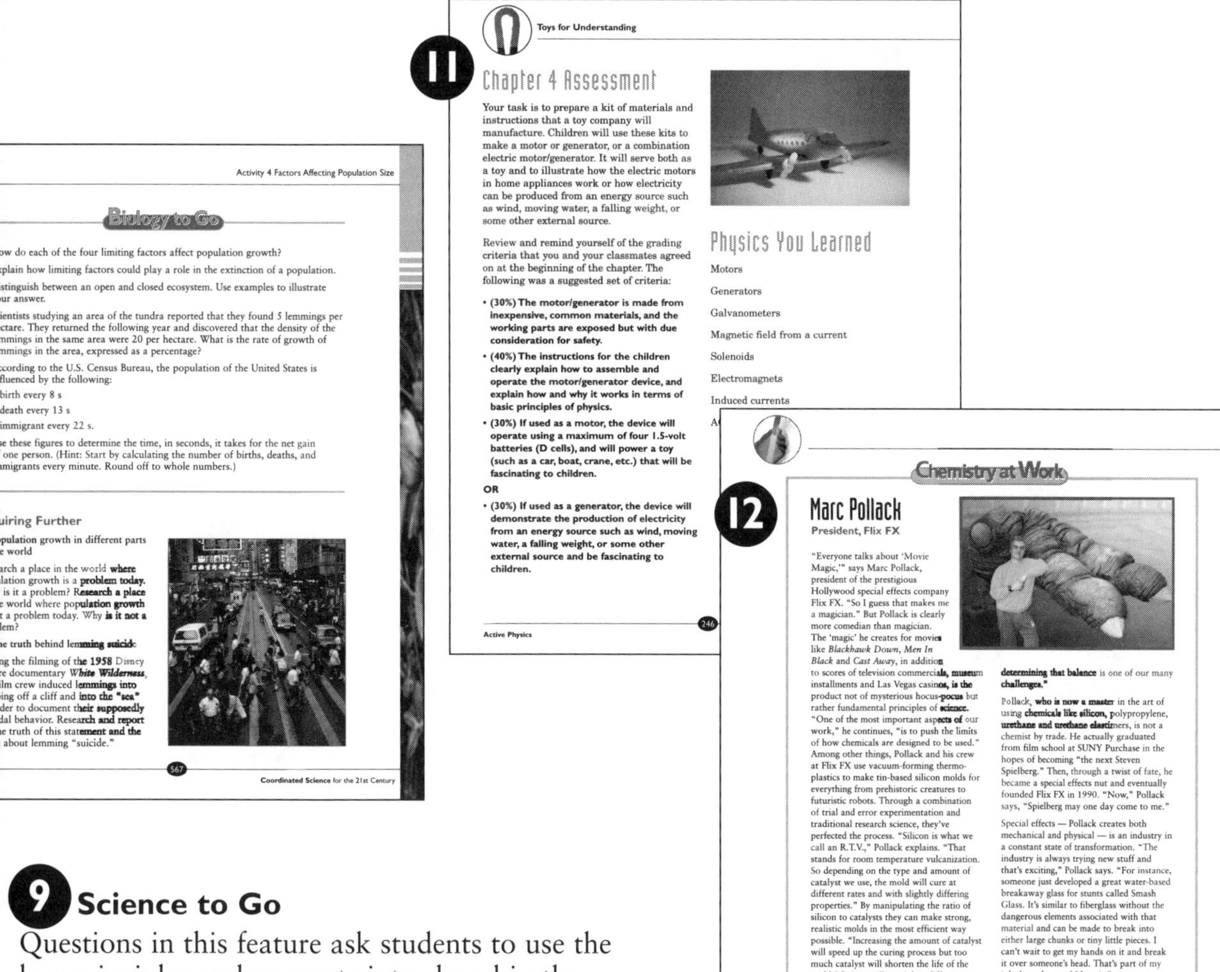

## 9 Science to Go

Questions in this feature ask students to use the key principles and concepts introduced in the activity. Students may also be presented with new situations to apply what they have learned. These questions provide a study guide, helping students review what is most important from the activity. Students will also be given suggestions for ways to organize their work and get ready for the challenge.

## 10 Inquiring Further

This feature stretches your students' thinking. It provides suggestions for deepening the understanding of the concepts and skills developed in the activity. If you're looking for more challenging or in-depth problems, questions, and exercises, you'll find them right here.

## 11 Chapter Assessment

How do your students measure up? Here is their opportunity to share what they have actually learned. Using the activities as a guide, they can now complete the challenge they were presented at the beginning of the chapter.

## 12 Science at Work

Science is an integral part of many fascinating careers. This feature introduces students to people working in fields that involve the principles of science.

# Table of Contents

Screen
Lens

# 'he National Science Education Standards (NSES)

*'oordinated Science for the **21st Century*** was developed using the Active Learning ıstructional Model. This model promotes the style of science instruction that is ncouraged by the NSES Standards.

## ;uide and facilitate learning

Focus and support inquiries while interacting with students.

Orchestrate discourse among students about scientific ideas.

Challenge students to accept and share responsibility for their own learning.

Recognize and respond to student diversity; encourage all to participate fully in science learning.

Encourage and model the skills of scientific inquiry as well as the curiosity openness to new ideas and data.

Promote the healthy scepticism that characterizes science.

## )esign and manage learning :nvironments that provide students vith time, space and resources ıeeded for learning science

Structure the time available so students are able to engage in extended investigations.

Create a setting for student work that is flexible and supportive of science inquiry.

Make available tools, materials, media, and technological resources accessible to students.

Identify and use resources outside of school.

## :ngage in ongoing assessment of ;tudent learning

Use multiple methods and systematically gather data about student understanding and ability.

Analyze assessment data to guide teaching decisions.

Guide students in self-assessment.

## Develop communities of science learners that reflect the intellectual rigor of scientific attitudes and social values conducive to science learning

- Display and demand respect for the diverse ideas, skills, and experiences of all members of the learning community.
- Enable students to have significant voice in decisions about content and context of work and require students to take responsibility for their own learning.
- Nurture collaboration among students.
- Structure and facilitate ongoing formal and informal discussion based on shared understanding of rules.
- Model and emphasize the skills, attitudes and values of scientific inquiry.

## Create authentic assessment standards

- Measure those features claimed to be measured.
- Give students adequate opportunity to demonstrate their achievement and understanding.
- Provide assessment that is authentic and developmentally appropriate, set in familiar context, and engaging to students with different interests and experiences.
- Assesses student understanding as well as knowledge.
- Improve classroom practice and plan curricula.
- Develop self-directed learners.

# Key NSES Recommendations

*Coordinated Science for the* ***21st Century*** will bring the key NSES recommendations into your classroom.

## Scenario-Driven

There are 15 chapters in *Coordinated Science for the 21st Century*. Each chapter begins with an engaging **Scenario.** This project-based assignment challenges the students and sets the stage for the learning activities and chapter assessments to follow. Chapter contents and activities are selectively aimed at providing the students with the knowledge and skills needed to address this introductory challenge, thus providing a natural content filter in the "less is more" curriculum.

## Flexibly Formatted

Chapters are designed to stand alone, so teachers have the flexibility of changing the sequence of presentation of the chapters, omitting an entire chapter, or not finishing all of the chapters.

## Multiple Exposure Curriculum

The thematic nature of the course requires students to continually revisit fundamental science principles throughout the year. Students extend and deepen their understanding of these principles as they apply them in new contexts. This repeated exposure fosters the retention and transferability of learning and promotes the development of critical thinking skills.

## Constructivist Approach

Students are continually asked to explore how they think about certain situations. As they investigate new situations, they are challenged to either explain observed phenomena using an existing paradigm or to develop a more consistent paradigm. This approach is especially critical in helping students abandon previously held notions in favor of the more powerful ideas and explanations offered by scientists.

## Authentic Assessment

For the culmination of each chapter, students are required to demonstrate the usefulness of their newly acquired knowledge by adequately meeting the challenge posed in the chapter introduction. Students are then evaluated on the degree to which they accomplish this performance task. In addition to the project-based assessment, the curriculum also includes other instruments for authentic assessments, as well as non-traditional procedures for evaluating and rewarding desirable behaviors and skills.

## ːooperative Grouping Strategies

Jse of cooperative groups is integral to the course as tudents work together in small groups to acquire the nowledge and information needed to address the eries of challenges presented through the chapter cenarios. Ample teacher guidance is provided to ssure that effective strategies are used in group ɔrmation, function, and evaluation.

## Math Skills Development/Graphing Calculators and Computer Spreadsheets

'he presentation and use of math in *Coordinated cience for the **21st Century*** varies substantially from raditional high school science courses. Math, rimarily algebraic expressions, equations, and graphs s approached as a way of representing ideas ymbolically. Students begin to recognize the sefulness of math as an aid in exploring and nderstanding the world around them. Finally, since nany of the students in the target audience are nsecure about their math backgrounds, the course ngages and provides instruction for the use of raphing calculators and computer spreadsheets to rovide math assistance.

## Minimal Reading Required

ecause it is assumed that the target audience reads nly what is absolutely necessary, the entire course is ctivity-driven. Reading passages are presented mainly vithin the context of the activities and are written at he ninth grade level.

## Jse of Educational Technologies

ːomputer software programs extend and enhance the earning opportunities.

- eLabs – Probes and sensors are used to collect data that is then graphed using Data Studio®.
- CPU – Simulated experiments in physical science direct students' inquiry process.
- GETIT – Research activities use catastrophic events to simulate real-life research practices.
- Test Generator – Crossplatformed software generates tests using a wide variety of assessment criteria.

## Problem Solving

For the curriculum to be both meaningful and relevant to the target population, problem-solving related to technological applications and related issues is an essential component of the course. Problem-solving ranges from simple numerical solutions where one result is expected, to more involved decision-making situations where multiple alternatives must be compared.

## Challenging Learning Extensions

Throughout the text, a variety of **Stretching Exercises** are provided periodically for more motivated students. These extensions range from more challenging design tasks, to enrichment readings, to intriguing and unusual problems. Many of the extensions take advantage of the frequent opportunities the curriculum provides for oral and written expression of student ideas.

# Cooperative Learning

Any classroom using *Coordinated Science for the **21st Century*** comes alive with cooperative learning opportunities. The group project-based approach gives students ample opportunitie

- face-to-face interactions
- respect for diversity
- development of interdependence
- understanding of individual accountability

## Benefits of Cooperative Learning

Cooperative learning requires you to structure a lesson so that students work with other students to jointly accomplish a task. Group learning is an essential part of balanced methodology. It should be blended with whole-class instruction and individual study to meet a variety of learning styles and to maintain a high level of student involvement.

Cooperative learning has been thoroughly researched and has been shown to:

- promote trust and risk-taking
- elevate self-esteem
- encourage acceptance of individual differences
- develop social skills
- permit a combination of a wide range of backgrounds and abilities
- provide an inviting atmosphere
- promote a sense of community
- develop group and individual responsibility
- reduce the time on a task
- result in better attendance
- produce a positive effect on student achievement
- develop key employability skills

As with any learning approach, some students will benefit more than others from cooperative learning. Therefore, you may question as to what extent you should use cooperative learning strategies. It is important to involve the student in helping decide which type of learning approaches they prefer, and to what extent each is used in the classroom. When students have a say in their learning, they will accept to a greater extent any method which you choose to use.

## Phases of Cooperative Learning Lessons

### Organizational Pre-Lesson Decisions

What academic and social objectives will be emphasized? In other words, what content and skills are to be learned and what interaction skills are to be emphasized or practiced?

What will be the group size? Or, what is the most appropriate group size to facilitate the achievement of the academic and social objectives? This will depend on: the amount of individual involvement expected (small groups promote more individual involvement), the task (diverse thinking is promoted by larger groups), the nature of the task or materials available and the time available (shorter time demands smaller groupings to promote involvement).

Who will make up the different groups? Teacher-selected groups usually have the best mix, but this can only happen after the teacher gets to know his/her students well enough to know who works well together. Heterogeneous groupings are most successful in that all can learn through active participation. The duration of the groups' existence may also have some bearing on deciding the membership of groups.

How should the room be arranged? Practicing routines where students move into their groups quickly and quietly is an important aspect. Having students face-to-face is also important. The teacher should still be able to move freely among the groups.

## Materials and/or Rewards to be Outlined n Advance

*tructure for Positive Interdependence*: When students eel they need one another, they are more likely to vork together—goal interdependence becomes mportant. Class interdependence can be promoted by etting class goals, which all teams must achieve in rder for class success.

*xplanation of the Academic Task*: Clear explanation f the academic outcomes is essential. An explanation f the relevance of the activity is also important. hecks for clear understanding can be done either efore the groups form or after, but they are necessary or delimiting frustrations.

*xplanation of Criteria for Success*: Groups should now how their level of success will be determined. llowing students to play a role in determining the riteria promotes involvement and a sense of vesting.

*pecification of Desired Social Behaviors*: Definition nd explanations of the importance of values of social kills will promote student practice and achievement of hese essential life skills.

*tructure for Individual Accountability*: The use of ndividual follow-up activities for tasks or social skills romotes individual accountability.

## Monitoring and Intervening During Group Work

Through careful monitoring of students' behaviors, ntervention can be used most beneficially. Students an be involved in the monitoring by being "a team bserver," but only when the students have a very lear understanding of the behavior being monitored.

Interventions can increase the chances for success in completing the activity and can also teach collaborative skills. These interventions should be used as necessary—they should not be interruptions. This means that the facilitating teacher should be moving among the groups as much as possible. During interventions, the problem should be turned back to the students as often as possible, taking care not to frustrate them.

## Evaluating the Content and Process of Cooperative Group Work

Assessment of the achievement of content objectives should be completed by both the teacher and the students. Students can go back to their groups after an assignment to review the aspects in which they experienced difficulties.

When assessing the accomplishment of social objectives, two aspects are important: how well things proceeded and where/how improvements might be attempted. Student involvement in this evaluation is a very basic aspect of successful cooperative learning programs.

## Organizing Groups

An optimum size of groups for most activities appears to be four; however, for some tasks, two may be more efficient. Heterogeneous groups organized by the teacher are usually the most successful. The teacher will need to decide what factors should be considered in forming the heterogeneous groups. Factors which can be considered are: academic achievement, cultural background, language proficiency, sex, age, learning style, and even personality type.

Initially, level of academic achievement may be the simplest way to form groups. Sort the students on the basis of marks on a particular task or on previous year's achievement. Then choose a student from each quartile to form a group. Once formed, groups should be flexible. Continually monitor groups for compatibility and make adjustments as required.

Students should develop an appreciation that it is a privilege to belong to a group. Remove from group work any student who is a poor participant or one who is repeatedly absent. These individuals can then be assigned the same tasks to be completed in the same time line as a group. You may also wish to place a ten percent reduction on all group work that is completed individually.

The chart on the next page presents a variety of group structures and their relative benefits.

# Group Structures and Their Functions*

| Structure | | Brief Description | Academic and Social Functions |
|---|---|---|---|
| **Team Building** | Round-robin | Each student in turn shares something with his/her teammates. | Expressing ideas and opinions, creating stories. Equal participation, getting acquainted with each other. |
| **Class Building** | Corners | Each student moves to a group in a corner or location as determined by the teacher through specified alternatives. Students discuss within groups, then listen to and paraphrase ideas from other groups. | Seeing alternative hypotheses, values, and problem solving approaches. Knowing and respecting differing points of view. |
| **Mastery** | Numbered heads together | The teacher asks a question, students consult within their groups to make sure that each member knows the answer. Then one student answers for the group in response to the number called out by the teacher. | Review, checking for knowledge comprehension, analysis, and divergent thinking. Tutoring. |
| | Color coded co-op cards | Students memorize facts using a flash card game or an adaption. The game is structured so that there is a maximum probability for success at each step, moving from short to long-term memory. Scoring is based on improvement. | Memorizing facts. Helping, praising. |
| | Pairs check | Students work in pairs within groups of four. Within pairs students alternate — one solves a problem while the other coaches. After every problem or so, the | Practicing skills. Helping, praising. |
| **Concept Development** | Three-step interview | pair checks to see if they have the same answer as the other pair. | Sharing personal information such as hypotheses, views on an issue, or conclusions from a unit. Participation, involvement. |
| | Think-pair-share | Students interview each other in pairs, first one way, then the other. Each student shares information learned during interviews with the group. | Generating and revising hypotheses, inductive and deductive reasoning, and application. Participation and involvement. |
| | Team word-webbing | Students think to themselves on a topic provided by the teacher; they pair up with another student to discuss it; and then share their thoughts with the class. | Analysis of concept into components, understanding multiple relations among ideas, and differentiating concepts. Role-taking. |
| **Multifunctional** | Roundtable | Students write simultaneously on a piece of paper, drawing main concepts, supporting elements, and bridges representing the relation of concepts/ideas. | Assessing print knowledge, practicing skills, recalling information, and creating designs. Team building, participation of all. |
| | Partners | Each student in turn writes one answer as a paper and a pencil are passed around the group. With simultaneous roundtables, more than one pencil and paper are used. | Mastery and presentation of new material, concept development. Presentation and communication skills. |
| | Jigsaw | Students work in pairs to create or master content. They consult with partners from other teams. Then they share their products or understandings with the other partner pair in their team. | Acquisition and presentation of new material review and informed debate. Independence, status equalization. |

* Adapted from Spencer Kagan (1990), "*The Structural Approach to Cooperative Learning*," Educational Leadership, December 1989/January 1990.

## Roles in Cooperative Learning

During a cooperative learning situation, students should be assigned a variety of roles related to the particular task at hand. Following is a list of possible roles that students may be given. It is important that students are given the opportunity of assuming a number of different roles over the course of a semester.

**Leader:**
Assigns roles for the group. Gets the group started and keeps the group on task.

**Organizer:**
Helps focus discussion and ensures that all members of the group contribute to the discussion. The organizer ensures that all of the equipment has been gathered and that the group completes all parts of the activity.

**Recorder:**
Provides written procedures when required, diagrams where appropriate and records data. The recorder must work closely with the organizer to ensure that all group members contribute.

**Researcher:**
Seeks written and electronic information to support the findings of the group. In addition, where appropriate, the researcher will develop and test prototypes. The researcher will also exchange information gathered among different groups.

**Encourager:**
Encourages all group members to participate. Values contributions and supports involvement.

**Checker:**
Checks that the group has answered all the questions and the group members agree upon and understand the answers.

**Diverger:**
Seeks alternative explanations and approaches. The task of the diverger is to keep the discussion open. "Are other explanations possible?"

**Active Listener:**
Repeats or paraphrases what has been said by the different members of the group.

**Idea Giver:**
Contributes ideas, information, and opinions.

**Materials Manager:**
Collects and distributes all necessary material for the group.

**Observer:**
Completes checklists for the group.

**Questioner:**
Seeks information, opinions, explanations, and justifications from other members of the group.

**Reader:**
Reads any textual material to the group.

**Reporter:**
Prepares and/or makes a report on behalf of the group.

**Summarizer:**
Summarizes the work, conclusions, or results of the group so that they can be presented coherently.

**Timekeeper:**
Keeps the group members focused on the task and keeps time.

**Safety Manager:**
Responsible for ensuring that safety measures are being followed, and the equipment is clean prior to and at the end of the activity.

## Assessment of Cooperative Group Work

Assessment should not end with a group mark. Students and their parents have a right to expect marks to reflect the students' individual contributions to the task. It is impossible for you as the instructor to continuously monitor and record the contribution of each individual student. Therefore, you will need to rely on the students in the group to assign individual marks as merited.

There are a number of ways that this can be accomplished. The group mark can be multiplied by the number of students in the group, and then the total mark can be divided among the students, as shown in the example that follows.

Activity:____________________

Group Mark: 8/10

Number in Group: 4

Total Marks: 32/40

Distribution of Marks

| Student's Name | Mark | Signature |
|---|---|---|
| Ahmed | 8/10 | ____________________ |
| Jasmin | 8/10 | ____________________ |
| Mike | 7/10 | ____________________ |
| Tabitha | 9/10 | ____________________ |

Another way to share group marks is to assign a factor to each student, which best represents his or her contribution to the task. The mark factors must total the number of students in the group. The group mark is then multiplied by this factor to arrive at each student's individual mark.

Activity:____________________

Group Mark: 8/10

Number in Group: 4

Mark Factors and Individual Marks

| Student's Name | Mark Factor | Individual Mark | Signature |
|---|---|---|---|
| Ahmed | 1.0 | 8/10 | ____________________ |
| Jasmin | 1.0 | 8/10 | ____________________ |
| Mike | 0.9 | 7.2/10 | ____________________ |
| Tabitha | 1.1 | 8.8/10 | ____________________ |
| Total Mark Factor | 4 | | |

In either case, students must sign to show that they are in agreement with the way the individual marks were assigned.

## Assessment Rubric

You may also wish to provide students with an **Assessment Rubric** similar to the one shown. Students can use this rubric to assess the manner in which the group worked together.

### Assessment Rubric for Group Work: Individual Assessment of the Group

Individual's name: ____________________

Names of group members: ____________________

Name of activity: ____________________

Circle the appropriate number: #1 is excellent, #2 is good, #3 is average, and #4 is poor.

| | | | | |
|---|---|---|---|---|
| 1. The group worked cooperatively. Everyone assumed a role and carried it out. | 1 | 2 | 3 | 4 |
| 2. Everyone contributed to the discussion. Everyone's opinion was valued. | 1 | 2 | 3 | 4 |
| 3. Everyone assumed the roles assigned to them. | 1 | 2 | 3 | 4 |
| 4. The group was organized. Materials were gathered, distributed, and collected. | 1 | 2 | 3 | 4 |
| 5. Problems were addressed as a group. | 1 | 2 | 3 | 4 |
| 6. All parts of the task were completed within the time assigned. | 1 | 2 | 3 | 4 |

Comments:

If you were to repeat the activity, what things would you change?

# Reading Strategies and Science

*Coordinated Science for the **21st Century*** is an activity-driven program that gives reading a high level of relevance. Reading passages are frequently presented within the context of th real-world, motivational activities. Even reluctant readers find a new-found reason to read.

## The Reading Process in the Science Classroom

The *Active Learning Instructional Model* is an activity-driven project based program that naturally lends itself to the established strategies for reading in the science content area. The curriculum was developed with the mandate and mission of "science for all students" and the text therefore has many of these literacy strategies embedded in the instructional model. Educators now understand that linking science and language education strengthens students' skills in both areas; the two disciplines are inherently interdependent. The natural synergy between language and science makes the purposeful linking of the two disciplines extremely valuable, especially when using a contextual guided inquiry approach. Science educators want students to be able to think scientifically, understand the reading and vocabulary of the science content presented, as well to as express themselves effectively. The structure of the ***Active Learning*** chapter format is based on the most current knowledge and research on cognitive learning and reading in the science content area.

The following is a summary of many of the specific strategies and techniques embedded in the *Active Learning Instructional Model* that will help teachers fuse science and language experiences in the classroom, and thus help increase student learning and performance in both content areas.

The reading process includes three essential phases: before reading, during reading, and after reading. Each of these phases is implicitly embedded into the *Active Learning Instructional Model.*

### Before Reading

- Purpose Setting
- Eliciting Prior Knowledge
- Previewing
- Predicting

### During Reading

- Monitoring Comprehension
- Using Context Clues
- Using Text Structure Clues

### After Reading

- Reflecting
- Summarizing
- Seeking Additional Information

### Before Reading

*Purpose Setting* – Teachers are asked to introduce each new chapter by whole class readings and discussions on the ***Chapter Scenario***, ***Chapter Challenge*** and ***Criteria*** to let students in on the big picture of their learning. These three beginning components of each chapter were developed to provide relevant background context, set the stage an motivation for the learning and then encourage students to become active participants in the assessing of that knowledge. The ***Chapter Scenario*** and ***Challenge*** addresses real issues relevant to students' lives and provide the purpose for the learning and reading that will follow. The ***Criteria*** for assessing the success of the ***Chapter Challenge*** must also be determined during the class discussion in which the students collectively develop a rubric. This componen of the curriculum is geared to let students in on – indeed help create – the criteria by which their challenge will be judged. When students agree to the rubric by which they will be measured, research has shown that the students will perform better and achieve more.

Seven to ten ***Activities*** (inquiry-based hands-on labs) then follow. Each ***Activity*** takes from one to two days t complete and will help the students develop the science content needed to complete their ***Chapter Challenge***.

*Eliciting Prior Knowledge* – A ***What Do You Think?*** question begins each ***Activity,*** as a means to uncover preconceived ideas and prior knowledge to help students in their learning about the specific content they will be exposed to in the ***Activity***. These tasks ask students to think about and then write in their journals their thoughts about the question for a few minutes prior to beginning an ***Activity***. The ***What Do You Think?*** question gives students the chance to verbalize what they think about the topic at hand. The questions are designed to elicit both common conceptions and misconceptions. After completing the ***Activity*** they can then review their prior ideas to see what new understandings on the subject they may have gained.

## During Reading

***For You To Do*** or ***Investigate*** is the guided inquiry-based lab in each ***Activity***. Students are asked to work in groups as they perform these labs. Within their groups students are asked to read the instructions, perform the tasks requested, and then to record and write up their results and data in logs or lab books. Reading in a group serves to provide support that is absent in independent reading. The collaborative group process helps to ensure that all types of learners participate in the activity and learning process. At this stage of the learning cycle, students do not read about what others have done in science; they read together in a group in order to **do** the science.

The ***For You To Do*** component of the curriculum requires students to use key parts of the During Reading process.

> *Focus attention* – Students must follow the step-by-step instructions, thus they are asked to carefully focus their attention on each step of the process.
>
> *Monitor comprehension* – As they perform the guided steps of the inquiry lab, students must attend to their own comprehension. In reading what to do, you must learn what to do.
>
> *Anticipate and predict* – Following the step-by-step sequence, students naturally begin to question: "What's next?"
>
> *Use text-structure* – The design, with lots of supportive graphics and text-structure clues, gives practice in using all the information at hand.

Short condensed readings (***For You To Read***, or ***Science Talk***, or **Digging Deeper**) follows the lab activities. *Activity Before Content (ABC)* is a central philosophical tenet of the *Active Learning Instructional Model*. These content readings

summarize the science principles learned in the inquiry. These readings extend the learning process by allowing students to integrate new information into what they already have learned.
In these sections science vocabulary may be defined and explained and mathematical equations are presented where appropriate. Insights increase as students use reading to extend their first-hand knowledge of science.

## After Reading

Each ***Activity*** concludes with ***Reflecting on the Activity and the Challenge***, which relates the activity the students have just completed to the bigger picture, their final chapter project. If students have lost sight of the bigger picture while completing the activities, the larger context of the ***Chapter Challenge*** is brought back into sight. In the section ***Preparing for the Chapter Challenge***, students see how each activity forms a piece of the completed puzzle that will become their ***Chapter Challenge***. In completing the ***Chapter Challenge***, each group will put together a public presentation of the content learned in the chapter in a new real–world context. The students must transfer the content learned from the activities into a new domain. The ***Chapter Challenge*** thus reinforces the readings and science learned from their activities. Each group's presentation also serves as an excellent vehicle for multiple exposures to content, as it repeatedly summarizes the content learned in the chapter.

# Assessment Strategies and Opportunities

The word that most aptly describes the assessment opportunities in *Coordinated Science for the **21st Century*** is "authentic." At the conclusion of each chapter, students demonstrate the usefulness of their newly-acquired knowledge by creating a culminating project, the **Chapter Challenge**.

## Classroom Assessment and the NSES Assessment Characteristics

In keeping with the discussion on assessment as outlined in the *National Science Education Standards* (NSES), four issues, which may present somewhat new considerations for teachers and students, are of particular importance.

### 1. Formative and Integrated

The *National Science Education Standards* (NSES) indicates that assessments should be seen as the ongoing process of gathering and analyzing evidence to determine appropriate instruction. That instruction and any assessment must be consistent in format and intent. In addition, students should also be involved in the assessment process.

### 2. Promoting Deep Learning

Another explicit focus of NSES is to promote a shift to deeper instruction on a smaller set of core science concepts and principles. Assessment can support or undermine that intent. It can support it by raising the priority of in-depth treatment of concepts, so that students recognize the relevance of core concepts. Assessment can undermine a deep treatment of concepts by encouraging students to parrot back large bodies of knowledge-level facts that are not related to any specific context in particular. In short, by focusing on a few concepts and principles, deemed to be of particularly fundamental importance, assessment can help to overcome a bias toward superficial learning. *This is an area that some students will find unusual, if their prior science instruction has led them to rely largely on memorization skills for success.*

### 3. Flexible and Celebrating Diversity

Students differ in many ways. Assessment that calls on students to give thoughtful responses must allow for those differences. Some students may initially find the open-ended character of ***Chapter Challenge*** disquieting. They may ask many questions to try to find out exactly what the finished product should look like. Teachers will have to give a consistent and repeated message to those students, expressed in many different ways, that this is an opportunity for students to be creative and to show what they know in a way that makes sense to them. This allows for the assessments to be adapted to students with differing abilities and diverse backgrounds.

### 4. Contextual

While the ***Chapter Challenges*** are intended to be flexible, they are also intended to be consistent with the manner in which instruction takes place. The *Active-Learning Instructional Model* is such that students have the opportunity to learn new material in a way that places it in context. Consistent with that, the ***Chapter Challenge*** calls for the new material to be expressed in context. Traditional tests are less likely to allow this kind of expression, and are more likely to be inconsistent with the manner of teaching that *Coordinated Science for the **21st Century*** is designed to promote.

## Authentic Assessment

At the culmination of each chapter, students demonstrate the usefulness of their newly acquired knowledge by completing the ***Chapter Challenge***. This authentic and performance-based assessment:

- Focuses on deep understanding of meaningful wholes rather than superficial knowledge of isolated parts
- Encourages the integration of knowledge into the learner's schema rather than the mere reproduction of information discovered by others
- Emphasizes motivational tasks rather than trivial tasks, so is seen to be meaningful in nonschool environments
- Promotes transfer of knowledge to real-life contexts.

## :hapter Challenge

Γhe *Chapter Challenge* is the cornerstone of *:oordinated Science for the **21st Century***. The hallenge provides the purpose for all the learning that akes place. It also provides the central assessment ool for the learning. Students demonstrate their nderstandings of the content learned in the chapter y creating a group project that they publicly present hat ties together the science knowledge gained in the ctivities. This performance-based assessment romotes: motivation, transfer of knowledge, and the olistic integration of the content learned.

*)pportunities to Meet Different Needs* – Another reat aspect of the ***Chapter Challenge*** lies in the wide ariety of tasks that are needed to complete the roject. These tasks give students with different talents pportunity to excel. Students who express themselves rtistically will have an opportunity to shine in some arts of the challenge. Students who can design and uild may take the lead in another part. Some hallenges have a major component devoted to vriting, while others require oral or visual resentations. All of the challenges require what's nportant: the demonstration of solid science nderstanding.

*Multiple Exposures to Content* – One of the main trengths of this project-based assessment is that tudents don't just get to create their own project, they et to view every other groups' project. Through this epeated public display, students get what many need multiple exposures to content. This content has been reated by their peers and is thus not seen as "boring" eview, but as a chance to view other's creations.

*)pportunities for Creativity* – In many science ourses, all students are expected to converge on the lentical solution. In *Coordinated Science for the **21st :entury***, each group is expected to create a unique ***:hapter Challenge***. Each group project must emonstrate correct science concepts, but there is mple room for creativity on the students' part.

## :hapter Challenge Assessment Rubrics

*tudent Participation* – The discussion of the criteria or grading the project and the creation of a grading ubric is an essential ingredient for student success. After the introduction of the challenge, students get to create their own class grading criteria. "What does an "A" look like?" "Should creativity be weighted more than delivery?" "How many scientific principles need to be included?" The criteria will be visited again at the end of the chapter, but at this point it provides the expectation level that students set for themselves.

*Sample Assessment Rubrics* – To give all involved ideas for what might be included in the class rubric, at least two sample rubrics are included at the beginning of each chapter in the Teacher Edition. Just as every ***Chapter Challenge*** is different, every suggested rubric emphasizes specific criteria appropriate for that challenge. There are even samples of three different types of rubrics. There are rubrics to evaluate: 1) the Group-Process Work, 2) the Scientific and Technological Thinking, and 3) Performance-Assessment Rubrics to evaluate the actual data-collection methods used within the investigations.

## Multiple Assessment Tools and Opportunities

Although the cornerstone of the program is the ***Chapter Challenge***, more traditional assessment tools, such as – activity logs, lab reports, and quizzes are also provided for and encouraged. The teacher's lesson plan book for a 21st Century class will look the same as for any other of their science classes, except for the addition of the ***Chapter Challenge*** assessment. Logs, data charts, and journal writing form an integral part of the assessment strategies in this guided-inquiry program. Every activity provides explicit instructions for what students are to record in their logbooks.

*Science to Go* – This section provides questions to check understanding of key principles in the activity. They are often completed as homework assignments and can serve as a study guide to review what is most important in each activity.

*Alternative Chapter Assessment* – For traditional written evaluation of key concepts, an alternative assessment is provided for each chapter. This test can serve as a study guide, a homework assignment, or as an additional method of evaluation to supplement the information gained from the ***Chapter Challenge***.

# Safety in the Science Classroom

*Coordinated Science for the* ***21st Century*** is an active, inquiry-based program. It is founded on the belief that science understanding comes best not from hearing or reading but from *doing*. However, with such doing comes the possibility of injury. Safety is a primary concer

## Safety Guidelines

A critical guideline with respect to safety is student ownership. No teacher can take sole responsibility for making each student safe at all times. The students themselves must also own the responsibility for safety at each and every moment.

Therefore, one primary goal is to ellicit student involvement in safety procedures. Begin a discussion of the necessary rules within your classroom. Solicit student feedback. You might even post these agreed-upon rules for all to see and revisit them periodically.

In addition, have students read the Safety Guidelines provided at the front of their Student Editions. After reading the rules, students and parents should sign the Safety Contract shown on the next page. It is provided as a Blackline Master in the Teacher Resources book.

**Coordinated Science** for the 21st Century

## Safety in the Science Classroom

Chemistry is a laboratory science. During this course you will be doing many activities in which safety is a factor. To ensure the safety of you and all students, the following safety rules will be followed. You will be responsible for abiding by these rules at all times. After reading the rules, you and a parent or guardian must sign a safety contract acknowledging that you have read and understood the rules and will follow them at all times. (See page xxxix.)

### General Rules

1. There will be no running, jumping, pushing, or other behavior considered inappropriate in the science laboratory. You must behave in an orderly and responsible way at all times.
2. Eating, drinking, chewing gum, or applying cosmetics is strictly prohibited.
3. All spills and accidents must be reported to your teacher immediately.
4. You must follow all directions carefully and use only materials and equipment provided by your teacher. Only activities approved by your teacher may be carried out in the chemistry laboratory.
5. No loose, hanging clothing is allowed in the laboratory; long sleeves must be rolled up; bulky jackets, as well as jewelry, must be removed.
6. Never work in the lab unless your teacher or an approved substitute is present.
7. Identify and know the location of a fire extinguisher, fire blanket, emergency shower, eyewash, gas and water shut-offs, and telephone.

### Equipment Rules

1. All equipment must be checked out and returned properly.
2. Do not touch any equipment until you are instructed to do so.
3. Do not use glassware that is broken or cracked. Alert your teacher to any glassware that is broken or cracked.

### Working with Chemicals

1. Never touch or smell chemical unless specifically instructed to do so by your teacher. Never taste chemicals.
2. Safety goggles must be worn at all times.
3. Carefully read all labels to make sure you are using the correct chemicals and use only the amount of chemicals instructed by your teacher.
4. Keep your hands away from your face and thoroughly wash with soap and water before exiting the classroom.
5. Contact lenses can absorb certain chemicals. Advise your teacher if you wear contact lenses.
6. Never add water to an acid and always add acid slowly to water.
7. Follow your teachers' instructions for the correct disposal of chemicals. Do not dispose of any chemical waste, including paper towels used for chemical spills, in the trash basket or down a sink drain.

### Flame Safety

1. Use extreme caution when using any type of flame. Keep your hands, hair, and clothing away from flames.
2. Long hair must be tied back at all times.
3. Keep all flammable materials away from open flames. Some winter jackets are extremely flammable and should be removed before entering the laboratory.
4. Always point the mouth of a test tube away from yourself or any other person when heating a substance.
5. Extinguish the flame as soon as you are finished.
6. Always use heat-resistant gloves when working with an open flame.

### Work Area

1. When working in the laboratory all materials should be removed from the workstation except for instructions and data tables. Materials should not be removed from the desktop to the floor as this is a hazard for someone walking with glassware or chemicals.
2. The work area should be kept clean at all times. After completing an activity, wipe down the area.
3. Notify your teacher of any spills immediately so they can be properly taken care of.

## Safety Contract

The following contract may be reproduced and must be signed by each student and a parent or guardian before participating in laboratory activities.

I have read **Safety in the Science Classroom** and understand the requirements fully. I recognize that there are risks associated with any science activity and acknowledge my responsibility in minimizing these risks by abiding by the safety rules at all times.

Please list any known medical conditions or allergies:

____________________________________________________________

____________________________________________________________

____________________________________________________________

____________________________________________________________

**I do / do not** wear contact lenses. (Circle one)

Emergency phone contact: ____________________________

Student signature ____________________________________ Date _______

Parent or guardian ___________________________________ Date _______

Teacher___________________________________________ Date _______

# Expanding the 5E Model

## A proposed 7E model emphasizes "transfer of learning" and the importance of eliciting prior understanding

SOMETIMES A CURRENT MODEL MUST BE AMENDED TO maintain its value after new information, insights, and knowledge have been gathered. Such is now the case with the highly successful 5E learning cycle and instructional model (Bybee 1997). Research on how people learn and the incorporation of that research into lesson plans and curriculum development demands that the 5E model be expanded to a 7E model.

**Arthur Eisenkraft**

The 5E learning cycle model requires instruction to include the following discrete elements: *engage*, *explore*, *explain*, *elaborate*, and *evaluate*. The proposed 7E model expands the *engage* element into two components—*elicit* and *engage*. Similarly, the 7E model expands the two stages of *elaborate* and *evaluate* into three components— *elaborate*, *evaluate*, and *extend*. The transition from the 5E model to the 7E model is illustrated in Figure 1.

These changes are not suggested to add complexity, but rather to ensure instructors do not omit crucial elements for learning from their lessons while under the incorrect assumption they are meeting the requirements of the learning cycle.

### Eliciting prior understandings

Current research in cognitive science has shown that eliciting prior understandings is a necessary component of the learning process. Research also has shown that expert learners are much more adept at the transfer of learning than novices and that practice in the transfer of learning is required in good instruction (Bransford, Brown, and Cocking 2000).

The *engage* component in the 5E model is intended to capture students' attention, get students thinking about the subject matter, raise questions in students' minds, stimulate thinking, and access prior knowledge. For example, teachers may engage students by creating surprise or doubt through a demonstration that shows a piece of steel sinking and a steel toy boat floating. Similarly, a teacher may plac an ice cube into a glass of water and have the class observe it float while the same ice cube placed in a second glass of liquid sinks. The corresponding conversation with the students may access their prior learning. The students should have the opportunity tc ask and attempt to answer, "Why is it that the toy boat does not sink?"

The *engage* component includes both accessing prior knowledge and generating enthusiasm for the subject matter. Teachers may excite students, get them interested and ready to learn, and believe they are fulfilling the engage phase of the learning cycle, while ignoring the need to find out what prior knowledge students bring to the topic. The importance of *elicitin* prior understandings in ascertaining what students know prior to a lesson is imperative. Recognizing tha students construct knowledge from existing knowledge, teachers need to find out what existing knowledge their students possess. Failure to do so may result in students developing concepts very different from the ones the teacher intends (Bransfor Brown, and Cocking 2000).

A straightforward means by which teachers may elici prior understandings is by framing a "What Do You Think" question at the outset of the lesson as is done consistently in some current curricula. For example, a common physics lesson on seat belts might begin wit a question about designing seat belts for a racecar traveling at a high rate of speed (Figure 2, p. xxxiv). "How would they be different from ones available on passenger cars?" Students responding to this question communicate what they know about seat belts and inform themselves, their classmates, and the teacher about their prior conceptions and understandings. There is no need to arrive at consensus or closure at this point. Students do not assume the teacher will tell them the "right" answer. The "What Do You Think" question is intended to begin the conversation.

The proposed expansion of the 5E model does not exchange the *engage* component for the *elicit* component; the *engage* component is still a necessary element in good instruction. The goal is to continue to excite and interest students in whatever ways possible and to identify prior conceptions. Therefore, the *elicit* component should stand alone as a reminder of its importance in learning and constructing meaning.

### Explore and explain

The *explore* phase of the learning cycle provides an opportunity for students to observe, record data, isolate variables, design and plan experiments, create graphs, interpret results, develop hypotheses, and organize their findings. Teachers may frame questions, suggest approaches, provide feedback, and assess understandings. An excellent example of teaching a lesson on the metabolic rate of water fleas (Lawson 2001) illustrates the effectiveness of the learning cycle with varying amounts of teacher and learner ownership and control (Gil 2002).

Students are introduced to models, laws, and theories during the *explain* phase of the learning cycle. Students summarize results in terms of these new theories and models. The teacher guides students toward coherent and consistent generalizations, helps students with distinct scientific vocabulary, and provides questions that help students use this vocabulary to explain the results of their explorations. The distinction between the *explore* and *explain* components ensures that concepts precede terminology.

### Applying knowledge

The *elaborate* phase of the learning cycle provides an opportunity for students to apply their knowledge to new domains, which may include raising new questions and hypotheses to explore. This phase may also include related numerical problems for students to solve. When students explore the heating curve of water and the related heats of fusion and vaporization, they can then perform a similar experiment with another liquid or, using data from a reference table, compare and contrast materials with respect to freezing and boiling points. A further elaboration may ask students to consider the specific heats of metals in comparison to water and to explain why pizza from the oven remains hot but aluminum foil beneath the pizza cools so rapidly.

The elaboration phase ties directly to the psychological construct called "transfer of learning" (Thorndike 1923). Schools are created and supported with the expectation that more general uses of knowledge will be found outside of school and beyond the school years (Hilgard and Bower 1975). Transfer of learning can range from transfer of one concept to another (e.g., Newton's law of gravitation and Coulomb's law of electrostatics); one school subject to another (e.g., math skills applied in

**FIGURE 1**

**The proposed 7E learning cycle and instructional model.**

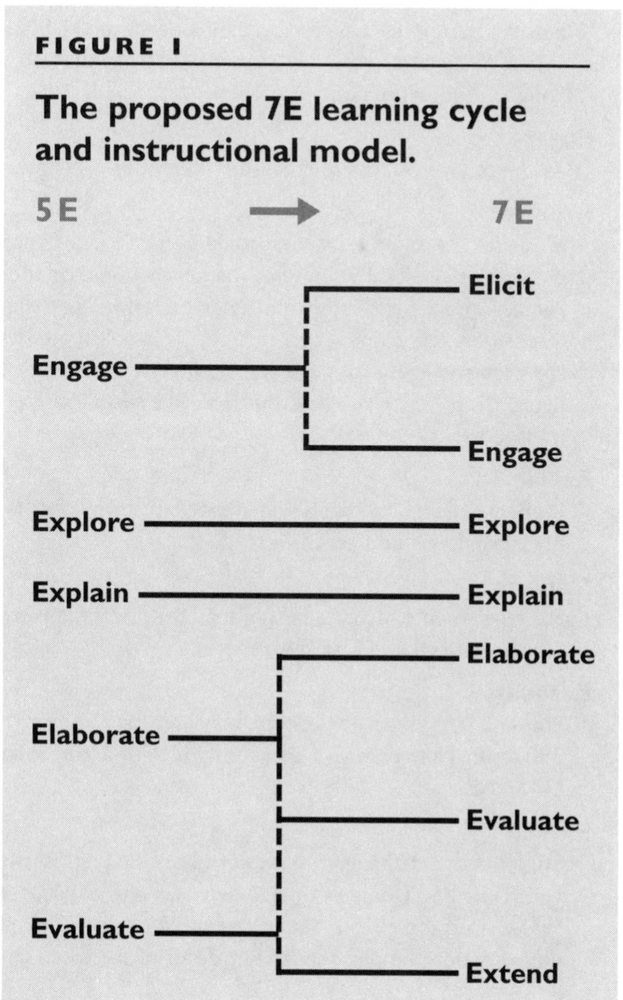

**FIGURE 2**

## Seat belt lesson using the 7E model.

**Elicit prior understandings**

- Students are asked, "Suppose you had to design seat belts for a racecar traveling at high speeds. How would they be different from ones available on passenger cars?" The students are required to write a brief response to this "What Do You Think?" question in their logs and then share with the person sitting next to them. The class then listens to some of the responses. This requires a few minutes of class time.

**Engage**

- Students relate car accidents they have witnessed in movies or in real life.

**Explore**

- The first part of the exploration requires students to construct a clay figure they can sit on a cart. The cart is then crashed into a wall. The clay figure hits the wall.

**Explain**

- Students are given a name for their observations. Newton's first law states, "Objects at rest stay at rest; objects in motion stay in motion unless acted upon by a force."

**Engage**

- Students view videos of crash-test dummies during automobile crashes.

**Explore**

- Students are asked how they could save the clay figure from injury during the crash into the wall. The suggestion that the clay figure will require a seat belt leads to another experiment. A thin wire is used as a seat belt. The students construct a seat belt from the wire and ram the cart and figure into the wall again. The wire seat belt keeps the clay figure from hitting the wall, but the wire slices halfway through the midsection.

**Explain**

- Students recognize that a wider seat belt is needed. The relationship of pressure, force, and area is introduced.

**Elaborate**

- Students then construct better seat belts and explain their value in terms of Newton's first law and forces.

**Evaluate**

- Students are asked to design a seat belt for a racing car that travels at 250 km/h. They compare their designs with actual safety belts used by NASCAR.

**Extend**

- Students are challenged to explore how air bags work and to compare and contrast air bags with seat belts. One of the questions explored is, "How does the air bag get triggered? Why does the air bag not inflate during a small fender-bender but does inflate when the car hits a tree?"

scientific investigations); one year to another (e.g., significant figures, graphing, chemistry concepts in physics); and school to nonschool activities (e.g., using a graph to calculate whether it is cost effective to join a video club or pay a higher rate on rentals) (Bransford, Brown, and Cocking 2000).

Too often, the elaboration phase has come to mean an elaboration of the specific concepts. Teachers may provide the specific heat of a second substance and have students perform identical calculations. This practice in transfer of learning seems limited to near transfer as opposed to far or distant transfer (Mayer 1979). Even though teachers expect wonderful results when they limit themselves to near transfer with large similarities between the original task and the transfer task, they know students often find elaborations difficult. And as difficult as near transfer is for students, the distant transfer is usually a much harder road to traverse. Students who are quite able to discuss phase changes of substances and their related freezing points, melting points and heats of fusion and vaporization may find it exceedingly difficult to transfer the concept of phase change as a means of explaining traffic congestion.

### Practicing the transfer of learning

The addition of the *extend* phase to the *elaborate* phase is intended to explicitly remind teachers of the importance for students to practice the transfer of learning. Teachers need to make sure that knowledge is applied in a new context and is not limited to simple elaboration. For instance, in another common activity students may be required to invent a sport that can be played on the moon. An activity on friction informs students that friction increases with weight.

Because objects weigh less on the moon, frictional forces are expected to be less on the moon. That elaboration is useful. Students must go one step further and extend this friction concept to the unique sports and corresponding play they are developing for the moon environment.

The *evaluate* phase of the learning cycle continues to include both formative and summative evaluations of student learning. If teachers truly value the learning cycle and experiments that students conduct in the classroom, then teachers should be sure to include aspects of these investigations on tests. Tests should include questions from the lab and should ask students questions about the laboratory activities. Students should be asked to interpret data from a lab similar to the one they completed. Students should also be asked to design experiments as part of their assessment (Colburn and Clough 1997).

Formative evaluation should not be limited to a particular phase of the cycle. The cycle should not be linear. Formative evaluation must take place during all interactions with students. The *elicit* phase is a formative evaluation. The *explore* phase and *explain* phase must always be accompanied by techniques whereby the teacher checks for student understanding.

Replacing *elaborate* and *evaluate* with *elaborate*, *extend*, and *evaluate* as shown in Figure 1, p. xxxiii, is a way to emphasize that the transfer of learning, as required in the extend phase, may also be used as part of the evaluation phase in the learning cycle.

## Enhancing the instructional model

Adopting a 7E model ensures that eliciting prior understandings and opportunities for transfer of learning are not omitted. With a 7E model, teachers will *engage* and *elicit* and students will *elaborate* and *extend*. This is not the first enhancement of instructional models, nor will it be the last. Readers should not reject the enhancement because they are used to the traditional 5E model, or worse yet, because they hold the 5E model sacred. The 5E model is itself an enhancement of the three-phrase learning cycle that included exploration, invention, and discovery (Karplus and Thier 1967.) In the 5E model, these phases were initially referred to as *explore*, *explain*, and *expand*. In another learning cycle, they are referred to as exploration, term introduction, and concept application (Lawson 1995).

The 5E learning cycle has been shown to be an extremely effective approach to learning (Lawson 1995; Guzzetti et al. 1993). The goal of the 7E learning model is to emphasize the increasing importance of eliciting prior understandings and the extending, or transfer, of concepts. With this new model, teachers should no longer overlook these essential requirements for student learning.

---

***Arthur Eisenkraft is a project director of Active Physics and currently the Distinguished Professor of Science Education and a Senior Research Fellow at the University of Massachusetts, Boston; and a past president of NSTA, e-mail: eisenkraft@worldnet.att.net.***

### REFERENCES

Bransford, J.D., A.L. Brown, and R.R. Cocking, eds. 2000. *How People Learn.* Washington, D.C.: National Academy Press.

Bybee, R.W. 1997. *Achieving Scientific Literacy.* Portsmouth, N.H.: Heinemann.

Colburn, A., and M.P. Clough. 1997. Implementing the learning cycle. *The Science Teacher* 64(5): 30–33.

Gil, O. 2002. Implications of inquiry curriculum for teaching. Paper presented at National Science Teachers Association Convention, 5–7 December, in Albuquerque, N.M.

Guzzetti B., T.E. Taylor, G.V. Glass, and W.S. Gammas. 1993. Promoting conceptual change in science: A comparative meta-analysis of instructional interventions from reading education and science education. *Reading Research Quarterly* 28: 117–159.

Hilgard, E.R., and G.H. Bower. 1975. *Theories of Learning.* Englewood Cliffs, N.J.: Prentice Hall.

Karplus, R., and H.D. Thier. 1967. *A New Look at Elementary School Science.* Chicago: Rand McNally.

Lawson, A.E. 1995. *Science Teaching and the Development of Thinking.* Belmont, Calif.: Wadsworth.

Lawson, A.E. 2001. Using the learning cycle to teach biology concepts and reasoning patterns. *Journal of Biological Education* 35(4): 165–169.

Mayer, R.E. 1979. Can advance organizers influence meaningful learning? *Review of Educational Research* 49(2): 371–383.

Thorndike, E.L. 1923. *Educational Psychology*, Vol. II: *The Psychology of Learning.* New York: Teachers College, Columbia University.

---

Reprinted with permission from *The Science Teacher* (70(6): 56–59), a journal for high school science educators published by the National Science Teachers Association (www.nsta.org).

# Unit 1

## Active Physics®

Chapter 1
PHYSICS IN ACTION

## Chapter 1-Physics in Action

# National Science Education Standards

### Chapter Challenge

PBS has decided to televise sporting events in a program that has educational value. Students are challenged to provide the voice-over on a sports video that explains the physics of the action as an audition for the job of sports broadcaster.

### Chapter Summary

To gain knowledge and understanding of physics principles necessary to meet this challenge, students work collaboratively on activities that investigate a variety of the forces that affect the changes in motion that are commonly observed in sports. These experiences engage students in the following content identified in the National Science Education Standards.

# Content Standards

### Unifying Concepts

- Systems, order, and organization
- Evidence, models and explanations
- Constancy, change and measurement

### Science as Inquiry

- Identify questions and concepts that guide scientific investigations
- Use technology and mathematics to improve investigations
- Communicate and defend a scientific argument
- Formulate and revise scientific explanations and models using logic and evidence

### Physical Science

- Motions and Forces
- Conservation of energy and increase in disorder

### History and Nature of Science

- Science as a human endeavor
- Nature of scientific knowledge
- Historical perspectives

# Key Physics Concepts and Skills

Chapter 1

| Activity Summaries | Physics Principles |
|---|---|
| **Activity 1: A Running Start and Frames of Reference**<br>Students measure the motion of a ball rolling down then up the sides of a bowl and find the ratio of the "running start" to the vertical distance. From this, they are introduced to the concept of inertia. | • Acceleration<br>• Gravity<br>• Galileo's Principle of Inertia<br>• Newton's First Law of Motion |
| **Activity 2: Push or Pull–Adding Vectors**<br>Students construct, calibrate, and use a simple force meter to explore the variables involved in throwing a shot put. They then connect their observations and data to a study of the laws of motion. | • Newton's Second Law of Motion<br>• Relationship of mass and force to acceleration<br>• Gravity |
| **Activity 3: Center of Mass**<br>By finding the balance points on objects with a variety of shapes, students are introduced to the effect motion of the athlete's center of mass has on balance and performance. | • Center of Mass<br>• Gravity |
| **Activity 4: Defy Gravity**<br>Students learn to measure hang time and analyze vertical jumps of athletes using slow-motion videos. This introduces the concept that work when jumping is force applied against gravity. | • Gravity<br>• Potential and kinetic energy<br>• Work<br>• Vertical accelerated motion |
| **Activity 5: Run and Jump**<br>Thinking about the direction in which they apply force to move in a desired way introduces students to the concept that a force has an equal and opposite force. They test this concept, then apply it to a variety of motions observed in sports. | • Force vectors<br>• Weight and gravity as forces<br>• Newton's Third Law of Motion |
| **Activity 6: The Mu of the Shoe**<br>Students measure the amount of force necessary to slide athletic shoes on a variety of surfaces. From this and the weight of the shoe, they learn to calculate friction coefficients. They then consider the effect of friction on an athlete's performance. | • Gravity<br>• Frictional force<br>• Normal force<br>• Coefficient of Sliding friction |
| **Activity 7: Concentrating on Collisions**<br>Students investigate the affect of a ball's velocity on its motion after a collision. They then apply these observations and what they now know about opposing forces in motion to describe collisions of balls and athletes in sporting events. | • Newton's Third Law of Motion<br>• Mass<br>• Velocity<br>• Momentum |
| **Activity 8: Conservation of Momentum**<br>Additional collisions between objects allow students to investigate what happens when the objects stay together or "stick" after the collision. | • Newton's Third Law of Motion<br>• Momentum = Mass × Velocity<br>• Velocity<br>• Law of Conservation of Momentum |
| **Activity 9: Circular Motion**<br>Students use an accelerometer to test the direction of acceleration when spinning in a chair. From this, they investigate the forces involved in the movement of turning objects and athletes. | • Inertia<br>• Centripetal acceleration<br>• Centripetal force |

# GETTING STARTED WITH EQUIPMENT NEEDED TO CONDUCT THE ACTIVITIES.

### Items needed—not supplied in Material Kits

Preparing the equipment needed for each activity in this chapter is an important procedure. There are some items, however, needed for the chapter that are not supplied in the It's About Time material kitpackage. Many of these items may already be in your school and would be an unnecessary expense to duplicate. Please read carefully the list of items to the right which are not found in the supplied kits and locate them before beginning activities.

**Items needed—not supplied by It's About Time:**

- Dried Peas
- Plastic beads
- Raw rice
- Eraser
- Teeter-totter
- VCR & Monitor
- Sports Content Video
- Computer
- MBL or CBL with Motion sensor
- Safety Helmet
- Knee Pads
- Elbow pads
- Chair with wheels
- Shoe
- Smooth surface (desk top, tile)
- Stack of books
- Wood board, 4 ft
- Rotating stool or chair

# Equipment List For Chapter 1 (Serves a Classroom of 30 Students)

| PART | ITEM | QTY | ACTIVITY |
|---|---|---|---|
| AH-9251 | Accelerometer | 6 | 9 |
| AH-9252-S2 | Accelerometer, Cork | 6 | 9 |
| AH-9250-S1 | Accelerometer-To-Cart | 6 | 9 |
| PH-1133 | Air Puck | 6 | 1A |
| BS-7203-S2 | Balls, Bocci | | 7,9 |
| BS-1400-S2 | Ball, Golf | 6 | 7 |
| NB-0004 | Ball, Nerf | 6 | 7, 9 |
| BS-6423 | Ball, Practice Golf | 6 | 7 |
| BS-7207-S2 | Balls, Set of 2 | 6 | 1, 2 |
| BS-0787-S2 | Ball, Super, 1" Diameter | 6 | 1 |
| BS-0790-S2 | Ball, Tennis | 6 | 2,7 |
| BS-5856-S2 | Bowl, Salad, | 6 | 1 |
| CM-1108-S2 | Calculator, Basic | 6 | 6A |
| CS-3246-S2 | C-Clamp | 6 | 2 |
| DS-7201-S2 | Dots, Adhesive, 3/4" | 168 | 3 |
| GC-0001 | Dynamics Cart | 12 | 8 |
| MS-1425-S2 | Marking Pen, Felt Tip | 6 | 1 |
| MC-5418-S2 | Modeling Clay | 6 | 8 |
| PM-0220-S2 | Plumb Bob W/String | 6 | 3 |
| PP-6128-S2 | Push Pins | 600 | 3 |
| SS-7209-S2 | Rough Horizontal Surface, 6" x 36" | 6 | 6 |
| RS-2723-S2 | Ruler, Flexible Plastic, 30 cm | 6 | 1,2,9 |
| RS-2826-S2 | Ruler, Metric, mm Marking | 6 | 1, 2, 6A, 9 |
| SS-0491 | Scale, Bathroom | 6 | 5A |
| SS-7210-S2 | Set of Shapes, A,B,C,D | 6 | 3 |
| SH-7708-S2 | Skateboard | 6 | 5 |
| SM-1676-S2 | Stick, Meter, 100 cm, Hardwood | 6 | 3, 4, 5 |
| SS-2304-S2 | Spring Scale, 0-20 Newton Range | 6 | 6,6A |
| SS-2303-S2 | Spring Scale, 0-10 Newton Range | 6 | 8 |
| RS-7211-S2 | Starting Ramp For Bocci Ball | 12 | 7 |
| TS-2662-S2 | Tape, Masking, 3/4" x 60 yds. | 12 | 1,4,7,8 |
| TT-6100-S2 | Ticker Tape Timer | 6 | 8 |
| RS-4462-S2 | Track W/Adjustable Outrun Slope | 6 | 1 |
| WS-0376-S2 | Weight, 100 g Slotted Mass | 60 | 5,6,8 |
| WS-6910-S2 | Washers, Metal, 3/4" | 24 | 2,5 |
| | | | |
| ITEMS NEEDED – NOT SUPPLIED BY IT'S ABOUT TIME | | | |
| | Dried Peas | 6 | 1A |
| | Plastic beads | 6 | 1A |
| | Raw rice | 6 | 1A |
| | Eraser | 6 | 1A |
| | Teeter-totter | 6 | 3A |
| | VCR & Monitor | 6 | 4 |
| | Sports Content Video | 6 | 4,7 |
| | Computer | 6 | 4A |
| | MBL or CBL with Motion sensor | 6 | 4A |
| | Safety Helmet | 6 | 5 |
| | Knee Pads | 6 | 5 |
| | Elbow pads | 6 | 5 |
| | Chair with wheels | 6 | 5 |
| | Shoe | 6 | 6,6A |
| | Smooth surface (desk top, tile) | 6 | 6 |
| | Stack of books | 6 | 6,6A |
| | Wood board, 4 ft | 6 | 6A |
| | Rotating stool or chair | 6 | 9 |

# Organizer for Materials Available in Teacher's Edition



| Activity in Student Text | Additional Material | Alternative / Optional Activities |
|---|---|---|
| ACTIVITY 1: A Running Start and Frames of Reference p. 6 | Performance Assessment Rubrics pgs. 28-29 | Activity 1 A: Can Objects Move Forever?, p. 27 |
| ACTIVITY 2: Push or Pull–Adding Vectors, p. 17 | | |
| ACTIVITY 3: Center of Mass p. 28 | Templates for Shapes A, B, C, and D, pgs. 52-53 | Activity 3 A: Alternative Method for Determining Center of Gravity, p. 51 |
| ACTIVITY 4: Defy Gravity p. 33 | Calculating Hang Time and Force During a Vertical Jump (Worksheet) pgs. 70-71 | Activity 4 A: High-Tech Alternative for Monitoring Vertical Jump Height, p. 69 |
| ACTIVITY 5: Run and Jump p. 47 | | Activity 5 A: Using a Bathroom Scale to Measure Forces, pgs. 80-81 |
| ACTIVITY 6: The Mu of the Shoe p. 52 | Assessment Rubric, Physics to Go Question 8, p. 90<br>Background Information for Activity 6 A: Alternative Activity for Measuring the Mu of the Shoe, pgs. 93-94 | Activity 6 A: Alternative Activity for Measuring the Mu of the Shoe, p. 92 |
| ACTIVITY 7: Concentrating on Collisions, p. 58 | | |
| ACTIVITY 8: Conservation of Momentum, p. 63 | | |
| ACTIVITY 9: Circular Motion p. 71 | | |

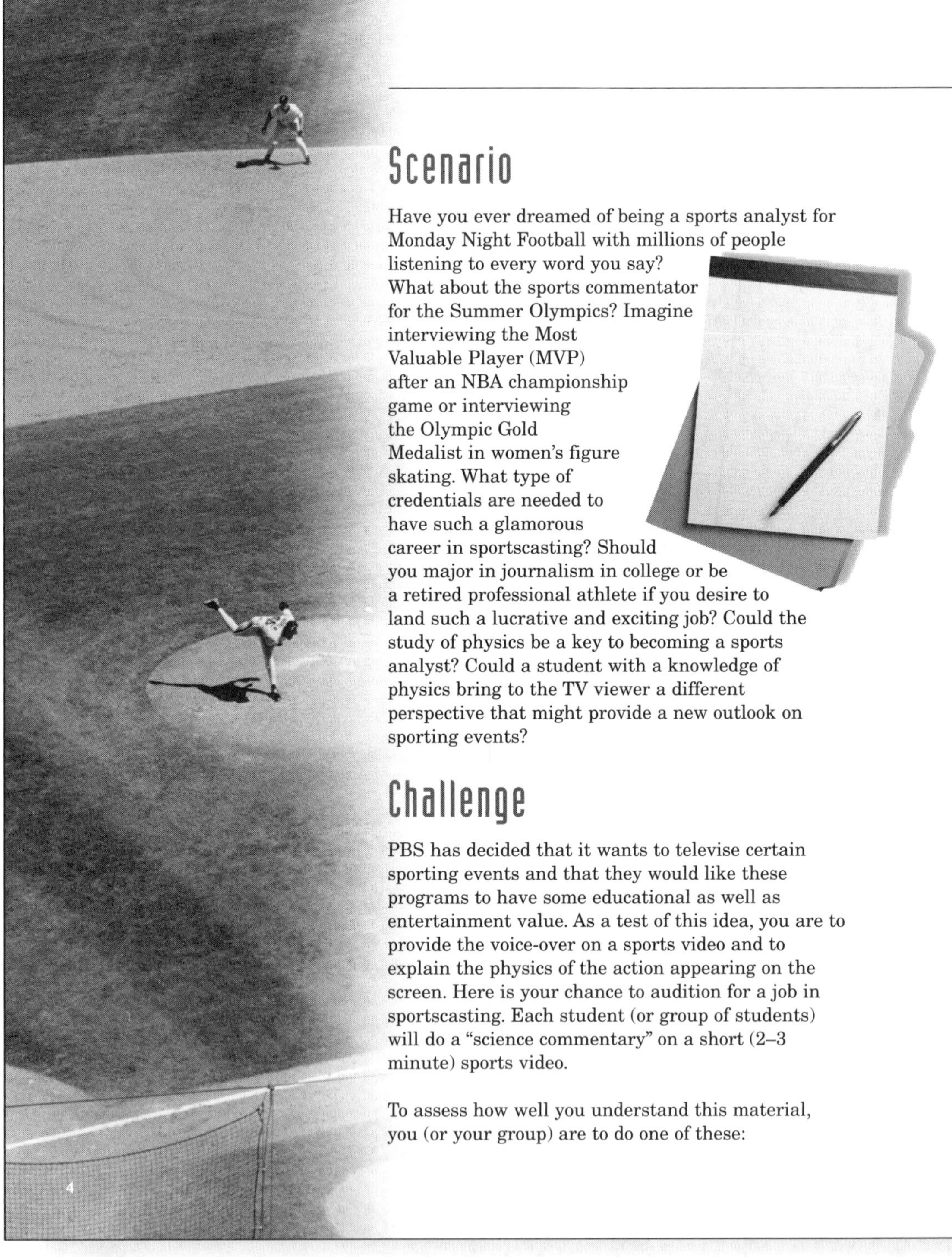

## Scenario

Have you ever dreamed of being a sports analyst for Monday Night Football with millions of people listening to every word you say? What about the sports commentator for the Summer Olympics? Imagine interviewing the Most Valuable Player (MVP) after an NBA championship game or interviewing the Olympic Gold Medalist in women's figure skating. What type of credentials are needed to have such a glamorous career in sportscasting? Should you major in journalism in college or be a retired professional athlete if you desire to land such a lucrative and exciting job? Could the study of physics be a key to becoming a sports analyst? Could a student with a knowledge of physics bring to the TV viewer a different perspective that might provide a new outlook on sporting events?

## Challenge

PBS has decided that it wants to televise certain sporting events and that they would like these programs to have some educational as well as entertainment value. As a test of this idea, you are to provide the voice-over on a sports video and to explain the physics of the action appearing on the screen. Here is your chance to audition for a job in sportscasting. Each student (or group of students) will do a "science commentary" on a short (2–3 minute) sports video.

To assess how well you understand this material, you (or your group) are to do one of these:

## Chapter and Challenge Overview

In this chapter Newton's Laws of Motion and also the concepts of force, inertia (mass), momentum, and the physics of rotation are introduced.

Students will be asked to produce a voice-over (or a script for a voice-over) to explain the physics behind a short sports video. You will select the video, but if the students are very ambitious, they can either find some footage themselves, shoot some scenes with a camcorder, or tape some sporting events from TV. The entire chapter will build toward this end, and the final evaluation of the student's progress will be based on the video voice-over.

There are a number of objectives in this chapter, one of which is to show the students that the laws of physics hold true not only in their science class and lab, but out in the world as well. The students should be able to look at a sporting event and realize what physical principle is involved. Hopefully this will carry over to everyday life, and the student will then be able to see the physics in the world around them.

Each class might start with a short video segment showing sports bloopers. They are commercially available and many of the students may have their own. After the class has covered some of the material, it is increasingly appropriate to discuss the physics that is being displayed in the blooper. Many of these bloopers are very humorous and the students look forward to the beginning of the class.

As you review the **Chapter Challenge** assignments, reassure the students that while they may feel incompetent now, by the end of the chapter they will have the necessary skills and vocabulary to respond adequately.

On the following pages of the Teacher's Edition there are suggestions on how to evaluate students on this material. It is very important at this time that the students be made aware of the method you are going to use and how you will evaluate their work. Have the students actively participate in deciding the criteria for evaluation.

The **Physics To Go** at the end of each section often contains more questions than should ever be assigned for homework. This section has been written in such a way as to give you a choice as to how much work, and the nature of the work the students will be expected to do each day out of class.

As you work with *Coordinated Science for the 21st Century*, be aware that the same physics concepts appear repeatedly in different contexts. It is not necessary for the students to achieve total understanding the first time that they encounter Newton's Laws of Motion, and the physics of rotation.

- **submit a written script**
- **narrate live**
- **dub onto the video soundtrack**
- **record on an audiocassette**

Your task is not to give a play-by-play description of the sporting event or give the rules of the game but rather to go a step beyond and educate the audience by describing to them the rules of nature that govern the event. This approach will give the viewer (and you) a different perspective on both sports and physics. The laws of physics cover not only obscure phenomena in the lab, but everyday events in the real world as well.

## Criteria

What criteria should be used to evaluate a voice-over dialogue or script of a sporting event? Since the intention is to provide an analysis of and interest in the physics of sports, the voice-over should include the use of physics terms and physics principles. All of these terms and principles should be used correctly. How many of these terms and principles would constitute an excellent job? Would it be enough to use one physics term correctly and explain how one physics principle is illustrated in the sport? Should use of one physics term and one physics principle be a minimum standard to get minimal credit for this assessment? Discuss in your small groups and your class and decide on reasonable expectations for the physics criteria for the assessment.

Since the assessment requires a product that will be a part of television, another aspect of the criteria for success would be the entertainment quality of the voice-over. Does a commentator who adds humor or drama receive a higher rating than someone who has similar physics content but has added no excitement or interest to the broadcast? How does one weigh the value of the entertainment quality and the value of relevant physics? What are reasonable expectations for the entertainment aspect of the voice-over? Discuss and decide as a class.

Although many people may be in the broadcast booth, a voice-over becomes the product of one person—the commentator or the scriptwriter. Although you will be working in cooperative groups during the chapter, each person will be responsible for a voice-over or script for a sporting event. As a team of two or three, you may wish to work together and share different aspects of the job, but the output of work per person should be the same. That is why one voice-over will be required of each person irrespective of whether individuals prefer to work independently or in groups.

5

# Assessment Rubric for Voice-Over Dialogue or Script

| | |
|---|---|
| **Meets the standard of excellence.** **5** | • A significant number of physics principles are consistently and correctly addressed.<br>• Physics concepts from the chapter are repeatedly integrated in the appropriate places.<br>• Physics terminology and equations are consistently incorporated as applicable.<br>• Correct estimates of the magnitude of physical quantities are frequently used.<br>• Additional research, beyond basic concepts presented in the chapter, is evident. Knowledge of the rules of the game are evident.<br>• The voice-over has great entertainment value. It contains humor and excitement. |
| **Approaches the standard of excellence.** **4** | • A significant number of physics principles are often correctly addressed.<br>• Physics concepts from the chapter are integrated in the appropriate places.<br>• Physics terminology and equations are incorporated as applicable.<br>• Correct estimates of the magnitude of physical quantities are frequently used.<br>• Knowledge of the rules of the game are evident.<br>• The voice-over has entertainment value. It contains some humor and excitement. |
| **Meets an acceptable standard.** **3** | • A sufficient number of physics principles are correctly addressed.<br>• Physics concepts from the chapter are integrated in the appropriate places.<br>• A limited amount of physics terminology and equations are incorporated as applicable.<br>• Correct estimates of the magnitude of physical quantities are occasionally used.<br>• Knowledge of the rules of the game are general.<br>• The voice-over has some entertainment value. |
| **Below acceptable standard and requires remedial help.** **2** | • Very few physics principles are addressed.<br>• Physics concepts from the chapter are not always integrated in the appropriate places.<br>• A limited amount of physics terminology is incorporated as applicable.<br>• Estimates of the magnitude of physical quantities are seldom used.<br>• Knowledge of the rules of the game is weak.<br>• The voice-over has a limited entertainment value. |
| **Basic level that requires remedial help or demonstrates a lack of effort.** **1** | • Physics principles are not addressed correctly.<br>• Physics concepts from the chapter are not integrated in the appropriate places.<br>• Physics terminology is not used.<br>• No attempt is made to include the magnitude of physical quantities.<br>• Knowledge of the rules of the game is lacking.<br>• The voice-over is difficult to follow and portions are missing. |

For use with *Physics in Action*, Chapter 1

# Assessment Rubric for Voice-Over Dialogue or Script

| | |
|---|---|
| **Meets the standard of excellence.** **5** | • Scientific vocabulary is used consistently and precisely.<br>• Sentence structure is consistently controlled.<br>• Spelling, punctuation, and grammar are consistently used in an effective manner.<br>• Scientific symbols for units of measurement are used appropriately in all cases. |
| **Approaches the standard of excellence.** **4** | • Scientific vocabulary is used appropriately in most situations.<br>• Sentence structure is usually consistently controlled.<br>• Spelling, punctuation, and grammar are generally used in an effective manner.<br>• Scientific symbols for units of measurement are used appropriately in most cases. |
| **Meets an acceptable standard.** **3** | • Some evidence that the student has used scientific vocabulary, although usage is not consistent or precise.<br>• Sentence structure is generally controlled.<br>• Spelling, punctuation, and grammar do not impede the meaning.<br>• Some scientific symbols for units of measurement are used. Generally, the usage is appropriate. |
| **Below acceptable standard and requires remedial help.** **2** | • Limited evidence that the student has used scientific vocabulary. Generally, the usage is not consistent or precise.<br>• Sentence structure is poorly controlled.<br>• Spelling, punctuation, and grammar impedes the meaning.<br>• Some scientific symbols for units of measurement are used, but most often, the usage is inappropriate. |
| **Basic level that requires remedial help or demonstrates a lack of effort.** **1** | • Limited evidence that the student has used scientific vocabulary and usage is not consistent or precise.<br>• Sentence structure is poorly controlled.<br>• Spelling, punctuation, and grammar impedes the meaning.<br>• No attention to using scientific symbols for units of measurement. |

**Maximum = 10 points**

For use with *Physics in Action*, Chapter 1

# What is in the Physics InfoMall for Chapter 1?

Chapter 1 deals with the physics of sports.

If you have had much experience with the *Physics InfoMall CD-ROM*, you have probably done a few searches, and no doubt some of the searches have resulted in "Too many hits." Surprisingly, searching the entire CD-ROM with the keyword "sport*" does not give "too many" hits, but provides some interesting hits. Note that the asterisk is a wild character; this searches for any word beginning with "sport."

If you do the search just mentioned, the first hit is a resource letter.

("Resource letter PS-1: Physics of sports," *American Journal of Physics, vol. 54, issue 7*) that discusses the published discussions on the physics of sports. According to this letter, "there is surprisingly little published information about the basic physics underlying most sports, even though the relevant physics is all classical." Included is a list of places you might find such information, including journals and books. The letter contains a list of specific references grouped by sport, such as Physics of basketball, *American Journal of Physics, vol. 49, issue 4*. Another interesting article is "Students do not think physics is 'relevant.' What can we do about it?," in the *American Journal of Physics, vol. 36, issue 12*.

Given that the physics in sports is classical, you might search for student difficulties learning classical physics in general. One article you might find is "Factors influencing the learning of classical mechanics," *American Journal of Physics 48, issue 12*. Knowledge of such factors affecting learning can be a valuable tool. Perform other searches that meet your needs, and the InfoMall is very likely to provide good information. And we have not even opened the Textbook Trove yet!

# ACTIVITY 1
## A Running Start and Frames of Reference

# Background Information

Two major ideas are introduced in this activity:

- Galileo's Principle of Inertia
- Newton's Second Law of Motion

Before attempting to identify causes and effects for generating, sustaining, and arresting motion, a pivotal question first must be answered: What kinds of motion require explanation?

Two distinct kinds of motion along a straight line often are encountered in nature: (1) motion with constant speed and (2) motion with uniform, or constant, acceleration.

Since the contributions of Galileo, physics has operated from the perspective that the first of these kinds of motion, constant speed, has no cause. Galileo devised a number of arguments and demonstrations, some of which are replicated in this activity, to support this notion.

The cause of all accelerated motion is force; some agent(s) must be pushing or pulling—exerting a force—on any object observed to be accelerating. Sources or kinds of forces abound. Every situation that involves acceleration has an associated net force. Observation: If an orange is dropped, it accelerates; assigned cause: the downward force due to gravity. When the orange hits the floor it stops; another acceleration, another force. The force which stops the orange is provided by the floor, upward. A magnet brought near another magnet will cause an acceleration; therefore, there must be a magnetic force.

Sometimes, we can also discover forces hiding in constant-speed linear motion. Drop a coffee filter: it accelerates downward for a bit, but the amount of acceleration falls off to zero, so that the coffee filter falls most of the way at constant speed. Did the force of gravity decrease or disappear? No, a coffee filter seems to weigh (a measure of the force of gravity) the same at every point in the descent path. Conclusion: there must be another force, the force of air resistance, acting in the opposite direction to gravity. The force of air resistance eventually balances out the gravitational force. It is possible for a combination of forces to have a net effect of zero.

So, it is the net force on an object that imparts the acceleration. Newton's First Law of Motion states the case: An object at rest tends to remain at rest, and an object in motion (in a straight line) tends to remain in motion unless acted upon by an outside (net, non-zero) force. This statement is more complete than the one provided to the students in **Activity 1**. Whenever speed, direction, or both speed and direction, are observed to change, a net force is the cause.

The First Law does not attempt to quantify the relationship between accelerations and the forces that cause them. Establishing the quantitative relationship requires experimental evidence which is the purpose of the next activity.

## *Active*-ating the Physics InfoMall

Note that this activity has students perform a simple experiment, and gradually leads them to concentrate on one aspect of the motion, then leads to predictions and generalizations. The importance of the prediction should not be overlooked; indeed, predictions force students to examine their understanding of a phenomena and actively engage thought. If you were to search the InfoMall to find more about the importance of predictions in learning, you would find that you need to limit your search. For example, a search for "prediction*" AND "inertia" resulted in several hits; the first hit is from *A Guide to Introductory Physics Teaching: Elementary Dynamics,* Arnold B. Arons' Book Basement entry. Here is a quote from that book: "Because of the obvious conceptual importance of the subject matter, the preconceptions students bring with them when starting the study of dynamics, and the difficulties they encounter with the Law of Inertia and the concept of force, have attracted extensive investigation and generated a substantial literature. A sampling of useful papers, giving far more extensive detail than can be incorporated here, is cited in the bibliography [Champagne, Klopfer, and Anderson (1980); Clement (1982); di Sessa (1982); Gunstone, Champagne, and Klopfer (1981); Halloun and Hestenes (1985); McCloskey, Camarazza, and Green (1980); McCloskey (1983); McDermott (1984); Minstrell (1982); Viennot (1979); White (1983), (1984)]." Note that students' preconceptions can have a large effect on how they learn something. It is important that they are forced to consciously acknowledge their preconceptions by making predictions.

Not surprisingly, among the list of hits from the search just mentioned is an article on Galileo,

"Galileo, yesterday and today," *American Journal of Physics* vol. 33, issue 9, 1965. This article provides an interesting insight into Galileo and his work, as well as several historical accounts of his work. Check it out; it might provide interesting additional reading for your students.

Of course, Newton had something to say about inertia, and another hit from the same search provides *Physics for Science and Engineering* in the Textbook Trove. See Chapter 4, Newton's Principles of Motion.

The search above was conducted initially to explore the importance of predictions, especially as related to the concept of inertia. As we can see, additional information was provided that was easily relevant to this topic. This is not unusual when searching the InfoMall — you will often find many interesting bits of information that may take you on unexpected, but enlightening, tangents.

In **Physics To Go, Question 2**, you are encouraged to find something about "curling." Sadly, the InfoMall has only one reference to this sport, and it is a short passage indicating that Lord Kelvin broke his leg while curling, and limped badly thereafter. While not directly related to anything in this section, it is another of those interesting articles one can find on the InfoMall.

# Planning for the Activity

## Time Requirements

- One class period.

## Materials Needed

### For each group:

- ball, set of 2
- ball, super, 1" diameter
- bowl, salad
- marking pen, felt tip
- ruler, flexible plastic, 30 cm
- ruler, metric, mm marking
- tape, masking, 3/4" x 60 yds.
- track w/adjustable outrun slope

## Advance Preparation and Setup

You may wish to consider using ball bearings or glass marbles rolling within flexible, transparent plastic tubing for the second part of **For You To Do** instead of a ball rolling on an adjustable track. If so, you may wish to procure the tubing and bearings in advance.

# Teaching Notes

*Coordinated Science for the 21st Century* uses a modified constructivist model. By confronting students' misconceptions and by having them do hands-on exploration of ideas, we seek to replace their misconceptions with correct perceptions of reality. In order to do this, a consistent scheme is integrated into the course activities to elicit the students' misconceptions early in any activity.

Students' current mental models are sampled by one or more **What Do You Think?** questions. Students are not expected to know a "right" answer. These questions are supposed to elicit from students their beliefs regarding a very specific prediction or outcome, and students should commit to a written specific answer in their logs.

When students have completed **For You To Do**, convene the entire class for a demonstration of objects moving at constant speed on low-friction surfaces. A demonstration is included at the end of this activity. Possible materials for demonstrating motion at constant speed for an object given a push start on a low-friction, flat surface such as a smooth counter top or the glass surface include:

- balloon air puck on a smooth, hard surface
- a piece of dry ice on a smooth, hard surface
- glider on air track
- puck on air table
- puck on raw rice in a ripple tank
- puck on plastic bead bearings in a ripple tank.

Then direct students to read the **For You To Read** and **Physics Talk** sections. Reserve some time for closure after students have completed the reading.

You may wish to direct the students' attention to the fact that several sports involve motion for which an initial speed does not involve running in the literal sense, but may involve an object, such as a shot put ball, being given an initial speed by an athlete.

# Activity Overview

## Student Objectives

### Students will:

- Understand and apply Galileo's Principle of Inertia.
- Understand and apply Newton's First Law of Motion.
- Recognize inertial mass as a physical property of matter.

ANSWERS FOR THE TEACHER ONLY

## What Do You Think?

The horizontal distance a basketball player travels while "hanging" is determined by the speed upon jumping; since the speed often is high, the trajectory is quite flat near the peak of flight, giving the illusion that the player "hangs" in the air.

Skaters maintain speed on ice due to very low friction between the blades and the ice.

Physics in Action

Activity 1

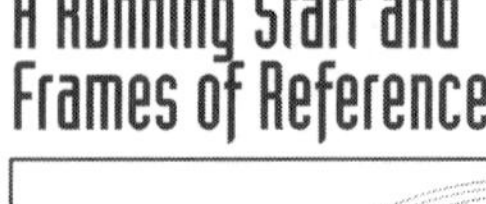

# A Running Start and Frames of Reference

**GOALS**

In this activity you will:

- Understand and apply Galileo's Principle of Inertia.
- Understand and apply Newton's First Law of Motion.
- Recognize inertial mass as a physical property of matter.

### What Do You Think?

Many things that happen in athletics are affected by the amount of "running start" speed an athlete can produce.

- **What determines the amount of horizontal distance a basketball player travels while "hanging" to do a "slam dunk" during a fast break?**
- **How do figure skaters keep moving across the ice at high speeds for long times while seldom "pumping" their skates?**

Record your ideas about these questions in your *Active Physics* log. Be prepared to discuss your responses with your small group and the class.

### For You To Do

1. Use a salad bowl and a ball to explore the question, "When a ball is released to roll down the inside surface of a salad bowl, is the motion of the ball up the far side of the bowl the 'mirror image' of the ball's downward motion?" Use a nonpermanent pen to mark a starting position for the ball near the top edge of the bowl. Use a

6

Active Physics

flexible ruler to measure, in centimeters, the distance along the bowl's curved surface from the bottom-center of the bowl to the mark.

a) Make a table similar to the one below in your log.

| Start | Trial | Starting Distance (cm) | Recovered Distance (cm) | Recovered Distance / Starting Distance |
|---|---|---|---|---|
| High | 1 | | | |
| | 2 | | | |
| | 3 | | | |
| Medium | 1 | | | |
| | 2 | | | |
| | 3 | | | |
| Low | 1 | | | |
| | 2 | | | |
| | 3 | | | |

b) Record the measured distance in your table as the High Starting Distance.

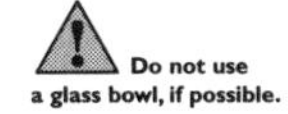

Do not use a glass bowl, if possible.

2. Prepare to observe and mark the position on the far side of the bowl where the ball stops when it is released from the starting position. Release the ball from the starting position and mark the position where it stops. Measure the distance from the bottom-center to the stop mark.

a) Record the distance in your table as the High Recovered Distance.

3. Repeat **Step 2** above two more times to see if the results are consistent.

a) Record the data for all three trials.

4. Mark two more starting positions on the surface of the bowl, one for Medium Starting Distance and another for Low Starting Distance.

a) Measure and record each of the new distances.

ANSWERS

# For You To Do

1. a) Students copy tables into their logs.
   b) Students record data.

2.–4. Students record data. When rolling the ball within the salad bowl, students should find the Recovered Distance to be very nearly equal to the Starting Distance.

b) Observe, mark, measure and record the recovered distances for three trials at each of the medium and low starting positions.

c) Complete the table by calculating and recording the value of the ratio of the Recovered Distance to the Starting Distance for each trial. (The ratio is the Recovered Distance divided by the Starting Distance.)

Example:
If the Recovered Distance is 6.0 cm for a Starting Distance of 10.0 cm, the value of the ratio is $\frac{6.0 \text{ cm}}{10.0 \text{ cm}} = 0.6$.

d) For each of the three starting distances, to what extent is the motion of the ball up the far side of the bowl the "mirror image" of the downward motion? Use data as evidence for your answer.

e) Does the fraction of the starting distance "recovered" when going up the far side of the bowl depend on the amount of starting distance? Describe any pattern of data that supports your answer.

5. Repeat the activity but roll the ball along varying slopes during its upward motion. Make a track that has the same slope on both sides, as shown below. Your teacher will suggest how high the ends of the track sections should be elevated. This time, concentrate on comparing the vertical height of the ball's release position to the vertical height of the position where the ball stops.

a) Measure and record the vertical height (not the distance along the track) from which the ball will be released at the top end of the left-hand section of track.

b) Prepare to observe and mark the position on the right-hand section of track where the ball stops when it is released from the starting position. Release the ball from the top end of the left-hand section of track and mark the position where it stops. Measure and record the vertical height of the position where the ball stops.



## Answers

# For You To Do *(continued)*

4. c)-e) The ratio of Recovered Distance to Starting Distance should be only slightly less than 1.00 and typically 0.90 or more. The actual value will, of course, depend on the coefficient of friction for the particular kind of ball and bowl used. You may expect that the error of measurement will be nearly as much as the observable difference in distances, indicating nearly complete "conservation" of distance. The ratio should remain essentially constant regardless of the starting height.

5. a)-c) Students record data and calculate the ratios of Recovered to Starting Distance.

Activity 1 A Running Start and Frames of Reference

c) Calculate the ratio of the recovered height to the starting height. How is this case, and the result, similar to what you did when using the salad bowl? How is it different?

6. Leave the left-hand starting section of track unchanged, but change the right-hand section of track so that it has less slope and is at least long enough to allow the ball to recover the starting height. The track should be arranged approximately as shown below.

a) Predict the position where the ball will stop on the right-hand track if it is released from the same height as before on the left-hand track. Mark the position of your guess on the right-hand track and explain the basis for your prediction in your log.

7. Release the ball from the same height on the left-hand section of track as before and mark the position where the ball stops on the right-hand section of track.

a) How well did you guess the position? Why do you think your guess was "on" or "off"?

b) Measure the vertical height of the position where the ball stopped and again calculate the ratio of the recovered height to the starting height. Did the ratio change? Why, do you think, did the ratio change or not change?

8. Imagine what would happen if you again did not change the left-hand starting section of track, but changed the right-hand section of track so that it would be horizontal, as shown below.

a) How far along the horizontal track would the ball need to roll to recover its starting height (or most of it)? How far do you think the ball would roll?

b) When rolling on the horizontal track, what would "keep the ball going"?



ANSWERS

## For You To Do *(continued)*

6. a) Make certain that students record their predictions.

7. a)-b) When rolling the ball down the adjustable track, students must shift attention from comparing distances traveled along the "down" and "up" paths to vertical distances "down" and "up." For symmetrical slopes, the former measurement—distance along either track—would serve, but, as the "up" slope is made less, the ball will roll farther on the "up" slope to gain nearly the same vertical height as the height from which it was released on the "down" slope.

8. a)-b) The intent is for students to realize, in accord with Galileo's ideal, that when the "up" track has no slope, the ball will roll "forever" in its attempt to gain the height from which it was released.

## FOR YOU TO READ

### Inertia

Italian philosopher Galileo Galilei (1564–1642), who can be said to have introduced science to the world, noticed that a ball rolled down one ramp seems to seek the same height when it rolls up another ramp. He also did a "thought experiment" in which he imagined a ball made of extremely hard material set into motion on a horizontal, smooth surface, similar to the final track in **For You To Do**. He concluded that the ball would continue its motion on the horizontal surface with constant speed along a straight line "to the horizon" (forever). From this, and from his observation that an object at rest remains at rest unless something causes it to move, Galileo formed the Principle of **Inertia**:

**Inertia is the natural tendency of an object to remain at rest or to remain moving with constant speed in a straight line.**

Isaac Newton, born in England on Christmas day in 1642 (within a year of Galileo's death), used Galileo's Principle of Inertia as the basis for developing his First Law of Motion, presented in **Physics Talk**. Crediting Galileo and others for their contributions to his thinking, Newton said, "If I have seen farther than others, it is because I have stood on the shoulders of giants."

### Running Starts

Running starts take place in many sporting activities. Since there seems to be this prior motion in many sports, there must be some advantage to it.

In sports where the objective is to maximize the speed of an object or the distance traveled in air, the prior motion may be essential. When a javelin is thrown, at the instant of release it has the same speed as the hand that is propelling it.

- The hand has a forward speed relative to the elbow, the elbow has a forward speed relative to the shoulder (because the arm is rotating around the elbow and shoulder joints), and the shoulder has a forward speed relative to the ground because the body is rotating and the body is also moving forward.
- The javelin speed then is the sum of each of the above speeds. If the thrower is not running forward, that speed does not add into the equation.

You can write a **velocity** equation to show the speeds involved.

$$v_{javelin} = v_{hand} + v_{elbow} + v_{shoulder} + v_{ground}$$

Motion captures everyone's attention in sports. Starting, stopping, and changing direction (**accelerations**) are part of the motion story, and they are exciting components of many sports. Ordinary, straight-line motion is just as important but is easily overlooked.

**PHYSICS TALK**

**Newton's First Law of Motion**

Isaac Newton included Galileo's Principle of Inertia as part of his **First Law of Motion**:

**In the absence of an unbalanced force, an object at rest remains at rest, and an object already in motion remains in motion with constant speed in a straight-line path.**

Newton also explained that an object's mass is a measure of its inertia, or tendency to resist a change in motion.

Here is an example of how Newton's First Law of Motion works:

Inertia is expressed in kilograms of mass. If an empty grocery cart has a mass of 10 kg and a cart full of groceries has a mass of 100 kg, which cart would be more difficult to move (have a greater tendency to remain at rest)? If both carts already were moving at equal speeds, which cart would be more difficult to stop (would have a greater tendency to keep moving)? Obviously in both cases, the answer is the cart with more mass.

Physics Words

**inertia:** the natural tendency of an object to remain at rest or to remain moving with constant speed in a straight line.

**acceleration:** the change in velocity per unit time.

**frame of reference:** a vantage point with respect to which position and motion may be described.

**FOR YOU TO READ**

**Frames of Reference**

In this activity, you investigated Newton's First Law. In the absence of external forces, an object at rest remains at rest and an object in motion remains in motion. If you were challenged to throw a ball as far as possible, you would probably now be sure to ask if you could have a running start. If you run with the ball prior to throwing it, the ball gets your speed before you even try to release it. If you can run at 5 m/s, then the ball will get the additional speed of 5 m/s when you throw it. When you do throw the ball, the ball's speed is the sum of your speed before releasing the ball, 5 m/s, and the speed of the release.

→

Physics in Action

It may be easier to understand this if you think of a toy cannon that could be placed on a skateboard. The toy cannon always shoots a small ball forward at 7 m/s. This can be checked with multiple trials. The toy cannon is then attached to the skateboard. A release mechanism is set up so that the cannon continues to shoot the ball forward at 7 m/s when the skateboard is at rest. When the skateboard is given an initial push, the skateboard is able to travel at 3 m/s. If the cannon releases the ball while the skateboard is moving, the ball's speed is now measured to be 10 m/s. From where did the additional speed come? The ball's speed is the sum of the ball's speed from the cannon plus the speed of the skateboard. 7 m/s + 3 m/s = 10 m/s.

You may be wondering if the ball is moving at 7 m/s or 10 m/s. Both values are correct — it depends on your **frame of reference**. The ball is moving at 7 m/s relative *to* the skateboard. The ball is moving at 10 m/s relative *to* the Earth.

Imagine that you are on a train that is stopped at the platform. You begin to walk toward the front of the train at 3 m/s. Everybody in the train will agree that you are moving at 3 m/s toward the front of the train. This is your speed *relative to* the train. Everybody looking into the train from the platform will also agree that you are moving at 3 m/s toward the front of the train. This is your speed *relative to* the platform.

Imagine that you are on the same train, but now the train is moving past the platform at 9 m/s. You begin to walk toward the front of the train at 3 m/s. Everybody in the train will agree that you are moving at 3 m/s toward the

front of the train. This is your speed *relative to* the train. Everybody looking into the train from the platform will say that you are moving at 12 m/s (3 m/s + 9 m/s) toward the front of the train. This is your speed relative *to* the platform.

Whenever you describe speed, you must always ask, "*Relative to what?*" Often, when the speed is relative to the Earth, this is assumed in

the problem. If your frame of reference is the Earth, then it all seems quite obvious. If your frame of reference is the moving train, then different speeds are observed.

In sports where you want to provide the greatest speed to a baseball, a javelin, a football, or a tennis ball, that speed could be increased if you were able to get on a moving platform. That being against the rules and inappropriate for many reasons, an athlete will try to get the body moving with a running start, if allowed. If the running start is not permitted, the athlete tries to move every part of his or her body to get the greatest speed.

### Sample Problem 1

A sailboat has a constant velocity of 22 m/s east. Someone on the boat prepares to toss a rock into the water.

a) Before being tossed, what is the speed of the rock with respect to the boat?

b) Before being tossed, what is the speed of the rock with respect to the shore?

c) If the rock is tossed with a velocity of 16 m/s east, what is the rock's velocity with respect to shore?

d) If the rock is tossed with a velocity of 16 m/s west, what is the rock's velocity with respect to shore?

***Strategy:*** Before determining a velocity, it is important to check the frame of reference. The rock's velocity with respect to the boat is different from the velocity with respect to the shore. The direction of the rock also impacts the final answer.

***Givens:***

$v_b$ = 22 m/s east

$v_r$ = 16 m/s (direction varies)

***Solution:***

a) With respect to the boat, the rock's velocity is 0 m/s.
The rock is moving at the same speed as the boat, but you wouldn't notice this velocity if you were in the boat's frame of reference.

b) With respect to shore, the rock's velocity is 22 m/s east.
The rock is on the boat, which is traveling at 22 m/s east. Relative to the shore, the boat and everything on it act as a system traveling at the same velocity.

c) With respect to the shore, the rock's velocity is now 38 m/s east.
It is the sum of the velocity values. Since each is directed east, the relative velocity is the sum of the two.

$v = v_b + v_r$

= 22 m/s east + 16 m/s east

= 38 m/s east

→

d) With respect to shore, the rock's velocity is now 6 m/s east.
Since the directions are opposite, the relative velocity is the difference between the two.

$$v = v_b - v_r$$
$$= 22 \text{ m/s east} - 16 \text{ m/s west}$$
$$= 6 \text{ m/s east}$$

### Sample Problem 2

A quarterback on a football team is getting ready to throw a pass. If he is moving backward at 1.5 m/s and he throws the ball forward at 10.0 m/s, what is the velocity of the ball relative to the ground?

***Strategy:*** Use a negative sign to indicate the backward direction. Add the two velocities to find the velocity relative to the ground.

***Givens:***

***Solution:***

Add the velocities.

$10.0 \text{ m/s} + (-1.5 \text{ m/s}) = 8.5 \text{ m/s}$

The ball is moving forward at 8.5 m/s relative to the ground.

## Reflecting on the Activity and the Challenge

Running starts can be observed in many sports. Many observers may not realize the important role that inertia plays in preserving the speed already established when an athlete engages in activities such as jumping, throwing, or skating from a running start. "Immovable objects," such as football linemen, illustrate the tendency of highly massive objects to remain at rest and can be observed in many sports. You should have no problem finding a great variety of video segments that illustrate Newton's First Law.

### Physics To Go

1. Provide three illustrations of Newton's First Law in sporting events. Describe the sporting event and which object when at rest stays at rest, or when in motion stays in motion. Describe these same three illustrations in the manner of an entertaining sportscaster.
2. Find out about a sport called curling (it is an Olympic competition that involves some of the oldest Olympians) and how this sport could be used to illustrate Newton's First Law of Motion.
3. When a skater glides across the ice on only one skate, what kind of motion does the skater have? Use principles of physics as evidence for your answer.
4. Use what you have learned in **Activity 1** to describe the motion of a hockey puck between the instant the puck leaves a player's stick and the instant it hits something. (No "slap shot" allowed; the puck must remain in contact with the ice.)
5. Why do baseball players often slide into second base and third base, but almost never slide into first base after hitting the ball? (The answer depends on both the rules of baseball and the laws of physics.)
6. Do you think it is possible to arrange conditions in the "real world" to have an object move, unassisted, in a straight line at constant speed forever? Explain why or why not.

ANSWERS

## Physics To Go

1. Answers will vary.
   Possible examples:
   An outfielder diving for a line drive. The outfielder continues in motion, sliding along the ground, his hat also continues in motion.

   A slap shot in hockey. The puck continues to move in a constant horizontal motion, once it has been set in motion by the player.
2. Curling, an Olympic competition, is similar to shuffleboard and involves sliding "stones" on ice.
3. A skater has either nearly constant speed, or very small uniform deceleration.
4. A hockey puck on ice has either nearly constant speed, or very small uniform deceleration.
5. A baseball player slides into second or third to decelerate to a stop at the base because if the base is overrun, the player would be "out" if tagged; at first base, a player can overrun the base without danger of being tagged out and the fastest way of beating a throw to first base is to run without sliding into the base.
6. It does not seem possible to eliminate friction to arrive at perpetual motion in the real world.

ANSWERS

## Physics To Go
*(continued)*

7. a) The ball will appear to go straight up and down.

   b) The little girl will see the ball travel in a parabola.

   c) The speed relative to the girl will be 2.5 m/s + 4.5 m/s = 7.0 m/s.

8. The relative velocity will be 4.2 m/s + 10.3 m/s = 14.5 m/s.

9. a) The velocity relative to the tracks is 5.6 m/s + 2.4 m/s = 8.0 m/s.

   b) The velocity relative to the tracks is 5.6 m/s − 2.4 m/s = 3.2 m/s.

   c) Since the two velocities are perpendicular, we must use the Pythagorean Theorem

$(5.6 \text{ m/s})^2 + (2.4 \text{ m/s})^2 = v^2$

$v = 6.1 \text{ m/s}$

Using the tangent button on the calculator or a vector diagram, the angle is 67°

(Students at this point should be able to make the diagram. Some will be able to use the Pythagorean Theorem. More emphasis on this will come in **Activity 2**.)

10. a) One vaulter is moving 4.3 m/s – 3.8 m/s = 0.5 m/s faster than the other. This is their relative speed.

    b) The one going faster has more kinetic energy. With all else being equal (skill, strength, etc.) the one going faster will also go higher.

11. The speed is 85 m/s – 18 m/s = 67 m/s.

7. You are pulling your little brother in his red wagon. He has a ball, and he throws it straight up into the air while you are pulling him forward at a constant speed.
   a) What will the path of the ball look like to your little brother in the wagon?
   b) What will the path of the ball look like to a little girl who is standing on the sidewalk watching you?
   c) If your brother throws the ball forward at a velocity of 2.5 m/s while you are pulling the wagon at a velocity of 4.5 m/s, at what speed does the girl see the ball go by?

8. A track and field athlete is running forward with a javelin at a velocity of 4.2 m/s. If he throws the javelin at a velocity relative to him of 10.3 m/s, what is the velocity of the javelin relative to the ground?

9. You are riding the train to school. Since the train car is almost empty, you and your friend are throwing a ball back and forth. The train is moving at a velocity of 5.6 m/s. Suppose you throw the ball to each other at the same speed, 2.4 m/s.
   a) What is the velocity of the ball relative to the tracks when the ball is moving toward the front of the car?
   b) What is the velocity of the ball relative to the tracks when it is moving toward the back of the car?
   c) What if you and your friend throw the ball perpendicular to the aisle of the train? What is the ball's velocity then?

10. Two athletes are running toward the pole vault. One is running at 3.8 m/s and the other is running at 4.3 m/s.
    a) What is their velocity relative to each other?
    b) If they leave the ground at their respective velocities, which one has the energy to go higher in the vault? Explain.

11. While riding a horse, a competitor shoots an arrow toward a target. The speed of the arrow as it reaches the target is 85 m/s. If the horse was traveling at 18 m/s, at what speed did the arrow leave the bow? (Assume the horse and arrow are traveling in the same direction.)

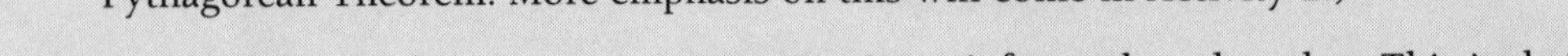

Activity 1 A

# Can Objects Move Forever?

## FOR YOU TO DO

In this exercise, you will observe the motion of various objects on a variety of surfaces to see whether an object might be able to move forever as Galileo concluded. Since you do not have an infinite time to work, nor an infinitely large room to work in, this question can only be approached by looking at limited examples and by trying to imagine the limitless consequences of what you see. Also, since space is limited, it is important to use as little space as possible for the process of getting the object started on the motion that is to be investigated. That leaves more room to see what happens when the object is "on its own."

1. The first object to start in motion is an eraser on a table top or on the floor. The spine of the eraser is to be in contact with the floor.

   a) Does the eraser sustain its motion on its own? Describe what happens.

2. Replace the table top with smooth glass or plastic.

   a) Describe the eraser's motion.

3. Put about 1/8 cup of raw rice on the same surface and try the eraser again.

   a) Describe the eraser's motion.

4. Remove the rice, and replace it with dried peas. Try the eraser again.

   a) Describe the eraser's motion.

5. Remove the peas, and pour on a thin, sparse layer of minute plastic beads. Be careful not to spill the beads on the floor, as they are very difficult to pick up and clean up. Try the eraser again.

   a) Describe the eraser's motion.

   b) Try other objects on the beaded surface (coins, small blocks of wood, objects with weights on them, etc.). Report the results.

6. Set a non-inflated air puck in motion along a table top.

   a) Describe what happens.

7. Fill the air reservoir and set the puck in motion again.

   a) Describe and explain what happens in each case.

For use with *Physics in Action*, Chapter 1, Activity 1: A Running Start and Frames of Reference

# Performance Assessment Rubrics

Part 1 = maximum 4
Part 2 = maximum 2
Part 3 = maximum 5
Part 4 = maximum 6

**1. Student records experimental data demonstrating that the distance a ball travels down a salad bowl is nearly the same distance that the ball travels up the same salad bowl.**

| Descriptor | Task accomplished | Task not accomplished |
|---|---|---|
| a) Measurement is taken with ruler. | | |
| b) Units of measurement are recorded in centimeters. | | |
| c) Three trials are used for high, medium, and low distances. | | |
| d) Release height is compared with recovery height. | | |

Maximum 4 marks if each of the sub tasks is accomplished.

**Total marks:** ________

**2. Student uses deductive reasoning to make a generalization about how the downward motion of the ball mirrors the upward motion of the ball. The greater the release height, the greater the recovery distance.**

| Descriptor | Task accomplished | Task not accomplished |
|---|---|---|
| a) Student explains that the recovery distance is dependent upon the start distance. | | |
| b) Student notes the constant ratio by comparing start distance of three places (high, medium, and low) to recovery distance. | | |

Maximum 2 marks if each of the sub tasks is accomplished.

**Total marks:** ________

For use with *Physics in Action*, Chapter 1, Activity 1: A Running Start and Frames of Reference

**3. Student uses experimental data to find a constant ratio of release distance to recovery distance.**

| Descriptor | Task accomplished | Task not accomplished |
|---|---|---|
| a) Vertical measurement is taken with a ruler. | | |
| b) Units of measurement are recorded in centimeters. | | |
| c) Release height is compared with recovery height. | | |
| d) Ratio of the recovery distance to start distance is calculated correctly. | | |
| e) Comparison is made between the ramp and salad bowl. Similarities and differences are identified. | | |

Maximum 5 marks if each of the sub tasks is accomplished.

**Total marks:** _________

**4. Student correctly decreases the slope on the right of the incline and notes that the release height affects the distance that the ball rolls along a vertical plane.**

| Descriptor | Task accomplished | Task not accomplished |
|---|---|---|
| a) Prediction of recovery distance. | | |
| b) Student notes that the ball travels a greater distance along a horizontal plane. | | |
| c) Student measures the vertical distance that the ball traveled up the ramp. | | |
| d) The ratio of the vertical start height to horizontal recovery height is calculated correctly. The ratio remains constant. | | |
| e) Gravity is identified as the force that caused the ball to move downward and slowed the ball as it moved upward. | | |
| f) Student concludes that the force of gravity remains constant for both downward and upward movement of the ball. | | |

Maximum 6 marks if each of the sub tasks is accomplished.

**Total marks:** _________

For use with *Physics in Action*, Chapter 1, Activity 1: A Running Start and Frames of Reference

# ACTIVITY 2
## Push or Pull – Adding Vectors

# Background Information

It is suggested that the "**Physics Talk:** Newton's Second Law of Motion" and "**For You To Read:** Weight and Newton's Second Law" sections of the student text for **Activity 1** be read again before proceeding in this section.

The unit of mass, or quantity of matter, in the International System of Units is the kilogram. One of seven base units from which all other units are derived, the kilogram originally was conceived as the quantity of matter represented by 1 liter of water at the temperature of maximum density, 4°C; today, the kilogram is defined by a carefully protected metal standard called the International Prototype Kilogram. When a balance which employs the force of gravity is used to measure the mass of an object by comparison to prototype masses, the resulting measurement is known as the "gravitational mass" of the object. Mass is also internationally recognized as a measure of the inertial resistance of an object to acceleration. When a standard force is used to compare an object's acceleration to the acceleration of a prototype mass as a means of measuring the mass of the object, the resulting measurement is known as the "inertial mass" of the object. It can be shown that 1 kilogram of gravitationally determined mass is equivalent to 1 kilogram of inertial mass.

A derived unit of force, the newton, is defined in terms of base units of mass, length and time using Newton's Second Law of Motion, $F = ma$. 1 newton (N) is the force which will cause 1 kilogram to accelerate at $1 m/s^2$, or $1 N = 1 (kg)m/s^2$.

The word "weight" denotes a force; the weight of an object is the product of its mass and the acceleration due to gravity, $9.81 m/s^2$. Since weight is the force due to gravity, weight is measured in newtons. One newton is roughly 1/4 lb., prompting the identification of the familiar 1/4 lb-burger as a "newton burger."

In summary, matter seems to have two distinct properties:
1. It exhibits a resistance to acceleration, property called "inertia."
2. It has the property of gravitation; matter is attracted to other matter.

It is clear why it is that all objects, irrespective of mass, have the same free fall acceleration at a given location. The more mass, the more gravitational force; but the more mass, the more difficult it is to accelerate the object. These two factors exactly compensate to produce the same acceleration for every freely falling object at a given location.

## *Active*-ating the Physics InfoMall

A big concept in this activity is the concept of force. Students' understanding of this concept has been studied extensively. An InfoMall search using "force" AND "misconception*" in only the Articles and Abstracts Attic produced many great references. The first such hit is the article containing the Force Concept Inventory. The second is "Common sense concepts about motion," *American Journal of Physics,* vol. 53, issue 11, 1985. in which it is mentioned that "(a) On the pretest (post-test), 47% (20%) of the students showed, at least once, a belief that under no net force, an object slows down. However, only 1% (0%) maintained that belief across similar tasks. (b) About 66% (54%) of the students held, at least once, the belief that under a constant force an object moves at constant speed. However, only 2% (1%) held that belief consistently." More results are reported in this article.

The third hit in this search is "Physics that textbook writers usually get wrong," in *The Physics Teacher*, vol. 30, issue 7, 1992. This article is good reading for any introductory physics teacher. The list of hits from this search is long. In fact, it had to be limited to just the Articles and Abstracts Attic to prevent the "Too many hits" warning. If you search the rest of the CD-ROM, you will find many other great hits, such as this quote from Chapter 3 of Arons' *A Guide to Introductory Physics Teaching: Elementary Dynamics*: "In the study of physics, the Law of Inertia and the concept of force have, historically, been two of the most formidable stumbling blocks for students, and, as of the present time, more cognitive research has been done in this area than in any other."

Newton's Second Law is discussed in virtually every physics textbook in existence, not to mention the InfoMall. Depending on the level at which you wish to present this Law, you may wish to examine the conceptual-level texts, the algebra-based texts, or even the calculus-bases textbooks on the InfoMall.

If you want more exercises to give to your students, searching the InfoMall is a bad idea — there are too many problems on the CD-ROM. Searching with keywords "force" AND "acceleration" AND "mass" in the Problems Place alone produces "Too many hits." However, you will find more than enough by simply going to the Problems Place and browsing a few of the resources you will find there. For example, *Schaum's 3000 Solved Problems in Physics* has a section on Newton's Laws of Motion. You will surely find enough problems there to keep any student busy for some time!

# Planning for the Activity

## Time Requirements

- One class period.

## Materials Needed

### For each group:

- balls, set of 2
- ball, tennis
- c-clamp
- ruler, flexible plastic, 30 cm
- ruler, metric, mm marking
- washers, metal, 3/4"

## Advance Preparation and Setup

Identify the particular combination of flexible rulers (or plastic strips) and weights (coins or metal washers) which will serve as force meters. Identify a means of preventing the weights from slipping off the bent plastic strip, such as a lightweight cardboard "lip" taped to the plastic strip.

Also identify the set of objects to be accelerated; either balls of about the same diameter but having different masses or laboratory carts which can be loaded to vary the mass would work. If possible, have at least three objects of different masses available for each group.

Try for yourself the calibration procedure and the use of the force meter to accelerate objects in advance of class. You may need to try different sizes of coins or metal washers to find a kind which will produce a reasonable amount of bend in the ruler for a 4-washer load while at the same time providing a reasonable amount of acceleration when the smallest and largest objects are pushed using the smallest and largest forces.

# Teaching Notes

Students can be expected to need practice to exert constant amounts of force on moving objects. Only semiquantitative comparisons of the amounts of acceleration (e.g., low, higher, even higher) which result from varying the amount of force (while mass is held constant) and from varying the amount of mass (when force is held constant) are intended.

Direct all students to silently read the **For You To Read** section. Then conduct a brief discussion of the assumption presented in the section. You may wish to point out that assumptions represent beliefs which may be argued, but not proven as "right" or "wrong." Another example of an assumption which could be used for the discussion is "There is a tooth fairy."

You may wish to see if students really believe that gravity treats all athletes equally by probing students about the "hang time" of basketball stars.

NOTES

# Activity 2 Push or Pull—Adding Vectors

**GOALS**

In this activity you will:

- Recognize that a force is a push or a pull.
- Identify the forces acting on an object.
- Determine when the forces on an object are either balanced or unbalanced.
- Calibrate a force meter in arbitrary units.
- Use a force meter to apply measured amounts of force to objects.
- Compare amounts of acceleration semiquantitatively.
- Understand and apply Newton's Second Law of Motion.
- Understand and apply the definition of the newton as a unit of force.
- Understand weight as a special application of Newton's Second Law.

### What Do You Think?

Moving a football one yard to score a touchdown requires strategy, timing, and many forces.

- **What is a force?**
- **Can the same force move a bowling ball and a ping-pong ball?**

Record your ideas about these questions in your *Active Physics* log. Be prepared to discuss your responses with your small group and the class.

### For You To Do

1. Use a flexible ruler as a "force meter". Use washers to make a scale of measurement for (that is, to calibrate) the meter inpennyweights. The force you are using to calibrate the meter is gravity, the force with which Earth pulls downward



ANSWERS

## For You To Do

1. Student activity.

# Activity Overview

In this activity students calibrate a crude "force meter" by deforming (bending) a plastic strip using washers. Students then use the force meter to accelerate the same object using different forces, and different objects using the same force.

## Student Objectives

### Students will:

- Recognize that a force is a push or a pull.
- Identify the forces acting on an object.
- Determine when the forces on an object are either balanced or unbalanced.
- Calibrate a force meter in arbitrary units.
- Use a force meter to apply measured amounts of force to objects.
- Compare amounts of acceleration semiquantitatively.
- Understand and apply Newton's Second Law of Motion, $F=ma$.
- Understand and apply the definition of the newton as a unit of force,

  $1 \text{ N} = 1 \text{ (kg)m/s}^2$.
- Understand weight as a special application of Newton's Second Law,

  Weight = $mg$.

ANSWERS FOR THE TEACHER ONLY

## What Do You Think?

In simple terms, a force is a push or a pull. Some forces, such as gravitational and magnetic forces can act on objects without having to be in contact with them. Many other forces, called mechanical forces, act when particles or objects contact each other Forces are very important in physics because they determine how matter interacts with other matter.

The same force could be used to move both a bowling ball and a ping-pong ball. The difference would be the amount by which each is accelerated by the force. The greater the mass, the less the acceleration experienced by an object when the same force is applied to it. Mass affects acceleration.

on every object near its surface. *Carefully* clamp the plastic strip into position as shown in the diagram on the previous page.

2. Draw a line on a piece of paper. Hold the paper next to the plastic strip so that the line is even with the edge of the strip. Mark the position of the end of the strip on the reference line and label the position as the "zero" mark.

3. Place one washer on the top surface of the strip near the strip's outside end. Notice that the strip bends downward and then stops. Hold the paper in the original position and mark the new position of the end of the strip. Label the mark as "1 pennyweight."

4. Repeat **Step 3** for two, three, and four coins placed on the strip. In each case mark and label the new position of the end of the strip.

   a) Copy the reference line and the calibration marks from the piece of paper into your log.

5. Practice holding one end of the "force meter" (plastic strip) in your hand and pushing the free end against an object until you can bend the strip by forces of 1, 2, 3, and 4-pennyweight amounts. To become good at this, you will need to check the amount of bend in the strip against your calibration marks as you practice.

6. Use the force meter to push an object such as a tennis ball with a continuous 1-pennyweight force. You will need to keep up with the object as it moves and to keep the proper bend in the force meter. You may need to practice a few times to be able to do this.

   a) In your log, record the amount of force used, a description of the object, and the kind of motion the object seemed to have.

7. Repeat **Step 6** three more times, pushing on the same object with steady (constant) 2, 3, and 4-pennyweight amounts of force.

   a) Record the results in your log for each amount of force.





ANSWERS

## For You To Do *(continued)*

2. – 3. Student activity.

4. a) Students record calibration in their logs. As each washer is added, the ruler deflects more. The force due to gravity of each washer is responsible for bending the ruler.

5. Student activity.

6. a) Students record observations in their logs.

7. a) The greater the force applied to the tennis ball, the greater its acceleration, as demonstrated by the increasing difficulty in keeping up with the ball to maintain the force on it.

8. Based on your observations, complete the statement: "The greater the constant, unbalanced force pushing on an object,..."
   a) Write the completed statement in your log.
9. Select an object that has a small mass. Use the force meter to push on the object with a rather large, steady force such as 3- or 4-pennyweight amounts.
   a) Record the amount of force used, a description of the object pushed (especially including its mass, compared to the other objects to be pushed) and the kind of motion the object seemed to have.
10. Repeat **Step 9** using the same amount of force to push objects of greater and greater mass.
    a) Record the results in your log for each object.
11. Based on your observations, complete the statement: "When equal amounts of constant, unbalanced force are used to push objects having different masses, the more massive object..."
    a) Write the completed statement in your log.

ANSWERS

## For You To Do *(continued)*

8. a) The greater the constant, unbalanced force pushing on an object, the greater the acceleration of that object.

9. a) Students record results in their logs.

10. a) As the mass of the objects increases, the acceleration decreases.

11. a) When equal amounts of constant, unbalanced forces are used to push objects having different masses, the more massive objects are accelerated less.

**Physics Words**

**Newton's Second Law of Motion:** if a body is acted upon by an external force, it will accelerate in the direction of the unbalanced force with an acceleration proportional to the force and inversely proportional to the mass.

**weight:** the vertical, downward force exerted on a mass as a result of gravity.

## PHYSICS TALK

### Newton's Second Law of Motion

Based on observations from experiments similar to yours, Isaac Newton wrote his **Second Law of Motion**:

**The acceleration of an object is directly proportional to the unbalanced force acting on it and is inversely proportional to the object's mass. The direction of the acceleration is the same as the direction of the unbalanced force.**

If 1 N (newton) is defined as the amount of unbalanced force that will cause a 1-kg mass to accelerate at 1 m/s$^2$ (meter per second every second), the law can be written as an equation:

$$F = ma$$

where $F$ is expressed in newtons (symbol N), mass is expressed in kilograms (kg), and acceleration is expressed in meters per second every second (m/s$^2$).

By definition, the unit "newton" can be written in its equivalent form: (kg)m/s$^2$.

Newton's Second Law can be arranged in three possible forms:

$$F = ma \qquad a = \frac{F}{m} \qquad m = \frac{F}{a}$$

## FOR YOU TO READ

### Weight and Newton's Second Law

Newton's Second Law explains what "weight" means, and how to measure it. If an object having a mass of 1 kg is dropped, its free fall acceleration is roughly 10 m/s$^2$.

Using Newton's Second Law,

$$F = ma$$

the force acting on the falling mass can be calculated as

$$F = ma$$
$$= 1 \text{ kg} \times 10 \text{ m/s}^2 \text{ or } 10 \text{ N}$$

The 10-N force causing the acceleration is known to be the gravitational pull of Earth on the 1-kg object. This gravitational force is given the special name **weight**. Therefore, it is correct to say, "The weight of a 1-kg mass is ten newtons."

What is the weight of a 2-kg mass? If dropped, a 2-kg mass also would accelerate due to gravity (as do all objects in free fall) at about 10 m/s$^2$. Therefore, according to Newton's Second Law, the weight of a 2-kg mass is equal to

$$2 \text{ kg} \times 10 \text{ m/s}^2 \text{ or } 20 \text{ N}$$

In general, to calculate the numerical value of an object's weight in newtons, it is necessary only to multiply the numerical value of its mass by the numerical value of the $g$ (acceleration due to gravity), which is about 10m/s$^2$.

$$\text{Weight} = mg$$

The preceding equation is the "special case" of Newton's Second Law that must be applied to any situation in which the force causing an object to accelerate is Earth's gravitational pull.

### Where There's Acceleration, There Must Be an Unbalanced Force

There are lots of different everyday forces. You just read about the force of gravity. There is also the force of a spring, the force of a rubber band, the force of a magnet, the force of your hand, the force of a bat hitting a ball, the force of friction, the buoyant force of water, and many more. Newton's Second Law tells you that accelerations are caused by unbalanced external forces. It doesn't matter what kind of force it is or how it originated. If you observe an acceleration (a change in velocity), then there must be an unbalanced force causing it.

When you apply a force, if the object has a small mass, the acceleration may be quite large for a given force. If the object has a large mass, the acceleration will be smaller for the same applied force. Occasionally, the mass is so large that you are not even able to measure the acceleration because it is so small.

If you push on a go-cart with the largest force possible, the cart will accelerate a great deal. If you push on a car with that same force, you

will measure a much smaller acceleration. If you were to push on the Earth, the acceleration would be too small to measure. Can you convince someone that a push on the Earth moves the Earth? Why should you believe something that you can't measure? If you were to assume that the Earth does not accelerate when you push on it, then you would have to believe that Newton's Second Law stops working when the mass gets too big. If that were so, you would want to determine how big is "too big." When you conduct such experiments, you find that the acceleration gets less and less as the mass gets larger and larger. Eventually, the acceleration gets so small that it is difficult to measure. Your inability to measure it doesn't mean that it is zero. It just means that it is smaller than your best measurement. In this way, you can assume that Newton's Second Law is always valid.

All of these statements are summarized in Newton's Second Law as you read in **Physics Talk**:

$$F = ma$$

or in forms that emphasize the acceleration and the mass

$$a = \frac{F}{m} \text{ and } m = \frac{F}{a}$$

### Sample Problem 1

A tennis racket hits a ball with a force of 150 N. While the 275-g ball is in contact with the racket, what is its acceleration?

***Strategy:*** Newton's Second Law relates the force acting on an object, the mass of the object, and the acceleration given to it by the force. Use the form of the equation that emphasizes acceleration to find the acceleration. The force unit, the newton, is defined as the amount of force needed to give a mass of 1.0 kg an acceleration of 1.0 m/s$^2$. Therefore, you will need to change the grams to kilograms.

***Givens:***

$F = 150.0$ N
$m = 275$ g

***Solution:***

$$275 \text{ g} = 0.275 \text{ kg}$$

$$a = \frac{F}{m} = \frac{150 \text{ N}}{0.275 \text{ kg}} = 545 \text{ m/s}^2$$

### Sample Problem 2

As the result of a serve, a tennis ball ($m_t = 58$ g) accelerates at 43 m/s$^2$.

a) What force is responsible for this acceleration?

b) Could an identical force accelerate a 5.0-kg bowling ball at the same rate?

***Strategy:*** Newton's Second Law states that the acceleration of an object is directly proportional to the applied force and indirectly proportional to the mass ($F = ma$).

***Givens:***

$a = 43$ m/s$^2$
$m_t = 58$ g $= 0.058$ kg
$m_b = 5.0$ kg

***Solution:***

a)

$$F = m_t a = 0.058 \text{ kg} \times 43 \text{ m/s}^2 = 2.494 \text{ N or } 2.5 \text{ N}$$

b) Since the mass of the bowling ball is much greater than that of the tennis ball, an identical force will result in a smaller acceleration.
(You can calculate the acceleration.)

$$a = \frac{F}{m_b}$$

$$= \frac{2.5\text{ N}}{5.0\text{ kg}}$$

$$= 0.50\text{ m/s}^2$$

## Adding Vectors

A vector is a quantity that has both magnitude and direction. Velocity is a vector. In the previous activity you found that the direction in which an object was traveling and the speed at which it was moving are equally important.

Force is also a vector because you can measure how big it is (its magnitude) and its direction. Acceleration is also a vector. The equation for acceleration reminds you that the force and the acceleration must be in the same direction.

Often, more than one force acts on an object. If the two forces are in the same direction, the sum of the forces is simply the addition of the two forces. A 30-N force by one person and a force of 40 N by a second person (pushing in the same direction) on the same desk provides a 70-N force on the desk. If the two forces are in opposite directions, then you give one of the forces a negative value and add them again. If one student pushes on a desk to the right with a force of 30 N and a second student pushes on the same desk to the left with a force of 40 N, the net force on the desk will be 10 N to the left. Mathematically, you would state that 30 N + (–40 N) = – 10 N where the negative sign denotes "to the left."

30 N
40 N
30 N + 40 N = 70 N

30 N –40 N
30 N + (–40 N) = –10 N

Occasionally, the two forces acting on an object are at right angles. For instance, one student may be kicking a soccer ball with a force of 30 N ahead toward the goal, while the second student kicks the same soccer ball with a force of 40 N toward the sideline. To find the net force on the ball and the direction the ball will travel, you must use vector addition. You can do this by using a vector diagram or the Pythagorean Theorem.

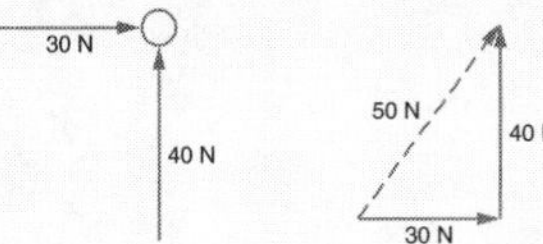


In the vector diagram shown above, the two force vectors are shown as arrows acting on the soccer ball. The magnitudes of the vectors are drawn to scale. The 30-N force may be drawn as 3.0 cm and the 40-N force may be drawn as 4.0 cm, if the scale is 10 N = 1 cm. To add the vectors, slide them so that the tip of the 30-N vector can be placed next to the tail of the 40-N vector (tip to tail method). The sum of the two vectors is then drawn from the tail of the 30-N vector to the tip of the 40-N vector. This *resultant* vector is measured and is found to be 5.0 cm, which is equivalent to 50 N. The angle is measured with a protractor and is found to be 53°.

Physics in Action

A second method of finding the resultant vector is to recognize that the 30-N and 40-N force vectors form a right triangle. The resultant is the hypotenuse of this triangle. Its length can be found by the Pythagorean Theorem.

$$a^2 + b^2 = c^2$$
$$30\text{ N}^2 + 40\text{ N}^2 = c^2$$
$$900\text{ N}^2 + 1600\text{ N}^2 = c^2$$
$$c = \sqrt{2500}\text{ N}^2$$
$$c = 50\text{ N}$$

The angle can be found by using the tangent function.

$$\tan\theta = \frac{\text{opposite}}{\text{adjacent}} = \frac{40\text{ N}}{30\text{ N}} = 1.33$$
$$\theta = 53°$$

Adding vector forces that are not perpendicular is a bit more difficult mathematically, but no more difficult using scale drawings and vector diagrams. Two other players are kicking a soccer ball in the direction shown in the diagram. The resultant vector force can be determined using the tip to tail approach.

The two arrows in the left diagram correspond to the two players kicking the ball at different angles. The diagram at the right shows the two vectors being added "tip to tail." The resultant vector (shown as a dotted line) represents the net force and is the direction of the acceleration of the soccer ball.

### Sample Problem 3

One player applies a force of 125 N in a north direction. Another player pushes with a force of 125 N west. What is the magnitude and direction of the resultant force?

***Strategy:*** Since the forces are acting at right angles, you can use the Pythagorean Theorem to find the resultant force. The direction of the force can be found using the tangent function.

***Givens:***

$F_1 = 125\text{ N}$

$F_2 = 125\text{ N}$

***Solution:***

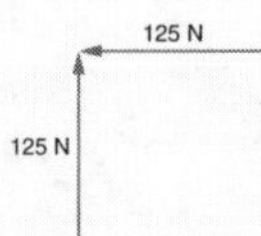


$$F_R^2 = F_1^2 + F_2^2$$
$$F_R = \sqrt{125\text{ N}^2 + 125\text{ N}^2}$$
$$= \sqrt{31{,}250\text{ N}^2}$$
$$= 177\text{ N}$$
$$\tan\theta = \frac{\text{opposite}}{\text{adjacent}} = \frac{125\text{ N}}{125\text{ N}} = 1$$
$$\theta = 45°$$

The resultant force is 177 N, 45° west of north.

### Reflecting on the Activity and the Challenge

What you learned in this activity really increases the possibilities for interpreting sports events in terms of physics. Now you can explain why accelerations occur in terms of the masses and forces involved. You know that forces produce accelerations. Therefore, if you see an acceleration occur, you know to look for the forces involved. You can apply this to the sport that you will describe.

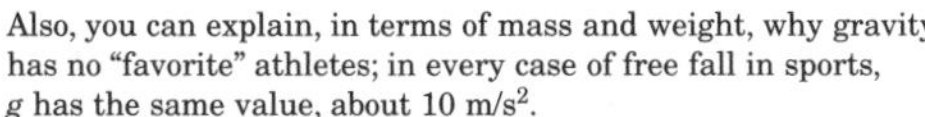

Also, you can explain, in terms of mass and weight, why gravity has no "favorite" athletes; in every case of free fall in sports, $g$ has the same value, about 10 m/s$^2$.

### Physics To Go

1. Copy and complete the following table using Newton's Second Law of Motion. Be sure to include the unit of measurement for each missing item.

| Newton's Second Law: | $F$ = | $m$ × | $a$ |
|---|---|---|---|
| Sprinter beginning 100-meter dash | ? | 70 kg | 5 m/s$^2$ |
| Long jumper in flight | 800 N | ? | 10 m/s$^2$ |
| Shot put ball in flight | 70 N | 7 kg | ? |
| Ski jumper going down hill before jumping | 400 N | ? | 5 m/s$^2$ |
| Hockey player "shaving ice" while stopping | −1500 N | 100 kg | ? |
| Running back being tackled | ? | 100 kg | −30 m/s$^2$ |

2. The following items refer to the table in **Question 1**:
   a) In which cases in the table does the acceleration match "$g$," the acceleration due to gravity 10 m/s$^2$? Are the matches to $g$ coincidences or not? Explain.
   b) The force on the hockey player stopping is given in the table as a negative value. Should the player's acceleration also be negative? What do you think it means for a force or an acceleration to be negative?
   c) The acceleration of the running back being tackled also is given as negative. Should the unbalanced force acting on him also be negative? Explain.

## Answers

## Physics To Go

1. See chart below.

2. a) The long jumper and the shot put ball both are cases of free fall; therefore the acceleration is g, the acceleration due to gravity.

   b) The negative sign is used to denote that the force and acceleration are in a direction opposite the motion.

   c) Since acceleration occurs in the direction of the causal force, yes the force should be shown as negative.

| Newton's Second Law: | $f$ = | $m$ × | $a$ |
|---|---|---|---|
| Sprinter beginning 100-meter dash | 350 N | 70 kg | 5 m/s$^2$ |
| Long jumper in flight | 800 N | 80 kg | 10 m/s$^2$ |
| Shot put ball in flight | 70 N | 7 kg | 10 m/s$^2$ |
| Ski jumper going down hill before jumping | 400 N | 80 kg | 5 m/s$^2$ |
| Hockey player "shaving ice" while stopping | –1500 N | 100 kg | -15 m/s$^2$ |
| Running back being tackled | –3000 N | 100 kg | –30 m/s$^2$ |

ANSWERS

## Physics To Go
*(continued)*

2. d) Students should be able to provide a plausible "voice-over" narration for an imagined video clip showing each event in the table.

3. 4.2 N / 0.30 kg = 14 $m/s^2$

4. 0.040 kg × 20 $m/s^2$ = 0.8 N

5. a) A bowling ball has greater inertia (mass) than a baseball; therefore, a bowling ball has a greater tendency to either remain at rest or remain in motion than does a baseball.

   b) More force is required to cause a bowling ball to accelerate than a baseball; therefore, throwing (accelerating) or catching (decelerating) a bowling ball involves much greater forces than throwing or catching a baseball when equal speeds are involved.

6. The sandwich would weigh 0.1 kg × 10 $m/s^2$ = 1 N.

7. Example: Weight
   = 150 lb. × 4.38 N/lb.
   = 657 N

   Mass
   = 657 N ÷ 10 $m/s^2$ = 65.7 kg

8. The component, or effectiveness, of the weight in the downhill direction, parallel to the slope of the hill, is 0.71 times the downward force of gravity (weight), or 7.1 N/kg; therefore, the acceleration is 7.1 N/kg = 7.1 $m/s^2$. This can be analyzed using either a scale drawing or trigonometry and should not be expected of all students at this level.

Physics in Action

d) In your mind, "play" an imagined video clip that illustrates the event represented by each horizontal row of the preceding table. Write a brief voice-over script for each video clip that explains how Newton's Second Law of Motion is operating in the event. Use appropriate physics terms, equations, numbers, and units of measurement in the scripts.

3. What is the acceleration of a 0.30-kg volleyball when a player uses a force of 42 N to spike the ball?

4. What force would be needed to accelerate a 0.040-kg golf ball at 20.0 $m/s^2$?

5. Most people can throw a baseball farther than a bowling ball, and most people would find it less painful to catch a flying baseball than a bowling ball flying at the same speed as the baseball. Explain these two apparent facts in terms of:
   a) Newton's First Law of Motion.
   b) Newton's Second Law of Motion.

6. Calculate the weight of a new fast-food sandwich that has a mass of 0.1 kg. Think of a clever name for the sandwich that would incorporate its weight.

7. In the United States, people measure body weight in pounds. Write down the weight, in pounds, of a person who is known to you. (This could be your weight or someone else's.)
   a) Convert the person's weight in pounds to the international unit of force, newtons. To do so, use the following conversion equation:
   Weight in newtons = Weight in pounds × 4.38 newtons/pound
   b) Use the person's body weight, in newtons, and the equation
   Weight = mg
   to calculate the person's body mass, in kilograms.

8. Imagine a sled (such as a bobsled or luge used in Olympic competitions) sliding down a 45° slope of extremely slippery ice. Assume there is no friction or air resistance (not really possible). Even under such ideal conditions, it is a fact that gravity could cause the sled to accelerate at a maximum of only 7.1 $m/s^2$. Why would the "ideal" acceleration of the sled not be $g$, 10 $m/s^2$? Your answer is expected only to suggest reasons why, on a 45° hill, the ideal free fall acceleration is "diluted" from 10 $m/s^2$ to about 7 $m/s^2$; you are not expected to give a complete explanation of why the "dilution" occurs.

Active Physics 26

9. If you were doing the voice-over for a tug-of-war, how would you explain what was happening? Write a few sentences as if you were the science narrator of that athletic event.
10. You throw a ball. When the ball is many meters away from you, is the force of your hand still acting on the ball?
11. Carlo and Sara push on a desk in the same direction. Carlo pushes with a force of 50 N, and Sara pushes with a force of 40 N. What is the total resultant force acting on the desk?
12. A car is stuck in the mud. Four adults each push on the back of the car with a force of 200 N. What is the total force on the car?
13. During a football game, two players try to tackle another player. One player applies a force of 50.0 N to the east. A second player applies a force of 120.0 N to the north. What is the total applied force? (Since force is a vector, you must give both the magnitude and direction of the force.)
14. In auto racing, a crash occurs. A red car hits a blue car from the front with a force of 4000 N. A yellow car also hits the blue car from the side with a force of 5000 N. What is the total force on the blue car? (Since force is a vector, you must give both the magnitude and direction of the force.)
15. A baseball player throws a ball. While the 700.0-g ball is in the pitcher's hand, there is a force of 125 N on it. What is the acceleration of the ball?
16. If the acceleration due to gravity at the surface of the Earth is approximately 9.8 m/s$^2$, what force does the gravitational attraction of the Earth exert on a 12.8-kg object?
17. A force of 30.0 N acts on an object. At right angles to this force, another force of 40.0 N acts on the same object.
    a) What is the net force on the object?
    b) What acceleration would this give a 5.6-kg wagon?
18. Bob exerts a 30.0 N force to the left on a box ($m$ = 100.0 kg). Carol exerts a 20.0 N force on the same box, perpendicular to Bob's.
    a) What is the net force on the box?
    b) Determine the acceleration of the box.
    c) At what rate would the box accelerate if both forces were to the left?

ANSWERS

## Physics To Go

*(continued)*

9. Students provide voice-over for tug-of-war.
10. No.
11. The resultant is 40 N + 50 N = 90 N.
12. The total force is 4 x 200 N = 800 N.
13. Application of the Pythagorean Theorem yields:

    $(50\ N)^2 + (120\ N)^2 = F^2$

    $F = 130\ N$

    Using the tangent button on the calculator or a vector diagram, the angle is 23° east of north.
14. Application of the Pythagorean Theorem yields:

    $(4000\ N)^2 + (5000\ N)^2 = F^2$

    $F = 6403\ N$

    Using the tangent button on the calculator or a vector diagram, the angle is 39°.
15. $F = ma$

    $a = F/m = (125\ N)/\ 0.7\ kg) = 179\ m/s^2$
16. $F = ma = (12.8\ kg)\ (9.8\ m/s^2) = 125\ N$
17. a) Application of the Pythagorean Theorem yields:

    $(40\ N)^2 + (30\ N)^2 = F^2$

    $F = 50\ N$

    Using the tangent button on the calculator or a vector diagram, the angle is 53°.

    b) $a = F/m = 50\ N/5.6\ kg = 8.9\ m/s^2$
18. a) Application of the Pythagorean Theorem yields:

    $(30\ N)^2 + (20\ N)^2 = F^2$

    $F = 36\ N$

    Using the tangent button on the calculator or a vector diagram, the angle is 34°.

    b) $a = F/m = 36\ N/100\ kg = 0.36\ N/kg = 0.36\ m/s^2$

    c) If both boxes were pushed toward the right, the new force would be 50 N. (30 N + 20 N = 50 N).

    The acceleration can then be calculated:

    $a = F/m = 50\ N/100\ kg = 0.5\ N/kg = 0.5\ m/s^2$

# ACTIVITY 3
## Center of Mass

# Background Information

The center of mass of an object is the only idea introduced in this activity.

Definition: The center of mass is the point at which the entire mass of an object may be thought of as being concentrated for purposes of analyzing the translational motion (motion along a path) or rotational motion (spinning motion) of the object.

For practical purposes, the location of the center of mass of an object having only one significant dimension—such as a straight stick, loaded teeter-totter, twirler's baton, screwdriver or wrench—corresponds to the object's balance point. For a two-dimensional object—such as a sheet of plywood cut into any shape—the location of the center of mass corresponds to the balance point located on either of the two large, flat surfaces of the object; to the extent that a two-dimensional object—such as a triangle cut from a sheet of plywood—may have significant thickness and, therefore, actually be three-dimensional, the center of mass would be located within the object, "in line" with the balance point, at the center of the thickness dimension.

For objects having simple three-dimensional shapes—such as homogeneous or symmetrically layered spheres (examples, in respective order: bowling ball, basketball), cubes, rectangular solids and cylinders—the center of mass is located within the object, at its center.

An alternative to balancing an object to locate the center of mass is to suspend the object from any point which is not the center of mass. When suspended, gravity serves to orient the object so that its center of mass is located directly below the point of suspension (this is an example that the Earth "views" an object near it as a "point mass" (located at the object's center of mass) and pulls the point mass as close to Earth as possible). A line extended straight downward from the point of suspension passes through the object's center of mass. The intersection of two such lines, corresponding to two points of suspension, locates the object's center of mass.

It is possible that the center of mass may not be located within the material of the object for some shapes. The "boomerang" shape is an example of such an object.

For purposes of applying Newton's Laws of Motion, an object is treated as if all of its mass is concentrated at the center of mass. The fact that objects behave this way in nature is verified by the observation that when a baton is thrown through the air as a twirling projectile, the baton's center of mass, if marked for high visibility, is seen to trace the familiar parabolic trajectory of a projectile. A twirling baton brings up another aspect of center of mass: when a force acting on an object is aligned with the object's center of mass, the object accelerates in accordance with Newton's second law; however, if the applied force is not aligned with the center of mass, the object also will rotate, or spin. The latter kind of case is not treated in *Coordinated Science for the 21st Century*.

Considerable emphasis in future activities will be placed on the center of mass of the human body. Except for contorted positions of body parts (e.g., the arched "Fosbury Flop" position in the high jump), the normal location of the body's center of mass is within the body at about the level of the navel.

## *Active*-ating the Physics InfoMall

The methods outlined in the *Active Physics* text are standard for finding the center of mass for objects. However, you may want demonstrations. A search of the Demo & Lab Shop produces many great, and tested, demonstrations. Just use keywords "Center of mass" and search only the Demo & Lab Shop. Of course, you can also find many problems in the Problems Place, if you wish, using the same keywords.

# Planning for the Activity

## Time Requirements

- One class period.

## Materials Needed

### For the class:

- hammer and catch box (demonstration, **Step 8**)

### For each group:

- dots, adhesive, 3/4"
- plumb bob w/string
- push pins
- set of shapes, A,B,C,D
- stick, meter, 100 cm, hardwood

## Advance Preparation and Setup

Cutouts of shapes A, B, C, and D need to be made for each group in the shapes of templates provided in the Additional Materials for this activity. (The templates are provided only for your convenience. Other shapes, in greater variety, may be used and the size may be scaled differently, if desired. If you depart from the shapes provided, be sure to include a "boomerang" shape for which the C of M will be outside the object.) The shapes may be cut from any thin, flat material such as corrugated cardboard or (more durable) plywood, plastic or metal; it would also be convenient to cut the shapes from a sheet of pegboard material to avoid need to drill holes for suspension.

Drill holes to serve to suspend the shapes, and plan how the shapes will be suspended from pins or nails from areas such as a bulletin board or pieces of wood mounted on laboratory table rods.

# Teaching Notes

Prepare a demonstration of one or more objects having complex shapes moving as spinning projectiles. For example, the centers of mass of shapes A, B, C, and D could be brightly marked and observed from a distance as two persons play catch with each of the objects. Even if an object spins, the center of mass will trace a parabolic trajectory. A baton having the center of mass marked with bright tape also could be used. When all students have completed **Steps 1** to **5** of **For You To Do**, convene the entire class to observe the motion of the center of mass as two persons play catch with the objects planned for the demonstration.

You may wish to recommend the opening montage on the *Active Physics* video as a possibility for tracing the motion of the center of mass moving as a projectile.

# Activity Overview

In this activity students locate the center of mass of various shaped objects by locating the center of gravity. This is done by balancing the object on a finger as well as by suspending the object and using a plumb bob. Students also estimate the location of their own center of mass.

## Student Objectives

- Locate the center of mass of oddly shaped two-dimensional objects.
- Infer the location of the center of mass of symmetrical three-dimensional objects.
- Measure the approximate location of the center of mass of the student's body.
- Understand that the entire mass of an object may be thought of as being located at the object's center of mass.

ANSWERS FOR THE TEACHER ONLY

## What Do You Think?

The center of mass is the point at which all of the mass of an object may be thought of as being concentrated. (As defined above: The center of mass is the point at which the entire mass of an object may be thought of as being concentrated for purposes of analyzing the translational motion (motion along a path) or rotational motion (spinning motion) of the object.)

The normal location of the body's center of mass is within the body at about the level of the navel.

Physics in Action

# Activity 3 Center of Mass

### GOALS

In this activity you will:

- Locate the center of mass of oddly shaped two-dimensional objects.
- Infer the location of the center of mass of symmetrical three-dimensional objects.
- Measure the approximate location of the center of mass of body.
- Understand that the entire mass of an object may be thought of as being located at the object's center of mass.

### What Do You Think?

The center of mass of a high jumper using the "Fosbury Flop" (arched back) technique passes below the bar as the jumper's body successfully passes over the bar.

- **What is "center of mass"? What does it mean?**
- **Where is your body's center of mass?**

Record your ideas about these questions in your *Active Physics* log. Be prepared to discuss your responses with your small group and the class.

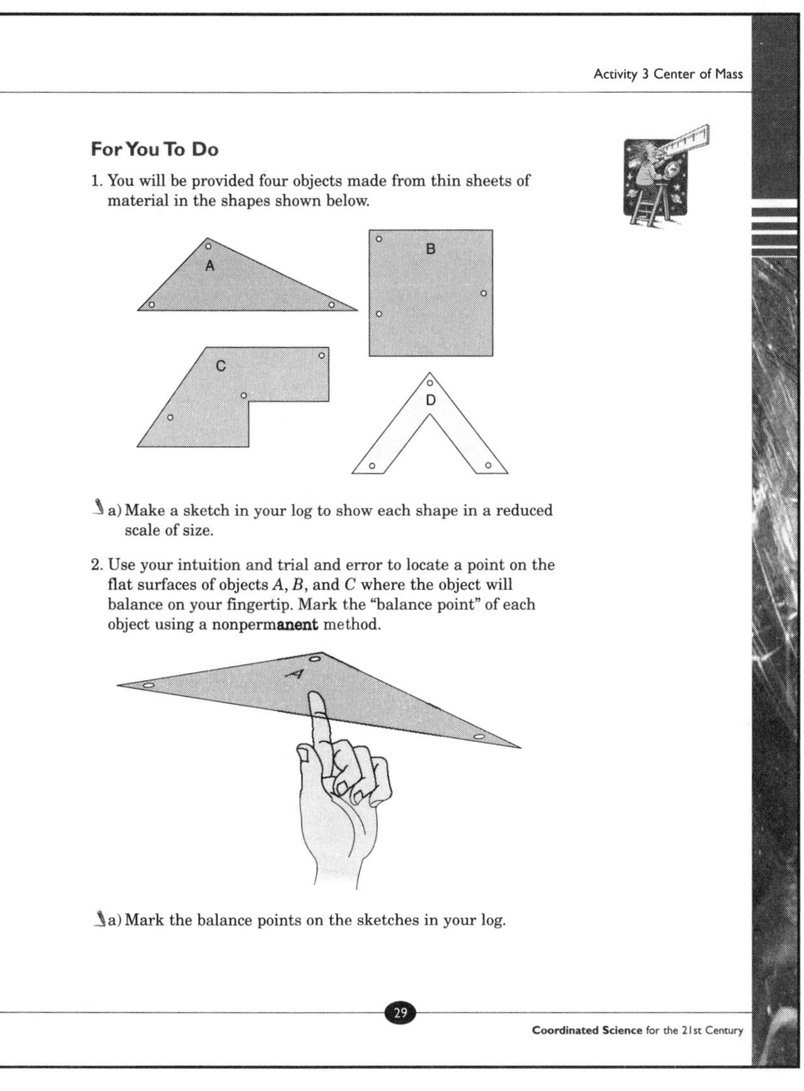



### For You To Do

1. You will be provided four objects made from thin sheets of material in the shapes shown below.



a) Make a sketch in your log to show each shape in a reduced scale of size.

2. Use your intuition and trial and error to locate a point on the flat surfaces of objects *A*, *B*, and *C* where the object will balance on your fingertip. Mark the "balance point" of each object using a nonpermanent method.



a) Mark the balance points on the sketches in your log.



ANSWERS

## For You To Do

1. a) You may wish to provide students with a copy of the templates at the end of this activity, rather than have them redraw each in their log.

2. a) Students should be able to locate the balance points for each shape.

3. To check on the balance points found by the above method for objects *A*, *B* and *C*, use one of the small holes in object *A* to hang it from a pin as shown, and, also as shown, hang a "plumb bob" (a weight on a string) from the same pin.

   a) Does the string pass over the balance point you marked for object *A* when you used your finger to balance the object? Should this happen? Write why you think it should or should not happen in your log.

   b) Use a different hole in object *A* to suspend it from the pin and again hang the plumb bob from the pin. Does the string pass over the balance point marked before? Should it? Write your responses in your log.

4. The intersection of the two lines made where the string passed over the surface of object *A* could have been used to predict the balance point without first trying to balance the object on your finger. Use the suspension and plumb bob method to check the correspondence of the two methods of finding the balance point for objects *B* and *C*.

   a) Record your findings about how well the two methods agree in your log.

5. Locate an "imaginary" balance point for object *D*. Tape a lightweight piece of paper between the "open arms" of object *D* and suspend the object and the plumb bob from the pin. Trace the path of the string across the piece of paper. Suspend the object from a different hole and trace the path of the string across the paper again.

   a) Do you agree that the intersection of the two lines on the paper mark the balance point of object *D*? What is special about this balance point? Write your answers in your log.

6. The above "balance points" that you found for two-dimensional, or "flat," objects *A*, *B*, *C*, and *D* were, in each case, the location of the object's "center of mass." Do a "thought experiment" (an experiment in your mind) to determine the location of the center of mass of each of the following objects:

   Shot put ball (solid steel) | Basketball
   Banana | Planet Earth
   Baseball bat | Hockey stick



ANSWERS

## For You To Do *(continued)*

3. a) Yes, the plumb bob should pass over the balance point. When suspended, gravity serves to orient the object so that its center of mass is located directly below the point of suspension. A line extended straight downward from the point of suspension passes though the object's center of mass.

   b) The string will pass over the balance point. The spot where the lines from **a)** and **b)** intersect represents the C of M.

4. a) Students will find that the two methods will produce similar results.

5. a) The C of M is not located on shape D. Students will find that the two lines cross as a point outside the shape.

6. a) Students answer will vary.

a) For each object, describe in your log how you decided upon the location of the center of mass.

7. The technique that was used to find the center of mass (C of M) relied on the fact that the C of M always lies beneath the point of support when an object is hanging. Similarly, when an object is balanced, the C of M is always above the point of support. To find your C of M, carefully balance on one foot and then the other. Try to keep your arms and legs in roughly identical positions as you shift your weight. Your C of M is located where a vertical meter stick from one foot and the other intersect. Locate this point. The actual C of M is inside your body, since nobody has zero thickness.

a) Record the location of your C of M.

8. Your teacher will balance a hammer on a finger to locate the hammer's C of M and make an obvious mark on the hammer at the C of M. As your teacher drops the hammer into a catch box on the floor, and it twists and turns, notice the movement of the C of M.

a) How does the movement of the C of M compare to the motion of the entire hammer?

**Physics Words**

**center of mass:** the point at which all the mass of an object is considered to be concentrated for calculations concerning motion of the object.

### Reflecting on the Activity and the Challenge

The **center of mass** is an important concept in any sports activity. The motion of the center of mass of a diver or gymnast is much easier to observe than the movements of the entire body. The sure-fire way of having a football player fall is to move his center of mass away from his support.

Think about the possibilities for using a transparent plastic cover on a TV monitor and using a pen to trace the motion of the center of mass of an athlete executing a free fall jump or dive. This could be used to simulate the light-pen technique used by TV commentators when they comment on football replays. This would seem a good way to add an interesting feature to your TV sports commentary.





ANSWERS

## For You To Do *(continued)*

7. a) The center of mass of a body is located inside the body at about the level of the navel.

8. a) The C of M moves directly down in a straight line, whereas the hammer twists and turns as it falls.

ANSWERS

## Physics To Go

1. If not directed toward the center of mass, part of the force will be used to make the object rotate, not accelerate along a line.
2. Referring to the above answer to **Question 1**, a player having a low center of gravity must be "hit" low, at the level of the center of mass, to have his state of rest or motion changed.
3. The body's center of mass has no support directly beneath it, so it falls.
4. Fosbury Flop: the center of mass is located behind the back, in the air outside the body.
5. The pushoff force is directed at an angle to the intended path of travel.
6. If the car were suspended from a crane twice, each time from a different point of attachment of the cable to the car, the intersection of lines representing, in each case of suspension, an extension of the cable through the car would locate the center of mass.
7. Students will probably find the center of mass by balancing the bat on their finger. Ask students to record what they did, and any problems they may have encountered.
8. When the support is moved away from the center of mass, the book will fall. By tackling below the center of mass, the support is moved away from under the center of mass, and the player will fall.

Physics in Action

### Physics To Go

1. When applying a force to make an object move, why is it most effective to have the applied force "aimed" directly at the object's center of mass?
2. "Center of gravity" means essentially the same thing as "center of mass." Why is it often said to be desirable for football players to have a low center of gravity?
3. Stand next to a wall facing parallel to the wall. With your right arm at your side pushing against the wall and with the right edge of your right foot against the wall at floor level, try to remain standing as you lift your left foot. Why is this impossible to do?
4. Think of positions for the human body for which the center of mass might be located outside the body. Describe each position and where you think the center of mass would be located relative to the body for each position.
5. An object tends to rotate (spin) if it is pushed on by a force that is not aimed at the center of mass. How do athletes use this fact to initiate spins before they fly through the air, as in gymnastics, skating, and diving events?
6. Could the suspension technique for finding the center of mass used in **For You To Do** be adapted to locate the center of mass of a three-dimensional object? If you had a crane that you could use to suspend an automobile from various points of attachment, how could you locate the auto's center of mass?
7. Find the center of mass of a baseball bat using the technique that you learned in class.
8. Carefully balance a light object (not too massive) over a table or catch box. Notice that the C of M is directly over the point of support. Move the support a little bit. Explain how this technique can be adapted to tackling in football.
9. Cut out a piece of cardboard in the shape of your state or a country. Find the geographic center of mass of your shape.

32

## Activity 3 A

# Alternative Method for Determining Center of Gravity

### FOR YOU TO DO

1. Locate the center of mass (often abbreviated C of M) of your body. For this you will need an equal arm "teeter-toter," meter stick, and two assistants. Your teacher will give you safety precautions.

Lay on your back on the teeter-toter as an assistant stabilizes each end to prevent extreme tipping. Adjust your position until balance is achieved without the assistants touching the system.

2. When at balance, have an assistant measure the distance, in meters, from the bottom of the heel of your shoe to the fulcrum (middle support) of the teeter-toter.

3. As the assistants again stabilize the teeter-toter, get off the teeter-toter.

a) Record the distance measured by the assistant.

Distance from heel to C of M = _____ m

4. Standing erect, measure the above distance from the floor upward to locate the height of your body's center of mass relative to a constant reference point such as your navel.

a) Record the location in your log so that you will be able to recover it easily (Example: three fingerwidths below navel). Actually, your center of mass is located inside your body (within your belly) when your body is in most positions. Usually you will need to know only how high above floor level it is located.

For use with *Physics in Action*, Chapter 1, Activity 3: Center of Mass

# Template for Shapes A and B

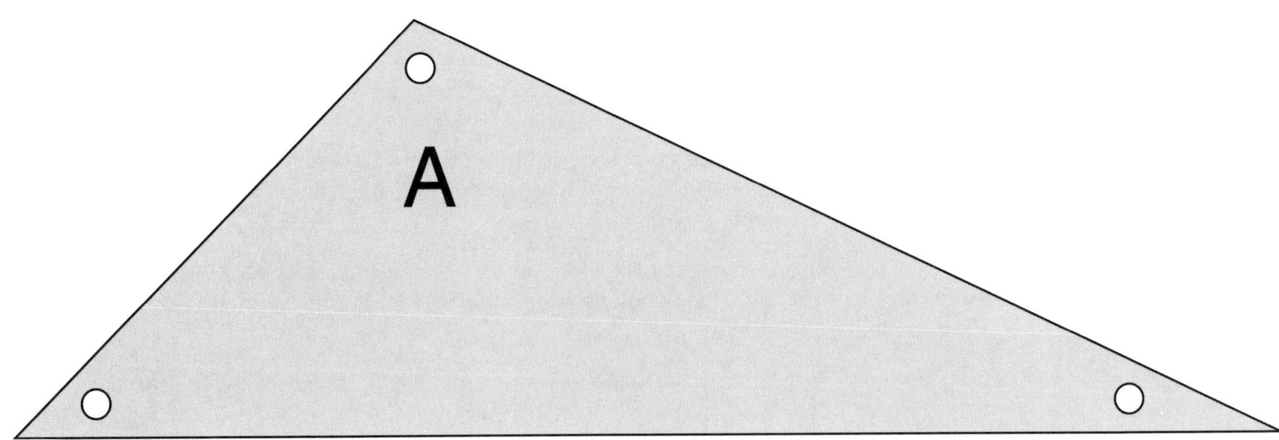


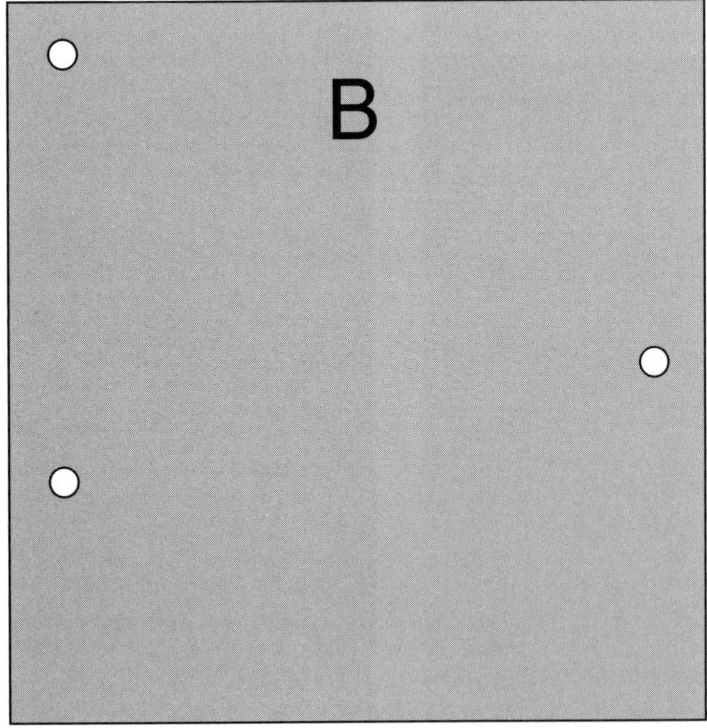


For use with *Physics in Action*, Chapter 1, Activity 3: Center of Mass

# Template for Shapes C and D

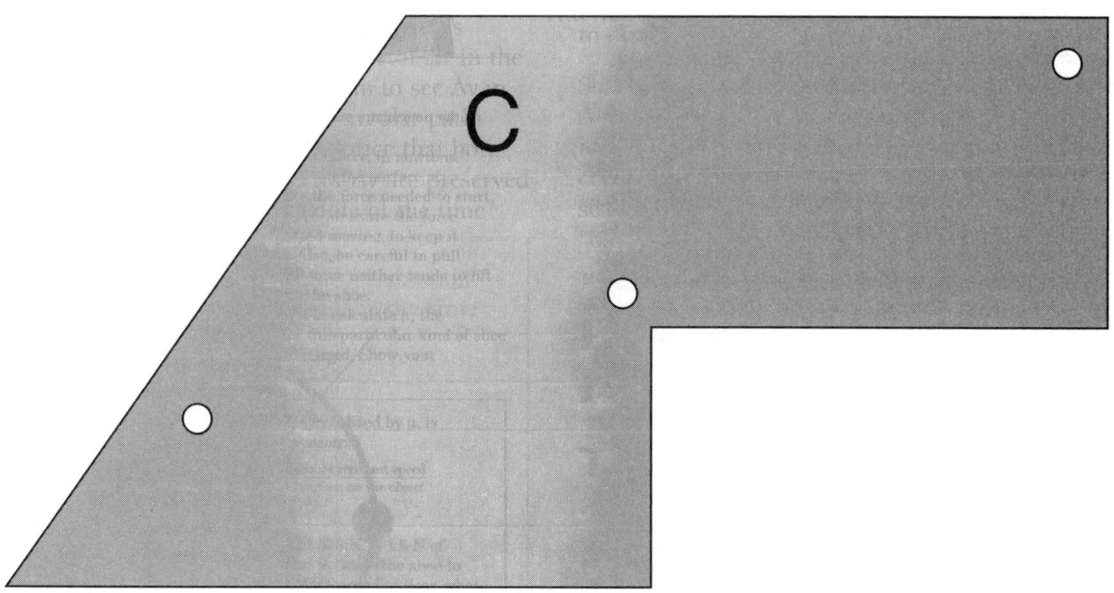


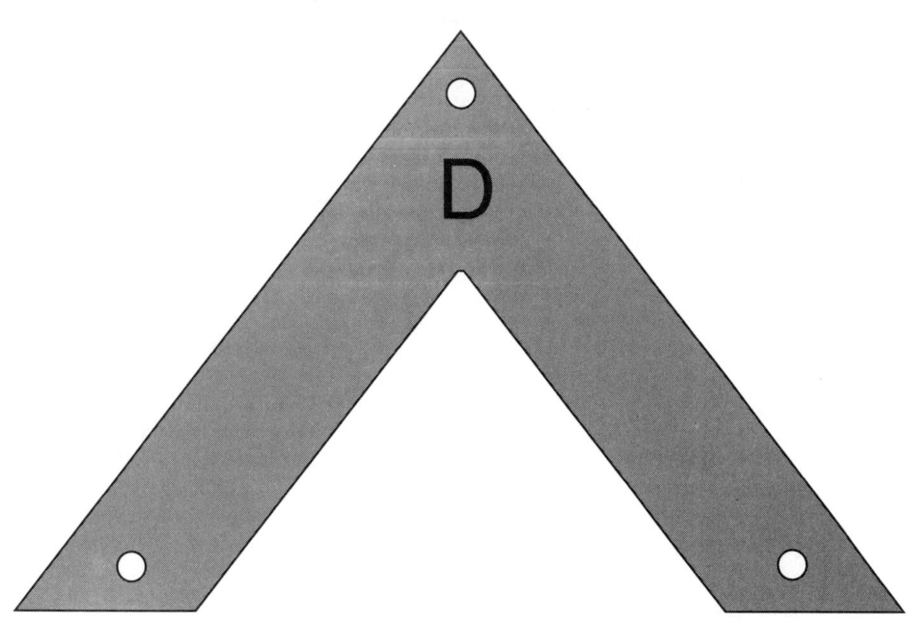


For use with *Physics in Action*, Chapter 1, Activity 3: Center of Mass

# ACTIVITY 4
## Defy Gravity

# Background Information

It is suggested that you read the **Physics Talk** and **Example Analysis** sections in the student text for this activity before proceeding in this section.

Work, the product force x distance, is expressed in joules. Work is equivalent to energy and, indeed, is transformed into kinetic energy and gravitational potential energy in the vertical jump.

Research has shown that the location of the center of mass within the jumper's body varies only slightly for the body positions assumed during the process of the vertical jump.

The force which lifts and accelerates the body's center of mass during a vertical jump is provided by muscles of the leg, ankles, and feet. The method of analysis used for this activity assumes that the muscular force is constant as the body rises from "ready" to "launch" positions; this is not entirely accurate—in a real jump, the force varies—but is a reasonable approximation of reality.

## *Active*-ating the Physics InfoMall

While "hang time" is discussed on the InfoMall, it is in the sense of how long a football stays in the air during a punt, and not how long a basketball player stays (or seems to stay) in the air.

Note that gravitational potential energy is mentioned in this activity, a topic we will encounter in another chapter. If you perform a search of the InfoMall for Work, Potential Energy, or Kinetic Energy, you will want to limit your search to only one or two stores at a time, or use additional keywords to restrict your search.

Should you desire additional problems for your students to work on, consult the Problems Place. For example, *Progressive Problems in Physics* has 16 problems on Work, and 25 on Energy.



# Planning for the Activity

## Time Requirements

- One class period.

## Materials Needed

### For the class:

- *Active Physics Sports* content video

(Segments: ice skater performing triple axel jump, basketball player "hanging" during slam dunk)

- VCR and TV monitor

### For each group:

- *Sports* Content Video
- masking tape, 3/4" x 60 yds
- stick, meter, 100 cm, hardwood
- VCR & monitor

## Advance Preparation and Setup

Reserve a VCR and TV monitor for showing segments of the *Active Physics Sports* video.

# Teaching Notes

View the slow-motion sequences of the jumping figure skater and the jumping basketball player to examine "hang time" and to see if either athlete remains suspended at the peak of flight. These are clear cases of free fall; since the basketball player has high horizontal running speed at take-off, the top of the trajectory is quite flat, giving illusion, when viewed in "real time" that the player "hangs" in the air.

Students may be expected to need help when applying their own data to replicate the calculations presented as an example in **Physics Talk**.

If a sonic ranger is available, monitor a jump from above and analyze the graphs of distance, speed, and acceleration versus time with the entire class. Directions for using a sonic ranger are provided at the end of this activity.

Suggest to students who have access to VCRs with slow-motion playback capability that they could record jumps during athletic contests and perform analysis similar to those conducted using the *Active Physics Sports* video.

# Activity 4 Defy Gravity

GOALS

In this activity you will:

- Measure changes in height of the body's center of mass during a vertical jump.
- Calculate changes in the gravitational potential energy of the body's center of mass during a vertical jump.
- Understand and apply the definition of work.
- Recognize that work is equivalent to energy.
- Understand and apply the joule as a unit of work and energy using equivalent forms of the joule.
- Apply conservation of work and energy to the analysis of a vertical jump, including weight, force, height, and time of flight.

### What Do You Think?

No athlete can escape the pull of gravity.

- **Does the "hang time" of some athletes defy the above fact?**
- **Does a world-class skater defy gravity to remain in the air long enough to do a triple axel?**

Record your ideas about these questions in your *Active Physics* log. Be prepared to discuss your responses with your small group and the class.

### For You To Do

1. Your teacher will show you a slow-motion video of a world-class figure skater doing a triple axel jump. The image of the skater will appear to "jerk," because a video camera completes one "frame," or one complete picture, every $\frac{1}{30}$ s. When the video is played at normal speed, you perceive the action as continuous; played at slow motion, the individual frames can be detected and counted. The duration of each frame is $\frac{1}{30}$ s.

# Activity Overview

In this activity students measure the positions of the C of M of a student during a vertical jump, and then analyze the amount of force and energy required by the student to perform the jump.

## Student Objectives

### Students will:

- Measure changes in height of the body's center of mass during a vertical jump.
- Calculate changes in the gravitational potential energy of the body's center of mass during a vertical jump.
- Understand and apply the definition of work, Work = *fd*.
- Recognize that work is equivalent to energy.
- Understand and apply the joule as a unit of work and energy using equivalent forms of the joule:

  $1\text{ J} = 1\text{ Nm} = 1\text{ (kg)m/s}^2 \times \text{(m)} = 1\text{ (kg)m}^2\text{/s}^2$
- Apply conservation of work and energy to analysis of a vertical jump, including weight, force, height, and time of flight.

ANSWERS FOR THE TEACHER ONLY

## What Do You Think?

The answer to both questions is the same. There is no evidence that athletes are able to defy gravity.

ANSWERS

## For You To Do

1. a) The skater is in the air for 15 frames.

   b) Time in air (s) = Number of frames × 1/30 s

   =15 × 1/30 s

   =15/30 s =1/2 s

   c) During the time frame as viewed on the video, the skater's position is constantly changing. There is no "hang" time.

2. a) The basketball player is in the air for 31 frames.

Time in air (s) = Number of frames × 1/30 s

=31 × 1/30 s

=31/30 s = 1 1/30 s

b) During the time frame as viewed on the video, the basketball player's position is constantly changing. There is no "hang" time. (Since the ball is moving upward before the player leaves the ground, and since on the way down his arms are extending and lifting the ball into the net, the illusion of hanging in the air may be created.)

3. a) Students' answers will vary according to their weight in pounds.

The following equations were presented on page 24.

Weight in newtons = Weight in pounds × 4.38 N/lb.

Weight (N) = Weight (lbs) × 4.38 N/1b

Weight = $mg$

Weight (N) = $m$(kg) × $g$(m/s²)

$$m \text{ (kg)} = \frac{\text{Weight (N or kg}\cdot\text{m/s}^2)}{g \text{ (m/s}^2)}$$

Physics in Action

a) Count and record in your log the number of frames during which the skater is in the air.

b) Calculate the skater's "hang time." (Show your calculation in your log.)

$$\text{Time in air (s)} = \text{Number of frames} \times \frac{1}{30}\text{ s}$$

c) Did the skater "hang" in the air during any part of the jump, appearing to "defy gravity"? If necessary, view the slow-motion sequence again to make the observations necessary to answer this question in your log. If your observations indicate that hanging did occur, be sure to indicate the exact frames during which it happened.

2. Your teacher will show you a similar slow-motion video of a basketball player whose hang time is believed by many fans to clearly defy gravity.

a) Using the same method as above for the skater, show in your log the data and calculations used to determine the player's hang time during the "slam dunk."

b) Did the player hang? Cite evidence from the video in your answer.

3. How much force and energy does a person use to do a vertical jump? A person uses body muscles to "launch" the body into the air, and, primarily, it is leg muscles that provide the force. First, analyze only the part of jumping that happens before the feet leave the ground. Find your body mass, in kilograms, and your body weight, in newtons, for later calculations. If you wish not to use data for your own body, you may use the data for another person who is willing to share the information with you. (See **Activity 2, Physics To Go, Question 7**, for how to convert your body weight in pounds to weight in newtons and mass in kilograms.)

a) Record your weight, in newtons, and mass, in kilograms, in your log.

34

Active Physics

4. Recall the location of your body's center of mass from **Activity 3**. Place a patch of tape on either the right or left side of your clothing (above one hip) at the same level as your body's center of mass. Crouch as if you are ready to make a vertical jump. While crouched, have an assistant measure the vertical distance, in meters, from the floor to the level of your body's center of mass (C of M).

   a) In your log, record the distance, in meters, from the floor to your C of M in the ready position.

Ready position

5. Straighten your body and rise to your tiptoes as if you are ready to leave the floor to launch your body into a vertical jump, but don't jump. Hold this launch position while an assistant measures the vertical distance from the floor to the level of your center of mass.

   a) In your log, record the distance, in meters, from the floor to your C of M in the launch position.

   b) By subtraction, calculate and record the vertical height through which you used your leg muscles to provide the force to lift your center of mass from the "ready" position to the "launch" position. Record this in your log as legwork height.

   Legwork height = Launch position – Ready position

Launch position

6. Now it's time to jump! Have an assistant ready to observe and measure the vertical height from the floor to the level of your center of mass at the peak of your jump. When your assistant is ready to observe, jump straight up as high as you can. (Can you hang at the peak of your jump for a while to make it easier for your assistant to observe the position of your center of mass? Try it, and see if your assistant thinks you are successful.)

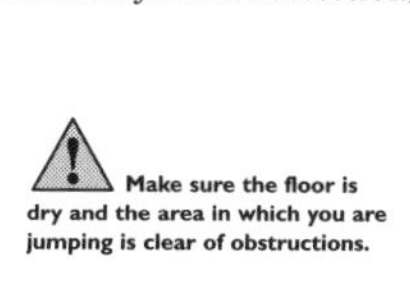


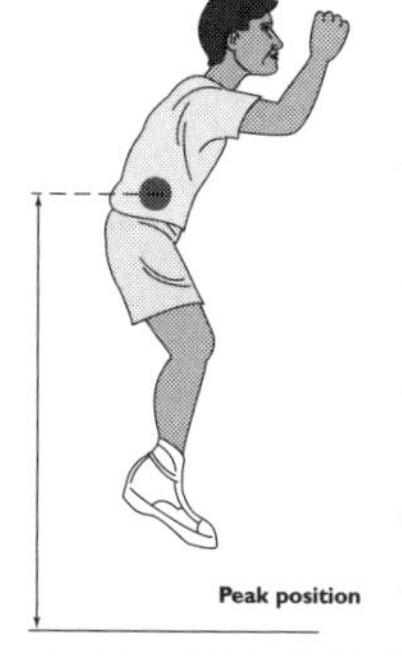

Peak position

ANSWERS

## For You To Do *(continued)*

4.-6. You may wish to provide the students with a copy of the Calculating Hang Time and Force during a Vertical Jump Worksheet provided after this activity. Expect to help students when applying their own data to replicate the calculations presented as an example in **Physics Talk**.

ANSWERS

## For You To Do *(continued)*

6. a) Answers will vary.

   b) Answers will vary.

7. a) Answers will vary.

8. Student activity.

**Physics Words**

**work:** the product of the displacement and the force in the direction of the displacement; work is a scalar quantity.

**potential energy:** energy that is dependent on the position of the object.

**kinetic energy:** the energy an object possesses because of its motion.

a) In your log, record the distance from the floor to C of M at peak position.

b) By subtraction, calculate and record the vertical height through which your center of mass moved during the jump.

Jump height = Peak position – Launch position

7. The information needed to analyze the muscular force and energy used to accomplish your jump—and an example of how to use sample data from a student's jump to perform the analysis—is presented in **Physics Talk** and the Example Analysis.

a) Use the information presented in the **Physics Talk** and Example Analysis sections and the data collected during above **Steps 4** through **6** to calculate the hang time and the total force provided by *your* leg muscles during your vertical jump. Show as much detail in your log as is shown in the Example Analysis.

8. An ultrasonic ranging device coupled to a computer or graphing calculator, which can be used to monitor position, speed, acceleration, and time for moving objects, may be available at your school. If so, it could be used to monitor a person doing a vertical jump. This would provide interesting information to compare to the data and analysis that you already have for the vertical jump. Check with your teacher to see if this would be possible.

## FOR YOU TO READ

### Conservation of Energy

In this activity you jumped and measured your vertical leap. You went through a chain of energy conversions where the total energy remained the same, in the absence of air resistance. You began by lifting your body from the crouched "ready" position to the "launch" position. The **work** that you did was equal to the product of the applied force and the distance. The work done must have lifted you from the ready position to the launch position (an increase in **potential energy**) and also provided you with the speed to continue moving up (the **kinetic energy**). After you left the ground, your body's potential energy continued to increase, and the kinetic energy decreased. Finally, you reached the peak of your jump, where all of the energy became potential energy. On the way down, that potential energy began to decrease and the kinetic energy began to increase.

When you are in the ready position, you have elastic potential energy. As you move toward the launch position, you have exchanged your elastic potential energy for an increase in gravitational potential energy and an increase in kinetic energy. As you rise in the air, you lose the kinetic energy and gain more gravitational potential energy. You can show this in a table.

| Energy→<br>Position↓ | Elastic potential energy | Gravitational potential energy = $mgh$ | Kinetic energy $= \frac{1}{2}mv^2$ |
|---|---|---|---|
| ready position | maximum | 0 | 0 |
| launch position | 0 | some | maximum |
| peak position | 0 | maximum | 0 |

The energy of the three positions must be equal. In this first table, the sum of the energies in each row must be equal. The launch position has both gravitational potential energy and kinetic energy. Using the values in the activity, the total energy at each position is 410 J.

| Energy→<br>Position↓ | Elastic potential energy | Gravitational potential energy = $mgh$ | Kinetic energy $= \frac{1}{2}mv^2$ |
|---|---|---|---|
| ready position | 410 J | 0 | 0 |
| launch position | 0 | 150 J | 260 J |
| peak position | 0 | 410 J | 0 |

→

**Physics in Action**

In the ready position, all 410 J is elastic potential energy. In the peak position, all 410 J is gravitational potential energy. In the launch position, the total energy is still 410 J but 150 J is gravitational potential energy and 260 J is kinetic energy.

Consider someone the same size, who can jump much higher. Since that person can jump much higher, the peak position is greater, and therefore the gravitational potential energy of the jumper is greater. In the example shown below, the gravitational potential energy is 600 J. Notice that this means the elastic potential energy of the jumper's legs must be 600 J. And when the jumper is in the launch position, the total energy (potential plus kinetic) is also 600 J.

| Energy→ / Position↓ | Elastic potential energy | Gravitational potential energy = $mgh$ | Kinetic energy $= \frac{1}{2}mv^2$ |
|---|---|---|---|
| ready position | 600 J | 0 | 0 |
| launch position | 0 | 150 J | 450 J |
| peak position | 0 | 600 J | 0 |

A third person of the same size is not able to jump as high. What numbers should be placed in blank areas to preserve the principle of conservation of energy?

Total energy must be conserved. Therefore, in the launch position the kinetic energy of the jumper must be 50 J. In the peak position, all the energy is in potential energy and must be 200 J.

The conservation of energy is a unifying principle in all science. It is worthwhile to practice solving problems that will help you to see the variety of ways in which energy conservation appears.

| Energy→ / Position↓ | Elastic potential energy | Gravitational potential energy = $mgh$ | Kinetic energy $= \frac{1}{2}mv^2$ |
|---|---|---|---|
| ready position | 200 J | 0 | 0 |
| launch position | 0 | 150 J | 50 J |
| peak position | 0 | 200 J | 0 |

A similar example to jumping from a hard floor into the air is jumping on a trampoline (or your bed, when you were younger). If you were to jump on the trampoline, the potential energy from the height you are jumping would provide kinetic energy when you landed on the trampoline. As you continued down, you would continue to gain speed because you would still be losing gravitational potential energy. The trampoline bends and/or the springs holding the trampoline stretch. Either way, the trampoline or springs gain elastic potential energy at the expense of the kinetic energy and the changes in potential energy.

| Energy → / Position ↓ | Elastic potential energy | Gravitational potential energy = $mgh$ | Kinetic energy $= \frac{1}{2}mv^2$ |
|---|---|---|---|
| High in the air position | 0 | 2300 J | 0 |
| Landing on the trampoline position | 0 | 500 J | 1800 J |
| Lowest point on the trampoline position | 2300 J | 0 | 0 |

A pole-vaulter runs with the pole. The pole bends. The pole straightens and pushes the vaulter into the air. The vaulter gets to his highest point, goes over the bar, and then falls back to the ground, where he lands on a soft mattress. You can analyze the pole-vaulter's motion in terms of energy conservation. (Ignore air resistance.)

A pole-vaulter runs with the pole. (*The vaulter has kinetic energy.*) The pole bends. (*The vaulter loses kinetic energy, and the pole gains elastic potential energy as it bends.*) The pole unbends and pushes the vaulter into the air. (*The pole loses the elastic potential energy, and the vaulter gains kinetic energy and gravitational potential energy.*) The vaulter gets to his highest point (*the vaulter has almost all gravitational potential energy*) goes over the bar, and then falls back to the ground (*the gravitational potential energy becomes kinetic energy*), where he lands on a soft mattress (*the kinetic energy becomes the elastic potential energy of the mattress, which then turns to heat energy*). The height the pole-vaulter can reach is dependent on the total energy that he starts with. The faster he runs, the higher he can go.

The conservation of energy is one of the great discoveries of science. You can describe the energies in words (elastic potential energy, gravitational potential energy, kinetic energy, and heat energy). There is also sound energy, →

light energy, chemical energy, electrical energy, and nuclear energy. The words do not give the complete picture. Each type of energy can be measured and calculated. In a closed system, the total of all the energies at any one time must equal the total of all the energies at any other time. That is what is meant by the conservation of energy.

If you choose to look at one object in the system, that one object can gain energy. For example, in the collision between a player's foot and a soccer ball, the soccer ball can gain kinetic energy and move faster. Whatever energy the ball gained, you can be sure that the foot lost an equal amount of energy. The ball gained energy, the foot lost energy, and the "ball and foot" total energy remained the same. The ball gained energy because work (force x distance) was done on it. The foot lost energy because work (force x distance) was done on it. The total system of "ball and foot" neither gained nor lost energy.

Physics provides you with the means to calculate energies. You may wish to practice some of these calculations now. Never lose sight of the fact that you can calculate the energies because the sum of all of the energies remains the same.

The equations for work, gravitational potential energy, and kinetic energy are given below.

The equation for work is:

$$W = F \cdot d$$

Work is done only when the force and displacement are (at least partially) in the same (or opposite) directions.

The equation for gravitational potential energy is:

$$PE_{gravitational} = mgh = wh$$

The $w$ represents the weight of the object in newtons, where $w = mg$. On Earth's surface, when dealing with $g$ in this course, consider it to be equal to 9.8 $m/s^2$. (Sometimes we use 10 $m/s^2$ for ease of calculations.)

The equation for kinetic energy is:

$$KE = \frac{1}{2}mv^2$$

### Sample Problem

A trainer lifts a 5.0-kg equipment bag from the floor to the shelf of a locker. The locker is 1.6 m off the floor.

a) How much force will be required to lift the bag off the floor?

b) How much work will be done in lifting the bag to the shelf?

c) How much potential energy does the bag have as it sits on the shelf?

d) If the bag falls off the shelf, how fast will it be going when it hits the floor?

***Strategy:*** This problem has several parts. It may look complicated, but if you follow it step by step, it should not be difficult to solve.

**Part (a):** Why does it take a force to lift the bag? It takes a force because the trainer must act against the pull of the gravitational field of the Earth. This force is called weight, and you can solve for it using Newton's Second Law.

**Part (b):** The information you need to find the work done on an object is the force exerted on it and the distance it travels. The distance was given and you calculated the force needed. Use the equation for work.

**Part (c):** The amount of potential energy depends on the mass of the object, the acceleration due to gravity, and the height of the object above what is designated as zero height (in this case, the floor). You have all the needed pieces of information, so you can apply the equation for potential energy.

**Part (d):** The bag has some potential energy. When it falls off the shelf, the potential energy becomes kinetic energy as it falls. When it strikes the ground in its fall, it has zero potential energy and all kinetic energy. You calculated the potential energy. Conservation of energy tells you that the kinetic energy will be equal to the potential energy. You know the mass of the bag so you can calculate the velocity with the kinetic energy formula.

***Givens:***

$m = 5.0 \text{ kg}$
$h = 1.6 \text{ m}$
$a = 9.8 \text{ m/s}^2$

***Solution:***

a)

$$\begin{aligned} F &= ma \\ &= (5.0 \text{ kg})(9.8 \text{ m/s}^2) \\ &= 49 \text{ kg} \cdot \text{m/s}^2 \text{ or } 49 \text{ N} \end{aligned}$$

b)

$$\begin{aligned} W &= F \cdot d \\ &= (49 \text{ N})(1.6 \text{ m}) \\ &= 78.4 \text{ Nm or } 78 \text{ J (Nm = J)} \end{aligned}$$

c)

$$\begin{aligned} PE_{gravitational} &= mgh \\ &= (5.0 \text{ kg})(9.8 \text{ m/s}^2)(1.6 \text{ m}) \\ &= 78 \text{ J} \end{aligned}$$

Should you be surprised that this is the same answer as **Part (b)**? No, because you are familiar with energy conservation. You know that the work is what gave the bag the potential energy it has. So, in the absence of work that may be converted to heat because of friction, which you did not have in this case, the work equals the potential energy.

d)

$$\begin{aligned} KE &= \frac{1}{2}mv^2 \\ v^2 &= \frac{KE}{\frac{1}{2}m} \\ &= \frac{78 \text{ J}}{\frac{1}{2}(5.0 \text{ kg})} \\ &= 31 \text{ m}^2/\text{s}^2 \\ v &= 5.6 \text{ m/s} \end{aligned}$$

## PHYSICS TALK

### Work

When you lifted your body from the ready (crouched) position to the launch (standing on tiptoes) position before takeoff during the vertical jump activity, you performed what physicists call work. In the context of physics, the word *work* is defined as:

**The work done when a constant force is applied to move an object is equal to the amount of applied force multiplied by the distance through which the object moves in the direction of the force.**

You used symbols to write the definition of work as:

$$W = F \cdot d$$

where $F$ is the applied force in newtons, $d$ is the distance the object moves in meters, and work is expressed in joules (symbol, J). At any time it is desired, the unit "joule" can be written in its equivalent form as force times distance, "(N)(m)."

The unit "newton" can be written in the equivalent form "(kg)m/s$^2$." Therefore, the unit joule also can be written in the equivalent form (kg)m$^2$/s$^2$. In summary, the units for expressing work are:

$$1 \text{ J} = 1 \text{ (N)(m)} = 1 \text{ (kg)m}^2/\text{s}^2$$

As you read, it is very common in sports that work is transformed into kinetic energy, and then, in turn, the kinetic energy is transformed into gravitational potential energy. This chain of transformations can be written as:

$$\text{Work} = KE = PE$$

$$Fd = \frac{1}{2}mv^2 = mgh$$

These transformations are used in the analysis of data for a vertical jump.

Example:
Calculation of Hang Time and Force During Vertical Jump

DATA: Body Weight = 100 pounds = 440 N
Body Mass = 44 kg
Legwork Height = 0.35 m
Jump Height = 0.60 m

Analysis:
Work done to lift the center of mass from ready position to launch position without jumping ($W_{R\ to\ L}$):

$$W_{R\ to\ L} = Fd = \text{(Body Weight)} \times \text{(Legwork Height)}$$
$$= 440\ \text{N} \times 0.35\ \text{m} = 150\ \text{J}$$

Gravitational Potential Energy gained from jumping from launch position to peak position ($PE_J$):

$$PE_J = mgh = \text{(Body Mass)} \times (g) \times \text{(Jump Height)}$$
$$= 44\ \text{kg} \times 10\ \text{m/s}^2 \times 0.60\ \text{m}$$
$$= 260\ \text{(kg)m}^2/\text{s}^2 = 260\ \text{(N)(m)} = 260\ \text{J}$$

The jumper's kinetic energy at takeoff was transformed to increase the potential energy of the jumper's center of mass by 260 J from launch position to peak position. Conservation of energy demands that the kinetic energy at launch be 260 J:

$$KE = \frac{1}{2}mv^2 = 260\ \text{J}$$

This allows calculation of the jumper's launch speed:

$$v = \sqrt{2(KE)/m} = \sqrt{2(260\ \text{J})/(44\ \text{kg})} = 3.4\ \text{m/s}$$

From the definition of acceleration, $a = \Delta v/\Delta t$, the jumper's time of flight "one way" during the jump was:

$$\Delta t = \Delta v/a = (3.4\ \text{m/s}) / (10\ \text{m/s}^2) = 0.34\ \text{s}$$

Therefore, the total time in the air (hang time) was $2 \times 0.34\ \text{s} = 0.68\ \text{s}$.

→

The total work done by the jumper's leg muscles before launch, $W_T$, was the work done to lift the center of mass from ready position to launch position without jumping, $W_{R\ to\ L}$ = 150 J, plus the amount of work done to provide the center of mass with 260 J of kinetic energy at launch, a total of 150 J + 260 J = 410 J. Rearranging the equation $W = F \cdot d$ into the form $F = W/d$, the total force provided by the jumper's leg muscles, $F_T$ was:

$$F_T = \frac{W_T}{\text{(Legwork Height)}}$$

$$= 410 \text{ J} / 0.35 \text{ m}$$

$$= 1200 \text{ N}$$

Approximately one-third of the total force exerted by the jumper's leg muscles was used to lift the jumper's center of mass to the launch position, and approximately two-thirds of the force was used to accelerate the jumper's center of mass to the launch speed.

## Reflecting on the Activity and the Challenge

Work, the force applied by an athlete to cause an object to move (including the athlete's own body as the object in some cases), multiplied by the distance the object moves while the athlete is applying the force explains many things in sports. For example, the vertical speed of any jumper's takeoff (which determines height and "hang time") is determined by the amount of work done against gravity by the jumper's muscles before takeoff. You will be able to find many other examples of work in action in sports videos, and now you will be able to explain them.

### Physics To Go

1. How much work does a male figure skater do when lifting a 50-kg female skating partner's body a vertical distance of 1 m in a pairs competition?
2. Describe the energy transformations during a bobsled run, beginning with team members pushing to start the sled and ending when the brake is applied to stop the sled after crossing the finish line. Include work as one form of energy in your answer.
3. Suppose that a person who saw the video of the basketball player used in **For You To Do** said, "He really can hang in the air. I've seen him do it. Maybe he was just having a 'bad hang day' when the video was taken, or maybe the speed of the camera or VCR was 'off.' How do I know that the player in the video wasn't a 'look-alike' who can't hang?" Do you think these are legitimate statements and questions? Why or why not?
4. If someone claims that a law of physics can be defied or violated, should the person making the claim need to provide observable evidence that the claim is true, or should someone else need to prove that the claim is not true? Who do you think should have the burden of proof? Discuss this issue within your group and write your own personal opinion in your log.
5. Identify and discuss two ways in which an athlete can increase his or her maximum vertical jump height.
6. Calculate the amount of work, in joules, done when:
   a) a 1.0-N weight is lifted a vertical distance of 1.0 m.
   b) a 1.0-N weight is lifted a vertical distance of 10 m.
   c) a 10-N weight is lifted a vertical distance of 1.0 m.
   d) a 0.10-N weight is lifted a vertical distance of 100 m.
   e) a 100-N weight is lifted a distance of 0.10 m.
7. List how much gravitational potential energy, in joules, each of the weights in **Question 6** above would have when lifted to the height listed for it.





ANSWERS

## Physics To Go

1. Work = $fd = (mg)d$
   = 50 kg × 10 m/s$^2$ × 1 m = 50 j.
2. Team members do work while running and pushing the sled to give it and their bodies kinetic energy before jumping on the sled, $fd = 1/2mv^2$. After the team has jumped on the sled, the total energy of the team + sled is equal to the kinetic energy gained during the pushing phase plus the gravitational potential energy = $mgh$, where $h$ is the vertical distance to the bottom of the hill. At the bottom of the hill, the kinetic energy of the sled should be equal to the kinetic energy gained during the pushing phase plus the loss in potential energy due to coming down the hill, $1/2mv^2 + mgh$. The brake must do enough work to cause the sled to lose all of its kinetic energy by exerting a force in the direction opposite the sled's motion.
3. It is apparent that the person wants to believe that the player can defy gravity and is attempting to justify that belief by rejecting scientific evidence. It could be said that the person is not reflecting open-mindedness, a desirable attribute in scientific pursuits.
4. The burden of proof rests with the person making the claim.
5. Increase the force the athlete is able to exert using muscles, lose weight without decreasing muscular force.
6. a) 1.0 N × 1.0 m = 1 J
   b) 1.0 N × 10 m = 10 J
   c) 10 N × 1.0 m = 10 J
   d) 0.10 N × 100 m = 10 J
   e) 100 N × 0.10 m = 10 J
7. All answers are the same as for #6 above.

ANSWERS

## Physics To Go
***(continued)***

8. All answers are the same as for #6 above.

9. $W = F \cdot d = (50.0\ \text{N})(43\ \text{m})$
$= 2150\ \text{J}\ (2200\ \text{J})$

10. $KE = 1/2\ mv^2 = 1/2\ (62\ \text{kg})$
$(8.2\ \text{m/s})^2 = 2084\ \text{J}\ (2100\ \text{J})$

11. a) $F = ma$
$a = F/m = 30.0\ \text{N}/5.0\ \text{kg}$
$= 6\ \text{m/s}^2$

b) $W = F \cdot d = (30.0\ \text{N})$
$(18.75\ \text{m}) = 563\ \text{J}$

12. a) $W = F \cdot d$
$d = W/F = 40{,}000\ \text{J}/3200\ \text{N}$
$= 12.5\ \text{m}\ (12\ \text{m})$

b) $F = ma$
$a = F/m = 3200\ \text{N}/1200\ \text{kg}$
$= 2.7\ \text{m/s}^2$

13. The work done is equal to the change in KE. The final KE is 0. The initial KE can be found.

$KE = 1/2\ mv^2 = 1/2\ (0.150\ \text{kg})$
$(40\ \text{m/s})^2 = 120\ \text{J}$

14. The change in KE is equal to the work done. Calculate the change in KE and then calculate the distance.

$KE = 1/2\ mv^2 = 1/2\ (64.0\ \text{kg})$
$(15.0\ \text{m/s})^2 = 7200\ \text{J}$

$W = F \cdot d$

$d = W/F = 7200\ \text{J}/417\ \text{N}$
$= 17.3\ \text{m}$

8. List how much kinetic energy, in joules, each of the weights in **Questions 6** and **7** would have at the instant before striking the ground if each weight were dropped from the height listed for it.
9. How much work is done on a go-cart if you push it with a force of 50.0 N and move it a distance of 43 m?
10. What is the kinetic energy of a 62-kg cyclist if she is moving on her bicycle at 8.2 m/s?
11. A net force of 30.00 N acts on a 5.00-kg wagon that is initially at rest.
    a) What is the acceleration of the wagon?
    b) If the wagon travels 18.75 m, what is the work done on the wagon?
12. Assume you do 40,000 J of work by applying a force of 3200 N to a 1200-kg car.
    a) How far will the car move?
    b) What is the acceleration of the car?
13. A baseball ($m$ = 150.0 g) is traveling at 40.0 m/s. How much work must be done to stop the ball?
14. A boat exerts a force of 417 N pulling a water-skier ($m$ = 64.0 kg) from rest. The skier's speed is now 15.0 m/s. Over what distance was this force exerted?

Activity 4 A

# High-Tech Alternative for Monitoring Vertical Jump Height

## FOR YOU TO DO

1. Place a computer motion sensor near the ceiling, pointing straight down.
2. Adjust the software so that the duration of the time axis is 5 s or less.
3. Activate the Distance versus Time graph.
4. Click the start button and, as soon as you hear the motion sensor clicking, jump. Try not to get closer than 50 cm to the sensor or you will get erroneous results.
5. Look at the resulting graph and try to find the following parts of the motion:
   - The initial bending of your knees in preparation for the jump.
   - The part of the motion when you were in the air.
   - The bending of your knees upon landing.

   a) Describe what each part of the jump looks like on the graph.
6. Use the software to zoom in on the part of the graph that contains the above-mentioned parts.
7. Switch to a Velocity versus Time graph.

   a) Describe your velocity while in the air.
8. Repeat the experiment for a higher and a lower jump. Compare and contrast your results.

For use with *Physics in Action*, Chapter 1, Activity 4: Defy Gravity

# Calculating Hang Time and Force During a Vertical Jump

Use a calculator to complete the following analysis of a vertical jump:

## DATA:

Calculate body weight.

Weight (N) = Weight (lb.) × 4.38 N/lb.

= ____________________ × 4.38 N/lb.

= ____________________

Body Weight = ____________________

Calculate body mass.

Weight (N) = $mg$

$m$ (kg) = $\dfrac{\text{weight } (\text{kg}\cdot\text{m/s}^2)}{g(\text{m/s}^2)}$

= $\dfrac{\text{\_\_\_\_\_\_\_\_\_\_\_\_\_\_\_\_\_\_\_\_}}{10 \text{ m/s}^2}$

= ____________________

Body Mass = ____________________

Calculate your legwork height.

Legwork Height = Launch position – Ready position

= ______________ – ______________

= ______________

Legwork Height = ____________________

Calculate your jump height.

Jump Height = Peak position – Launch position

= ____________ – ______________

= ____________

Jump Height = ____________________

Calculate the work done to lift the center of mass from ready position to launch position without jumping ($W_{R\ to\ L}$)

$W_{R\ to\ L} = fd$ = (Body Weight) × (Legwork Height)

= ____________ N × ____________ m

= ______________________________ N·m or J (joules)

Calculate the gravitational potential energy gained from jumping from launch position to peak position ($PE_J$).

$PE_J = mgh$ = (Body Mass) × (g) × (Jump Height)

= _________ kg × 10 m/s$^2$ × _________ m

= ______________ kg·m$^2$/s$^2$

= ______________ N·m or J (joules)

Conservation of energy demands that the kinetic energy at launch is equal to the gravitational potential energy at peak position.

$KE = PE$

= ______________________________ J

(Insert the figure you calculated for *PE* above.)

For use with *Physics in Action*, Chapter 1, Activity 4: Defy Gravity

Calculate the jumper's launch speed by writing the following equation in a different form.

$KE = 1/2mv^2$

$v = \sqrt{2(KE)/m}$

$= \sqrt{2\_\_\_\_\_\_\_\_ / \_\_\_\_\_\_\_\_}$

(Insert the *KE* from above) (Insert the figure for mass from data above)

= ____________________ m/s

Calculate the jumper's time of flight "one way" during the jump by using the following equation (acceleration is change in velocity divided by the time). The change in velocity is the jumper's final velocity subtracted from the jumper's launch velocity. In this case, the final velocity at the top of the jump will be zero, the velocity before the jumper begins to come down. The acceleration is the acceleration due to gravity ($a = g$).

$a = \Delta v / \Delta t$

$g = \Delta v / \Delta t$

This equation can be rearranged in the following form to find the jumper's time of flight one way.

$\Delta t = \Delta v / g$

$\Delta t$ = ____________________ $/10 \text{ m/s}^2$

(Insert the value of jumper's launch speed)

= ____________________ s

Calculate the total time in the air (hang time) by multiplying by two, to account for the time going up and the time coming down.

Hang time = 2 × jumper's flight one way

= 2 × ____________________

(Insert the value of time for one-way trip.)

= ____________________

Calculate the total work done by the jumper's leg muscles before launch, $W_T$. This was the work done to lift the center of mass from ready position to launch position without jumping ,$W_{R\ to\ L}$, plus the amount of work done to provide the center of mass with kinetic energy (*KE*) at launch.

$W_T = W_{R\ to\ L}, + KE$

= __________ + __________

= ____________________ J

Calculate the total force provided by the jumper's leg muscles, $f_T$ by rearranging the following equation

$W = fd$

$f = W/d,$

$f_T = W_T$ ____________

(Legwork Height)

= ____________

= ____________

# ACTIVITY 5
## Run and Jump

# Background Information

It is recommended that you read the section **Physics Talk:** Newton's Third Law of Motion in the student text for this activity before proceeding in this section.

The explanation of forces involved in walking given in the teacher's **Background Information** for **Activity 4** will serve to explain the forces involved with walking and running brought up in this activity. You may wish to review the **Background Information** for **Activity 4** before proceeding.

The pairs of equal and opposite forces identified during earlier activities to explain friction and walking are examples of Newton's Third Law of Motion, often stated as: "For every action there is an equal and opposite reaction." Another equal and opposite pair of forces arises during this activity when a student standing on a skateboard sets himself into motion by using a leg and foot to push off from the wall.

Inevitably, forces exist in equal and opposite pairs, and often the force which we identify as the force responsible for motion is not the correct one. For example, a person who says, "I pushed down on the trampoline with a mighty force, and my force launched me upward in a high jump," is mistaken; it was the equal and opposite reaction force provided by the trampoline that launched the person upward.

## *Active*-ating the Physics InfoMall

In addition to looking for information on Newton's Third Law (look at problem 4.14 in *Schaum's 3000 Solved Problems in Physics*, in the Problems Place), perform a search using "force diagrams" as the keywords, and the first hit is a great one! It is, again, from Arons' *A Guide to Introductory Physics Teaching: Elementary Dynamics*, Chapter 3. Section 3.12 is on Newton's Third Law and Free Body Diagrams. Arons mentions common problems and suggests solutions, including suggestions of what not to do.

Arons also notes that "Students do not really begin to understand the concept of force until they become able to apply the third law correctly and draw proper, isolated force diagrams of interacting objects," in his article "Thinking, reasoning, and understanding in introductory physics courses," in *The Physics Teacher*, vol. 19, issue 3, 1981. Check out this article.

This same search produces the warning that "Introductory textbooks are liberally decorated with diagrams, but they fail to convey to students the essential role of diagrams in problem solving or, indeed, to distinguish the roles of different kinds of diagrams" from "Toward a modeling theory of physics instruction," in the *American Journal of Physics*, vol. 55, issue 5, 1987. It is clear that the practice and ability to draw force diagrams are important.

**Stretching Exercise:** Add the word "elevator" to the search above (so now it is "force diagrams" AND "elevator*") for some nice discussions related to the **Stretching Exercise.**

# Planning for the Activity

## Time Requirements

- One class period.

## Materials Needed

### For each group:

- chair with wheels
- elbow pads
- knee pads
- safety helmet
- skateboard
- stick, meter, 100 cm, mardwood
- washers, metal, 3/4"
- weight, 100 g slotted mass

## Advance Preparation and Setup

If you do not have access to a skateboard and safety equipment ask the students if you may borrow theirs for use in class.

# Teaching Notes

SAFETY PRECAUTIONS: Close supervision is needed to prevent injury or damage to equipment and surrounding items when students set themselves into motion by pushing off from the wall.

You may wish to emphasize that the reaction forces which we "feel" throughout a day are not limited to the reaction forces felt while running or walking, but include the reaction force which occurs whenever we touch something.

You may wish to do **Activity 5 A**, Using a Bathroom Scale to Measure Forces, following this activity in the Teacher's Edition. If you do, be sensitive to the fact that many students will not wish to step on scales in front of their peers. Be sure that the students are given the opportunity to volunteer for this activity. Also, forewarn students that the reading of the bathroom scale "goes crazy" when the applied force increases or decreases by great amounts in short time intervals because the scales inertia causes it to "overshoot" maximum or minimum reading. Therefore, the scale reading must be observed very soon after a dramatic change is made in the force applied to the scale. Also point out to students that a specific value of force is not expected to be read on the scale during the push-off phase of the vertical jump; only the nature of the change in force—e.g., greater, no change, less—needs to be observed.

NOTES

# Activity 5 Run and Jump

## GOALS

In this activity you will:

- Understand the definition of acceleration.
- Understand meters per second per second as the unit of acceleration.
- Use an accelerometer to detect acceleration.
- Use an accelerometer to make semiquantitative comparisons of accelerations.
- Distinguish between acceleration and deceleration.

## What Do You Think?

The men's high jump record is over 8 feet.

**• Pretend that you have just met somebody who has never jumped before. What instructions could you provide to get the person to jump up (that is, which way do you apply the force)?**

Record your ideas about this question in your *Active Physics* log. Be prepared to discuss your responses with your small group and the class.

## For You To Do

1. Carefully stand on a skateboard or sit on a wheeled chair near a wall. By touching only the wall, not the floor, cause yourself to move away from the wall to "coast" across the floor. Use words and diagrams to record answers to the following questions in your log:

a) When is your motion accelerated? For what distance does the accelerated motion last? In what direction do you accelerate?





# Activity Overview

In this activity students analyze the forces involved in running, stopping, and jumping. First they push against a wall while standing on a skateboard. Then they perform a thought experiment about the forces involved.

## Student Objectives

### Students will:

- Understand the definition of acceleration.
- Understand meters per second per second as the unit of acceleration.
- Use an accelerometer to detect acceleration.
- Use an accelerometer to make semi-quantitative comparisons of accelerations.
- Distinguish between acceleration and deceleration.

ANSWERS FOR THE TEACHER ONLY

## What Do You Think?

Jumping involves the downward force exerted by the feet on the jumping surface and the equal and opposite reaction force exerted by the jumping surface on the feet and, in turn, the body.

ANSWERS

## For You To Do

1. a) Your motion is accelerated when you push away from the wall. The acceleration lasts for a short distance after which you move at a constant velocity, and then slow down. The direction of acceleration is away from the wall.

ANSWERS

## For You To Do

***(continued)***

1. b) Motion is at a constant speed just after the initial acceleration, when the force is acting on you. If you neglected friction you would keep moving until another force acted on you to slow you down or stop you.

 c) The force is supplied by the wall. The force must be acting in the direction of motion, away from the wall.

 d) You push on the wall, in a direction towards the wall.

 e) The two forces are equal, but opposite in direction.

2. a)-b) You push your foot on the ground backwards from yourself. How much you can push your foot parallel to the surface of the sidewalk depends on how much frictional force can be sustained by the interaction of the sole of the shoe and the sidewalk's surface. If the shoe does not slip on the surface, the sidewalk surface's equal and opposite reaction to the rearward force of friction causes your body to accelerate forward.

 c) On the slippery ice surface, the force of friction is greatly reduced. You are unable to apply much of a backwards force on the ice surface because your shoe will slide on the surface. In turn the opposite reaction of the sidewalk will also be minimal with the result that you go nowhere, there is no force to push you forward.

3. a)-b) Students generate a force diagram similar to the one shown.

**Physics Words**

**Newton's Third Law of Motion:** forces come in pairs; the force of object A on object B is equal and opposite to the force of object B on object A.

b) When is your motion at constant speed? Neglecting the effects of friction, how far should you travel? (Remember Galileo's Principle of Inertia when answering this question.)

c) Newton's Second Law, $F = ma$, says that a force must be active when acceleration occurs. What is the source of the force, the push or pull, that causes you to accelerate in this case? Identify the object that does the pushing on your mass (body plus skateboard) to cause the acceleration. Also identify the direction of the push that causes you to accelerate.

d) Obviously, you do some pushing, too. On what object do you push? In what direction?

e) How do you think, on the basis of both amount and direction, the following two forces compare?
- The force exerted by you on the wall
- The force exerted by the wall on you

2. Do a "thought experiment" about the forces involved when you are running or walking on a horizontal surface. Use words and sketches to answer the following questions in your log:

a) With each step, you push the bottom surface of your shoe, the sole, horizontally backward. The force acts parallel to the surface of the ground, trying to scrape the ground in the direction opposite your motion. Usually, friction is enough to prevent your shoe from sliding across the ground surface.

b) Since you move forward, not rearward, there must be a force in the forward direction that causes you to accelerate. Identify where the forward force comes from, and compare its amount and direction to the rearward force exerted by your shoe with each step.

c) Would it be possible to walk or run on an extremely slippery skating rink when wearing ordinary shoes? Discuss why or why not in terms of forces.

3. Think about the vertical forces acting on you while you are standing on the floor.

a) Copy the diagram of a person at left in your log.

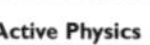



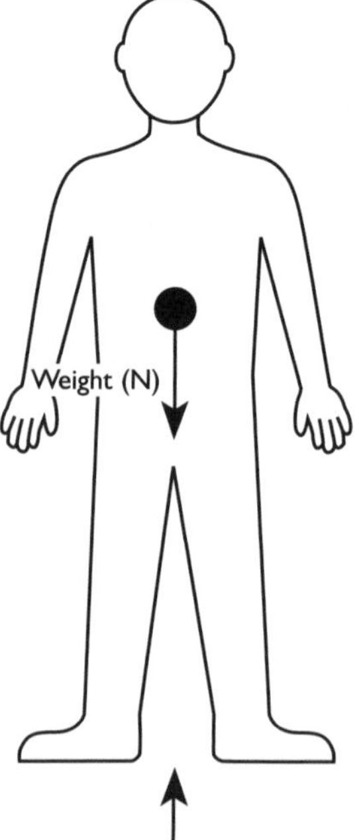

b) Identify all the vertical forces. Use an arrow to designate the size and direction of the force. Draw the forces from the dot.

c) How can you be sure that the force with which you push on the floor and the floor pushes on you are equal?

4. Set up a meter stick with a few books for support as shown.

5. Place a washer in the center of the meter stick.

a) In your log, record what happens.

6. Remove the washer and replace it with 100 g (weight of 100 g = 1.0 N). Continue to place 1.0 N weights on the center of the meter stick. Note what happens as you place each weight on the stick.

a) Measure the deflection of the meter stick for each 1.0 N of weight and record the values for these deflections.

b) How does the deflection of the meter stick compare to the weight it is supporting? In your log, sketch a graph to show this relationship.

c) Write a concluding statement concerning the washer and the deflection of the meter stick.

**PHYSICS TALK**

**Newton's Third Law of Motion**

Newton's **Third Law of Motion** can be stated as:

**For every applied force, there is an equal and opposite force.**

If you push or pull on something, that something pushes or pulls back on you with an equal amount of force in the opposite direction. This is an inescapable fact; it happens every time.

Chapter 1

ANSWERS

## For You To Do

***(continued)***

3. c) The forces must be equal because you are not moving.

4. Student activity.

5. a) Nothing happens.

6. a)-c) The more weight that is added to the meter stick, the greater the deflection of the stick. As the meter stick is deflected, the restoring forces in the wood build up until they exert an upward force equal to the downward force of the weight.

ANSWERS

## Physics To Go

1. Yes, the forces are equal and opposite.

2. The restoring forces within the material from which the chair is made build up until the upward force exerted by the chair equals the downward force caused by your weight; if someone sits in your lap, the chair "bends" (or is otherwise deformed) more, resulting in a higher equal and opposite reaction force.

3. The forces on the ball and the bat are equal and opposite; sometimes the force exerted by the ball on the bat is enough to break the wood of the bat.

4. The forces on the players are equal and opposite, but the smaller player experiences a greater acceleration, which can have more harmful effects on the human body than a lesser acceleration.

5. The forces are equal and opposite; the hockey player is more likely than the boards to complain about the pain involved.

6. Gloves having padding which compresses and/or webbing which deforms when the ball hits the glove. The "softness" of a glove reduces the force which the glove exerts on the ball to a lower amount than a stationary hand would need to exert to stop the ball. The lower force causes the ball to decelerate at a lower rate, also reducing the reaction force which the ball exerts on the glove during stopping. A sure way to reduce the forces during a collision is to increase the amount of time that the objects exert forces on each other during the collision; that is one reason why airbags reduce injuries in automobile collisions.

### Reflecting on the Activity and the Challenge

According to Newton's Third Law, each time an athlete acts to exert a force on something, an equal and opposite force happens in return. Countless examples of this exist as possibilities to include in your video production. When you kick a soccer ball, the soccer ball exerts a force on your foot. When you push backward on the ground, the ground pushes forward on you (and you move). When a boxer's fist exerts a force on another boxer's body, the body exerts an equal force on the fist. Indeed, it should be rather easy to find a video sequence of a sport that illustrates all three of Newton's Laws of Motion.

### Physics To Go

1. When an athlete is preparing to throw a shot put ball, does the ball exert a force on the athlete's hand equal and opposite to the force the hand exerts on the ball?

2. When you sit on a chair, the seat of the chair pushes up on your body with a force equal and opposite to your weight. How does the chair know exactly how hard to push up on you—are chairs intelligent?

3. For a hit in baseball, compare the force exerted by the bat on the ball to the force exerted by the ball on the bat. Why do bats sometimes break?

4. Compare the amount of force experienced by each football player when a big linebacker tackles a small running back.

5. Identify the forces active when a hockey player "hits the boards" at the side of the rink at high speed.

6. Newton's Second Law, $F = ma$, suggests that when catching a baseball in your hand, a great amount of force is required to stop a high-speed baseball in a very short time interval. The great amount of force is needed to provide the great amount of deceleration required. Use Newton's Third Law to explain why baseball players prefer to wear gloves for catching high-speed baseballs. Use a pair of forces in your explanation.

7. Write a sentence or two explaining the physics of an imaginary sports clip using Newton's Third Law. How can you make this description more exciting so that it can be used as part of your sports voice-over?

8. Write a sentence or two explaining the concept that a deflection of the ground can produce a force. How can you make this description more exciting so that it can be used as part of your sports voice-over?

### Stretching Exercises

Ask the manager of a building that has an elevator for permission to use the elevator for a physics experiment. Your teacher may be able to help you make the necessary arrangements.

1. Stand on a bathroom scale in the elevator and record the force indicated by the scale while the elevator is:

   a) At rest.
   b) Beginning to move (accelerating) upward.
   c) Seeming to move upward at constant speed.
   d) Beginning to stop (decelerating) while moving upward.
   e) Beginning to move (accelerating) downward.
   f) Seeming to move downward at constant speed.
   g) Beginning to stop (decelerating) while moving downward.

2. For each of the above conditions of the elevator's motion, the Earth's downward force of gravity is the same. If you are accelerating up, the floor must be pushing up with a force larger than the acceleration due to gravity.

   a) Make a sketch that shows the vertical forces acting on your body.
   b) Use Newton's Laws of Motion to explain how the forces acting on your body are responsible for the kind of motion—at rest, constant speed, acceleration, or deceleration—that your body has.

ANSWERS

## Physics To Go *(continued)*

7.–8. Students may wish to use parts of the voice-overs generated for these questions for their **Chapter Challenge.**

ANSWERS

## Stretching Exercises

1. For the answers below, $m$ is the person's mass, $g$ is the acceleration due to gravity, $mg$ is the person's weight, and $a$ is the acceleration of the elevator.

   a) $mg$

   b) $mg + ma$

   c) $mg$

   d) $mg - ma$

   e) $mg - ma$

   f) $mg$

   g) $mg + ma$

## Activity 5 A

# Using a Bathroom Scale to Measure Forces

### FOR YOU TO DO

1. Stand on a bathroom scale. Preferably, the scale should be calibrated in newtons. If not, see page 25, of the Student Edition, **Activity 2, Physics to Go**, for how to convert, the calibration of the scale to newtons.

a) Reproduce the sketch of you standing on a bathroom scale in your log. Use a dot to show the approximate location of your center of mass.

b) Add an arrow pointing downward from your center of mass to show your weight, in newtons. Label the arrow: Weight = ___ N (enter the amount).

c) Since, when standing on the scale, you are not moving, the scale must be pushing upward on you with a force which balances your weight. Decide upon the amount and direction of the second force which must be acting on you. Starting from your center of mass in the sketch, draw the arrow to represent the force. Label the arrow: Force of Scale = ___ N (enter the amount).

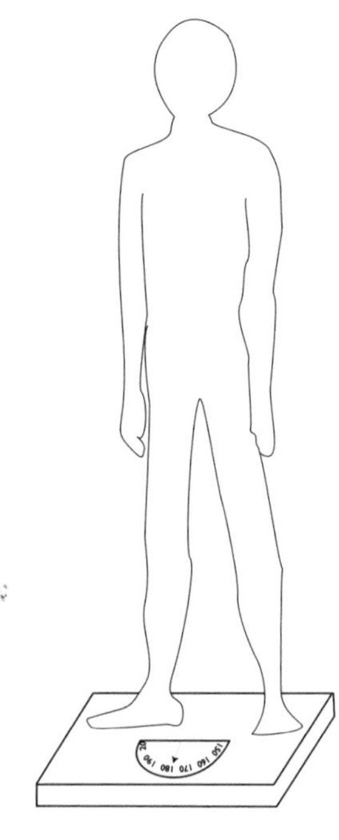

2. Stand on the bathroom scale again, but stand in a crouched position.

a) Reproduce the position of you crouching on a bathroom scale sketch in your journal.

b) Without moving any part of your body while in a crouched position are you able to "bear down" to increase the scale reading to more than when you are standing upright on the scale? Explain why or why not, and show the forces acting on your center of mass when crouched on the scale.

3. Crouch on the scale and rise to full standing position at very low, constant speed. Accelerate as little as possible as you move; you will need to accelerate somewhat to start moving upward, but, after that, try to move your center of mass at low, constant speed upward.

a) Describe in your log that you are moving your center of mass upward at constant speed from the positions used in the above steps.

b) Even though the scale reading may "wiggle" somewhat due to small, not-on-purpose accelerations as you move your center of mass upward at constant speed, how does the scale reading compare to when you were at rest in both crouched and standing positions? In your sketch, draw arrows to represent the forces acting on your center of mass as it moves upward at constant speed.

For use with *Physics in Action*, Chapter 1, Activity 5: Run and Jump

4. Explain how Newton's First Law of Motion applies to the above situations when you were:

a) At rest, standing upright on the scale.

b) At rest, crouched on the scale.

c) Standing on the scale while moving your center of mass upward at constant speed.

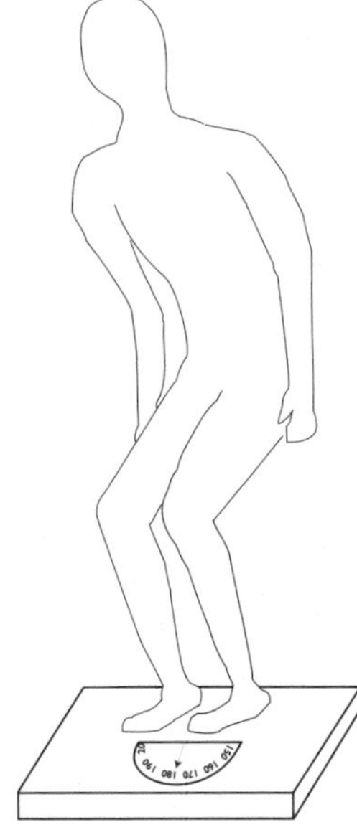

5. Again stand on the scale in crouched position. This time, you will accelerate your center of mass to jump upward. Design your jump to go slightly forward so that you land on the floor, not the scale, when you come back down – this is to prevent damage to the scale by landing on it.

Have an assistant ready to observe how the scale reading changes as you are accelerating your center of mass upward, but before you "launch" into the air. This observation will be "tricky" because bathroom scales cannot rapidly respond to changing forces. Also, inertia causes the scale to "overshoot" the maximum reading when the force on the scale is suddenly increased. Therefore, observation of how the scale reading changes will need to be made a fraction of a second after you begin rising from crouched position.

a) Sketch the situation in your log.

b) Jump as your assistant observes the scale. Compared to when your center of mass is at rest or moving with constant upward speed, is the downward force you exert on the scale when accelerating your center of mass upward more or less? Since the scale couldn't "keep up" with the action to allow reading a specific value of force, report in your journal whether the force was "going up" or "going down" as you were pushing on the scale before launching into your jump.

c) In your sketch, draw arrows to represent the forces acting on your center of mass as it accelerates upward before launch, including (i) the downward pull of gravity on your body (your weight) and (ii) the upward push of the scale. You do not have a value, in newtons, to determine how long to make the arrow representing the upward push of the scale, so indicate, according to how the scale reading was changing, that the amount of force is either greater or less than your weight.

d) Which of Newton's Laws of Motion best applies to this situation, the First Law or the Second Law? How does the law which you choose apply?

# ACTIVITY 6
## The Mu of the Shoe

# Background Information

When it is desired to accelerate an object (change its speed, its direction, or both), a net force must act on the object. If you are standing still on a sidewalk and want to get moving, you must somehow cause a force to be exerted on your body's mass. You accomplish application of a force by pushing your foot in the backward direction parallel to the surface of the sidewalk. How much you can push your foot parallel to the surface of the sidewalk depends on how much frictional force can be sustained by the interaction of the sole of the shoe and the sidewalk's surface. If the shoe does not slip on the surface, the sidewalk surface's equal and opposite reaction to the rearward force of friction causes your body to accelerate forward; if there is ice on the sidewalk, the available force will be reduced, and your shoes may just slide on the surface with the result that you go nowhere.

The maximum frictional force, $F$, that can be generated between the surfaces of two materials in contact is expressed by the equation $F = \mu N$ where $N$ is the "normal force" (the word "normal" in this context means "perpendicular") that is perpendicularly pushing the two surfaces together, and $\mu$ is the "coefficient of friction" for the pair of materials from which the surfaces are made. In the above example, the normal force, $N$, would be equal to your weight. The value of $\mu$ depends on the quality of the two materials in contact. For a given pair of materials, such as leather shoe soles on a concrete sidewalk surface, there are two kinds of $\mu$, the "coefficient of static (starting) friction" and the "coefficient of sliding friction" (sometimes called the coefficient of "kinetic"—meaning "moving"—friction). The larger of the two, the "coefficient of static (stationary) friction" applies when the surfaces are at rest with respect to each other; the value of $F$ resulting from calculations using the coefficient of static friction is the minimum force required to "tear the surfaces loose" to cause them to begin sliding across one another. The second kind of $\mu$, the smaller of the two kinds, applies when the surfaces are moving with respect to each other in a sliding mode; the value of $F$ resulting from calculations using the coefficient of sliding friction is the minimum force required to cause the surface to slide across one another at constant speed. It is the second kind, the coefficient of sliding friction, that is measured in this activity

The frictional force generated between the shoe and the sidewalk—due to pushing the foot rearward as gravity and the upward restoring force of the sidewalk squeeze the sole of the shoe and the surface of the sidewalk together—is answered by a corresponding equal forward push by the sidewalk on the foot. The latter force, the forward push by the sidewalk, is the push you "feel" and which causes you to accelerate forward.

It is not always the case that the normal force, $N$, is equal to the weight of the object. In the case of an object on a sloped surface, the normal force is less than the weight, equaling the component, or effectiveness, of the object's weight in the direction perpendicular to the sloped surface. Other examples of a cases where $N$ is not equal to the weight of an object bearing on a surface would include the frictional force between belts riding on pulleys in machines; in such cases, tensioned springs usually are used to force the surfaces together to provide sufficient $N$ to prevent sliding, or, intentionally as when stopping a machine, to reduce tension to cause $N$ to be reduced to an amount where a belt will slide on a pulley.

## *Active*-ating the Physics InfoMall

Discussions of friction can be found throughout the InfoMall. If you choose to do a search using the keyword "friction," you will need to limit your search to only a few stores at a time. If you look in the Articles and Abstracts Attic, one of the titles that may interest you is "Twas the class before Christmas," from *The Physics Teacher*, vol. 24, issue 9, 1986. At the very least, the problems involving friction can be amusing.

Try searching the Demo & Lab Shop with the keyword "friction." You will want to look at these yourself, so no examples are included here. Choose the demonstration that best suits your style and situation.

And you will not be surprised to find that there are many, many problems you can find involving friction in the Problems Place.

# Planning for the Activity

## Time Requirements

- One class period.

## Materials Needed

### For each group:

- rough horizontal surface, 6" x 36"
- shoe
- smooth surface (desk top, tile)
- scale, 0-20 newton range
- stack of books
- weight, 100 g slotted mass
- wood board, 4 ft

## Advance Preparation and Setup

You will need a variety of athletic shoes and samples of floor materials. If enough students wear athletic shoes to school, use their shoes as samples; if not, arrange to have one shoe per group available. Floor materials may include your classroom floor, a table top to simulate a floor, samples of floor materials from a retail store, or, if they are different from the floor in your classroom, floors in other areas of your school such as the gymnasium. It is desired to have contrasting degrees of "roughness" represented in the samples. Surfaces which would be of most interest to your students would be best to use.

# Teaching Notes

It is recommended that a brief discussion of the symbolism used in physics be conducted with the class after students have read the section "What is Mu?" and before beginning **For You To Do.** Students may feel it is "cool" to ask other students not taking physics about the "$\mu$" of their athletic shoes.

Students may wish to measure $\mu$ for more kinds of shoes and floor materials called for in the instructions. The instructions include only the minimum number of samples needed to acquire meaningful data; encourage students to test more samples of shoes and/or floor materials if time allows.

Coefficients of friction for many pairs of materials are listed in the *Handbook of Chemistry and Physics* and in many traditional physics textbooks. You may wish to have listings available for students to examine.

In addition to athletic shoes which are specialized for particular sports, the variety of waxes used by skiers to control friction in various conditions may be of high interest to some students.

This activity can also be done with a smart pulley, computer, and weights hanging over the smart pulley.

You may wish to do the **Activity 6 A**: Alternative Activity for Measuring the Mu of the Shoe presented after this activity in the Teacher's Edition as a **Stretching Exercise** with the class, after you have completed the activity in the textbook. The explanation for this activity is also presented following the activity.

# Activity Overview

In this activity students investigate the effect of different surfaces, and different weights on the coefficient of friction.

## Student Objectives

### Students will:

- Understand and apply the definition of the coefficient of sliding friction, $\mu$.
- Measure the coefficient of sliding friction between the soles of athletic shoes and a variety of floor surface materials.
- Calculate the effects of frictional forces on the motion of objects.

ANSWERS FOR THE TEACHER ONLY

## What Do You Think?

Vast amounts of engineering knowledge and research about friction are applied to the design of athletic footgear.

Physics in Action

# Activity 6 The Mu of the Shoe

**GOALS**

In this activity you will:

- Understand and apply the definition of the coefficient of sliding friction, $\mu$.
- Measure the coefficient of sliding friction between the soles of athletic shoes and a variety of floor surface materials.
- Calculate the effects of frictional forces on the motion of objects.

**What Do You Think?**

A shoe store may sell as many as 100 different kinds of sport shoes.

- **Why do some sports require special shoes?**

Record your ideas about this question in your *Active Physics* log. Be prepared to discuss your responses with your small group and the class.

**For You To Do**

1. Take an athletic shoe. Use a spring scale to measure the weight of the shoe, in newtons.
   a) Record a description of the shoe (such as its brand) and the shoe's weight, in your log.

Active Physics 52

ANSWERS

## For You To Do

1. a) Answers will vary depending on the brand of shoe used and the size of the shoe.

2. Place the shoe on one of two horizontal surfaces (either rough or smooth) designated by your teacher to be used for testing. Attach the spring scale to the shoe as shown below so that the spring scale can be used to slide the shoe across the surface while, at the same time, the amount of force indicated by the scale can be read.

a) Record in your log a description of the surface on which the shoe is to slide.

b) Measure and record the amount of force, in newtons, needed to cause the shoe to slide on the surface at constant speed. Do not measure the force needed to start, or "tear the shoe loose," from rest. Measure the force needed, after the shoe has started moving, to keep it sliding at low, constant speed. Also, be careful to pull horizontally so that the applied force neither tends to lift the shoe nor pull downward on the shoe.

c) Use the data you have gathered to calculate μ, the coefficient of sliding friction for this particular kind of shoe on the particular kind of surface used. Show your calculations in your log.

> The coefficient of sliding friction, symbolized by μ, is calculated using the following equation:
>
> $$\mu = \frac{\text{force required to slide object on surface at constant speed}}{\text{perpendicular force exerted by the surface on the object}}$$
>
> Example:
>
> Brand X athletic shoe has a weight of 5 N. If 1.5 N of applied horizontal force is required to cause the shoe to slide with constant speed on a smooth concrete floor, what is the coefficient of sliding friction?
>
> $$\mu_{x \text{ on concrete}} = \frac{1.5\text{ N}}{5.0\text{ N}} = 0.30$$





ANSWERS

## For You To Do *(continued)*

2. a)-c) Students' answers will vary depending on the surface and the shoe used.

ANSWERS

## For You To Do

***(continued)***

3\. a)-c) Taking into account possible errors in measurement, students should find that the value of $\mu$ is not affected by the weight of the shoe.

4\. a) Student sketch (see bottom right).

b)-c) Students should recognize that the value of $\mu$ will be different for different surfaces. The "rougher" the surface, the greater the coefficient of friction. For example, the coefficient of friction for rubber on dry concrete is 140 times greater than rubber on ice.

d) The previous step in the activity indicates that the weight of the shoe should not make a difference in the coefficient of friction.

Physics in Action

3\. Add "filler" to the shoe to approximately double its weight and repeat the above procedure for measuring the $\mu$ of the shoe.

a) Calculate $\mu$ for this surface, showing your work in your log.

b) Taking into account possible errors of measurement, does the weight of the shoe seem to affect $\mu$? Use data to answer the question in your log.

c) How do you think the weight of an athlete wearing the shoe would affect $\mu$? Why?

4\. Place the shoe on the second surface designated by your teacher and repeat the procedure.

a) Make another sketch to show the forces acting on the shoe.

b) Calculate $\mu$.

c) How does the value of $\mu$ for this surface compare to $\mu$ for the first surface used? Try to explain any difference in $\mu$.

d) Would it make any difference if you used the empty shoe or the shoe with the filler to calculate $\mu$ in this activity? Explain your answer.

### Reflecting on the Activity and the Challenge

Many athletes seem more concerned about their shoes than most other items of equipment, and for good reason. Small differences in the shoes (or skates or skis) athletes wear can affect performance. As everyone knows, athletic shoes have become a major industry because people in all "walks" of life have discovered that athletic shoes are great to wear, not only on a track but, as well, just about anywhere. Now that you have studied friction, a major aspect of what makes shoes function well when need exists to be "sure-footed," you are prepared to do "physics commentary" on athletic footgear and other effects of friction in sports. Your sports commentary may discuss the $\mu$ of the shoe, the change in friction when a playing field gets wet, and the need for friction when running.

54

Active Physics

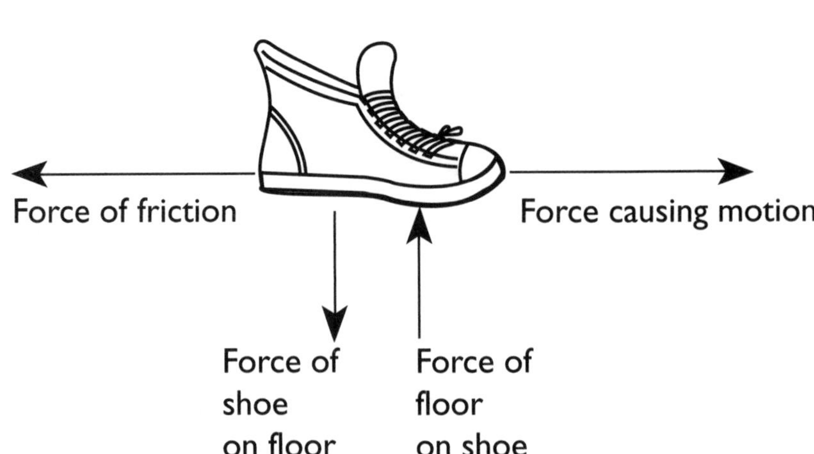

Activity 6 The Mu of the Shoe

**PHYSICS TALK**

**Coefficient of Sliding Friction, μ**

There are not enough letters in the English alphabet to provide the number of symbols needed in physics, so letters from another alphabet, the Greek alphabet, also are used as symbols. The letter μ, pronounced "mu," traditionally is used in physics as the symbol for the "coefficient of sliding friction."

The coefficient of sliding friction, symbolized by μ, is defined as the ratio of two forces:

$$\mu = \frac{\text{force required to slide object on surface at constant speed}}{\text{perpendicular force exerted by the surface on the object}}$$

Facts about the coefficient of sliding friction:

- **μ does not have any units because it is a force divided by a force; it has no unit of measurement.**
- **μ usually is expressed in decimal form, such as 0.85 for rubber on dry concrete (0.60 on wet concrete).**
- **μ is valid only for the pair of surfaces in contact when the value is measured; any significant change in either of the surfaces (such as the kind of material, surface texture, moisture, or lubrication on a surface, etc.) may cause the value of μ to change.**
- **Only when sliding occurs on a horizontal surface, and the pulling force is horizontal, is the perpendicular force that the sliding object exerts on the surface equal to the weight of the object.**

Answers

## Physics To Go

1. In football, players change the length of shoe cleats or sometimes wear shoes without cleats to improve footing in bad weather.

2. Downhill skiers use wax to reduce friction.

3. A common misconception is that the coefficient of friction—and, therefore, the force of friction—depends only on the shoe; it depends as much as on the nature of the surface beneath the shoe. No, the athlete cannot be assured that the same amount of frictional force will be present when the same shoe is used on a court having a different surface.

4. $F = 0.03 \times 600 \text{ N} = 18 \text{ N}$

5. Normal force, $F$N= Weight of vehicle = $mg$ = 1000 kg × 10 m/s$^2$ = 10,000 N

Frictional (stopping) force = $\mu$N = 0.55 × 10,000 N = 5500 N

Work to stop vehicle = fd = 5500 N × 100 m = 550,000 joules

Work to stop vehicle = $KE$ of vehicle before brakes were applied:

$550{,}000J = 1/2\ mv^2$

Therefore,

$v = \sqrt{2\ (550{,}000\ J)/m}$

$= \sqrt{1{,}100{,}000\ J/1000\ \text{kg}}$

$= \sqrt{1100\ \text{m}^2/\text{s}^2}$

= 33 m/s = 75 miles/hr

The driver has a problem because the laws of physics will prevail in court.

Physics in Action

### Physics To Go

1. Identify a sport and changing weather conditions that probably would cause an athlete to want to increase friction to have better footing. Name the sport, describe the change in conditions, and explain what the athlete might do to increase friction between the shoes and ground surface.

2. Identify a sport in which athletes desire to have frictional forces as small as possible and describe what the athletes do to reduce friction.

3. If a basketball player's shoes provide an amount of friction that is "just right" when she plays on her home court, can she be sure the same shoes will provide the same amount of friction when playing on another court? Explain why or why not.

4. A cross-country skier who weighs 600 N has chosen ski wax that provides $\mu = 0.03$. What is the minimum amount of horizontal force that would keep the skier moving at constant speed across level snow?

5. A racecar having a mass of 1000 kg was traveling at high speed on a wet concrete road under foggy conditions. The tires on the vehicle later were measured to have $\mu = 0.55$ on that road surface. Before colliding with the guardrail, the driver locked the brakes and skidded 100 m, leaving visible marks on the road. The driver claimed not to have been exceeding 65 miles per hour (29 m/s). Use the equation:

Work = Kinetic Energy

to estimate the driver's speed upon hitting the brakes. (Hint: In this case, the force that did the work to stop the car was the frictional force; calculate the frictional force using the weight of the vehicle, in newtons, and use the frictional force as the force for calculating work.)

6. Identify at least three examples of sports in which air or water have limiting effects on motion similar to sliding friction. Do you think forces of "air resistance" and "water resistance" remain constant or do they change as the speeds of objects (such as athletes, bobsleds, or rowing sculls) moving through them change? Use examples from your own experience with these forms of resistance as a basis for your answer.
7. If there is a maximum frictional force between your shoe and the track, does that set a limit on how fast you can start (accelerate) in a sprint? Does that mean you cannot have more than a certain acceleration even if you have incredibly strong leg muscles? What is done to solve this problem?
8. How might an athletic shoe company use the results of your experiment to "sell" a shoe? Write copy for such an advertisement.
9. Explain why friction is important to running. Why are cleats used in football, soccer, and other sports?
10. Choose a sport and describe an event in which friction with the ground or the air plays a significant part. Create a voice-over or script that uses physics to explain the action.

ANSWERS

## Physics To Go
***(continued)***

6. Any sports which involve objects moving through air or water will involve fluid resistance.
7. Yes, the maximum frictional force between your shoe and the track does place a limit on acceleration. Sprinters use blocks when beginning a race, thereby providing a surface that is not horizontal, and does not depend on the weight of the runner.
8. See the **Assessment Rubric** following this activity in the Teacher's Edition.
9. Without friction it would be impossible to walk or run. Cleats increase the friction between the shoe and the ground by increasing the amount of surface in contact.
10. Students provide a voice-over related to friction.

# Assessment Rubric: Physics To Go Question 8

How might an athletic shoe company use the results of your experiment to sell a shoe? Write a copy for an advertisement.

| Descriptors | Levels of Attainment | | | |
|---|---|---|---|---|
| | poor | average | good | excellent |
| **1. Physics concepts are accurately presented.**<br>• Newton's Third Law of Motion is explained in terms of running: For every applied force, there is an equal and opposite force.<br>• The force required to cause the shoe to move on different surfaces is explained: *Coefficient of friction is expressed as a ratio of the force needed to move the shoe at a constant speed, by the force exerted by the surface of the shoe.* | 1 | 2 | 3 | 4 |
| **2. Physics concepts explained in everyday language.**<br>• Examples are provided for each concept.<br>• Presentation style does not talk down to the consumer. | 1 | 2 | 3 | 4 |
| **3. The need for buying specialized sports shoes is established.**<br>• Examples are provided illustrating why different shoes are used for different sporting events. *Presentation identifies the amount of friction provided by different surfaces and the advantages of solid traction for different sports.* | 1 | 2 | 3 | 4 |
| **4. Design and appeal.**<br>• Organization of information is short and snappy.<br>• Message is clearly identified with a target audience.<br>• Presentation designed for a specific media: i.e., visual images utilize color effectively and enhance message for television presentation. | 1 | 2 | 3 | 4 |

For use with *Physics in Action*, Chapter 1, Activity 6: The Mu of the Shoe

NOTES

## Activity 6 A

# Alternative Activity for Measuring the Mu of the Shoe

### FOR YOU TO DO

"Step right up. Have the Mu of your shoe measured right here! Takes only a minute." Can the coefficient of sliding friction be measured in a quicker, easier way than the method used in **For You To Do.** Well, it can.

1. For a quick-and-easy way to measure, all you need is a sample of "floor" material (such as a wooden board) which can be arranged into a ramp, a carpenter's square (if not available, a ruler can be substituted), a calculator, and, of course, a shoe. Use a stack of books (or some other method) to raise one end of the floor sample to form a sloped ramp and place the shoe on the ramp facing "downhill."

2. Adjust the slope of the ramp to find the slope at which the shoe, given only a "nudge" to start it sliding, continues sliding down the ramp "on its own" (without further pushing) at low, constant speed:

   - If the shoe does not slide down the ramp on its own after it is tapped, increase the slope of the ramp.
   - If the shoe accelerates while sliding down the ramp after it is tapped, decrease the slope of the ramp.

3. When you have found the proper slope for the ramp, use a carpenter's square (or a ruler) to measure the vertical rise and corresponding horizontal run of the ramp.

   a) Record the rise and the run in your log.

4. The coefficient of sliding friction is:

$$\mu = \frac{\text{Length of vertical rise of ramp}}{\text{Length of horizontal run of ramp}}$$

   a) Calculate the coefficient of sliding friction of the shoe.
   This method gives the same result as the method used in **For You To Do.** Your teacher may be able to help you understand why the two methods are equivalent.

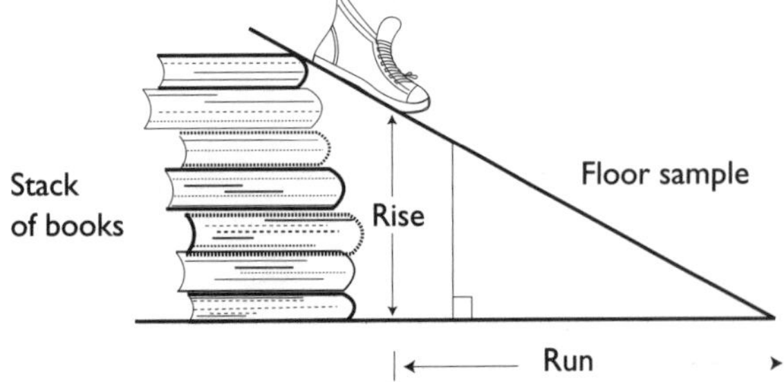


For use with *Physics in Action*, Chapter 1, Activity 6: The Mu of the Shoe

# Background Information for Activity 6 A: Alternative Activity for Measuring the Mu of the Shoe

Below is the theoretical basis for the method of measuring $\mu$ presented in the **Alternative Activity 6 A.** Highly able students who have had experience with geometry may be able to understand the principles underlying the method. It is recommended that the procedure presented in the alternative activity be read carefully before proceeding to read the below explanation.

The purpose of this explanation is to prove that $\mu$ = (Rise) ÷ (Run) when the slope of the inclined plane made from a sample of floor material is adjusted so that the shoe, or other object represented by mass *m*, whose coefficient of friction with the floor material is desired to be measured—slides down the slope at low, constant speed when given a "tap" start. (A tap start is required to overcome the force of "static," or starting, friction which is greater than the force of sliding friction.) Referring to the below diagram, side lengths A and B of triangle ABC correspond, respectively, to the Rise and Run distances. Therefore, it is to be proven that $\mu$ = A/B.

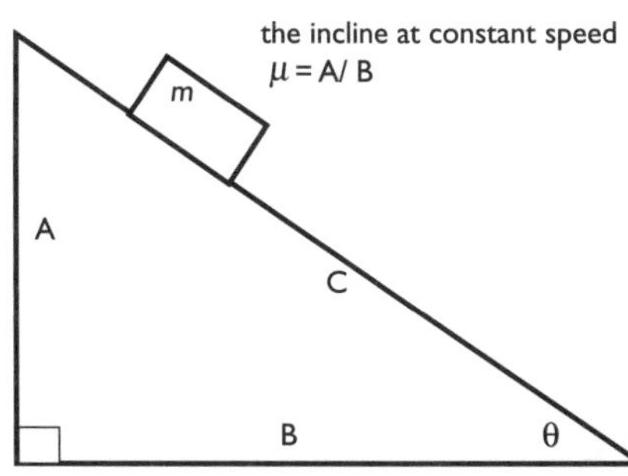


The weight, *W*, of mass *m* is a force which acts straight downward, as shown in the below diagram. The amount of force due to the object's weight, *W* = *mg*, is represented by the length of vector *W*.

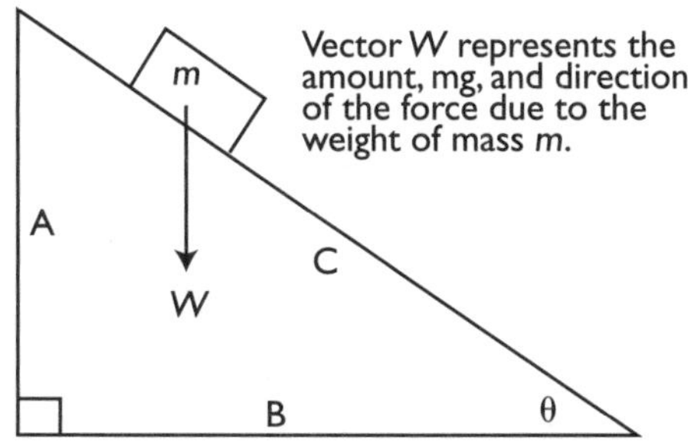


The weight vector, *W*, can be resolved into, or thought of as having the same effect as, two vectors called "components" of vector *W*: (1) a vector *R* which has a direction normal, or perpendicular, to the surface represented by side C of triangle ABC and (2) a vector *P* which has a direction parallel to side C. Vector *R* is the force which mass m exerts perpendicular to the surface on which sliding occurs, and vector *P* is the force which causes mass m to slide on the surface at constant speed.

The below diagram shows vectors *R* and *P* as components of vector *W*.

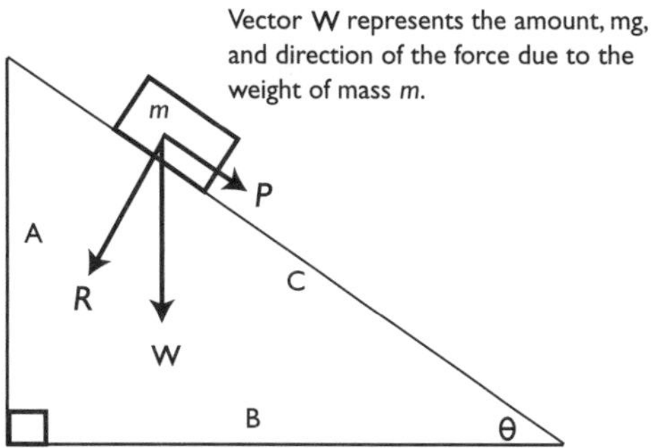


Vector mathematics requires that component vectors, in this case vectors *P* and *R*, must, when added together as vectors, equal the vector of which they are components, in this case vector *W*. On the right-hand side of the above diagram, vectors *P* and *R* are shown repositioned, but not altered in length or direction, for head-to-tail vector addition. As shown, when the component vectors are placed head-to-tail with one of the component vectors having its tail end corresponding to the tail end of vector *W*, the head end of the second component vector drawn in the head-to-tail vector addition process arrives at the head end of vector *W*. The fact that the component vectors "close" on the head end of *W* when placed head-to-tail shows, according to vector algebra, that the components add together to equal vector *W*. Since the directions of the component vectors *P* and *R* were specified as, respectively, parallel and normal to triangle side C, only the lengths of *P* and *R* shown in the diagram would satisfy the requirement of the vectors to close on the head end of vector *W*.

Referring to the above diagram, triangle ABC is similar to triangle *PRW* formed for the vector addition process. This is true by the theorem that two triangles which each contain a right angle and which have two mutually perpendicular sides are similar. Therefore, the sides of the two triangles exist in equal proportions, and:

A/B = *P*/*R*

Since, by definition, $\mu$ is the ratio of the force required to slide mass m at constant speed on the surface to the force which *m* exerts perpendicular to the surface:

$\mu = P/R = A/B$

Therefore, it is proven: the Rise divided by the Run, A/B, when mass *m* slides at constant speed is equal to $\mu$.

For those familiar with trigonometry, A/B in triangle ABC is the tangent of angle θ.

Therefore, as an alternative to measuring the Rise and Run to determine $\mu$, θ can be measured instead, and $\mu$ can be determined using a calculator or table by the relation:

$\mu = \tan\theta$

All of the forces active as mass *m* slides on the incline at constant speed are shown in the below diagram:

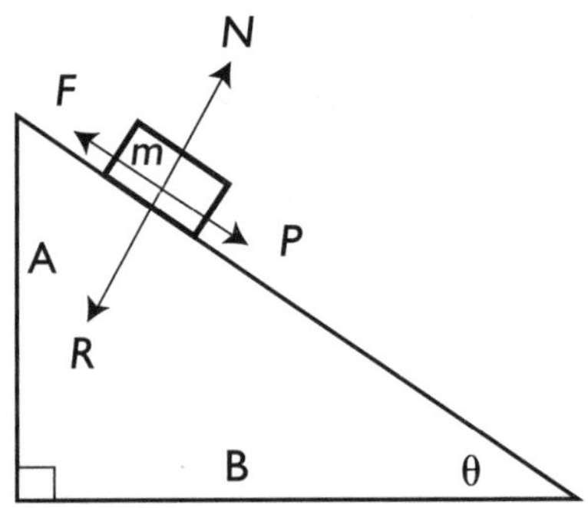


Where the new forces introduced to the diagram, *F* and *N*, are, respectively, the frictional force and the restoring force of the material represented by side C of triangle ABC. It is important to observe that, since mass *m* moves at constant speed, the sum—vector sum, that is—of the four forces acting on mass *m* is zero, or, in other words, all of the forces acting on mass *m* are balanced.

For students who wish to do the activity , samples of floor materials will need to be able to be arranged on an adjustable slope; boards 2 or 3 feet in length could be used as a base for samples to be arranged on a slope.

NOTES

# ACTIVITY 7
## Concentrating on Collisions

# Background Information

This **Background Information** will serve for both this and the next activity because both involve the same topic, momentum. Momentum is introduced in this activity, and conservation of momentum is implied; the next activity, **Activity 8**, focuses on the Law of Conservation of Momentum.

Before proceeding in this section it is recommended that you read the following sections in this chapter of the student text: **For You To Read:** Mass, Velocity and Momentum in **Activity** 7, and **For You To Read:** Conservation of Momentum in **Activity 8**. Both selections are assumed as background for the below discussion of momentum.

A corollary to Newton's Third Law is that action/reaction pairs of forces always act for the same amount of time as each other. In any interaction, forces may vary in complicated ways during the time of interaction; nevertheless, the average force during the interaction often can be used to provide an accurate, but greatly simplified, analysis of the net result of an interaction such as a collision.

If $F$ is used to represent the average force on one object during a collision and $\Delta t$ is used to represent the duration of the collision, one can produce the following mathematical path to the quantity of ultimate interest for this discussion, the quantity called momentum. Newton's Second Law, written in rearranged form, may be applied to the object involved in the collision as:

$$F\Delta t = m\Delta v = m(v_f - v_i) = (mv_f - mv_i) = \Delta(mv)$$

$$F\Delta t = \Delta(mv)$$

where $v_i$ is the speed with which the object entered into the collision and $v_f$ is the speed of the object after the collision.

In the above equation $F\Delta t = \Delta(mv)$, the product of force and time, $F\Delta t$, is called the "impulse," and the quantity $mv$ is called the "momentum." The equation can be read literally as, "When during a collision an object of mass $m$ experiences an impulse $F\Delta t$, the object experiences a change in momentum $\Delta(mv)$. Indeed, it is the impulse that causes the object's change in momentum.

The force-time product, impulse, is expressed in the unit newton-seconds; dimensional analysis shows that Ns is equivalent to the unit of momentum, which, as a mass-velocity product, must be (kg)m/s:

Ns = [(kg)(m/s$^2$)] (s) = (kg)m/s (True, the unit of impulse equals the unit of momentum.)

But, according to Newton's Third Law, whatever happens to one of a pair of objects during a collision should, in terms of force, happen equally, but in the opposite direction, to the other object. Also, the objects involved in a collision touch each other for equal amounts of time, the duration of mutual contact. Thus, whatever change in momentum one object experiences, the other object must experience an equal and opposite change in momentum. For example, if one object gains momentum in a head-on collision, the other object must lose an equal amount of momentum. It must always be true that the combined momenta of the two objects before the collision are preserved, or conserved, after the collision. It is a far-reaching law of physics that momentum is conserved in all interactions, no matter how complex, convoluted, and intense the interactions may be.

This chapter considers only collisions that are "head-on," so the motions all are along the same line before and after the collision. However, the caution is made that when dealing with momentum in collision events, one must take care to keep track of the directions of travel of the objects both before and after the collision. For example, for a collision in which an object of mass $m$ leaves the collision traveling at the same rate but in an opposite direction is not one for which the momentum change is zero. In this case, the momentum change is $2mv$ or $-2mv$, depending on how positive and negative signs are attributed to direction of travel.

Similarly, if two objects enter a collision with equal magnitudes of momentum, the opposite signs mean that the sum of momenta is zero. As a result, when the collision is completed, one possible outcome is to have both objects at rest at the point of impact. The only other possibility is that both objects will rebound with equal and opposite momenta, but not necessarily equal and opposite speeds because the masses may be unequal.

Depending on the nature of the objects involved in a collision, kinetic energy may be conserved to extents ranging from not at all (as when the objects stop upon colliding) to almost completely (as when extremely hard objects such as ball bearings or gas molecules collide and rebound). Whether or not kinetic energy is conserved, momentum is conserved, which makes the Law of Conservation of Momentum a very powerful tool.

## *Active*-ating the Physics InfoMall

The physics of collisions is yet another of those topics discussed in almost every textbook around. To keep InfoMall searches interesting and related to this book, search the Articles and Abstracts Attics using keywords "collisions" AND "sport*". Several articles result, including "Batting the ball," from the *American Journal of Physics*, vol. 31, issue 8, 1963. You will also find plenty on billiard balls, kicking footballs, and using tennis rackets.

You may wish to investigate the known misconceptions students have regarding momentum. Try a search with "momentum" AND "misconcept*". One of the hits is "Verification of fundamental principles of mechanics in the computerized student laboratory," in *American Journal of Physics*, vol. 58, issue 10, 1990. This article is also great for using computers for teaching.

# Planning for the Activity

## Time Requirements

- One class period.

## Materials Needed

### For each group:

- balls, bocci
- ball, golf
- ball, Nerf®
- bBall, practice golf
- ball, tennis
- *Sports* Content Video
- starting ramp for bocci ball
- tape, masking, 3/4" x 60 yds.

## Advance Preparation and Setup

A pair of (nearly) matched bocci balls is recommended for each group. Contacting bocci establishments in your area perhaps will result in donations of balls, and putting out a call to your school community—parents, faculty, etc.—also should be productive. If you simply cannot obtain bocci balls, hard balls, such as croquet balls, could be substituted.

You will also need a soccer ball, a golf ball, and a tennis ball for each group. You may be able to borrow the balls from your school's athletic department.

# Teaching Notes

Discuss with the students what is happening in the humorous illustration on page 58. What will happen to the soccer ball? Students who enjoy drawing, may wish to draw the "next frame" in this action. You may wish to return to the illustration at the end of the activity to compare students responses to their original predictions.

Bocci balls, with their large inertias, are less sensitive to annoyances like small imperfections in the floor or a little grit in the path. They also have the advantage of a large diameter, so it is easier to create collisions that are nearly head on. They also demand a lot of attention. Their disadvantage is their large mass!

SAFETY PRECAUTION: Moving bocci balls can hurt people (especially toes and fingers) and property. Instruct students to use extreme care and provide close supervision when bocci balls are in use.

Contact sports are, of course, rich sources of collisions to be analyzed in terms of momentum.

Need to assign positive and negative values to velocity and momentum is implied in the discussion in **For You To Read.** The unit of momentum, (kg)m/s is new, but is sufficiently straightforward that it should not be a problem for students; somehow, the unit of momentum seems to have escaped being collected up under someone's name.

# Activity Overview

In this activity students investigate the principle of momentum by rolling balls down ramps and staging collisions. They will infer the relative masses of two balls by staging and observing collisions between them.

## Student Objectives

### Students will:

- Understand and apply the definition of momentum: Momentum = $mv$.
- Conduct semi-quantitative analyses of the momentum of pairs of objects involved in one-dimensional collisions.
- Infer the relative masses of two objects by staging and observing collisions between the objects.

ANSWERS FOR THE TEACHER ONLY

## What Do You Think?

Both mass and speed have roles in collisions between two objects. The mass of each player and their speeds must be considered.

Physics in Action

# Activity 7 Concentrating on Collisions

**GOALS**

In this activity you will:

- Understand and apply the definition of momentum.
- Conduct semiquantitative analyses of the momentum of pairs of objects involved in one-dimensional collisions.
- Infer the relative masses of two objects by staging and observing collisions between the objects.

**What Do You Think?**

In contact sports, very large forces happen during short time intervals.

- **A football player runs toward the goal line, and a defensive player tries to stop him with a head-on collision. What factors determine whether the offensive player scores?**

Record your ideas about this question in your *Active Physics* log. Be prepared to discuss your responses with your small group and the class.

**For You To Do**

1. You will stage a head-on collision between two matched bocci balls. Set up a launch ramp for one ball, and find a level area clear of obstructions nearby where the other bocci ball can be at rest.

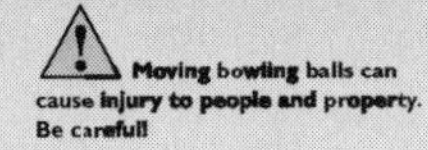

**Moving bowling balls can cause injury to people and property. Be careful!**

58

Active Physics

ANSWERS

## For You To Do

1. Student activity.

2. Temporarily remove the "target" bocci ball. Find a point of release within the first one-fourth of the ramp's total length that gives the ball a slow, steady speed across the floor. Mark the point of release on the ramp with a piece of tape.

3. Replace the target ball. Adjust the aim until a good approximation of a head-on collision is obtained. Stage the collision.

a) Record the results in your log. Use a diagram and words to describe what happened to each ball.

4. Repeat the above type of collision, but this time move the release point up the ramp to at least double the ramp distance.

a) Describe the results in your log.

b) How did the results of the collision change from the first time?

c) Identify a real-life situation that this collision could represent.

5. Arrange another head-on collision between the balls, but this time have both balls moving at equal speeds before the collision. Using a second, identical ramp, aim the second ramp so that the second ball's path is aligned with the first ball's path. Mark a release point on the second ramp at a height equal to the mark already made on the first ramp. This should ensure that the balls will have low, approximately equal speeds. On a signal, two persons should release the balls simultaneously from equal ramp heights.

a) Describe the results in your log.

b) Identify a real-life situation that this collision could represent.





ANSWERS

## For You To Do

***(continued)***

2. Student activity.

3. a) Assume the bocci balls have reasonably well-matched masses. A moving ball striking a stationary ball of equal mass should result in the moving ball stopping upon colliding and the stationary ball moving away from the collision at about the same speed that the incoming ball had before the collision. bocci balls have a high coefficient of restitution, and the first ball can be expected to come nearly to a stop, while the other leaves at something close to the speed of the incident ball.

4. a)-b) As the speed of the released ball increases, the speed at which the stationary ball moves away will increase.

   c) This type collision may occur when a forward collides with a goaltender, or the head of a golf club collides with a golf ball.

5. a) Balls of equal mass and equal speed colliding head-on should result in both balls rebounding at speeds less than or equal to their speeds before the collision.

   b) This type of collision may occur between two football players. However, since people are less elastic than bocci balls, they won't bounce off each other so far. However, you can expect both players to bounce in such a manner that each finds himself on his back.

ANSWERS

## For You To Do
***(continued)***

6. a) In collisions between a nerf ball and a bocci ball it is generally the case that the nerf ball "loses," or undergoes the greatest change in velocity. When a stationary nerf ball is hit head-on by a moving bocci ball, the nerf ball will leave the collision at a speed higher than the incoming speed of the bocci ball, and the bocci ball will keep moving in its original direction of motion after the collision, but with a slightly reduced speed.

7. a) When a stationary bocci ball is hit head-on by a moving nerf ball, it should occur, for ordinary incoming speeds of the nerf ball, that the bocci ball will move away from the collision at a relatively low speed and the nerf ball will rebound from the collision.

8. a) Students will stage a collision between a golf ball and a wiffle golf ball and determine from observing the balls after the collision that the golf ball is much more massive.

   b) Students determine mass of both balls. Their results from the previous method will be verified.

**Physics Words**
**momentum:** the product of the mass and the velocity of an object; momentum is a vector quantity.

6. Repeat **Steps 1, 2, and 3**, but replace the stationary bocci ball with a nerf ball.
   a) Be sure to write all responses, including identification of a similar situation in real life, in your log.

7. Repeat **Steps 1, 2,** and **3**, but in this case have the nerf ball roll down the ramp to strike a stationary bocci ball.
   a) Be sure to write all responses, including identification of a similar situation in real life, in your log.

8. Using your observations, determine the relative mass of a golf ball compared to a wiffle ball by staging collisions between them.
   a) Which ball has the greater mass? How many times more massive is it than the other ball? Describe what you did to decide upon your answer.
   b) Use a scale or balance to check your result. Comment on how well observing collisions between the balls worked as a method of comparing their masses.

### FOR YOU TO READ
#### Momentum

Taken alone, neither the masses nor the velocities of the objects were important in determining the collisions you observed in this activity. The crucial quantity is **momentum** (mass × velocity). A soccer ball has less mass than a bocci ball, but a soccer ball can have the same momentum as a bocci ball if the soccer ball is moving fast. A soccer ball moving very fast can affect a stationary bocci ball more than a soccer ball moving very slowly. This is similar to the damage small pieces of sand moving at very high speeds can cause (such as when a sand blaster is used to clean various surfaces).

Sportscasters often use the term *momentum* in a different way. When a team is doing well, or "on a roll," that team has momentum.

A team can gain or lose momentum, depending on how things are going. This momentum clearly does not refer to the mass of the entire team multiplied by the team's velocity.

Other times, sportscasters use the term momentum to mean exactly how it is defined in the activity (mass × velocity), when they say things such as, "Her momentum carried her out of bounds."

### Reflecting on the Activity and the Challenge

You already have identified several real-life situations that involve collisions, and many such situations happen in sports. Some involve athletes colliding with one another as in hockey and football. Others cases include athletes colliding with objects, such as when kicking a ball. Still others include collisions between objects such as a golf club, bat, or racquet with a ball. Some spectacular collisions in sports provide fun opportunities for demonstrating your knowledge about collisions during voice-over commentaries. Use the concept of momentum when describing collisions in your sports video.

### Physics To Go

1. Sports commentators often say that a team has momentum when things are going well for the team. Explain the difference between that meaning of the word momentum and its specific meaning in physics.
2. Suppose a running back collides with a defending linebacker who has just come to a stop. If both players have the same mass, what do you expect to see happen in the resulting collision?
3. Describe the collision of a running back and a linebacker of equal mass running toward each other at equal speeds.

ANSWERS

## Physics To Go

1. To say that a team or a candidate for election has momentum is to say that things are going well, are "on the rise;" momentum in physics is defined as mass times velocity.
2. You would expect the defending linebacker to be set in motion, and the speed of the running back to be greatly reduced.
3. The two players will bounce back from each other, at least to the extent that they will end up on their backsides.

ANSWERS

## Physics To Go
***(continued)***

4. a) The heavier bat has, for the same speed, more momentum than the lighter bat and will transfer more momentum to the ball, hitting the ball farther.

   b) For the same effect, the lighter bat would need to be swung at a speed 38/30 = 1.3 times faster than the heavy bat.

5. It is difficult to change the momentum of a massive person.

6. The relative speeds of the players determines who gets knocked backward; if the small player moves fast enough, the big player can get knocked backward.

7. $100 \text{ kg} \times 10 \text{ m/s}$
   $=(0.10 \text{ kg})\ v,$
   $v = 10{,}000 \text{ m/s}$

8. Before the collision one puck is moving and the other is stationary; after the collision, the pucks have "traded" conditions, with the puck which originally was moving being stationary and the puck which originally was stationary moving at about the same speed which the other puck had before the collision.

4. Suppose that you have two baseball bats, a heavy (38-ounce) bat and a light (30-ounce) bat.
   a) If you were able to swing both bats at the same speed, which bat would allow you to hit the ball the farthest distance? Explain your answer.
   b) How fast would you need to swing the light bat to produce the same hitting effect as the heavy bat? Explain your answer.

5. Why do football teams prefer offensive and defensive linemen who weigh about 300 pounds?

6. What determines who will get knocked backward when a big hockey player checks a small player in a head-on collision?

7. A 100.0-kg athlete is running at 10.0 m/s. At what speed would a 0.10-kg ball need to travel in the same direction so that the momentum of the athlete and the momentum of the ball would be equal?

8. Use the words *mass*, *velocity*, and *momentum* to write a paragraph that gives a detailed "before and after" description of what happens when a moving shuffleboard puck hits a stationary puck of equal mass in a head-on collision.

9. Describe a collision in some sport by using the term *momentum*. Adapt this description to a 15-s dialogue that could be used as part of the voice-over for a video.

NOTES

# ACTIVITY 8
## Conservation of Momentum

# Background Information

The **Background Information** for this activity is presented in the **Background Information** for **Activity** 7, Concentrating on Collisions, because the same topic, momentum, is involved in both activities.

### *Active*-ating the Physics InfoMall

In addition to the references to **Activity** 7, you may wish to examine the Problems Place for even more exercises in momentum conservation. Remember, *Schaum's 3000 Solved Problems in Physics* has the problem and the solution. It can be a source for you, as well as a way to provide your students with solved problems for them to study!

# Planning for the Activity

## Time Requirements

- One class period.

## Materials Needed

### For each group:

- dynamics cart
- modeling clay
- spring scale, 0-10 newton range
- tape, masking, 3/4" x 60 yds
- ticker tape timer
- weight, 100 g slotted mass

## Advance Preparation and Setup

Two matters need to be considered in advance: (1) the pairs of objects to be used by each group in the collisions, including how the masses will be varied and how the masses will be caused to stick together upon colliding, and (2) how the velocity will be measured before and after the collision.

Regarding the objects to be collided, two possibilities seem to exist, air track gliders or laboratory carts. At a minimum, three collisions are desired, involving mass ratios of 1:1, 2:1 and 1:2 (the moving mass is listed first in the ratios listed – see the data table in **For You To Do** for details). Masses of 1 and 2 kg are recommended, but certainly not required. Instead, one laboratory cart could collide with another identical laboratory cart to provide a 1:1 mass ratio, and a 2:1 (and 1:2) ratio could be obtained by loading one cart to double its mass. Similarly, air track gliders could be rigged to provide 1:1 and 2:1 mass ratios.

The colliding objects need to stick together upon colliding to move as a single object after the collision. Stick-on patches are very convenient for this purpose, but double-stick tape or modeling clay also works.

Whatever objects are used, have the students measure and use their masses, in kilograms, to call attention to and engage students in using the unit of momentum, (kg)m/s. If the masses are not to be 1 and 2 kg, have students change the values listed in the data table in **For You To Do** to the values to be used in your class.

The speed, in meters per second, must be measured before and after the collision. It is necessary to measure the speed of the incoming mass before the collision and the speed of the combined masses after the collision. One way to accomplish this would be to use a sonic ranging device to monitor the speed of Object 1 (the mass moving before the collision) before, during, and after the collision. Other possibilities for measuring the speeds include stop action video, strobe photography, a ticker-tape timer or a spark timer. The particular method to be used depends on the equipment available at your school. Whatever method of measuring speeds is used, students should record speeds in m/s.

For better data if friction is involved (as when using laboratory carts) it would be best to use the speed values which occur just before and just after the collision; this would help to avoid changes in speed (deceleration) as a source of error.

You may wish to provide additional mass ratios for students to use for staging additional collisions. A 3:1 mass ratio run "both ways," 3:1 and 1:3, gives particularly interesting results; if organized on bifilar supports to collide head-on as pendulums, hard wooden or metal spheres having a 3:1 mass ratio provide a cyclically repeating sequence of collisions.

"Nonsticky" collisions present problems for measuring speed because both masses move simultaneously at different speeds. This is difficult, but not impossible to overcome with ordinary equipment and may present an interesting challenge to interested students.

# Teaching Notes

Monitor the class as students begin to stage collisions. They may need assistance choosing appropriate push-off speeds for the incoming mass and getting equipment to function.

Sample data is not provided because it cannot be anticipated what masses and speeds will be used in your class. However, conservation of momentum allows you to predict with ease what the velocity of the combined masses after the collision should be compared to the velocity of Object 1 before the collision:

Conservation of momentum applied to this kind of collision:

$m_1 v_{before} = (m_1 + m_2) v_{after}$

Solving the above equation for $v_{after}$:

$v_{after} = v_{before}\, m_1 / (m_1 + m_2)$

The final equation above will allow you to predict before class begins what the relative speeds before and after each collision should be. Another way of saying the same thing is that $v_{after} / v_{before} = m_1 / (m_1 + m_2)$.

Allowing for errors of measurement, students should find that the momenta before and after each collision are equal (in the Analysis table, the momentum of Object 2 will be zero, unless an additional collision is staged which is not included in the written procedure; this is pointed out only to avoid confusion).

It is important to discuss the collision analyzed as an example in **For You To Read** because it is of a general kind where both objects are moving before and aftera head-on collision. The same method of analysisis needed to solve some of the problems in **Physics To Go.**

Assigning positive and negative values to directions of velocities is very useful, if not necessary, for solving many collision problems.

NOTES

# Activity 8 Conservation of Momentum

**GOALS**

In this activity you will:

- Understand and apply the Law of Conservation of Momentum.
- Measure the momentum before and after a moving mass strikes a stationary mass in a head-on, inelastic collision.

**What Do You Think?**

The outcome of a collision between two objects is predictable.

- **What determines the momentum of an object?**
- **What does it mean to "conserve" something?**

Record your ideas about these questions in your *Active Physics* log. Be prepared to discuss your responses with your small group and the class.

**For You To Do**

1. From the objects provided arrange to have a head-on collision between two objects of equal mass. Before the collision, have one object moving and the other object at rest. Arrange for the objects to stick together to move as a single object after the collision. Stage a head-on, sticky collision between equal masses. Measure the velocity, in meters per second, of the moving mass before the collision and the velocity of the combined masses after the collision.

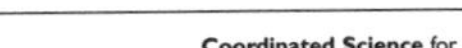



## Activity Overview

In this activity students stage collisions between objects to investigate the conservation of momentum.

## Student Objectives

### Students will:

- Understand and apply the Law of Conservation of Momentum.
- Measure the momentum before and after a moving mass strikes a stationary mass in a head-on, inelastic collision.

ANSWERS FOR THE TEACHER ONLY

## What Do You Think?

An object's mass and velocity determine its momentum, $mv$.

To conserve means to keep the same amount.

ANSWERS

## For You To Do

Sample data is not provided because it cannot be anticipated what masses and speeds will be used in your class. However, conservation of momentum allows you to predict with ease what the velocity of the combined masses after the collision should be compared to the velocity of Object 1 before the collision. (See **Teaching Notes.**)

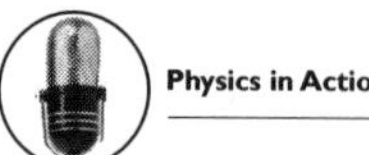



a) Prepare a data table in your log similar to the one shown below. Provide enough horizontal rows in the table to enter data for at least four collisions.

**Sticky Head-on Collisions: One Object Moving before Collision**

| Mass of Object 1 (kg) | Mass of Object 2 (kg) | Velocity of Object 1 before Collision (m/s) | Velocity of Object 2 before Collision (m/s) | Mass of Combined Objects after Collision (kg) | Velocity of Combined Objects after Collision (m/s) |
|---|---|---|---|---|---|
| 1.0 | 1.0 | | 0.0 | 2.0 | |
| 2.0 | 1.0 | | 0.0 | 3.0 | |
| 1.0 | 2.0 | | 0.0 | 3.0 | |
| | | | 0.0 | | |

b) Record the measured values of the velocities in the first row of the data table.

2. Stage other sticky head-on collisions using the masses listed in the second and third rows of the data table. Then stage one or more additional collisions using other masses. Measure the velocities before and after each collision.

a) Enter the measured values in the data table.

3. Organize a table for recording the momentum of each object before and after each of the above collisions.

a) Prepare a table similar to the following example in your log:

**Momentum of Object before and after Collisions**
**Momentum = Mass × Velocity**

| Before the Collision | | After the Collision |
|---|---|---|
| Momentum of Object 1 kg (m/s) | Momentum of Object 2 kg (m/s) | Momentum of Combined Objects 1 and 2 kg (m/s) |
| | | |
| | | |
| | | |
| | | |
| | | |



ANSWERS

## For You To Do *(continued)*

Sample data is not provided because it cannot be anticipated what masses and speeds will be used in your class. However, conservation of momentum allows you to predict with ease what the velocity of the combined masses after the collision should be compared to the velocity of Object 1 before the collision. (See **Teaching Notes.**)

Activity 8 Conservation of Momentum

b) Calculate the momentum of each object before and after each of the above collisions and enter each momentum value in the table.

c) Calculate and compare the total momentum before each collision to the total momentum after each collision.

d) Allowing for minor variations due to errors of measurement, write in your log a general conclusion about how the momentum before a collision compares to the momentum afterward.

### FOR YOU TO READ

#### The Law of Conservation of Momentum

In this activity, you investigated another conservation principle that is a hallmark of physics—the conservation of momentum. If you sum all of the momenta before a collision or explosion, you know that the sum of all the momenta after the collision will be the same.

If the momentum before a collision is 500 kg·m/s, then the momentum after the collision is 500 kg·m/s. A football player stops to catch a pass. The player is not moving and therefore has momentum equal to zero. If an opponent that has a momentum of 500 kg·m/s then hits the player, both players move off with (a combined) 500 kg·m/s of momentum. Any time you see a collision in sports, you can explain that collision using the conservation of momentum.

Conservation of momentum is an experimental fact. Physicists have compared the momenta before and after a collision between pairs of objects ranging from railroad cars slamming together to subatomic particles impacting one another at near the speed of light. Never have any exceptions been found to the statement, "The total momentum before a collision is equal to the total momentum after the collision if no external forces act on the system." This statement is known as the Law of Conservation of Momentum. In all collisions between cars and trucks, between protons and protons, between planets and meteors, the momentum before the collision equals the momentum after.



ANSWERS

## For You To Do (continued)

Sample data is not provided because it cannot be anticipated what masses and speeds will be used in your class. However, conservation of momentum allows you to predict with ease what the velocity of the combined masses after the collision should be compared to the velocity of Object 1 before the collision. (See **Teaching Notes**.)

**Physics in Action**

A single cue ball hits a rack of 15 billiard balls and they all scatter. It would seem like everything has changed. Physicists have discovered that in this collision, as in all collisions and explosions, nature does keep at least one thing from changing—the total momentum. The sum of the momenta of all of the billiard balls immediately after the collision is equal to the momentum of the original cue ball. Nature loves momentum. Irrespective of the changes you visually note, the total momentum undergoes no change whatsoever. The objects may move in new directions and with new speeds, but the momentum stays the same. There aren't many of these conservation laws that are known.

Conservation of momentum can be shown to emerge from Newton's Laws. Newton's Third Law states that if object *A* and object *B* collide, the force of object *A* on *B* must be equal and opposite to the force of object *B* on *A*.

$$F_{A \text{ on } B} = -F_{B \text{ on } A}$$

The negative sign shows mathematically that the equally sized forces are in opposite directions.

Newton's Second Law states that $F = ma$. Also, acceleration, $a$, equals the change in velocity divided by the change in time ($a = \Delta v/\Delta t$):

$$m_B a_B = -m_A a_A$$

$$m_B \frac{\Delta v_B}{\Delta t} = -m_A \frac{\Delta v_A}{\Delta t}$$

$$m_B \frac{(v_f - v_i)_B}{\Delta t} = -m_A \frac{(v_f - v_i)_A}{\Delta t}$$

Since the change in time must be the same for both objects (*A* acts on *B* for as long as *B* acts on *A*), then $\Delta t$ can be eliminated from both sides of the equation.

Combining the initial velocities ($v_i$) on one side of the equation and the final velocities ($v_f$) on the other side of the equation:

$$m_A v_{iA} + m_B v_{iB} = m_A v_{fA} + m_B v_{fB}$$

$$(m_A v_A)_{before} + (m_B v_B)_{before} = (m_A v_A)_{after} + (m_B v_B)_{after}$$

Newton's Laws have yielded the conservation of momentum. The momentum of object *A* *before* the collision plus the momentum of object *B* *before* the collision equals the momentum of object *A* *after* the collision plus the momentum of object *B* *after* the collision

This equation not only works in one-dimensional collisions, but works equally well in the extraordinarily complex two-dimensional collisions of billiard balls and three-dimensional collisions of bowling.

Solving conservation of momentum problems is easy. Calculate each object's momentum

66

Active Physics

before the collision. Calculate each object's momentum after the collision. The totals before the collision must equal the total after the collision.

There are a variety of collisions involving two objects. In each collision, momentum is conserved and the same equation is used. The equation gets simpler when one of the objects is at rest and has zero momentum. You may wish to draw two sketches for each collision—one showing each object before the collision and one showing each object after the collision. By writing the momenta you know directly on the sketch, the calculations become easier.

Collision Type 1: One moving object hits a stationary object and both stick together and move off at the same speed:

before the collision

after the collision

Collision Type 2: Two stationary objects explode and move off in opposite directions.

Collision Type 3: One moving object hits a stationary object. The first object stops, and the second object moves off.

Collision Type 4: One moving object hits a stationary object, and both move off at different speeds.

Collision Type 5: Two moving objects collide, and both objects move at different speeds after the collision.

Collision Type 6: Two moving objects collide, and both objects stick together and move off at the same speed.

### Sample Problem 1

A 75.00-kg ice skater is moving to the east at 3.00 m/s toward his 50.00-kg partner, who is moving toward him (west) at 1.80 m/s. If he catches her up and they move away together, what is their final velocity?

***Strategy:*** This is a problem involving the Law of Conservation of Momentum. The momentum of an isolated system before an interaction is equal to the momentum of the system after the interaction. As you are working through this problem, remember that the $v$ in this expression is velocity and that it has direction as well as magnitude. Make east the positive direction, and then west will be negative.

***Givens:***

$m_b = 75.00$ kg
$m_g = 50.00$ kg
$v_b = 3.00$ m/s
$v_g = -1.80$ m/s

$m_b = 75.0$ kg $m_g = 50.0$ kg

$v_b = 3.00$ m/s $v_g = -1.80$ m/s

→

Physics in Action

**Solution:**

$(m_b v_b)_{before} + (m_g v_g)_{after} = [(m_b + m_g)v_{bg}]_{after}$

$(75.00 \text{ kg})(3.00 \text{ m/s}) + (50.00 \text{ kg})(-1.80 \text{ m/s}) = (75.00 \text{ kg} + 50.00 \text{ kg})v_{bg}$

$$v_{bg} = \frac{225 \text{ kg·m/s} - 90.0 \text{ kg·m/s}}{125.00 \text{ kg}}$$

$= 1.08 \text{ m/s}$

## Sample Problem 2

A steel ball with a mass of 2 kg is traveling at 3 m/s west. It collides with a stationary ball that has a mass of 1 kg. Upon collision, the smaller ball moves away to the west at 4 m/s. What is the velocity of the larger ball?

***Strategy:*** Again, you will use the Law of Conservation of Momentum. Before the collision, only the larger ball has momentum. After the collision, the two balls move away at different velocities.

***Givens:***

before the collision

$m_1 = 2$ kg $m_2 = 1$ kg

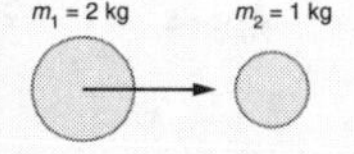

$v_{b1} = 3$ m/s $v_{b2} = 0$ m/s

after the collision

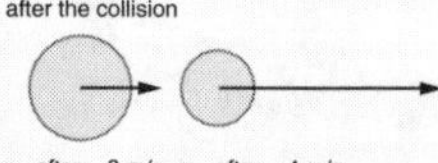

$v_{a1}$ after = ? m/s $v_{a2}$ after = 4 m/s

$m_1 = 2$ kg $v_{b2} = 0$ m/s
$m_2 = 1$ kg $v_{a2} = 4$ m/s
$v_{b1} = 3$ m/s

**Solution:**

$(m_1 v_1)_b + (m_2 v_2)_b = (m_1 v_1)_a + (m_2 v_2)_a$

$(2 \text{ kg})(3 \text{ m/s}) + (1 \text{ kg})(0 \text{ m/s}) = (2 \text{ kg})v_{a1} + (1 \text{ kg})(4 \text{ m/s})$

$6 \text{ kg·m/s} = (2v_{a1}) \text{ kg} + 4 \text{ kg·m/s}$

$v_{a1} = 1 \text{ m/s}$

68

Active Physics

### Reflecting on the Activity and the Challenge

The Law of Conservation of Momentum is a very powerful tool for explaining collisions in sports and other areas. The law works even when, as often happens in sports, one of the objects involved in a collision "bounces back," reversing the direction of its velocity and, therefore, its momentum, as a result of a collision. When describing a collision between a bat and ball, or a collision between two people, you can describe how the total momentum is conserved.

### Physics To Go

1. A railroad car of 2000 kg coasting at 3.0 m/s overtakes and locks together with an identical car coasting on the same track in the same direction at 2.0 m/s. What is the speed of the cars after they lock together?
2. In a hockey game, an 80.0-kg player skating at 10.0 m/s overtakes and bumps from behind a 100.0-kg player who is moving in the same direction at 8.00 m/s. As a result of being bumped from behind, the 100.0-kg player's speed increases to 9.78 m/s. What is the 80.0-kg player's velocity (speed and direction) after the bump?
3. A 3-kg hard steel ball collides head-on with a 1-kg hard steel ball. The balls are moving at 2 m/s in opposite directions before they collide. Upon colliding, the 3-kg ball stops. What is the velocity of the 1-kg object after the collision? (Hint: Assign velocities in one direction as positive; then any velocities in the opposite direction are negative.)
4. A 45-kg female figure skater and her 75-kg male skating partner begin their ice dancing performance standing at rest in face-to-face position with the palms of their hands touching. Cued by the start of their dance music, both skaters "push off" with their hands to move backward. If the female skater moves at 2.0 m/s relative to the ice, what is the velocity of the male skater? (Hint: The momentum before the skaters push off is zero.)



ANSWERS

## Physics To Go

1. $m(3.0\text{ m/s}) + m(2.0\text{ m/s}) = (2m)v$

   $(3.0\text{ m/s} + 2.0\text{ m/s}) = 2mv$

   $v = m(5.0\text{ m/s}) / 2m$
   $= (5.0\text{ m/s})/2 = 2.5\text{ m/s}$

2. $(80.0\text{ kg})(10.0\text{ m/s}) + (100\text{ kg})(8.00\text{ m/s}) = (80\text{ kg})v + (100\text{ kg})(9.78\text{ m/s})$

   $800\text{ (kg)m/s} + 800\text{ (kg)m/s} = (80\text{ kg})v + 978\text{ (kg)m/s}$

   $1600\text{ (kg)m/s} = (80\text{ kg})v + 978\text{ (kg)m/s}$

   $622\text{ (kg)m/s} = (80\text{ kg})v$

   $v = [622\text{ (kg)m/s}] / 80\text{ kg} = 7.78\text{ m/s}$

3. The direction of travel of the 3-kg ball before the collision is assigned as positive:

   $(3\text{ kg})(2\text{ m/s}) + (1\text{ kg})(-2\text{ m/s}) = (1\text{ kg})v$

   $6\text{ (kg)m/s} - 2\text{ (kg)m/s} = (1\text{ kg})v$

   $4\text{ (kg)m/s} = (1\text{ kg})v$

   $v = [4\text{ (kg)m/s}] / (1\text{ kg}) = 4\text{ m/s}$

   The 1-kg ball bounces back after the collision at twice the speed it had coming into the collision.

4. The direction of the female skater after pushoff is assigned as positive:

   $0 = (45\text{ kg})(2\text{ m/s}) + (75\text{ kg})v$

   $0 = 90\text{ (kg)m/s} + (75\text{ kg})v$

   $-90\text{ (kg)m/s} = (75\text{ kg})v$

   $v = [-90\text{ (kg)m/s}] /(75\text{ kg}) = -1.2\text{ m/s}$

The male skater moves at 1.2 m/s in the direction opposite the female skater.

Chapter 1

5. A 0.35-kg tennis racquet moving to the right at 20.0 m/s hits a 0.060-kg tennis ball that is moving to the left at 30.0 m/s. The racquet continues moving to the right after the collision, but at a reduced speed of 10.0 m/s. What is the velocity (speed and direction) of the tennis ball after it is hit by the racquet?

6. A stationary 3-kg hard steel ball is hit head-on by a 1-kg hard steel ball moving to the right at 4 m/s. After the collision, the 3-kg ball moves to the right at 2 m/s. What is the velocity (speed and direction) of the 1-kg ball after the collision? (Hint: Direction is important.)

7. A 90.00-kg hockey goalie, at rest in front of the goal, stops a puck ($m$ = 0.16 kg) that is traveling at 30.00 m/s. At what speed do the goalie and puck travel after the save?

8. A 45.00-kg girl jumps from the side of a pool into a raft ($m$ = 0.08 kg) floating on the surface of the water. She leaves the side at a speed of 1.10 m/s and lands on the raft. At what speed will the girl-raft system begin to travel across the pool?

9. Two cars collide head on. Initially, car A ($m$ = 1700.0 kg) is traveling at 10.00 m/s north and car B is traveling at 25.00 m/s south. After the collision, car A reverses its direction and travels at 5.00 m/s while car B continues in its initial direction at a speed of 3.75 m/s. What is the mass of car B?

10. A proton ($m = 1.67 \times 10^{-27}$ kg) traveling at $2.50 \times 10^5$ m/s collides with an unknown particle initially at rest. After the collision the proton reverses direction and travels at $1.10 \times 10^5$ m/s. Determine the change in momentum of the unknown particle.

11. You shoot a 0.04-kg bullet moving at 200.0 m/s into a 20.00-kg block initially at rest on an icy pond.
    a) What is the velocity of the bullet-block combination?
    b) The coefficient of friction between the block and the ice is 0.15. How far would the block slide before coming to rest?

12. Write a 15- to 30-s voice-over that highlights the conservation of momentum in a sport of your choosing.

ANSWERS

## Physics To Go *(continued)*

5. To the right is assigned as the positive direction:

$$(0.35\text{ kg})(20\text{ m/s}) + (0.060\text{kg})(30\text{ m/s}) = (0.35\text{ kg})(10\text{ m/s}) + (0.060\text{ kg})v$$
$$7.0\text{ (kg)m/s} + 1.8\text{ (kg)m/s} = 3.5\text{ (kg)m/s} + (0.060\text{ kg})v$$
$$5.3\text{ (kg)m/s} = (0.060\text{ kg})v$$
$$v = [5.3\text{ (kg)m/s}]/(0.060\text{ kg}) = 88\text{ m/s to the right}$$

6. To the right is assigned as the positive direction:

$$0 + (1\text{ kg})(4\text{ m/s}) = (3\text{ kg})(2\text{ m/s}) + (1\text{ kg})v$$
$$-2\text{ (kg)m/s} = (1\text{ kg})v$$
$$v = -2\text{ m/s}$$

The 1-kg ball rebounds from the collision, moving to the left at half the speed it had coming into the collision; its speed after the collision also is observed to be the equal and opposite of the 3-kg ball's speed after the collision. If after this collision each ball hit a bumper and the balls came back to collide again, this would take us back to the beginning of **Problem 3** above, and if they rebounded after that collision, we'd be back to this problem again, and so on...

ANSWERS

## Physics To Go *(for Questions 7–12)*

7. $\text{Momentum}_{before} = \text{Momentum}_{after}$

$(m_{goalie}v_{goalie})_{before} + (m_{puck}v_{puck})_{before} = [(m_g + m_p)v_{gp}]_{after}$

$(90.00 \text{ kg})(0 \text{ m/s}) + (0.16 \text{ kg})(30.00 \text{ m/s}) = (90.16 \text{ kg})(v_{gp})$

$v = 0.05 \text{ m/s}$

8. $\text{Momentum}_{before} = \text{Momentum}_{after}$

$(m_{girl}v_{girl})_{before} + (m_{raft}v_{raft})_{before} = [(m_g + m_r)v_{gr}]_{after}$

$(45.00 \text{ kg})(1.10 \text{ m/s}) + (0.08 \text{ kg})(0 \text{ m/s}) = (45.08 \text{ kg})\ v_{gr}$

$v_{gr} = 1.098 \text{ m/s } (1.1 \text{ m/s})$

9. $\text{Momentum}_{before} = \text{Momentum}_{after}$

$(m_A v_A)_{before} + (m_B v_B)_{before} = (m_A v_A)_{after} + (m_B v_B)_{after}$

$(1700.0 \text{ kg})(10.0 \text{ m/s}) + m_B(-25 \text{ m/s}) = (1700.0 \text{ kg})(-5.00 \text{ m/s}) + m_B(-3.75 \text{ m/s})$

$m_B = (25{,}500 \text{ kg m/s})/(21.25 \text{ m/s})$

$m_B = 1200 \text{ kg}$

10. The change in momentum of the proton must be equal and opposite to the change in momentum of the unknown particle.

Change in momentum = $m\Delta v = (1.67 \times 10^{-27} \text{ kg})(1.10 \times 10^5 \text{ m/s}) - (-2.50 \times 10^5 \text{ m/s})$ $= 6.0 \times 10^{-22} \text{ kg m/s}$.

The unknown particle changed momentum by exactly this much but in the opposite direction.

11. a) $(m_A v_A)_{before} + (m_B v_B)_{before} = (m_A v_A)_{after} + (m_B v_B)_{after}$

$(0.04 \text{ kg})(200.0 \text{ m/s}) + (20.00 \text{ kg})(0 \text{ m/s}) = (20.04 \text{ kg})(v)$

$v = 0.4 \text{ m/s}$

b) The change in KE of the block as it slides will be equal to the work done. This will allow the student to find the distance.

$\Delta KE = 1/2\ mv^2 = 1/2\ (20.04 \text{ kg})(0.4 \text{ m/s})^2 = 1.6 \text{ J}$

$W = F \cdot d$

$d = W/F = 1.6 \text{ J}/F$

Find the force using the coefficient of kinetic friction.

$F_f = \mu F_N = (0.15)(20.04 \text{ kg})(9.8 \text{ m/s}^2) = 29 \text{ N}$

$d = W/F = 1.6 \text{ J}/F = 1.6 \text{ J}/29 \text{ N} = 0.06 \text{ m}$

12. Student answers will vary. The voice-over should include a collision of a player with a player or a player with object. It should also include a statement of the conservation of momentum.

# ACTIVITY 9
## Circular Motion

# Background Information

It already has been established that if an object is moving along a straight line at constant velocity (constant speed, always in the same direction, along the line of motion) all of the forces on the object are balanced; in other words, the net force on the object is zero. If a sudden, momentary force is applied to the same object in a direction exactly sideways, the speed of the object neither increases nor decreases because the sideways force has no effectiveness in either the same or opposite direction as the object's motion. The result of applying a sudden, momentary, sideways force to the object is to cause the object to turn in the direction of the applied force; after the force has been removed, the object would be found moving at the same constant speed as it was before application of the force, but it would be moving along a different direction line, its direction of motion having been changed by the force. Such a force is called a "deflecting force," and the result of a deflecting force—applied exactly sideways to the motion of an object—is to cause the object to change, or deflect, to a new direction of motion.

If instead of being applied suddenly and momentarily, a deflecting force is applied continuously, always in the same amount, and continuously adjusted in a direction to act exactly sideways to the object's motion, the object moves in a circular path at constant speed. In this case the force is called a "centripetal force" which is defined as "the force required to keep a mass moving in a circular path at constant speed." It can be experimentally determined that the relationship between the centripetal force, the object's mass and speed, and the radius of the object's circular path is:

$$F_c = mv^2/r$$

where $F_c$ is the centripetal force in newtons, $m$ is the object's mass in kilograms, $v$ is the objects speed in meters/second, and $r$ is the radius of the circular path in meters. A convenient alternate equation for centripetal force can be used when the period, T, of the objects circular motion (T is the time, in seconds, for the object to travel once around the circle) is known but the speed is not known:

$$F_c = m4\pi^2 R/T^2$$

According to Newton's Second Law of Motion, when there is an unbalanced force acting on an object, the object must be accelerating. This certainly must apply to an object moving in a circular path under the influence of an unbalanced, centripetal force. However, it is puzzling to contemplate that an object moving at constant speed, even on a circular path, somehow can be construed to be accelerating, because it is neither gaining nor losing speed. Consider the below "strobe" diagram showing an object moving at constant speed along a circular path at four instants equally separated in time:

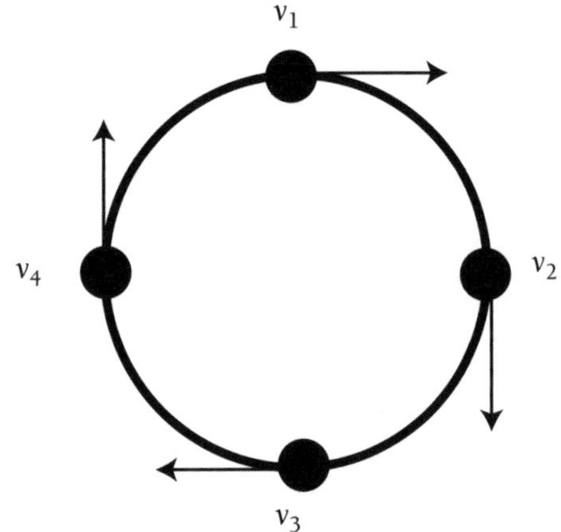


Diagram of object moving in a circular path.

The definition of acceleration, $a = \Delta v/\Delta t = (v_2 - v_1)/\Delta t$ will be applied to the above diagram to attempt to discover any basis for existence of acceleration. Since the instantaneous velocities $v_1$ and $v_2$ in the diagram are vectors which do not share a common direction, the quantity $\Delta v$ in the defining equation for acceleration must be found by treating $v_1$ and $v_2$ as vectors during the subtraction process; that is, a vector subtraction method must be applied to find the difference $v_2 - v_1 = \Delta v$. To do so, the negative of vector $v_1$ (a vector having the same length but opposite direction as $v_1$) will be added to vector $v_2$ using the tip-to-tail method of adding vectors:

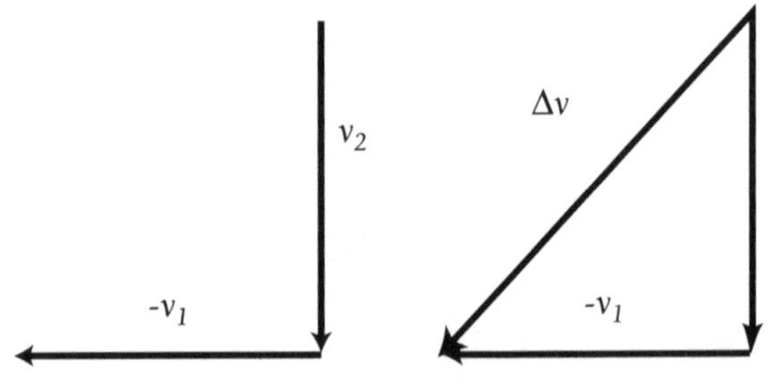


Clearly in the above diagram a change in velocity, $\Delta v$, occurred during the time interval, $\Delta t$, that the object moved at constant speed 1/4 of the distance around its circular path. If a numerical values were assigned to time and to the object's speed, a value for $\Delta v$ could be determined by comparing the length of the $\Delta v$ vector to the velocity vectors, and the average acceleration for the time interval could be calculated from $a = \Delta v/\Delta t$. Therefore, there is an acceleration associated with a centripetal force, but the nature of the acceleration is that it does not alter the object's speed, but does alter the direction of the object's velocity. It is informative to move the vector $\Delta v$ in the above diagram into the original diagram to see $\Delta v$ in relationship to the motion along the circular path. This is done in the below diagram. Notice that both the length and direction of the vector $\Delta v$ are preserved and that $\Delta v$ is positioned at the middle of the time interval from which it was derived:

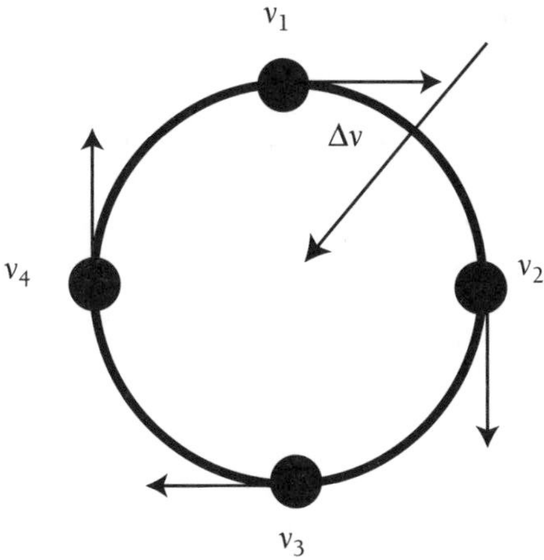


Object in circular motion with vector

Interestingly and not accidentally, $\Delta v$ points toward the center of the circular path. Since the acceleration calculated from $\Delta v$ using the defining equation $a = \Delta v/\Delta t$ would have the same direction as $\Delta v$, the acceleration also points toward the center of the circle. Indeed, it is a centripetal acceleration, caused, as one may suspect by the centripetal force, which also points toward the center of the circular path. Now it can be reasoned that the equations presented above for centripetal force are nothing more that Newton's Second Law, $f = ma$, applied to the special case of circular motion. The terms on the right-hand side of the equation other than the mass, $m$, are the centripetal acceleration, $a_c$:

$$F_c = ma_c = m(v^2/r) = m(4\pi^2 r/t^2)$$

In summary, the velocity, acceleration, and force vectors of an object moving at constant speed on a circular path constantly change in direction and remain constant in amount. Each vector rotates once during each trip of the object around the circular path.

## *Active*-ating the Physics InfoMall

When teaching about circular motion, you will almost certainly have to dispel ideas about centrifugal forces. Perhaps you noticed, while performing some of the searches mentioned in the previous activities, the article "Centrifugal force: fact or fiction?," in *Physics Education*, vol. 24, issue 3, 1989. If not, you may want to check it out now.

Search the InfoMall using keywords "circular motion" AND "misconcept*" for a list of articles discussing known problems students have with this common concept. You may also find useful demonstrations by searching the Demo & Lab Shop using the keywords "circular motion."

# Planning for the Activity

## Time Requirements

- One class period.

## Materials Needed

### For each group:

- accelerometer
- accelerometer, cork
- accelerometer-to-cart
- balls, bocci
- ball, nerf
- rotating stool or chair
- ruler, flexible plastic, 30 cm
- ruler, metric, mm marking

## Advance Preparation and Setup

A ball will be needed to serve as the subject of each group's inquiry. The ball needs to be sufficiently massive so that, when rolling at low speed across the floor, students can use a force meter to push sideways to the ball's motion to cause it to turn on a curve of observable, reasonable radius. The best choice of ball may be the bocci balls; if not available another kind of ball must be substituted. Try this yourself in advance of class to be sure that the ball's mass, the ball's speed, and the amount of force used, combine to produce a nice, observable curve of the ball's path.

If you have not done an earlier activity in *Sports*, yet, you will need to take the time to construct a cork accelerometer.

# Teaching Notes

The cork accelerometer will indicate an acceleration toward the center of the circular path in which the accelerometer moves.

SAFETY PRECAUTION: If a rotating stool or chair is used to place a student holding an accelerometer in a state of circular motion, provide close supervision and maintain safe conditions; the effect can be observed without a rotating stool or chair if students simply twirl around while holding an accelerometer in the hands.

If students have not already done so, some time may be required to familiarize them with the use of the accelerometer.

Students can be expected to need practice at keeping alongside the ball while applying a constant force always sideways (at a right angle to) the ball's motion; in fact, to do so extremely well perhaps is nearly impossible, but "close" will do well enough for students to observe the tendency for the ball to move in a circular path.

Some students may raise the question, "How can there be an acceleration when the object moves at constant speed?" You may wish to see the explanation in the **Background Information** for the Teacher for this activity to decide how you will deal with that question.

Students also may ask about, or bring up, "centrifugal force." This is addressed in the **Background Information** for the Teacher. A first response to a question about centrifugal force would be to refer students to **Physics To Go, Question 2.**

Students may have an inclination to "run together," or treat as the same phenomenon, circular motion with spinning motion. They are separate, but related, phenomena. This activity applies to the former, an object whose center of mass is moving along a circular path. When a figure skater does an in-place spin, the skater's center of mass does not move; it is a different phenomenon. It is possible, however, to treat part of the skater, such as an extended foot at the end of the spin, as an object in circular motion to which the ideas in this activity could be applied.

# Activity 9 Circular Motion

**GOALS**

In this activity you will:

- Understand that a centripetal force is required to keep a mass moving in a circular path at constant speed.
- Understand that a centripetal acceleration accompanies a centripetal force, and that, at any instant, both the acceleration and force are directed toward the center of the circular path.
- Apply the equation for circular motion.
- Understand that centrifugal force is the reaction to centripetal force.

**To avoid becoming too dizzy, limit your spins while standing to about four.**

### What Do You Think?

Racecars can make turns at 150 mph.

- **What forces act on a racecar when it moves along a circular path at constant speed on a flat, horizontal surface?**

Record your ideas about this question in your *Active Physics* log. Be prepared to discuss your responses with your small group and the class.

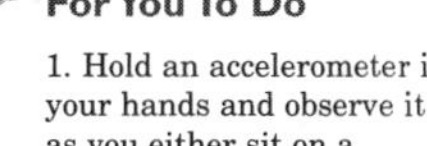

### For You To Do

1. Hold an accelerometer in your hands and observe it as you either sit on a rotating stool or spin around while standing. What is the direction of the acceleration indicated by the accelerometer? (You can find out how the cork indicates acceleration by holding it and noting its behavior as you accelerate forward.)





# Activity Overview

In this activity students use an accelerometer to identify the direction of centripetal acceleration. They then use a force meter to provide the centripetal force needed to deflect an rolling ball moving in a straight line into a curved path.

## Student Objectives

### Students will:

- Understand that a centripetal force is required to keep a mass moving in a circular path at constant speed.
- Understand that a centripetal acceleration accompanies a centripetal force, and that, at any instant, both the acceleration and force are directed toward the center of the circular path.
- Apply the equation $F_c = ma_c = m(v^2/r)$ to calculations involving circular motion.
- Understand that centrifugal force is the reaction to centripetal force.

ANSWERS FOR THE TEACHER ONLY

## What Do You Think?

The forces acting on a race car include gravity downward, the force of the road up and centrifugal force.

ANSWERS

## For You To Do

1. a) Students provide a sketch similar to the one shown.

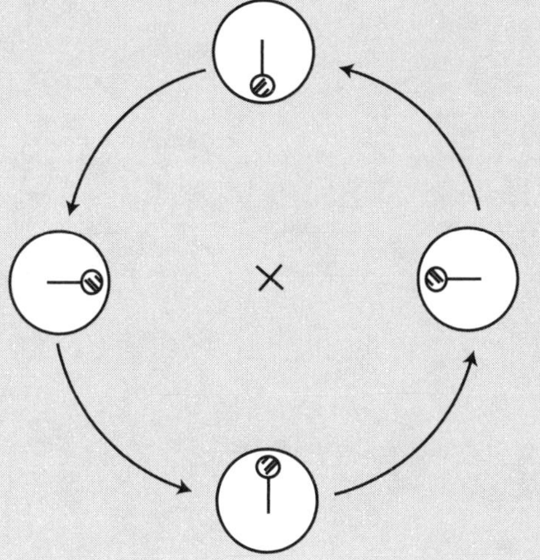

Answers

# For You To Do
***(continued)***

2. a) Students should indicate that the greater the force applied to a mass, the greater the acceleration. The acceleration occurs in the direction of the unbalanced force.

3. a) Students provide a sketch similar to the one shown.

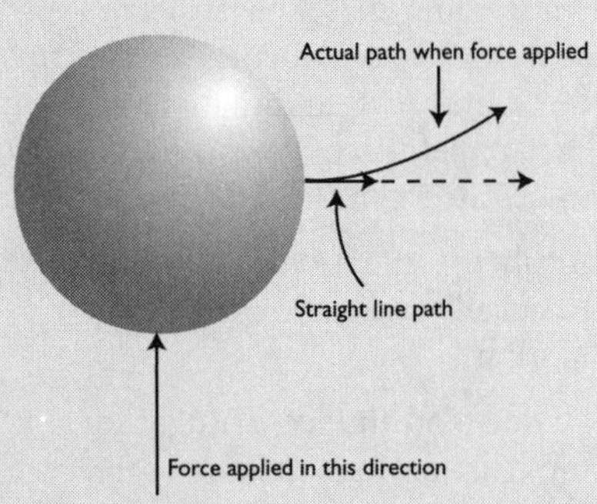


b) The speed of the ball does not change.

c) A constant force sideways needs to be applied to the ball.

d) If you stop pushing on the ball, the ball will follow a straight path. Students should provide a sketch similar to the one shown at right.

Physics in Action

a) Make a sketch in your log to simulate a snapshot photo taken from above as the accelerometer was moving along a circular path. Show the circular path, the accelerometer "frozen" at one instant, the cork "frozen" in leaning position, and an arrow to represent the velocity of the accelerometer at the instant represented by your sketch.

2. Review in your textbook and your log how you used a force meter to apply a constant force to objects to cause the objects to accelerate in **Activity 2**.

a) Based on the results of **Activity 2**, write a brief statement in your log that summarizes how the amount and direction of acceleration of an object depends on amount and direction of the force acting on the object.

3. Start a ball rolling across the floor. While it is rolling, catch up with the ball and use the force meter to push exactly sideways, or perpendicular, to the motion of the ball with a fixed amount of force. Carefully follow alongside the ball and, as will be necessary, keep adjusting the direction of push so that it is always perpendicular to the motion of the ball.

a) Make a top view sketch in your log that shows:
- a line to represent the straight-line path of the ball before you began pushing sideways on the ball
- a dashed line to represent the straight-line path on which the ball would have continued moving if you had not pushed sideways on it
- a line of appropriate shape to show the path taken by the ball as you pushed perpendicular to the direction of the ball's motion with a constant amount of force.

b) When you pushed on the ball exactly sideways to its motion, did you cause the ball to move either faster or slower? Explain your answer.

c) Assuming that friction could be eliminated to allow the ball to continue moving at constant speed, describe what you would need to do to make the ball keep moving on a circular path.

d) If you stop pushing on the ball, how does the ball move? Try it, and use a sketch and words in your log to describe what happens.

Active Physics 72

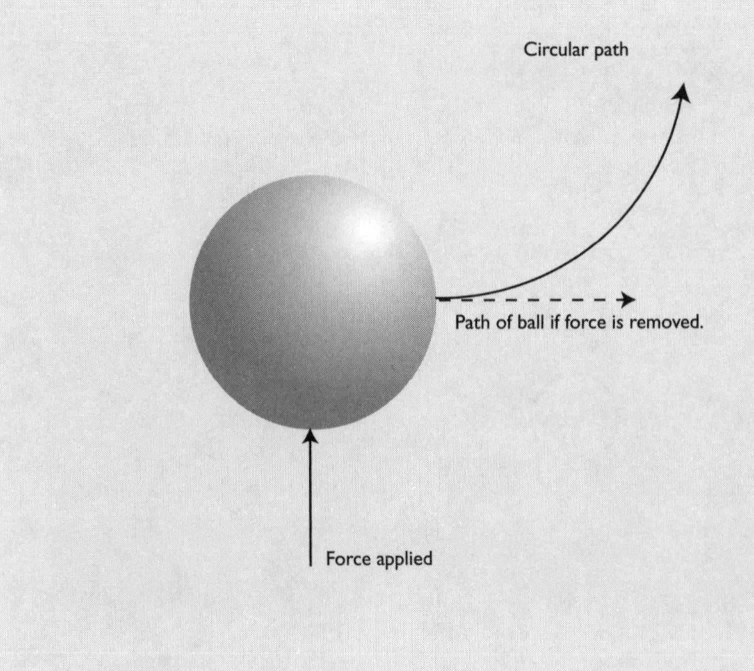

4. Review each of the items listed below. Copy each item into your log and write a statement to discuss how each item is related to an object moving along a circular path. If an item does not apply to circular motion, explain why.
   a) Galileo's Principle of Inertia
   b) Newton's First Law of Motion
   c) Newton's Second Law of Motion

**Physics Words**

**centripetal acceleration:** the inward radial acceleration of an object moving at a constant speed in a circle.

**FOR YOU TO READ**

**The Unbalanced Force Required for Circular Motion**

During the above activities you saw two things that are related by Newton's Second Law of Motion, $F = ma$. First, the accelerometer showed that when an object moves in a circular path there is an acceleration that at any instant is toward the center of the circle. This acceleration has a special name, **centripetal acceleration.** The word centripetal means "toward-the-center"; therefore, centripetal acceleration refers to acceleration toward the center of the circle when an object moves in a circular path.

You also saw that a centripetal force, a toward-the-center force, causes circular motion. When a centripetal force is applied to a moving object, the object's path curves; without the centripetal force, the object follows the tendency to move in a straight line. Therefore, a centripetal force, when applied, is an unbalanced force, meaning that it is not "balanced off" by another force.

Newton's Second Law seems to apply to circular motion just as well as it applies to accelerated motion along a straight line, but with a strange "twist." It is a clearly correct application of $F = ma$ to say that a centripetal force, $F$, causes a mass, $m$, to experience an acceleration, $a$. However, the strange part is that when an object moves along a circular path at constant speed, acceleration is happening with no change in the object's speed. The force changes the direction of the velocity.

Velocity describes both the amount of speed and the direction of motion of an object. Thinking about the velocity of an object moving with constant speed on a circular path, it is true that the velocity is changing from one instant to the next not in the amount of the velocity, but with respect to the direction of the velocity. The diagram shows an object moving at constant speed on a circular path. Arrows are used to represent the velocity of the object at several instants during one trip around the circle.



## ANSWERS

# For You To Do

***(continued)***

4. a) Inertia is the natural tendency of an object to remain at rest or to remain moving with constant speed in a straight line.

   When there is no centripetal force acting on an object, it will move in a straight line.

   b) In the absence of an unbalanced force, an object at rest remains at rest and an object already in motion remains in motion with constant speed on a straight line path.

   In the absence of a sideways, towards-the-center force (centripetal force), the object going in a circle continues to move straight (at a tangent to the circle).

   c) The acceleration of an object is directly proportional to the unbalanced force acting on it and is inversely proportional to the object's mass. The direction of the acceleration is the same as the direction of the unbalanced force.

   When a centripetal force is applied to an object, it moves in a circular path. The acceleration is toward the center of the circle. It is definitely an application of $F = ma$ because the force acts (toward the center of the circle) on the mass giving it an acceleration (toward the center of the circle).

Physics in Action

Physicists have shown that a special form of Newton's Second Law governs circular motion:

$$F_C = ma_C = \frac{mv^2}{r}$$

where $F_C$ is the centripetal force in newtons,
$m$ is the mass of the object moving on the circular path in kilograms,
$a_C$ is the centripetal acceleration in m/s$^2$,
$v$ is the velocity in m/s, and
$r$ is the radius of the circular path in meters.

### Sample Problem

Find the centripetal force required to cause a 1000.0-kg automobile travelling at 27.0 m/s (60 miles/hour) to turn on an unbanked curve having a radius of 100.0 m.

***Strategy:*** This problem requires you to find centripetal force. You can use the equation that uses Newton's Second Law to calculate $F_C$.

***Givens:***

$m$ = 1000.0 kg

$y$ = 27.0 m/s

$r$ = 100.0 m/s

***Solution:***

$$F_C = \frac{mv^2}{r}$$

$$= \frac{(1000.0 \text{ kg } (27.0 \text{ m/s})^2}{100.0 \text{ m}}$$

$$= \frac{(1000.0 \text{ kg} \times 730 \text{ m}^2\text{/s}^2)}{100.0 \text{ m}}$$

$$= 7300 \text{ N}$$

If the force of friction is less than the above amount, the car will not follow the curve and will skid in the direction in which it is travelling at the instant the tires "break loose."

Activity 9 Circular Motion

### Reflecting on the Activity and the Challenge

Both circular motion and motion along curved paths that are not parts of perfect circles are involved in many sports. For example, both the discus and hammer throw events in track and field involve rapid circular motion before launching a projectile. Track, speed skating, and automobile races are done on curved paths. Whenever an object or athlete is observed to move along a curved path, you can be sure that a force is acting to cause the change in direction. Now you are prepared to provide voice-over explanations of examples of motion along curved paths in sports, and in many cases you perhaps can estimate the amount of force involved.

### Physics To Go

1. For the car used as the example in the **For You To Read**, what is the minimum value of the coefficient of sliding friction between the car tires and the road surface that will allow the car to go around the curve without skidding? (Hint: First calculate the weight of the car, in newtons.)
2. If you twirl an object on the end of a string, you, of course, must maintain an inward, centripetal force to keep the object moving in a circular path. You feel a force that seems to be pulling outward along the string toward the object. But the outward force that you detect, called the "centrifugal force," is only the reaction to the centripetal force that you are applying to the string. Contrary to what many people believe, there is no outward force acting on an object moving in a circular path. Explain why this must be true in terms of what happens if the string breaks while you are twirling an object.
3. A 50.0-kg jet pilot in level flight at a constant speed of 270.0 m/s (600 miles per hour) feels the seat of the airplane pushing up on her with a force equal to her normal weight, 50.0 kg × 10 m/s$^2$ = 500 N. If she rolls the airplane on its side and executes a tight circular turn that has a radius of 1000.0 m, with how much force will the seat of the airplane push on her? How many "*g*'s" (how many times her normal weight) will she experience?



Chapter 1

ANSWERS

## Physics To Go

1. $\mu$ = (frictional force)/(weight)
 = 7300 N / $mg$
 = 7300 N / (1,000 kg × 10 m/s$^2$)
 = 7300 N / 10,000 N = 0.73

2. If there were an outward force acting on the object, it would be expected to fly radially outward when the string breaks; when the string breaks, the object flies tangent to the circular path, indicating that there is no outward force.

3. $F_C = m(v^2/r)$
 = (50 kg) (270 m/s)$^2$/(1000 m)
 = (50 kg)(73,000 m$^2$/s$^2$)/(1000 m)
 = 3600 N

 The pilot's normal weight is $mg$ = 50 kg × 10 m/s$^2$ = 500 N

 Therefore, the pilot "pulls" 3600 N / 500 N = 7.2 g's during the turn; that is, she feels 7.2 times her normal weight. Assuming an inside turn with the top of the pilots head facing the center of the circle, the blood in her brain would tend to keep going straight ahead, tangent to the circle, draining from her brain and causing her to lose consciousness. However, her automatic pressure suit would inflate to squeeze against her legs, pushing blood upward to her brain to keep her from "blacking out."

ANSWERS

## Physics To Go

***(continued)***

4. For the discus event in track and field, assume mass of disc = 1 kg, radius of twirling action before throw = 1 m, and speed during rotation before throw = 10 m/s:

   $F_C = m(v^2/r)$
   $= (1 \text{ kg})\ (10 \text{ m/s})^2/(1 \text{ m})$
   $= (1 \text{ kg})(100 \text{ m}^2/\text{s}^2)/(1 \text{ m}) = 100 \text{ N}$

   The athlete must "hold on" to the disc, using the throwing hand to provide an inward force of 100 N while twirling prior to release of the disc.

5. Viewed from a helicopter above the event, the passenger would be observed to keep going in a straight line as the car turns. The seat would slide under the passenger, and the door on the passenger side would hit the right shoulder of the passenger, providing the centripetal force thereafter to push the passenger into the same curve as the car.

6. The tilt-a-whirl ride is fun (but sometimes sickening) because two circular motions are "superimposed" on the body. One motion carries the body around in a large circular path corresponding to the circular path around which the chair moves; the second, superimposed, motion is provided as the chair spins as it revolves on the circular path, causing the body simultaneously to move in a circle of small radius. Sometimes the two centripetal forces add together, and sometimes they are in different directions. All of the effects on the body are too numerous to understand or explain, but one thing is for sure, the hot dogs in the stomach don't know which way to go!

7. a) They need a frictional force to turn. On a wet field, without friction, they continue moving in the same direction.

   b) Friction supplies the centripetal force. On a wet field, without friction, the players continue moving in the same direction, obeying Newton's First Law.

   c) Student work.

Physics in Action

4. Imagine a video segment of an athlete or an item of sporting equipment moving on a circular path in a sporting event. Estimate the mass, speed, and radius of the circle. Use the estimated values to calculate centripetal force and identify the source of the force.

5. Below are alternate explanations of the same event given by a person who was not wearing a seat belt when a car went around a sharp curve:

   a) "I was sitting near the middle of the front seat when the car turned sharply to the left. The centrifugal force made my body slide across the seat toward the right, outward from the center of the curve, and then my right shoulder slammed against the door on the passenger side of the car."

   b) "I was sitting near the middle of the front seat when the car turned sharply to the left. My body kept going in a straight line while, at the same time due to insufficient friction, the seat slid to the left beneath me, until the door on the passenger side of the car had moved far enough to the left to exert a centripetal force against my right shoulder."

   Are both explanations correct, or is one right and one wrong? Explain your answer in terms of both explanations.

6. People seem to be fascinated with having their bodies put in a state of circular motion. Describe an amusement park ride based on circular motion that you think is fun, and describe what happens to your body during the ride.

7. a) Explain why football players fall on a wet field while changing directions during a play.

   b) Include the concepts centripetal force and Newton's Laws in a revised explanation.

   c) In a new revision, make the explanation exciting enough to include in your sports video voice-over.

76

# PHYSICS AT WORK

## Dean Bell

**TELEVISION PRODUCER USES SPORTS TO TEACH MATH AND PHYSICS**

Dean Bell is an award-winning filmmaker and television writer, director, and producer. His show *Sports Figures* is a highly acclaimed ESPN educational television series designed to teach the principles of physics and mathematics through sports. His approach has been to tell a story, pose a problem, and then follow through with its mathematical and scientific explanation. "But always," he says, "you must make it fun. It has to be both educational and entertaining."

Dean began his career as a filmmaker after college. He landed the apprentice film editor's position on a Woody Allen film. From there, he worked his way up in the field, from assistant editor, to editor and finally writer, director, and producer.

"I've always been a fan of educational TV," he states, "although I never thought that was where my career would take me. It's one of life's little ironies that I've ended up producing this type of show. You see, my father worked in scientific optics and was very science oriented. He was always delighted in finding out how things worked, and was even on the Mr. Wizard TV show a few times."

Dean writes the script for each segment, working together with top educational science consultants. "We spend a day coming up with ideas and then researching each subject thoroughly. Our job is to illustrate the relationship between a sports situation and the related mathematical or physics principles.

"At the end of the day," says Dean, "it really is nice to be working on a show that means something and that is so worthwhile. I'm still getting ahead in my career as a film and TV producer, but now I'm also an educator."

## Chapter 1 Assessment

Your big day has arrived. You will be meeting with the local television station to audition for a job as a "physics of sports" commentator. Whether you will get the job will be decided on the quality of your voice-over.

With what you learned in this chapter, you are ready to do your science commentary on a short sports video. Choose a videotape from a sports event, either a school event or a professional event. Each of you will be responsible for producing your own commentary, whether or not you worked in cooperative groups during the activities. You are not expected to give a play-by-play description, but rather describe the rules of nature that govern the event. Your viewers should come away with a different perspective of both sports and physics. You may produce one of the following:

- **a written script**
- **a live narrative**
- **a video soundtrack**
- **an audiocassette**

Review the criteria by which your voice-over dialogue or script will be evaluated. Your voice-over should:

- **use physics principles and terms correctly**
- **have entertainment value**

After reviewing the criteria, decide as a class the point value you will give to each of these criteria:

- **How important is the physics content? How many physics terms and principles should be illustrated to get the minimum credit? The maximum credit?**
- **What value would you place on the entertainment aspect? How do you fairly assess the excitement and interest of the broadcast?**

## Physics You Learned

Galileo's Principle of Inertia

Newton's First Law of Motion

Newton's Second Law of Motion

Newton's Third Law of Motion

Weight

Center of mass; center of gravity

Friction between different surfaces

Momentum

Law of Conservation of Momentum

Centripetal acceleration

Centripetal force

# Alternative Chapter Assessment

**Multiple Choice**: Select the letter of the choice that best answers the question or best completes the statement.

1. Which of the following best illustrates Newton's First Law of Motion?

   a) A collision between a running back and a linebacker in football.

   b) An ice hockey puck sliding along the ice after being hit by a player's stick.

   c) A bowling pin being struck by a bowling ball.

   d) A volleyball being "spiked" across the net.

2. A small ball rests on a circular turntable, rotating clockwise at a constant speed as illustrated in the below diagram. Which of the below pair of arrows best describes the direction of the acceleration and the net force acting on the ball at the point indicated in the diagram?

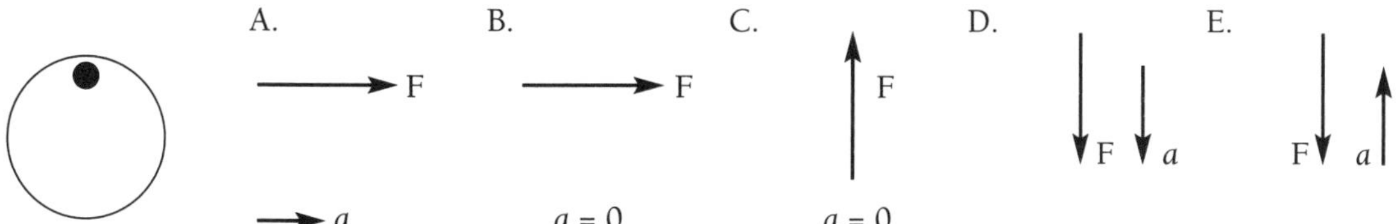


3. A person sliding into second base continues to slide past the base due to:

   a) inertia

   b) friction

   c) weight

   d) gravity

4. Newton's First Law of Motion states that an object at rest stays at rest unless acted upon by a:

   a) balanced force

   b) net force

   c) weak force

   d) strong force

5. In the absence of air, a penny and a feather dropped from the same height:

   a) fall at different rates

   b) float

   c) fall at equal rates

   d) do not have momentum

6. An object rolling across a level floor without any horizontal net force acting on it will:

a) slow down

b) speed up

c) keep moving forever

7. An object falling to Earth in the absence of air resistance:

a) falls with a constant speed of 9.8 m/s

b) falls with constant acceleration of 9.8m/s2

c) slows down

8. A constant net force acting on an object causes the object to move with constant:

a) speed

b) velocity

c) acceleration

d) momentum

9. Which of the following is NOT one of Newton's Laws of Motion?

a) An object in motion stays in motion unless acted upon by an unbalanced force.

b) A constant net force acting on an object produces a change in the object's motion.

c) For every action, there is an equal and opposite reaction.

d) Energy is neither created not destroyed; it simply changes form.

10. Newton's First Law is known as the law of:

a) impetus

b) inertia

c) acceleration

d) resistance

**True or Replace False Word**: Determine whether the word in bold print makes each statement true or false. If a statement is true, write "true" in the answer space. If a statement is false, write in the answer space a replacement for the word in bold print which would make the statement become true.

11. If a net force acts on an object, the object will change speed, direction or **neither**.

Answer: __________

12. **Gravity** is the tendency of an object to resist any change.

Answer: __________

13. A net force acting on an object causes the object to move with constant **velocity**.

 Answer: __________

14. **Forces** that are equal in amount and opposite in direction are balanced forces.

 Answer: __________

15. A bowling ball has more **inertia** than a tennis ball.

 Answer: __________

**Short Answer**: Write a brief response to each item. Show your work for responses which require calculations.

16. Describe how Galileo's experiments with balls and ramps led to Newton's First Law of Motion.

17. Using Newton's First and Second Laws of Motion, explain why a ball thrown into the air follows a parabolic path.

18. On the diagrams below, draw arrows to show the vertical forces acting on a person jumping into the air. Use diagram A to show the forces acting during the push-off, and use diagram B to show the forces acting as the person lands on the ground. Write a paragraph comparing the forces in each situation, including a discussion on the relative sizes of the forces.

A. B.

19. Describe the collision between a bat and a ball in terms of conservation of momentum.

20. Imagine a tug-of-war between two people. Draw a sketch indicating all of the forces acting in the tug-of-war. Using Newton's Laws and your picture, explain how and why one side will win.

21. For each of the following forces, identify the "reaction" force from Newton's Third Law:

 a) Volleyball hitting the floor.

 Answer: __________

 b) Softball bat hitting a softball.

 Answer: __________

c) Punter kicking a football.

Answer: __________

22. Explain why sprinters prefer to use longer spikes on their shoes than long distance runners, even though both run on same track surface.

23. A 75-kg ice hockey forward moving at 5.0 m/s collides with a stationary 85-kg defenseman and they become entangled. With what velocity will the pair move across the ice?

24. Explain how a soccer player can cause a ball to spin as a result of kicking it. What can the player do to change the direction of spin on the ball?

25. Two objects that have the same mass are dropped from the top of a 20-meter high building. One object is larger and flatter than the other object. Which hits the ground first? Use the terms gravity, acceleration, and air resistance correctly in your discussion.

## Multiple Choice

1. b
2. d
3. a
4. b
5. c
6. c
7. b
8. c
9. d
10. b

## True/Replace False Word

11. both
12. inertia
13. acceleration
14. true
15. true

## Short Answer

16. Galileo used two ramps, initially in a V-shape, and demonstrated that a ball rolled from a specific height would roll up the opposite side until it reached the original height. If the angle of the second ramp was decreased, the ball still rolled to the same height but this ball traveled further on the ramp. He reasoned that the ball will roll until it reaches the height from which it was released, therefore, if the second ramp were horizontal the ball would continue to roll in the horizontal direction since it would not be able to reach this height. This supports the idea that the "natural" state of an object's motion is not necessary to be at rest.

17. When a ball is thrown into the air the primary force acting on it is its weight. Since weight is a force that acts "down", from Newton's Second Law the acceleration of the ball will be down, therefore the vertical velocity of the ball will be changing. If we assume no air resistance, horizontally the ball will continue in its original state of motion according to Newton's First Law, minimal air resistance would at least show very little change in horizontal motion. During equally spaced time intervals the horizontal displacement will be constant and the vertical displacement will be changing, resulting in a parabolic path for the ball.

18. In each diagram the vertical forces that need to be shown are the person's weight (acting down) and the normal force from the ground (acting up). In both cases the normal force will exceed the weight since both require an acceleration in the upward direction.

19. When a bat collides with a ball the total momentum of the system will remain the same as long as no other forces acted. The bat will slow down upon collision, thus decreasing the momentum of the bat. The momentum that the bat lost will be gained by the ball, thus changing the momentum of the ball, since the bat was originally traveling in the opposite direction from the ball, the ball's motion will change such that its direction will change.

20. Consider the entire tug-of-war as one object. The forces acting are as follows:
"outside forces" — The weight of each person (down) and the normal force from the ground (up). Each side will have a friction force acting (horizontally) as a reaction to each person pushing sideways on the ground.
"internal forces" — Each side applies a force to the rope, the rope applies a force back on each side.
"Outside" forces change the motion of objects, the internal forces have no effect on the total motion of the object. The vertical forces (weights and normal forces) will add to zero, the horizontal friction forces will add to show a net force in one direction that will cause the entire tug-of-war to accelerate in that direction.

21. a) Floor applies a force on volleyball.

    b) Softball applies a force on bat.

    c) Football applies force on punter's foot.

22. Sprinters require larger accelerations since they wish to reach maximum velocity in a short period of time. The spikes allow for the sprinter to push harder on the ground so the ground can push pack with a larger force.

23. $(75*5.0) + (85*0) = (75+85)v$

$$v = 2.3 \text{ m/s}$$

24. To cause a ball to spin a force must be applied off-center. To change the direction of the spin the player need apply the force at different points on the ball, e.g., to have the ball spin to the left, the force needs to be applied to the right of center.

25. The larger and flatter object will be influenced by the air as it falls. The force of air resistance will oppose the force of gravity acting on the object causing the net downward force to be less. Therefore, the downward acceleration of the object will be less than 9.8 m/s/s. It will take longer to fall the 20 m.

Chapter 2
SAFETY

## Chapter 2-Safety

# National Science Education Standards

### Chapter Challenge

Dangers inherent in travel provide the context for this chapter. Students are challenged to design or build a safety device, or system, for protecting automobile, airplane, bicycle, motorcycle, or train passengers. New laws, increased awareness, and improved safety systems are explored as students work on this challenge. They are also encouraged to design improvements to existing systems and to find ways to minimize harm caused by accidents.

### Chapter Summary

To meet this challenge, students engage in collaborative activities that explore motions and forces and the principles of design technology. These experiences engage students in the following content from the National Science Education Standards.

# Content Standards

### Unifying Concepts

- Systems, order and organization
- Evidence, models and explanations
- Constancy, change, and measurement

### Science as Inquiry

- Identify questions and concepts that guide scientific investigations
- Use technology and mathematics to improve investigations
- Formulate and revise scientific explanations and models using logic and evidence
- Communicate and defend a scientific argument

### History and Nature of Science

- Nature of scientific knowledge

### Physical Science

- Motions and forces

### Science in Personal and Social Perspectives

- Personal and community health
- Natural and human-induced hazards

### Science and Technology

- Understandings about science and technology
- Ability to apply technology

# Key Physics Concepts and Skills

| Activity Summaries | Physics Principles |
|---|---|
| **Activity 1: Response Time**<br>Using a response timer, students explore the time required for a driver to respond to a hazard. This activity introduces students to the process of beginning with their own ideas and predictions, then implementing an investigation that results in both qualitative and quantitative data. | • Series circuits<br>• Switches<br>• Response time |
| **Activity 2: Speed and Following Distance**<br>Strobe, or multiple exposure photos of a moving vehicle are used to discuss speed and acceleration. Students then use a sonic ranger to measure how fast they walk and obtain a computer generated graph of their speed. Information about speed is then connected to response time with a discussion of tailgating. | • Average speed<br>• Using data as basis for predictions<br>• Speed, distance, and time relationships |
| **Activity 3: Accidents**<br>Following an investigation crashing cars against barriers, students use advertisements and consumer reports to learn about safety devices on automobiles. Each is analyzed to determine the type of collision-related injuries it prevents, and to identify if the device could in fact increase injuries in a unique setting. | • Physical properties of matter<br>• Effect of forces on motion |
| **Activity 4: Life (and Death) before Seat Belts**<br>Using a lump of clay on a motion cart to represent a person in a car, students explore "objects in motion stay in motion." They then relate this to actual automobile collisions. | • Acceleration<br>• Inertia |
| **Activity 5: Life (and Fewer Deaths) after Seat Belts**<br>Students focus on the design and materials used in seat belt construction as they study force and pressure. They investigate how increasing surface area decreases the pressure exerted. They relate this to the challenge by finding ways to increase the area of impact in a collision. | • Inertia<br>• Newton's Laws of Motion<br>• Force and pressure<br>• Newton as a unit of measure |
| **Activity 6: Why Air Bags?**<br>A model of an air bag is used in an investigation of what happens on impact when objects of different mass are dropped from different heights. They observe the amount of damage in each case and relate this to the concept of "impulse" and how spreading out the time of the impulse reduces damage. | • Inertia<br>• Force and pressure<br>• Impulse |
| **Activity 7: Automatic Triggering Devices**<br>In this inquiry investigation, students design a device that will trigger an air bag to inflate. These simulations allows them to apply concepts of inertia and impulse as they test ideas that help them address the chapter challenge. | • Inertia<br>• Force and pressure<br>• Impulse |
| **Activity 8: The Rear-End Collision**<br>Students investigate the effect of rear-end collisions on passengers by using a model of the neck muscles and bones of the vertebral column. They then read to learn more about Newton's Second Law of Motion and consider how they can apply this information in designing a safety device that prevents movement of the head in a collision. | • Collisions<br>• Newton's Second Law of Motion<br>• Momentum |
| **Activity 9: Cushioning Collisions (Computer Analysis)**<br>Using a force probe, students investigate the effectiveness of different types of systems designed to minimize the impact of collisions. The systems include sand canisters around bridge supports and padded car interiors. This investigation provides an opportunity to develop deeper understanding of the concepts of acceleration, velocity, and momentum. | • Inertia<br>• Impulse<br>• Momentum<br>• Change in Momentum<br>• Conservation of Momentum |

# GETTING STARTED WITH EQUIPMENT NEEDED TO CONDUCT THE ACTIVITIES.

### Items needed—not supplied in Material Kits

Preparing the equipment needed for each activity in this chapter is an important procedure. There are some items, however, needed for the chapter that are not supplied in the It's About Time material kitpackage. Many of these items may already be in your school and would be an unnecessary expense to duplicate. Please read carefully the list of items to the right which are not found in the supplied kits and locate them before beginning activities.

**Items needed—not supplied by It's About Time:**

- **Electrical Clock Measuring 1/100 seconds**
- **Digital Photogates**
- **MBL or CBL With Sonic Ranger**
- **Concrete block or similar barrier**
- **Camcorder on tripod**
- **VCR having single frame advance mode**
- **Video monitor**
- **Baseball glove**
- **VCR & TV monitor**

# Equipment List For Chapter 2 (Serves a Classroom of 30 Students)

| PART | ITEM | QTY | ACTIVITY |
|---|---|---|---|
| AC-6003-T2 | Alligator Clip Leads | 18 | 1, 7 |
| SS-6206-T2 | Balance Platform | 6 | 4A, 8, 9 |
| BH-6600-T2 | Ball, Heavy | 6 | 6 |
| WS-6731-T2 | Bare Copper, #22 Wire, 1ft. | 6 | 5 |
| BS-1596-T2 | Battery, D-Cell | 6 | 1, 7 |
| BS-6701-T2 | Brochures on Auto Safety Features | 6 | 3 |
| LS-6960-T2 | Bulb Base for Miniature Screwbase | v | 1, 7 |
| LS-6959-T2 | Bulb, Miniature Light | 18 | 1, 7p |
| MC-5418-T2 | Clay, Modeling. | 18 | 4, 4A, 5 |
| CB-1060-T2 | Clear Bag, Inflatable | 36 | 6 |
| MS-6726-T2 | Cushioning Materials Set, Variety of | 36 | 9 |
| DU-0026-T2 | Duct Tape, Roll | 6 | 8 |
| GC-0001 | Dynamics Carts | 12 | 4, 5, 6, 7, 8 |
| SS-6710-T2 | Landing Surface Materials of 3 Hardnesses | 6 | 6 |
| RS-6719-T2 | Ribbon, Various Widths | 48 | 5 |
| RS-2826 | Ruler | 6 | 1 |
| BH-6068-T2 | Single Battery Holder, D-Cell | 6 | 1, 7 |
| SS-2303-T2 | Spring Scale, 0-10 Newton Range | 6 | 8 |
| SR-6736-T2 | Starting Ramp for Lab Cart | 6 | 4, 5, 7, 8, 9 |
| SM-1676-T2 | Stick, Meter, 100 cm, Hardwood | 6 | 8 |
| SS-7778 | Stopwatches | 12 | 1 |
| SS-9019-T2 | Switch, Spst. | 12 | 1, 7 |
| TS-2662-T2 | Tape, Masking, 3/4 x 36 yds. | 6 | 8, 9 |
| WS-6732-T2 | Wood Piece, 1" x 2" x 2" | 6 | 8 |
| WS-6733-T2 | Wood Piece, 1" x 3" x 10" | 6 | 8 |
| WS-6734-T2 | Wood Piece, 2" x 4" x 1' | 6 | 8 |
| | | | |
| ITEMS NEEDED – NOT SUPPLIED BY IT'S ABOUT TIME | | | |
| | Electrical Clock Measuring 1/100 seconds | 6 | 1A |
| | Digital Photogates | 12 | 1A |
| | MBL or CBL With Sonic Ranger | 6 | 2, 9 |
| | Concrete block or similar barrier | | 2, 7 |
| | Camcorder on tripod | | 4 |
| | VCR having single frame advance mode | | 4 |
| | Video monitor | | 4 |
| | Baseball glove | | 4A |
| | VCR & TV monitor | | 3, 9 |

# Organizer for Materials Available in Teacher's Edition

| Activity in Student Text | Additional Material | Alternative / Optional Activities |
|---|---|---|
| ACTIVITY 1:<br>Response Time<br>p. 82 | Assessment: Group Work, p.156<br><br>Assessment: Scientific and Technological Thinking, p 157 | Activity 1A: High-Tech Alternative, pgs. 158-159 |
| ACTIVITY 2:<br>Speed and Following Distance<br>p. 88 | Assessment: Graphing Skills, p.172 | |
| ACTIVITY 3:<br>Accidents<br>p. 96 | Assessment: Participation in Discussion, p.181 | |
| ACTIVITY 4:<br>Life (and Death) before Seat Belts<br>p. 101 | | Activity 4 A: Dropping a Clay Ball to Investigate Inertia, pgs. 191-192<br><br>Activity 4 B: Low-Tech Alternative p. 193 |
| ACTIVITY 5:<br>Life (and Fewer Deaths) after Seat Belts<br>p. 107 | Assessment, p. 204 | |
| ACTIVITY 6:<br>Why Air Bags?<br>p. 113 | Assessment, p. 215 | |
| ACTIVITY 7:<br>Automatic Triggering Devices<br>p. 119 | Assessments for Activity 7, p. 224 and for Scientific and Technological Thinking, p. 224 | |
| ACTIVITY 8:<br>The Rear-End Collision<br>p. 124 | | |
| ACTIVITY 9:<br>Cushioning Collisions (Computer Analysis)<br>p. 131 | | |

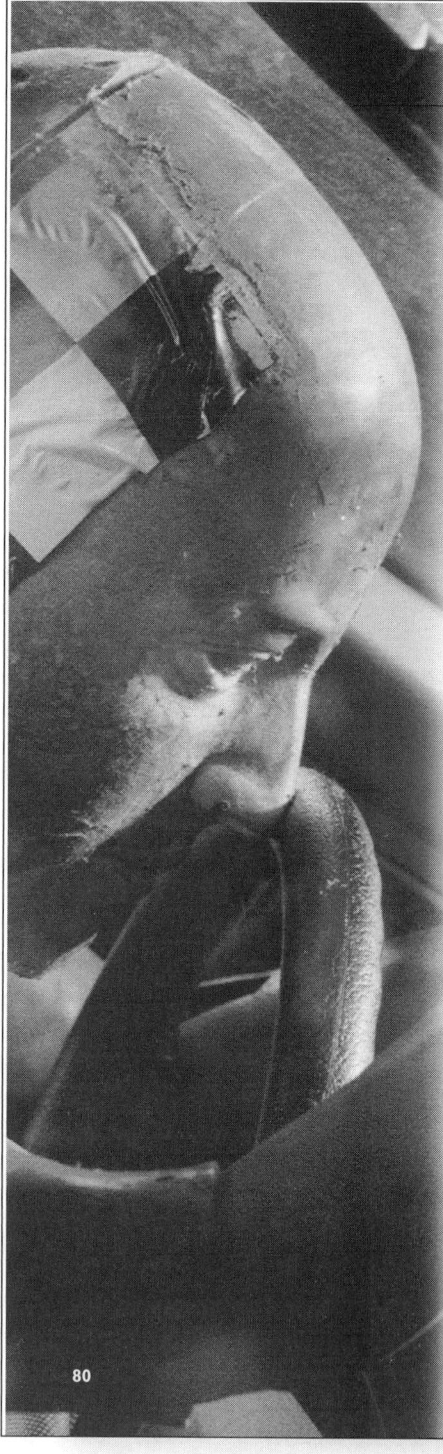

## Scenario

Probably the most dangerous thing you will do today is travel to your destination. Transportation is necessary, but the need to get there in a hurry, and the large number of people and vehicles, have made transportation very risky. There is a greater chance of being killed or injured traveling than in any other common activity. Realizing this, people and governments have begun to take action to alter the statistics. New safety systems have been designed and put into use in automobiles and airplanes. New laws and a new awareness are working together with these systems to reduce the danger in traveling.

What are these new safety systems? You are probably familiar with many of them. In this chapter, you will become more familiar with most of these designs. Could you design or even build a better safety device for a car or a plane? Many students around the country have been doing just that, and with great success!

## Challenge

Your design team will develop a safety system for protecting automobile, airplane, bicycle, motorcycle, or train passengers. As you study existing safety systems, you and your design team should be listing ideas for improving an existing system or designing a new system for preventing accidents. You may also consider a system that will minimize the harm caused by accidents.

80

## Chapter and Challenge Overview

Divide the class into design teams consisting of three or four students. As there is a diversity of skills required – design, construction, writing, speaking, etc. – all students should be involved in the process. You may want to allow some time early in the week for group meetings for the purpose of brainstorming as well as an opportunity for you to check on the progress of the teams.

Encourage broad thinking. The projects do not have to be limited to the vehicle, but can include the roadway, traffic control, etc. Students may use their experiences with skateboarding, cycling, in-line skating or other athletic pursuits to help them get started.

Explain the two-day presentation format. On the first day a poster session is conducted during which students informally explain their projects to classmates and answer their questions. Meanwhile, the students are writing down questions about the projects, and placing them into envelopes, provided by you for each project. The formal presentations the next day may be quite brief, since most students will have seen the projects the day before. After a few sentences addressing the points listed in the student text, randomly draw a student question from the appropriate envelope for the team to answer. Use as many as time might allow.

Scoring the project might be based on assigning credit for each of the items listed in the student text with, perhaps, greater emphasis on how the physics concepts are utilized and explained. Discuss with the students the relative credit weighting of the project. Use as a starting point, the criteria mentioned in the student text, with the total points being 100. An example might be Part 1 – 30, Part 2 – 20, Part 3 – 20, with a teacher assigned (or peer assessment) of 30 points, adding to a total of 100.

The intention in this exercise is to motivate students into learning about the safety of various modes of transportation. As they develop their project, while studying this chapter, they should be revising and changing their safety system.

Your final product will be a working model or prototype of a safety system. On the day that you bring the final product to class, the teams will display them around the room while class members informally view them and discuss them with members of the design team. During this time, class members will ask questions about each others products. The questions will be placed in envelopes provided to each team by the teacher. The teacher will use some of these questions during the oral presentations on the next day.

The product will be judged according to the following three parts:

1. The quality of your safety feature enhancement and the working model or prototype.
2. The quality of a five-minute oral report that should include:
   - **the need for the system**
   - **the method used to develop the working model**
   - **the demonstration of the working model**
   - **the discussion of the physics concepts involved**
   - **the description of the next-generation version of the system**
   - **the answers to questions posed by the class**
3. The quality of a written and/or multimedia report including:
   - **the information from the oral report**
   - **the documentation of the sources of expert information**
   - **the discussion of consumer acceptance and market potential**
   - **the discussion of the physics concepts applied in the design of the safety system**

## Criteria

You and your classmates will work with your teacher to define the criteria for determining grades. You will also be asked to evaluate your own work. Discuss as a class the performance task and the points that should be allocated for each part. A starting point for your discussions may be:

- **Part 1 = 40 points**
- **Part 2 = 30 points**
- **Part 3 = 30 points**

Since group work is made up of individual work, your teacher will assign some points to each individual's contribution to the project. If individual points total 30 points, then parts 1, 2 and 3 must be changed so that the total remains at 100.

81

## ssessment Rubric for Challenge: Group Work in Designing Safety eature Content

al = 9 marks

ow level – indicates minimum effort or effectiveness.

verage – acceptable standard has been achieved, but the group could have worked more effectively if better organized.

ood – this rating indicates a superior effort. Although improvements might have been made, the group was on task all f the time and completed all parts of the activity.

| escriptor | Values | | |
|---|---|---|---|
| . The group worked cooperatively to design a safety feature.<br>Comments: | 1 | 2 | 3 |
| . The group was organized. Materials were collected and the problems were addressed by the entire group.<br>Comments: | 1 | 2 | 3 |
| . Data was collected and recorded in an organized fashion in data tables in their logs.<br>Comments: | 1 | 2 | 3 |

# Assessment Rubric for Challenge: Safety Feature and Working Model

| Descriptor | 5 | 4 | 3 | 2 | 1 |
|---|---|---|---|---|---|
| **Skills Required for Working Model/Prototype** | | | | | |
| understands the need to control variables | | | | | |
| has run at least three trials with safety feature | | | | | |
| demonstrates or explains why there is a need for several trials | | | | | |
| has rebuilt or modified safety feature as necessary | | | | | |
| uses appropriate materials in the construction of the safety feature | | | | | |
| care and attention has been given in assembling the working model | | | | | |
| working model functions appropriately during demonstration | | | | | |
| group has worked efficiently as a team in assembling the model | | | | | |
| **Oral Report** | | | | | |
| understands and explains the need for the system | | | | | |
| describes the method used to develop the working model | | | | | |
| demonstrates the working model | | | | | |
| discusses the physics concepts illustrated by the safety feature | | | | | |
| describes the next generation of the system | | | | | |
| answers questions posed by the class | | | | | |
| **Written and/or Multimedia Report** | | | | | |
| contains the points included in the oral report | | | | | |
| spelling, punctuation, grammar, and sentence structure are correctly used | | | | | |
| science vocabulary and symbols are used correctly | | | | | |
| documents sources of expert information | | | | | |
| discusses consumer acceptance and marketing potential | | | | | |
| data is presented in tables and graphs as appropriate | | | | | |

For use with *Safety*, Chapter 2

# What is in the Physics InfoMall for Chapter 2?

**Chapter 2**, *Safety* deals with the physics of safety systems in automobiles, airplanes, and bicycles. The Physics InfoMall CD-ROM contains an enormous amount of material related to the physics of many phenomena, and safety is one of them. At first, it seems like a good idea to see what the InfoMall has to say about "safety systems." So the first thing you may want to do is perform a search on the entire CD-ROM for "safety system*" (the asterisk is a wild character asking the search engine to look for any words that share the same beginning, such as "system", "systems", or even "systematic"). The only result from this search that is relevant to our needs is "The science of traffic safety," by Leonard Evans, and found in *The Physics Teacher*, volume 26, issue 7. Early in this article, Evans states "No one who lives in a motorized society can fail to be concerned about the enormous human cost of traffic crashes; as many young males are killed in traffic crashes as by all other causes combined. The United States Department of Transportation maintains a file containing information on all fatal traffic crashes in the United States since 1975. This data file now documents over half a million fatalities, and of course, injuries are enormously more numerous. Recent research, discussed below, should demonstrate that the study of phenomena related to traffic safety presents problems of intellectual challenge similar in character and difficulty to those encountered in physics. Traffic safety means the safety of the overall traffic system, as distinct from more specific properties of individual components, such as laboratory crash tests of vehicles." This passage lends support to the **Scenario** described at the beginning of this chapter. Following this article is a list of references that you may also find useful in teaching traffic safety. While this article is specific to traffic safety, it is a great beginning to this chapter.

# ACTIVITY 1
## Response Time

## Background Information

**Background Information** for most activities is provided for the interest and insight of the teacher only. It is not intended to be part of the classroom instruction.

Reaction time can be understood by grouping physiological processes into three categories: input of sensory information, coordination by the central nervous system, and the response by motor nerves and their effectors, muscles and/or glands. The simplest reaction pathway is that of a reflex arc. Sensory receptors identify environmental stimuli causing a sensory nerve cell to become excited. The sensory nerve transmits an electrochemical impulse to the spinal cord. Here an intermediary nerve cell transmits the sensory impulse to a motor nerve cell. The impulse is carried by the motor nerve cell to a muscle (or in some cases a gland). The contraction of the muscle signals the response. A knee-jerk response provides an excellent example of this simple nerve pathway. The impulse is carried between three nerve cells: sensory nerve, interneuron, and motor nerve cell, toward the muscle. Surprisingly, no integration is required by the brain. These reactions occur without thinking.

Reactions that require integration by the central nervous system, such as those that occur when driving, take considerably longer to occur. A moose running in front of a vehicle is identified by visual receptors within the eye. Sensory impulses are carried toward the brain by the optic nerve. Here the information is accumulated and the driver is made aware of the problem. Multiple nerve connections carry the impulses toward the motor area of the brain. A conscious decision is made to lift the foot from the accelerator peddle and push down on the brake. Because the sensory nerves are connected with motor nerves through a maze of circuits within the brain, the reaction time is much longer than that of a reflex arch. Each time an impulse passes between connecting nerve cells, the speed of transmission is slowed.

Conscious decisions, such as braking for a moose, depend upon a number of variables. The time it takes to catch sight of the moose may well be the largest variable. Any distraction or driver fatigue will increase reaction times. Most impulses travel at approximately 100 m/sec along a nerve cell, but the time required for the impulse to travel between two different nerve cells varies greatly. Transmitter chemicals diffuse between connecting nerve cells. Because diffusion takes much more time than the movement of an impulse along a nerve cell, the connections between nerve cells slows reaction time. Not surprisingly, the complexity of integration of sensory impulses by the brain to create a visual image and the number of nerve cells involved also affects response time. The greater the number of interconnecting nerves, the slower is the processing time. Moving images require greater time to process and interpret than still images.

To accurately determine response times, we must consider how the reaction is measured. The removal of the foot from the driver's pedal takes considerably more time than just pushing down on the brake. The distance the leg moves, the amount of muscle required, and the health of the muscle also affect reaction rates.

As people age reaction rates are said to decline. The buildup of pigmented Nissl Bodies with nerve cells, slows the transmission of nerve impulses. In addition, the production of transmitter chemicals, the things that allow impulses to travel between nerves, decreases with age. Older people also tend to have less healthy muscles, further increasing the time it takes to respond to a stimulus. But age is not the major factor when considering reaction rates. The alertness of the driver is far more important.

In this activity students are asked to wire a series circuit. A series circuit has all of the current from the battery traveling through every part of the circuit. If either switch is open, the current is not able to traverse the entire circuit. In the reaction time circuit, one switch begins in the closed position and the other in the open position. One student is able to complete the circuit by closing the switch and lighting the bulb. The other student will then turn the light off by opening the other switch.

The reaction time graph is created by using the equation for free fall motion

$d = 1/2at^2$

where
$a$ is the acceleration due to gravity (9.8 m/s$^2$),
$t$ is the elapsed time and
$d$ is the distance fallen. Since all objects fall at the same rate, there is no need to be concerned with the mass of the ruler.

Solving the above equation for time:

$t = \sqrt{2d/a}$

allows us to compute the reaction tome for any given distance. The students will be introduced to this equation later in the course. To provide the equation with no evidence of constant acceleration would not help their understanding at this point. If, on the other hand, they have studied acceleration previously, you may use this equation to provide a reinforcement of this concept.

## *Active*-ating the Physics InfoMall

This activity is primarily about reaction times. There are several good items from the InfoMall that relate to this. If you choose to search the InfoMall CD-ROM for this activity, choose "reaction time" rather than "response time," as the latter will find mostly items that describe mechanical or chemical systems rather than people. All of the items found for **Activity 1**, and many for the following activities, were found searching for "reaction time" in all stores on the InfoMall at the same time (select stores to be searched using "compound search" and choose "select..." below "search in databases:").

The InfoMall has several methods for measuring a person's reaction time. The methods used in *Active Physics* can also be found on the InfoMall. Although not related to driving, the effect of reaction time in analyzing the Kennedy assassination can be found in the Articles and Abstracts Attic, *American Journal of Physics*, volume 44, issue 9, "A Physicist Examines the Kennedy Assassination Film."

Some good places to look for the effect on driving are the following:

## For You To Do

**Step 1:** Testing the reaction time for the foot may be different than for the hand. This is discussed briefly in Articles and Abstracts Attic, *The Physics Teacher*, volume 8, issue 4, "Problems for Introductory Physics," problem 49. This can be found most easily by scrolling down to near the bottom of the article and then searching up, rather than down. Included are some questions to consider about the effect of reactions time on driving.

**Step 2:** The reaction time for visual stimuli can differ from the time for audible stimuli. This is discussed, along with methods for measuring the difference, in Articles and Abstracts Attic, *The Physics Teacher*, volume 28, issue 6, "Speed of Sound in a Parking Lot." Reaction time is discussed as a source for error in the measurement of the speed of sound for this particular activity, but the difference in audible versus visual stimuli is measured using a meter stick.

**Steps 7 and 8:** Alternate methods for measuring reaction times can also be found on the InfoMall.

A circuit using an oscilloscope can be found in the Demo and Lab Shop, *Laboratory Manual to Accompany Physics Including Human Applications*, by Fuller, Fuller, & Fuller, The Oscilloscope, Application I: Reaction Time Measurement. A graphic showing how the circuit should be set up is included.

Using a clock (if it uses a large sweep second hand and displays time in small increments) to measure reaction time is discussed in the Demo and Lab Shop, *Demonstration Handbook for Physics*, Mechanics, Kinematics, Reaction Time, and scroll down to Mb-1. Also, the use of a meter stick is discussed here, as it is in many other places.

**Step 9:** *Laboratory Manual to accompany Physics Including Human Applications*, by Fuller, Fuller, & Fuller; Human senses, Part III: reaction time. This shows how a meter stick can be used to measure reaction time.

## Physics To Go

Methods mentioned in this activity include using a ruler or a dollar bill. These methods, including graphics are also discussed in the following places on the InfoMall:

Book Basement, *A Guidebook for Teaching Physics*, by Yurkewicz, Motion, topic IV: Uniform Acceleration. Activity #9 is about using a dollar bill, and Activity #10 is about using a meter stick.

The same methods are mentioned in the Demo and Lab Shop, *Physics Demonstrations and Experiments for High School*, Part II - Lab Experiments, #2 Using Acceleration of Gravity to Calculate Reaction Time," and uses a dollar bill and a meter stick.

The use of a dollar bill is also mentioned in Teacher Treasures, *Demonstration Guide for High School Physics*, and scroll down to "$ Bill & Reaction Time."

# Planning for the Activity

## Time Requirements

Allow about 40 minutes to construct the electrical circuit and complete **Steps 1** to 6. An additional 20 minutes to complete the remaining steps and record data may be required.

## Materials Needed

### For each group:

- ruler
- alligator clip leads
- battery, d-cell
- bulb base for miniature screwbase
- bulb, miniature light
- ruler
- single battery holder, d-cell
- stopwatches
- switch, spst.

## Advance Preparation and Setup

Search your school for a response timer formerly used in drivers' education. These units, about the size of an old movie projector, used to be quite common and included apparatus for testing peripheral vision and color blindness. It may well be in a closet somewhere in your building.

Should a response timer from a drivers' education class be unavailable, a usable circuit can be rigged if a clock measuring hundredths of a second is available. Set it up in series with its power source, a switch, and a normally closed foot switch. Response time is measured by having one student watch the clock, with a foot resting on the floor near the foot switch. When the student's partner starts the clock, the one being tested stops it by pressing the foot switch.

# Teaching Notes

*Coordinated Science for the 21st Century* uses a modified constructivist model. By confronting students' misconceptions and by having them do hands-on exploration of ideas, we seek to replace their misconceptions with correct perceptions of reality. In order to do this, a consistent scheme is integrated into the course activities to elicit the students' misconceptions early in any activity. Students' current mental models are sampled by one or more **What Do You Think?** questions. Students are not expected to know a "right" answer. These questions are supposed to elicit from students their beliefs regarding a very specific prediction or outcome, and students should commit to a written specific answer in their logs.

In **Activity 1** the term "response time" is used instead of the more common "reaction time" to differentiate the behavior from the reflex reaction.

Discuss the response-time circuit. The circuit is not complicated, but will provide the students with the experience of wiring a circuit. This is an opportunity to point out the characteristics of a simple series circuit. Do not begin an extensive lesson on circuit theory unless the class is really excited by the topic. Let students follow the direction in the text, and answer the questions. The sophistication of the circuit can be improved with the use of specialized normally on and normally off switches, if available.

You may be tempted to skip building the circuit and get right into measuring response time accurately. Do not skip this step. The qualitative estimates before building the circuit and using the circuit provide the foundation for understanding the short time intervals measured next.

Encourage students to work cooperatively by assigning tasks prior to beginning the activity. The following tasks are designed for groups of four students:

- **Organizer:** helps focus discussion and ensures that all members of the group contribute to the discussion. The organizer ensures that all of the equipment has been gathered and that the group completes all parts of the activity.
- **Recorder:** provides written procedures when required, diagrams where appropriate and records data. The recorder must work closely with the organizer to ensure that all group members contribute.
- **Researcher:** seeks written and electronic information to support the findings of the group. In addition, where appropriate, the researcher will develop and test prototypes. The researcher will also exchange information gathered among different groups.
- **Diverger:** seeks alternative explanations and approaches. The task of the diverger is keep the discussion open. "Are other explanations possible?"

The technique using two stopwatches in **Step 8** can become a popular game. Count your stopwatches before the students leave.

Students often believe that age is the largest factor when determining response time. Many will indicate that because of their much faster response times, they are better equipped to travel at higher speeds than older people. Variations due to age increase response time by as little a 0.01 sec.

The identification of sensory information has the greatest impact upon reaction time. Therefore, the alertness of the driver has the greatest impact upon stopping distances. The driver's experience may also play an important part in response time.

See the High Technology Alternative to **Activity 1**: Response Time. The use of photogates and 0:00 sec timers will be required if you choose to do this Alternative Activity.

# Activity Overview

This activity addresses questions of response time in its relation to the problem of bringing a car to rest.

# Student Objectives

## Students will:

- Identify the parts of the process of stopping a car.
- Measure reaction time.
- Wire a series circuit.

ANSWERS FOR THE TEACHER ONLY

# What Do You Think?

Response times will vary. The most important factor is speed, and distance is proportional to the square of the speed. The activity will provide a basis for this relationship. The identification of sensory information has the greatest impact on reaction time. Therefore, the alertness of the driver has the greatest impact on stopping distances. The driver's experience may also play an important part in response time. **What Do You Think?** is designed to provoke a discussion of the effects of listening to loud music on response time, as well as the distraction of talking or the effect of fatigue, alcohol, or drugs.

Safety

## Activity 1 Response Time

**GOALS**

In this activity you will:

- Identify the parts of the process of stopping a car.
- Measure reaction time.
- Wire a series circuit.

### What Do You Think?

Many deaths that occur on the highway are drivers and passengers in vehicles that did not cause the accident. The driver was not able to respond in time to avoid becoming a statistic.

- **How long would it take you to respond to an emergency?**

Record your ideas about this question in your *Active Physics* log. Be prepared to discuss your responses with your small group and the class.

### For You To Do

1. To stop a car, you must move your foot from the gas pedal to the brake pedal. Try moving your right foot between imaginary pedals.

Active Physics 82

a Estimate how long it takes to move your foot between the imaginary pedals. Record your estimate.

2. The first step in stopping a car happens even before you move your foot to the brake. It takes time to see or hear something that tells you to move your foot. Test this by having a friend stand behind you and clap. When you hear the sound, move your foot between imaginary pedals.

a) Estimate how long it took you to respond to the loud noise. Record your estimate.

3. Create a simple electric circuit to test your response time. Your group will need a battery in a clip, two switches, a flashlight bulb in a socket, and connecting wires. Connect the wires from one terminal of the battery to the first switch, then to the second switch, to the light bulb, and back to the battery.

**Have your teacher approve your circuit before proceeding to Step 4.**

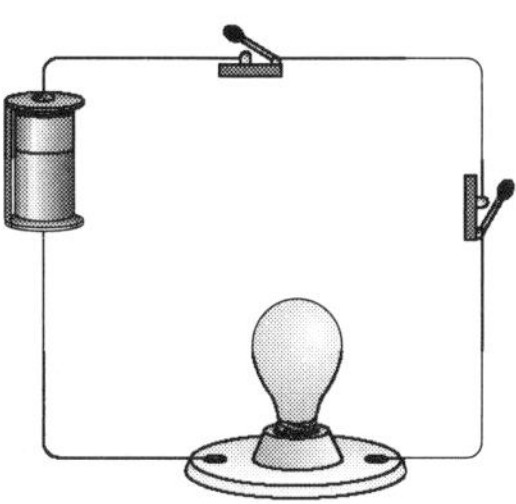

4. Close one switch wh'le the other 's open. Close the other switch. Take turns turning the light off and on with each person operating only one switch.

a) Record what happens in each case.

5. Try to keep the light on for exactly one second, then five seconds. You can estimate one second by saying "one thousand one."

a) How quickly do you think you can turn the light off after your partner turns it on? The time the bulb is lit is your response time. Record an estimate of your response time in your log.

Chapter 2

ANSWERS

## For You To Do

1. a) A reasonable estimate would be about half a second.

2. a) Estimates will vary. A reasonable estimate would be about half a second.

3. Students set up circuit.

4. a) The light bulb glows when both switches are closed.

5. a) Estimates will vary. Most students will be able to turn the light off in a quarter of a second or less. This is considerably less than the time it takes to move the foot from the gas pedal to the brake pedal.

ANSWERS

## For You To Do

***(continued)***

6. a) If the light bulb were replaced by a clock, the accuracy of the measurement would be improved. Students should also note that it takes a lot longer to move your foot and press a pedal than it takes to move a finger that is already posed for action. Students might suggest replacing the switch with a foot pedal.

   b) By averaging the results of a number of trials, the accuracy can be improved.

7. a) Response times will vary. Response times will probably be between 0.5 s and 0.8 s.

8. a) Expect the response times using this method to be slower than for previous trials.

9. a) Expect the distances to vary. If students are anticipating the release by closing their fingers periodically, and their fingers are close enough together, results may be as low as 2 cm. Suggest students average a number of trials to obtain a more reasonable result.

   b) If interpreting the graph of distance vs. time for a body falling from rest: $d = 1/2gt^2$ is difficult for your students, they could use the formula: $t = 0.45\sqrt{d}$ .

6. Find your response time using the electric circuit.
   a) How could you improve the accuracy of the measurement?
   b) How would repeating the investigation improve the accuracy?

7. Test your response time with the other equipment set up in your classroom. Use a standard reaction time meter, such as one used in driver education. You will need to follow the directions for the model available in your class.
   a) Record your response time.

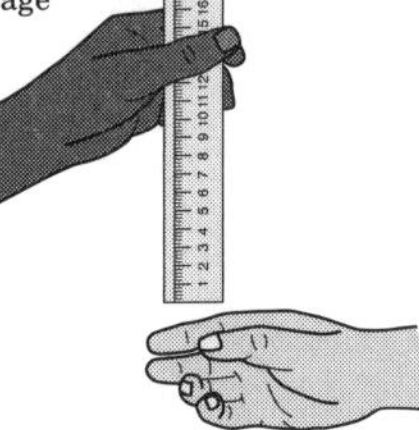

8. Use two stopwatches. One person starts both stopwatches at the same time, and hands one to her lab partner. When the first person stops her watch, the lab partner stops his. The difference in the two times is the response time.
   a) Record your response time.

9. Use a centimeter ruler. Hold the centimeter ruler at the top, between thumb and forefinger, with zero at the bottom. Your partner places thumb and forefinger at the lower end, but does not touch the ruler. Drop the ruler. Your partner must stop the ruler from falling by closing thumb and forefinger.
   a) The position of your partner's fingers marks the distance the ruler fell while her nervous system was responding. Record the distance in your log.
   b) The graph at the top of the next page shows the relationship between the distance the ruler fell and the time it took to stop it. Use the graph to find and record your response time.

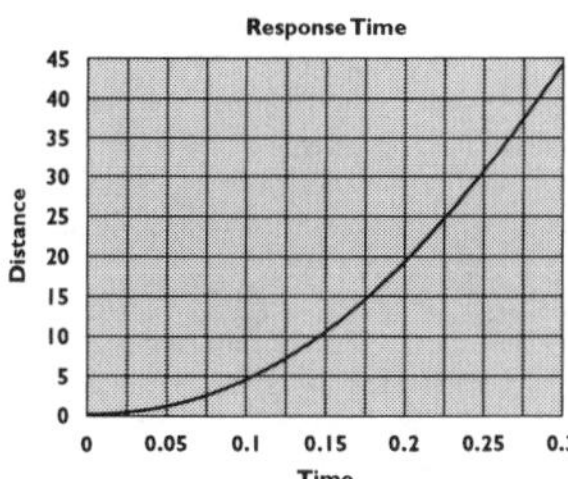


10. Compare the measures of your response obtained from each strategy.
   a) Explain why they were not all the same.
   b) What measure do you think best reports your response time? Why?
11. Compare the measures you obtained with those of other students.
   a) Record the results for the fastest, slowest, and average response times.
   b) Why do you think response times vary for people of the same age? Discuss this with your group and then record your answer.

### Reflecting on the Activity and the Challenge

The amount of time people require before they can act has a direct impact on their driving. It takes time to notice a situation and more time to respond. A person who requires a second to respond to what he or she sees or hears is more likely to have an accident than someone who responds in half a second. Your **Chapter Challenge** is to design and build an improved safety device for a car. You may be able to design a car that helps drivers to stay alert and helps them become more aware of their surroundings. Anything that you can do to decrease a driver's response time will make the car safer.



ANSWERS

## For You To Do
***(continued)***

10.a) Students may suggest that some of the methods permit them to be ready and poised to respond better than others. Estimating the point on the ruler where the fingers are located may cause errors.

b) Students may suggest that the standard reaction timer used in drivers' education might be the most accurate. Accept any reasonable explanation.

11.a) Students compare classroom data.

b) Students answers will vary. Students may consider: physiological difference in nervous systems, the amount and health of the muscles being used, the alertness of the subject, the time of day the test was conducted, use of medications such as cough syrups.

ANSWERS

## Physics To Go

1. Encourage the students to continue the discussion of safe driving at home.
2. Students may find that response time for much older and very young family members may be slower. On the other hand, parents and siblings at home may be more motivated to produce excellent response times and therefore may be more alert than the subjects tested in class.
3. The length of a dollar bill is 15.7 cm. The free fall time is under 0.2 s, which makes it nearly impossible to catch the bill unless the hand is lowered or the release is anticipated. The grasping action with the thumb and forefinger is a much more familiar one than between the forefinger and middle finger. Refer the students to the graph, or provide them with a simple mathematical equation they can use to quantify their results.
4. The speed at which the race car driver is traveling requires a very quick response time.
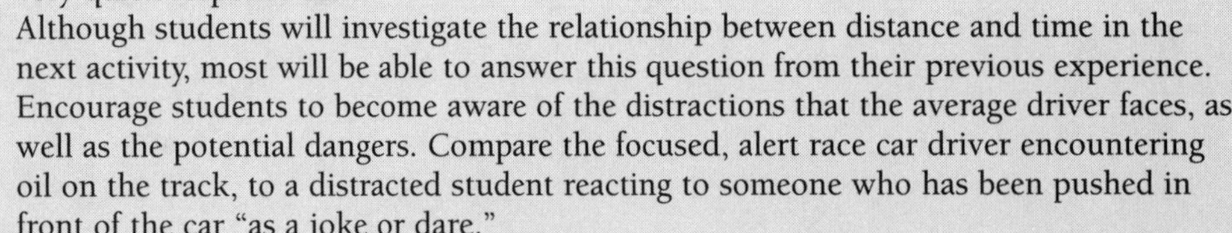
Although students will investigate the relationship between distance and time in the next activity, most will be able to answer this question from their previous experience. Encourage students to become aware of the distractions that the average driver faces, as well as the potential dangers. Compare the focused, alert race car driver encountering oil on the track, to a distracted student reacting to someone who has been pushed in front of the car "as a joke or dare."
5. Students should be assigned these questions routinely throughout the chapter. By answering this question, they have completed a part of their chapter challenge. Suggest that the students read the **Reflecting on the Activity and the Challenge** before they proceed with the answer. Answers will vary greatly. Some students who are familiar with video games may show "superhuman" responses.

Safety

### Physics To Go

1. Test the response time of some of your friends and family with the centimeter ruler. Bring in the results from at least three people of various ages.
2. How do the values you found in **Question 1** compare with those you obtained in class? What do you think explains the difference, if any?
3. Take a dollar bill and fold it in half lengthwise. Have someone try to catch the dollar bill between his or her forefinger and middle finger. Most people will fail this task.
   a) Explain why it is so difficult to catch the dollar bill.
   b) Repeat the dollar bill test, letting them catch it with their thumb and forefinger.
   c) Explain why catching it with thumb and forefinger may have been easier. Try to include numbers in your answer such as the length of the dollar, the time for the dollar to fall, and average response time.
4. Does a racecar driver need a better response time than someone driving around a school? Explain your answer, giving examples of the dangers each person encounters.
5. Apply what you learned from this activity to describe how knowing your own response time can help you be a safer driver.

86

Active Physics

Activity 1 Response Time

### Stretching Exercises

1. Build a device with a red light and a green light. If the red light turns on, you must press one button and measure the response time. If the green light turns on, you must press a second button and measure the response time. Have your teacher approve your design before proceeding. How do response times to this "decision" task compare with the response times measured earlier?
2. Use the graph for response time to construct a response-time ruler with the distance measurement converted to time. You can now read response times directly.
3. Do you think some groups of people have better or worse response times than others? Consider groups such as basketball players, video game players, taxi drivers, or older adults. Plan an investigation to collect data that will help you find an answer. Include in your plan the number of subjects, how you will test them, and how you will organize and interpret the data collected. Have your teacher approve your plan before you proceed.

87



ANSWERS

## Stretching Exercises

1. This assignment may interest students who are enrolled in a technology program. It may be a way of getting a special reaction timer for use next year.
2. The students might want to mark a time scale on a strip of masking tape which could be affixed to a ruler or paint stirrer.
3. Students should be reminded to be courteous to their subjects.

# Assessment: Group Work

The following rubric can be used to assess group work during the first six steps of the activity. Each member of the group can evaluate the manner in which the group worked to solve problems in the activity.

**Maximum value = 12**

1. Low level — indicates minimum effort or effectiveness.
2. Average — acceptable standard has been achieved, but the group could have worked more effectively if better organized.
3. Good — this rating indicates a superior effort. Although improvements might have been made, the group was on task all of the time and completed all parts of the activity.

| Descriptor | | Values | |
|---|---|---|---|
| 1. The group worked cooperatively to design a circuit that would measure reaction times.<br>Comments: | 1 | 2 | 3 |
| 2. A plan was established before beginning and a light bulb was used to test the circuit.<br>Comments: | 1 | 2 | 3 |
| 3. The group was organized. Materials were collected and the problems were addressed by the entire group.<br>Comments: | 1 | 2 | 3 |
| 4. Data was collected and recorded in an organized fashion in data tables and in journals.<br>Comments: | 1 | 2 | 3 |

For use with *Safety*, Chapter 2, Activity 1: Response Time

## Assessment: Scientific and Technological Thinking

Scientific and Technological Thinking can be assessed using the rubric below. Allow one mark for each check mark.

**Maximum value = 10**



| Descriptor | Yes | No |
|---|---|---|
| 1. A complete circuit is constructed. | | |
| 2. A light bulb is used to check for battery life and the functioning circuit. | | |
| 3. A switch is used to time reaction rates. | | |
| 4. Controls, such as the distance of the hand from the switch, are maintained throughout the experiment. | | |
| 5. Proper units are used to measure reaction times. | | |
| 6. A clock or timing device is integrated into the circuit to provide accuracy of measurement. | | |
| 7. Timing devices are tested and/or modified prior to collecting final data. | | |
| 8. Response time can be determined by using a distance vs time graph. | | |
| 9. Students can identify variables used in the experiment that would alter response time. | | |
| 10. Student is able to relate response times to the need for safe driving. | | |

For use with *Safety*, Chapter 2, Activity 1: Response Time

## Activity 1 A

# Response Time: High-Tech Alternative

### FOR YOU TO DO

1. Photogates can be used along with an electronic timer to determine reaction times. Hide the first photogate within a cardboard box, so that observation and reaction times can be accurately monitored. Use the setup below. The subject observes the timer clock for the "GO" signal.

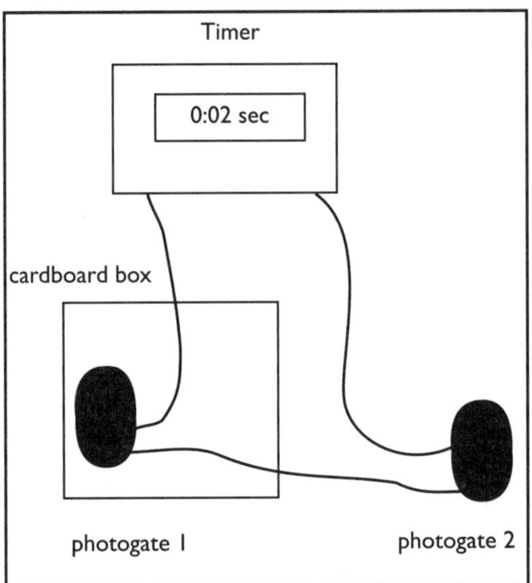


2. The tester moves his or her hand inside of the cardboard box and activates the first photocell.

3. The subject sees the clock begin to move and quickly moves his or her hand in front of the second photogate to stop the clock.

a) Complete multiple trials and record your data. Calculate the mean time taken to stop the electronic timer.

b) Explain why the mean time was determined.

c) Repeat the procedure but this time use your non-dominant hand. Account for any differences in reaction rate.

d) Why must the test subject's hand be held a specified distance from the timer?

e) When braking, the right foot is moved from the gas to the brake. What advantage is gained from only using one foot to control both the accelerator pedal and the brake?

f) A faster reaction time could be obtained by holding the left foot just millimeters above the brake pedal. As the left foot tramps on the brake, the right foot would leave the gas pedal. Explain why the practice of two-foot driving (using both the right and left foot) is discouraged.

For use with *Safety*, Chapter 2, Activity 1: Response Time

Activity 1 A

# Response Time: High-Tech Alternative

## Time Requirements

Approximately 40 minutes is required to complete the experiment.

## Materials needed

- electrical clock (measuring 1/100 sec)
- 2 photogates



ANSWERS

## For You To Do

3. a)-b) Small variations can be expected. Sometimes the subject may anticipate when the experimenter was about to throw the switch. Other times small distractions may have increased the reaction time. By taking an average, variables that increase and decrease reaction times can be eliminated.

c) In general, the dominant hand responds faster. The more neural circuits are used, the faster is the response time.

d) The further the hand is from the photogate, the greater the time it takes to reach the photocell.

e) You can't accelerate and brake at the same time. Not only would this increase wear on the brakes, it would tend to throw the car into a spin. By using one foot, the problem of simultaneously braking and accelerating is eliminated.

f) Simultaneous pushing down on the gas pedal and the brake would increase the braking distance. Because the engine would be pulling or pushing the car forward, while the brakes are applied — the effectiveness of the braking system would be reduced. The car is not able to do both at the same time.

For use with *Safety*, Chapter 2, Activity 1: Response Time

# ACTIVITY 2
## Speed and Following Distance

# Background Information

Kinematics is the study of motion. Every person will have experienced kinematics. From the moment we are able to crawl, we have a basic understanding of kinematics. As we grow and gain more experience, we are able to recognize objects as moving "fast" or "slow". We can make comparisons between the speed of a hare with the speed of a tortoise. In physics, we observe an object in motion, and then, using measurement and graphs, we are able to analyze the motion of that object.

To do this analysis, tools of measurement must be established which are appropriate for the object in motion and the speed at which it's moving. For example, a geologist who studies the movement of plates within the Earth's crust would measure the distances in inches (or cm) and the time in years or even thousands of years. When measuring the speed of an electron in a particle accelerator, we would use distances measured in meters or kilometers and times in millionths of seconds.

Understanding speed is critical in understanding motion and two tools used by scientists to achieve this understanding are mathematics, and graphical analysis. In this activity the students will be analyzing motion both mathematically and graphically.

A mathematical analysis is using the formula $v = d/t$.

Speed (symbol for speed is $v$) is the distance an object moves in a given time.

Average speed is the total distance traveled/total time. For example, you can travel from one city to the next in two hours, a total distance of 100 miles. Your average speed is (calculated mathematically)

$v =$ total distance/total time

$v =$ 100 miles/2 hours

$v =$ 50 mph

However, if on the return trip you had a flat tire, and spent 30 minutes fixing your tire, the total time has now changed to 2.5 h and the average speed is now 100 miles/2.5 hours or 40 mph. Instantaneous speed, on the other hand, is the speed that you are traveling at a given moment. For example on the return trip, even though your average speed was 40 mph, your instantaneous speed at a given time may have been 65 mph. Instantaneous speed is the speed at which you happen to be traveling when you look down at your speedometer.

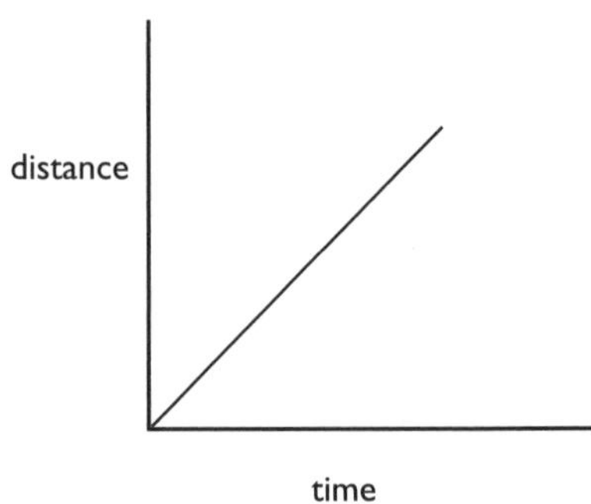


A graphical analysis of uniform motion (we will be using constant motion and uniform motion interchangeably) will involve collection of data, and then plotting that data onto a graph. Putting the information onto a distance-time graph will produce a straight line. (A straight line indicates uniform or constant motion.) The slope of that line ($\Delta d/\Delta t$) will give us the speed. ($\Delta d$ ($d_2 - d_1$) = meters (m) (or miles) and $\Delta t$ ($t_2 - t_1$) = seconds (s) (or hours), therefore the unit for the slope is m/s (or mph).) For distance, $d_2$ usually represents the final distance. In most situations $d_1$ is the starting point and is most often indicated by zero. Similarly, $\Delta t$ represents change in time, where $t_2$ is the final or end time and $t_1$ is the initial time.

Although the following is beyond the intent of the activity, some extension is presented here for the teacher.

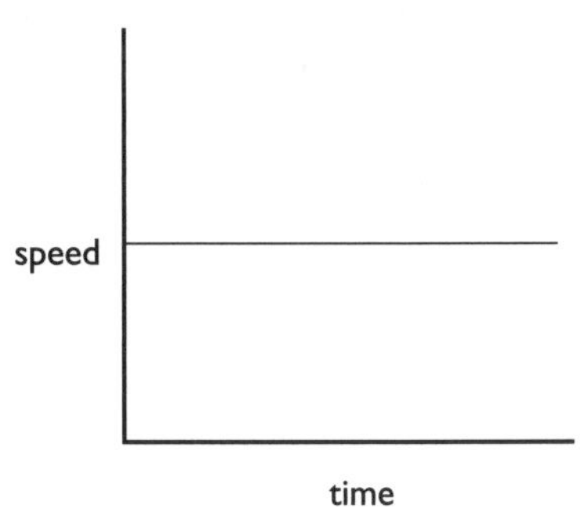


The speed-time graph of the same object can be obtained by taking the speed from the distance-time graph, at various times, and plotting them against time on a speed-time graph. The resulting line will be a horizontal straight line, which reinforces the concept of constant motion.

To find the distance an object travels while experiencing constant motion, we find the area under the graph. Area = length of side x the width of the side ($A = l \times w$). Therefore, $d = v \times t$.

Mathematically the formula $v = d/t$ will give the rearranged formula to solve for $d$ as $d = v \times t$.

In later grades, students will move from the study of speed to the study of velocity. The concept is essentially the same, with the difference being speed is the only magnitude or how fast something is moving whereas velocity is the speed and the direction in which the object is moving.

## *Active*-ating the Physics InfoMall

The effect of reaction time on following distance is discussed in Articles and Abstracts Attic, *The Physics Teacher,* volume 8, issue 4, "Problems for Introductory Physics," problem 49. This can be found most easily by scrolling down to near the bottom of the article and then searching up, rather than down. Included are some questions to consider about the effect of reactions time on driving.

# Planning for the Activity

## Time Requirements

During the class each group of students will require approximately 20 – 30 minutes to do the experiment and gather their data.

## Materials Needed

### For each group:

- MBL or CBL with sonic ranger

## Advance Preparation and Setup

Become familiar with the sonic rangers and the software that controls their output. Instructions come with the equipment, but you should use it yourself before putting it into the classroom. It will expedite the experiment if you are aware of the near and far limits of range. Prepare a list of the keystrokes required by your software to erase one graph and get ready to gather new data. Some programs ask several questions before recording data. You should know the best responses so that there will be little delay. Check as well on an appropriate sampling rate for your machine, since too much detail will cloud the issues at the moment. The sampling rate should be relatively slow.

In preparing for the class, set up the ranger and computer ready to go. Put masking tape on the floor to mark the range and perhaps another tape to indicate the lane so that students will stay "on track."

# Teaching Notes

After the students have read and reacted to the **What Do You Think?** question, perhaps enhanced with a personal story about being tailgated by a poor driver, get right to the sonic rangers to perform the exercises. If you do not have access to separate computer setups for each group of students, do this as a demonstration using student volunteers.

Make sure the computer equipment is in a secure and stable environment.

Even if you have only one setup available, it is likely that most students will remain engaged by the activity. It can be presented as a human-sized video game where the challenge is to create the straightest line, the smoothest curve, the steepest slope, etc. One student can be asked to create a graph while the others look away, after which another student can describe the motion that caused the graph. There are many variations, including having students copy graphs created by other students.

While it is fairly easy to obtain good speed-time graphs using the sonic ranger, an uneven gait or even a baggy sweatshirt can produce some irregularity. Teach the students to focus on the general trends. Students may accidentally collect data or construct graphs of displacement or acceleration instead of speed. You can use this error to your mutual advantage by asking them to explain what they see.

The graphs will quickly give students a secure understanding of the meaning of constant speed, and they can identify the different graphs almost at once. This provides preparation for algebraic analysis.

A common misconception held by students is that they go from zero to the speed they are walking or running instantly. In fact, there will always be some acceleration. Illustrate this with a set of data that has enough points to show that the first part of the graph is not straight.

Another common misconception of students that has been well documented is the confusion between distance and velocity. Look at the following pair of strobe pictures:

O     O     O     O     O
   X     X        X           X

Students may think that the X and O are traveling the same speed when they are aligned. They have the same speed when the distance between adjacent Xs is identical to the spacing between adjacent Os.

The sonic ranger has been shown to be extremely effective for students to gain an understanding of motion graphs. It is well worth the equipment investment.

The definition of velocity as the change in distance divided by the change in time can be written as an algebraic equation and can be solved for any of the variables.

$$d = vt$$

$$v = d/t$$

$$t = d/v$$

Some students deficient in algebra skills will need some help with this. You should emphasize the following points while students are learning this relationship:

The units of distance and time and velocity should always be presented with the numbers.

Any distance and any time can be used for velocity. (Cars move at miles/hour, km/h, or m/s; glaciers move at meters per year.)

To measure velocity, you need a ruler and a stopwatch.

Average velocity should be distinguished from initial velocity and final velocity. The average velocity for a trip does not give any indication of the initial and final velocity.

Before assigning the **Physics To Go Questions 4 to 7**, work a few examples with the students. Some students will need help in manipulating the speed equation. You may wish to provide them with the equations required to calculate distance and time.

NOTES

# Activity Overview

This activity should provide the student with a feel and definition for the notion of speed.

## Student Objectives

### Students will:

- Define speed.
- Identify constant and changing speeds.
- Interpret distance-time and speed-time graphs.
- Contrast average and instantaneous speeds.
- Calculate the distance traveled at constant speed.

ANSWERS FOR THE TEACHER ONLY

## What Do You Think?

The proper interval between your car and the vehicle in front is two seconds. (See the answer for **Physics To Go, Step 8** for more information.)

# Activity 2 Speed and Following Distance

**GOALS**

In this activity you will:

- **Define** speed.
- **Identify** constant and changing speeds.
- **Interpret** distance-time and speed-time graphs.
- **Contrast** average and instantaneous speeds.
- **Calculate** the distance traveled at constant speed.

**What Do You Think?**

In a rear-end collision, the driver of the car in back is always found at fault.

- **What is a safe distance between your car and the car in front of you?**
- **How do you decide?**

Record your ideas about these questions in your *Active Physics* log. Be prepared to discuss your responses with your small group and the class.

**For You To Do**

1. A strobe photo is a multiple-exposure photo in which a moving object is photographed at regular time intervals. The strobe photo below shows a car traveling at 30 mph.

a) Copy the sketch in your log.

2. Think about the difference between the motion of a car traveling at 30 mph and one traveling at 45 mph.

a) Draw a sketch of a strobe photo, similar to the one above, of a car traveling at 45 mph.

b) Are the cars the same distance apart? Were they farther apart or closer together than at 30 mph?

c) Draw a sketch for a car traveling at 60 mph. Describe how you decided how far apart to place the cars.

3. The following sketch shows a car traveling at different speeds.

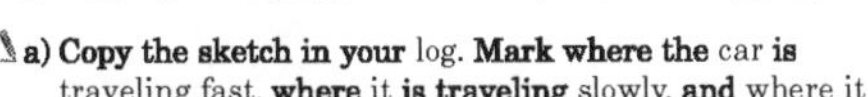

a) Copy the sketch in your log. Mark where the car is traveling fast, where it is traveling slowly, and where it is traveling at a constant speed. How did you know?

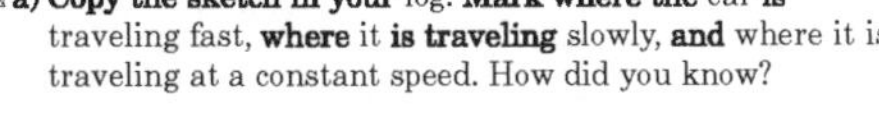

4. A sonic ranger connected to a computer will produce a graph that shows an object's motion. Use the sonic ranger setup to obtain the following graphs to print or sketch in your log.

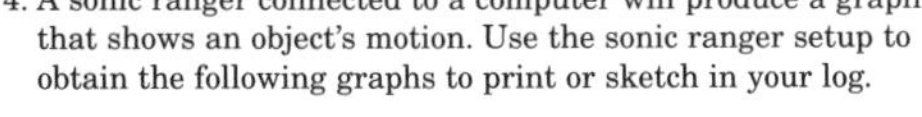

Make sure the path of motion is clear of any hazards.

a) Sketch a graph of a person walking toward the sonic ranger at a normal speed.

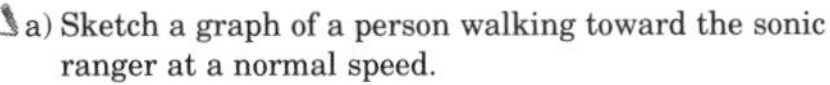

b) Sketch a graph of a person walking away from the sonic ranger at a normal speed.

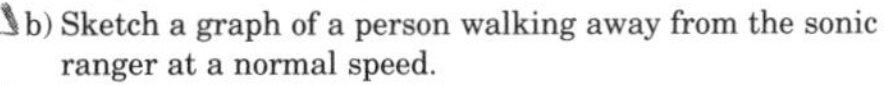

c) Sketch a graph of a person walking both directions at a very slow speed.

d) Sketch a graph of a person walking both directions at a fast speed.

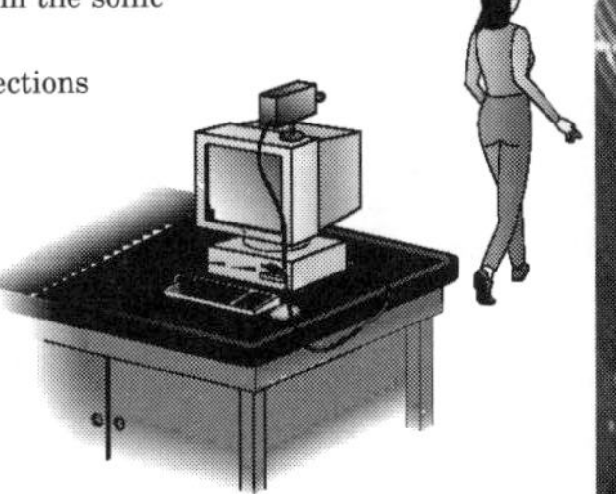

5. Predict what the graph will look like if you walk toward the system at a slow speed and away at a fast speed. Test your prediction.

a) Record your prediction in your log.

b) Based on your measurements, how accurate was your prediction?



Answers

## For You To Do

1. a) If the students do not feel comfortable sketching cars, suggest that they use O or X symbols instead. Check to see that the students understand that the spaces between the cars are even.

2. a) The sketch should show the cars with larger, even gaps between them.

   b) The marks are a greater distance apart.

   c) The cars should be twice the distance apart as they were for 30 mph.

3. a) The car is traveling slowly for the first three intervals, then fast for the next two intervals, and slower and slower for the last two intervals.

4. a)-d) The slope will be positive for motion away from the detector and negative for motion towards the detector.

5. a)-b) The prediction of a steeper slope for high speed than for low speed will be confirmed.

ANSWERS

## For You To Do

*(continued)*

6. a) For motion away from the sonic ranger, the total distance will be the maximum *y*-value of the graph. For motion towards the ranger the final *y*-value must be subtracted from the starting *y*-value.

   b) The total time is measured on the *x*-axis for the points corresponding to those for distance.

   c) Students divide the value they obtained in **6. a)** by the one in **6. b)** to obtain their average speed. Check that the units used are m/s.

Safety

6. Repeat any of the motions in **Steps 4** or **5** for a more thorough analysis.

   a) From your graph, determine the total distance you walked.

   b) How long did it take to walk that distance?

   c) Divide the distance you walked by the time it took. This is your average speed in meters per second (m/s).

Physics Words

**speed:** the change in distance per unit time; speed is a scalar, it has no direction.

**PHYSICS TALK**

**Speed**

The relationship between **speed**, distance, and time can be written as:

$$\text{Speed} = \frac{\text{Distance traveled}}{\text{Time elapsed}}$$

If your speed is changing, this gives your average speed. Using symbols, the same relationship can be written as:

$$v_{av} = \frac{\Delta d}{\Delta t}$$

where $v_{av}$ is average speed

$\Delta d$ is change in distance or displacement.

$\Delta t$ is change in time or elapsed time.

**Sample Problem I**

You drive 400 mi. in 8 h. What is your average speed?

***Strategy:*** You can use the equation for average speed.

$$v_{av} = \frac{\Delta d}{\Delta t}$$

***Givens:***

$$\Delta d = 400 \text{ mi.}$$

$$v_{av} = \frac{\Delta d}{\Delta t} = \frac{400 \text{ mi.}}{8 \text{ h}} = 50 \text{ mph (miles per hour)}$$

Your average speed is 50 mph. This does not tell you the fastest or slowest speed that you traveled. This also does not tell you how fast you were going at any particular moment.

### Sample Problem 2

Elisha would like to ride her bike to the beach. From car trips with her parents, she knows that the distance is 30 mi. She thinks she can keep up an average speed of about 15 mph. How long will it take her to ride to the beach?

***Strategy:*** You can use the equation for average speed.

$$v_{av} = \frac{\Delta d}{\Delta t}$$

However, you will first need to rearrange the terms to solve for elapsed time.

$$\Delta t = \frac{\Delta d}{v_{av}}$$

***Solution:***

$$\Delta t = \frac{\Delta d}{v_{av}} = \frac{30 \text{ mi.}}{15 \text{ mph}} = 2 \text{ h}$$



Chapter 2

## FOR YOU TO READ

### Representing Motion

One way to show motion is with the use of strobe photos. A strobe photo is a multiple-exposure photo in which a moving object is photographed at regular time intervals. The sketches you used in **Steps 1, 2,** and **3** in **For You To Do** are similar to strobe photos. Here is a strobe photo of a car traveling at the average speed of 50 mph.

Another way to represent motion is with graphs. The graph below shows a car traveling at the average speed of 50 mph.

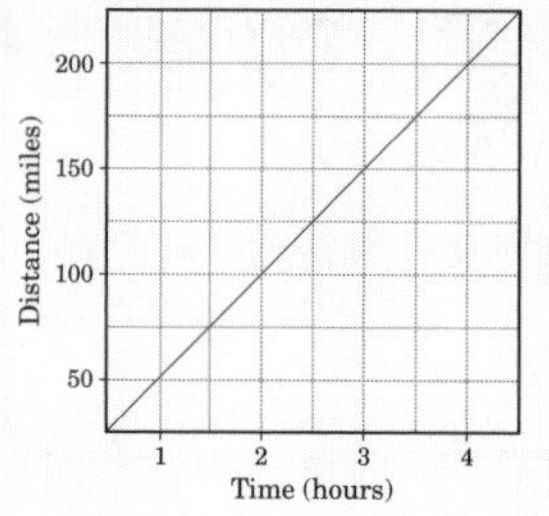


### Kilometers and Miles

Highway signs and speed limits in the USA are given in miles per hour, or mph. Almost every other country in the world uses kilometers to measure distances. A kilometer is a little less than two-thirds of a mile. Kilometers per hour (km/h) is used to measure highway driving speed. Shorter distances, such as for track events and experiments in a science class, are measured in meters per second, m/s.

You will use mph when working with driving speeds, but meters per second for data you collect in class. The good news is that you do not need to change measures between systems. It is important to be able to understand and compare measures.

To help you relate the speeds with which you are comfortable to the data you collect in class, the chart below gives *approximate* comparisons.

| | | | |
|---|---|---|---|
| School zone | 25 mph | 40 km/h | 11 m/s |
| Residential street | 35 mph | 55 km/h | 16 m/s |
| Suburban interstate | 55 mph | 90 km/h | 25 m/s |
| Rural interstate | 75 mph | 120 km/h | 34 m/s |

### Reflecting on the Activity and the Challenge

You now know how reaction time and speed affect the distance required to stop. You should be able to make a good argument about tailgating as part of the **Chapter Challenge**. If your car can be designed to limit tailgating or to alert drivers to the dangers of tailgating, it will add to improved safety.

### Physics To Go

1. Describe the motion of each car moving to the right. The strobe pictures were taken every 3 s (seconds).

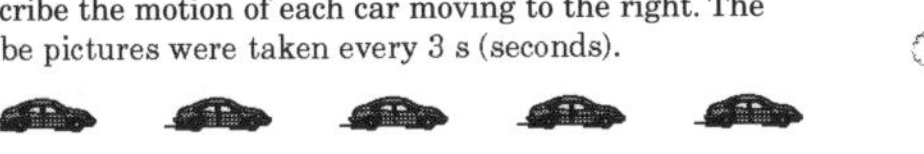


2. Sketch strobe pictures of the following:
   a) A car starting at rest and reaching a final constant speed.
   b) A car traveling at a constant speed then coming to a stop.

3. For each graph below, describe the motion of the car:

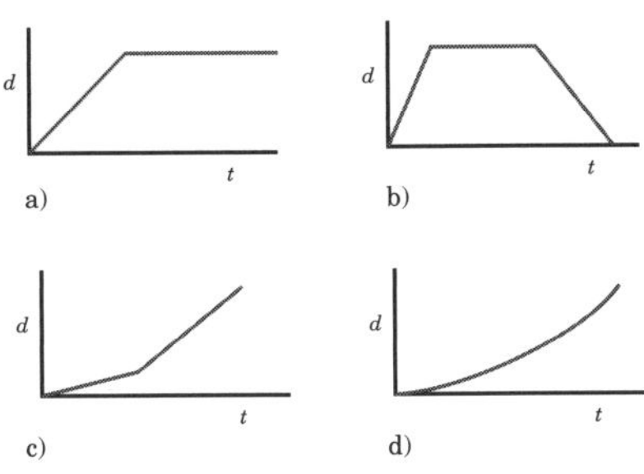




Chapter 2

## Answers

## Physics To Go

1. a) The car is moving at a constant speed every three seconds.
   b) The car speeds up, slows down, speeds up, and then slows again.

2. a) Answers will vary, but there should be a measurable increase in the distance of the first few sketches, and then the same distance while the car travels at a constant speed.
   E.g.: ΔΔ Δ Δ Δ Δ Δ Δ Δ Δ
   b) The reverse of the above situation, moving at a constant speed then slowing down.
   E.g.: Δ Δ Δ Δ Δ Δ Δ Δ Δ ΔΔ

3. a) A car travels at a constant speed, and then stops.
   b) A car travels at a rapid, constant speed, stops, and then returns at a slower constant speed.
   c) A car travels at a slow constant speed, and then increases to travel at a faster constant speed.
   d) A car accelerates as it travels.

   Note: If students have difficulty interpreting this graph, suggest they return to the question after completing the next activity.

ANSWERS

## Physics To Go
***(continued)***

4. Using the formula $v = d/t$, therefore; $d = vt$, then; $d = 110 \text{m/s} \times 20 \text{ s}$
$d = 2200 \text{ m}$ or 2.2 km
The driver traveled 2200 m, or 2.2 km.

5. a) $v = d/t$ $v = 215$ miles/4.5 hours
$v = 48$ mph. Her average speed was about 48 mph.

b) This question cannot be answered with the data provided. We do not know her instantaneous speed at any one time. We can only assume that she was *probably* doing about 48 mph.

6. $v = d/t$ $v = 5$ miles/2 hours
$v = 2.5$ mph You would need to keep up an average speed of 2.5 mph.

7. a) At 55 mph, the car will be moving at about 25 m/s, therefore to find the distance,
$v = d/t$, then $d = vt$
$d = 25$ m/s x response time
(response times will vary according to student)

For example, for a response time of 0.8 s: $d = vt$
$d = 25 \text{ m/s} \times 0.8 \text{ s}$ $d = 20 \text{ m}$

b) Assume a response time of 0.8 s.At 40 mph, the car will be moving at about 19 m/s, therefore to find the distance,
$v = d/t$, then $d = vt$
$d = 19 \text{ m/s} \times 0.8 \text{ s}$ $d = 15 \text{ m}$
A greater speed produces a greater stopping distance with the same reaction time.

c) You will travel double the distance.

8. Since distance traveled is directly proportional to time for constant speed, the two-second rule should work at any speed, and the distance can be described by the travel time. The premise for this rule is that the two cars have similar braking ability. The lead vehicle does not "stop on a dime." The two-second distance is not an adequate stopping distance, it is meant to cover the driver's response time and the time that the lead car has slowed before the driver's car begins to slow. Exceptions can be made for poor road, tire, or brake conditions.

9. Students' answers will vary depending on their response time. For a response time of 0.8s:
a) 9 m b) 20 m c) 27 m

4. A racecar driver travels at 110 m/s (that's almost 250 mph) for 20 s. How far has the driver traveled?

5. A salesperson drove the 215 miles from New York City to Washington, DC, in 412 hours.
a) What was her average speed?
b) How fast was she going when she passed through Baltimore?

6. If you planned to walk to a park that was 5 miles away, what average speed would you have to keep up to arrive in 2 hours?

7. Use your average response time from **Activity 1** to answer the following:
a) How far does your car travel in meters during your response time if you are moving at 55 mph (25 m/s)?
b) How far does your car travel during your response time if you are moving at 35 mph (16 m/s)? How does the distance compare with the distance at 55 mph?
c) Suppose you are very tired and your response time is doubled. How far would you travel at 55 mph during your response time?

8. According to traffic experts, the proper following distance you should leave between your car and the vehicle in front of you is two seconds. As the vehicle in front of you passes a fixed point, say to yourself "one thousand one, one thousand two." Your car should reach the point as you complete the phrase. How can the experts be sure? Isn't two seconds a measure of time? Will two seconds be safe on the interstate highway?

9. You calculated the distance your car would move during your response time. Use that information to determine a safe following distance at:
a) 25 mph
b) 55 mph
c) 75 mph

10. Apply what you learned in this activity to write a convincing argument that describes why following a car too closely (tailgating) is dangerous. Include the factors you would use to decide how close counts as "tailgating."

### Stretching Exercises

Measure a distance of about 100 m. You can use a football field or get a long tape or trundle wheel to measure a similar distance. You also need a watch capable of measuring seconds. Determine your average speed traveling that distance for each of the following:

a) a slow walk
b) a fast walk
c) running
d) on a bicycle
e) another method of your choice

ANSWERS

## Physics To Go

***(continued)***

10. Students answers will vary. Tailgating increases the potential hazards of driving. During the time from when you see the brake light of the lead car flash on until you hit the brake and begin to slow, your car is still traveling at top speed while the lead car has been slowing down from the time the light first flashed. If the two cars have equivalent braking ability, there had better be an extra distance equal to the product of your speed and response time between the two cars. Since safe driving means minimizing risks, following distances must be commensurate with speed and response time, in addition to road conditions.

    Student answers could also indicate that road conditions will increase the stopping time, physical conditions of the driver will increase the response time, distractions in and outside the car will increase the response time, etc. Also, the condition of the car in front, which is not in the control of the person tailgating, must be taken into consideration as an intangible. This could be that the car in front has faulty tail lights, or brakes that are very good, or that the driver in front is slowing down using the engine (gearing down) rather than brakes.

Chapter 2

# Assessment: Graphing Skills

The following assessment rubric provides insight into the attainment of graphing skills and monitors communication by way of mathematical expression. Place a check mark (√) in the appropriate box. Two check marks will be required for one point. A sample conversion scale is provided below the chart.

| Descriptor | Yes | No |
|---|---|---|
| **Analysis and Communication**<br>• data is recorded in an organized data table<br>• manipulated and responding variables are identified in data table<br>• appropriate units are recorded for distance (meters), time (seconds) and speed (m/sec)<br>• multiple trials are used and averages are calculated<br>• average speed ($v$) is calculated from formula $d/t$<br>• distance or time can be calculated from $v = d/t$<br>• distance can be calculated by pacing<br>• student is able to explain why increasing the speed will require greater distance between cars to allow for reaction time | | |
| **Graphing Skills**<br>• graph has a title<br>• the $x$-axis and $y$-axis are clearly labeled<br>• units of measurement are provided for distance and time<br>• manipulated variable (time) is plotted along the $x$-axis<br>• responding variable (distance) is plotted along the $y$-axis<br>• $x$-axis and $y$-axis are drawn to proper scale<br>• student is able to plot distance/time coordinates<br>• a best fit line is used to connect coordinates for a line graph<br>• distance and time relationships taken from the sonic ranger can be interpreted from computer-generated graph<br>• distance traveled can be determined from graphs where a constant velocity is provided<br>• time of travel can be determined from graphs where a constant velocity is provided<br>• average speed can be determined from distance/time graph by calculating the slope of the line | | |

**Conversion**

20 check marks = 10 points

18 + check marks = 9 points

16 + check marks = 8 points, etc.

For use with *Safety*, Chapter 2, Activity 2: Speed and Following Distance

NOTES

# ACTIVITY 3
## Accidents

# Background Information

Most of the physics involved with this chapter on *Safety*, involves an understanding of Inertia and Momentum.

**Symbols**

| | |
|---|---|
| $v$ = speed | $m$ = mass (kg) |
| $d$ = distance | $F_{net}$ = acceleration force |
| $t$ = time | $F_A$ = applied force |
| $a$ = acceleration | $F_f$ = force of friction |
| $F$ = force (N) | |

Newton's First Law of Motion states (Inertia) that an object in motion or at rest will remain in motion or at rest unless acted upon by an outside unbalanced force. An object at rest staying at rest is fairly obvious for students to understand. Even understanding that an object in motion remaining in motion should be easily understood once an understanding that friction is acting on things on Earth in one form or another.

Therefore, an object in motion, say an automobile, will continue in motion. However, students will observe that the automobile slows down (decelerates). Enter into a discussion on what slows the automobile. Friction (in the engine, axles, wheels, transmission, and tires on the road) is the outside unbalanced force acting on the automobile which stops it from moving at a constant speed. Therefore in order to keep it moving at a constant speed, you must be applying a force to it.
$F_{net}$ (the accelerating force )
= $F_A$ (the force applied by the engine) +
$F_f$ (force of friction always acting opposite to the direction of motion).

$$F_{net} = F_A + F_f$$

When there is no acceleration (therefore, constant motion), there is no net force. This means that the force applied is the same as the force of friction; only opposite in direction. As long as the force applied and the force of friction are the same magnitude but opposite in direction, the object will continue to move at a constant speed.

Safety, then is a discussion on how inertia affects the movement of a body in an automobile. While the automobile is moving, everything and everyone in the automobile are moving at the same speed as the automobile. When the automobile stops, everything that is attached to the automobile (bumper, seats, steering wheel, etc.) are stopping or experiencing the deceleration as well. However, anything not attached to the automobile, (people, dogs, tape cases, hockey sticks, etc.) will continue to move according to Newton's First Law. In this chapter, we will be doing an analysis of the inertia of the objects inside a vehicle and how to prevent injury.

## *Active*-ating the Physics InfoMall

The articles mentioned above are good for this activity. One of the safety devices mentioned in this activity is the air bag. A quick search for "air bags" found several interesting hits, including "Resource letter PE-1: Physics and the environment, *American Journal of Physics*, vol. 42, 267-273 (1974).

# Planning for the Activity

## Time Requirements

This activity is centered primarily on reading and answering the questions in **For You To Do**. After the students have answered the questions, there should be a class discussion, to investigate the understanding students have now about accidents and to help the students to open their minds to the seriousness of accidents. As the majority of students are entering the most dangerous time of their lives, in terms of learning to drive, and being involved in accidents, the discussion should be serious. Allow more time if a film and a discussion are added to the lesson.

## Materials Needed

### For each group:

- brochures on auto safety features

# Teaching Notes

If students have a tough time with the discussion from **What Do You Think?** you may try using brainstorming on these activities. See **Assessment** for a possible rubric for this activity.

While discussing accidents, be aware that some students may have already been in a serious accident, or know of someone close to them, who has been injured or killed in an accident. Be sensitive to the student who sits quietly and doesn't want to participate in the discussion.

# Activity Overview

This activity centers around the students own experiences with safety in transportation. The students will be exploring their own ideas, misconceptions and evaluating quantitatively their ideas in the "test".

## Student Objectives

### Students will:

- Evaluate their own understandings of safety.
- Evaluate the safety features on selected vehicles.
- Compare and contrast the safety features on selected vehicles.
- Identify safety features in selected vehicles.
- Identify safety features required for other modes of transportation (in-line skates, skate boards, cycling, etc.).

Safety

## Activity 3 Accidents

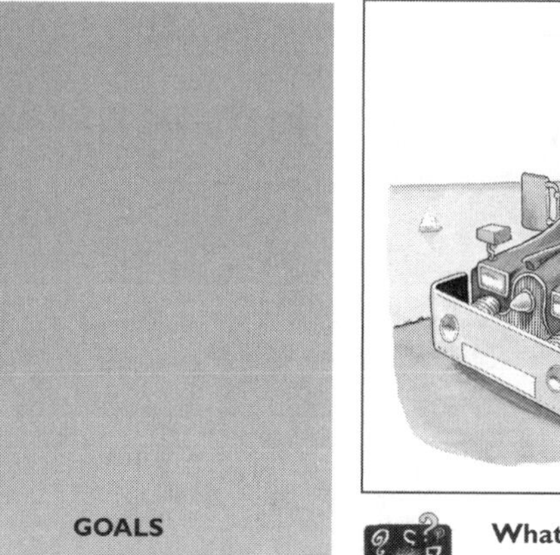

**GOALS**

In this activity you will:

- Evaluate your own understandings of safety.
- Evaluate the safety features on selected vehicles.
- Compare and contrast the safety features on selected vehicles.
- Identify safety features in selected vehicles.
- Identify safety features required for other modes of transportation (in-line skates, skateboards, cycling, etc.).

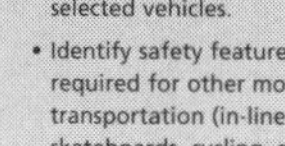

**What Do You Think?**

Chances are you will not be able to avoid being in an accident at some time in the future.

- **How can you protect yourself from serious injury, or even death, should an accident occur?**
- **What do you think is the greatest danger to you or the people in an accident?**

Record your ideas about these questions in your *Active Physics* log. Be prepared to discuss your responses with your small group and the class.

Active Physics 96

ANSWERS FOR THE TEACHER ONLY

## What Do You Think?

Students answers will vary. Most students will come up with the obvious – seat belts, air bags, roll cages, etc. However, some students may come up with ideas, which will convey their misconceptions of accidents, such as, bend over, and get into a ball, hang on to the dashboard.

Again, students will relay the most obvious, flying through the windshield, automobile rolling over you if you get thrown out, etc.; be prepared for answers such as being burned in the automobile, the automobile explodes on impact; look for sensible answers, that will steer the students toward an understanding of the physics, i.e., inertia, forces, momentum, without having to talk about the details of the "physics". Some of the answers which students may not think of might include: other people in the automobile flying around during the collision, being knocked unconscious as you enter a lake or slough, objects not tied or secured flying around inside the automobile.

Try not to limit the discussion to only cars and trucks. Expand if appropriate to motorcycles, snowmobiles, motorized tricycles and quads. Also, if it is brought up by the students, a discussion about the safety of bicycles could develop.

### For You To Do

1. Many people think that they know the risks involved with day-to-day transportation. The "test" below will check your knowledge of automobile accidents. The statements are organized in a true and false format. Record a T in your log for each statement you believe is true and an F if you believe the statement is false. Your teacher will supply the correct answers for discussion at the end of the activity.

a) More people die because of cancer than automobile accidents.

b) Your chances of surviving a collision improve if you are thrown from the car.

c) The fatality rate of motorcycle accidents is less than that of cars.

d) A large number of people who are belted into their cars are killed in a burning or submerged car.

e) If you don't have a child restraint seat, you should place the child in your seat belt with you.

f) You can react fast enough during an accident to brace yourself in the car seat.

g) Most people die in traffic accidents during long trips.

h) A person not wearing a seat belt in your car poses a hazard to you.

i) Traffic accidents occur most often on Monday mornings.

j) Male drivers between the ages of 16 and 19 are most likely to be involved in traffic accidents.

k) Casualty collisions are most frequent during the winter months due to snow and ice.

l) More pedestrians than drivers are killed by cars.

m) The greatest number of roadway fatalities can be attributed to poor driving conditions.

n) The greatest number of females involved in traffic accidents are between the ages of 16 and 20.

o) Unrestrained occupant casualties are more likely to be young adults between the ages of 16 and 19.



Answers

## For You To Do

1. a) False, more people die because of automobile accidents.

   b) False, 30% of the people who die in car crashes die because they are thrown. One report indicates that at least 50% of these fatalities need not have occurred.

   c) False, motorcycles are less common, but they are much more dangerous.

   d) False, very few people are killed because of burning or drowning. One study indicates that it is less than 0.1%.

   e) False, the child can be sandwiched by your mass. The abdomen of small children is particularly vulnerable to impact.

   f) False, you don't have enough time to react even at 50 km/h.

   g) False, most people die within 40 kilometers of home.

   h) True, the person becomes a projectile once the car stops.

   i) False, most traffic accidents occur on Friday afternoon or evening traffic.

   j) True, that accounts for the higher insurance rates.

   k) False, the months May, June, and August have the greatest number of casualty collisions.

   l) False, pedestrians only account for 6.1% of traffic fatalities.

   m) False, more accidents take place when road conditions are dry.

   n) True, young female drivers, like their male counterparts, are responsible for the greatest number of accidents. Inexperience may be the major contributing factor.

   o) True.

2. Calculate your score. Give yourself two points for a correct answer, and subtract one point for an incorrect answer. You might want to match your score against the descriptors given below.

   21 to 30 points: Expert Analyst

   14 to 20 points: Assistant Analyst

   9 to 13 points: Novice Analyst

   8 points and below: Myth Believer

   a) Record your score in your log. Were you surprised about the extent of your knowledge? Some of the reasons behind these facts will be better understood as you continue to travel through this chapter.

Obtain permission from the cars' owners before proceeding.

3. Survey at least three different cars for safety features. The list on the next page will allow you to evaluate the safety features of each of the cars. Place a check mark in the appropriate square.

   Number 1 indicates very poor or nonexistent, 2 is minimum standard, 3 is average, 4 is good, and 5 is very good.

   For example, when rating air bags: a car with no air bags could be given a 1 rating, a car with only a driver-side air bag a 2, a car with driver and passenger side air bags a 3, a car with slow release driver and passenger-side air bags a 4, and a car which includes side-door air bags to the previous list a 5. You may add additional safety features not identified in the chart. Many additional features can be added!

   a) Copy and complete the table in your log.

   b) Which car would you evaluate as being safest?

ANSWERS

## For You To Do *(continued)*

2. a) Students scores will vary. A short discussion can follow with the expectation that the students will be able to better understand the physics of the accidents after the next few chapters.

3. a) Students should be able to come up with at least five more safety features. These might include side impact beams in doors, shoulder belts for all outboard seats, laminated windshield glass, ABS brakes, tempered shatterproof glass, etc.

   b) Students answers will vary. Look for reasonable arguments for their evaluation. They should include factors such as speed, forces on the body, analysis of the effect on the body from different kinds of collisions.

| Car Tested: Make and Model _____ | Year _____ | | | | |
|---|---|---|---|---|---|
| Safety Feature | Rating | | | | |
| Padded front seats | 1 | 2 | 3 | 4 | 5 |
| Padded roof frame | 1 | 2 | 3 | 4 | 5 |
| Head rests | 1 | 2 | 3 | 4 | 5 |
| Knee protection | 1 | 2 | 3 | 4 | 5 |
| Anti-daze rear-view mirror that brakes on impact | 1 | 2 | 3 | 4 | 5 |
| Child proof safety locks on rear doors | 1 | 2 | 3 | 4 | 5 |
| Padded console | 1 | 2 | 3 | 4 | 5 |
| Padded sun visor | 1 | 2 | 3 | 4 | 5 |
| Padded doors and arm rests | 1 | 2 | 3 | 4 | 5 |
| Steering wheel with padded rim and hub | 1 | 2 | 3 | 4 | 5 |
| Padded gear level | | | | | |
| Padded door pillars | | | | | |
| Air bags | | | | | |

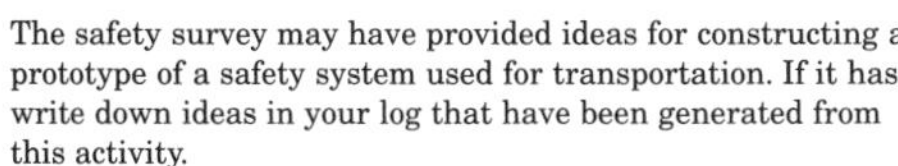

## Reflecting on the Activity and the Challenge

Serious injuries in an automobile accident have many causes. If there are no restraints or safety devices in a vehicle, or if the vehicle is not constructed to absorb any of the energy of the collision, even a minor collision can cause serious injury. Until the early 1960s, automobile design and construction did not even consider passenger safety. The general belief was that a heavy car was a safe car. While there is some truth to that statement, today's lighter cars are far safer than the "tanks" of the past.

The safety survey may have provided ideas for constructing a prototype of a safety system used for transportation. If it has, write down ideas in your log that have been generated from this activity.

ANSWERS

## Physics To Go

1. Looking at the chart from page 97, these are possible answers-
   - Padded front seats F, R
   - Padded roof frame T
   - Head rests F, R
   - Knee protection F, S
   - Anti-daze rear-view mirror F, S
   - Child-proof safety locks
   - Padded console F, R, S, T
   - Padded sun visor F, S, T
   - Padded doors and arm rests S
   - Steering wheel-padded F
   - Padded gear level F, S
   - Padded door pillars S, T
   - Air bags F, S, T

2. Safety features for cycling could be: padded handle bars, better brakes, padded seat and top bar, lots of lighting, and reflectors, safety flags for better visibility, safety training, like defensive driving, helmet, padding for knees, hands and elbows and shoulders.

3. Many of the same for bikes, but emphasis on padding and helmets.

4. As above, with an emphasis on padding again.

5. Students answers will vary. Look for a thorough evaluation, using the previous list, and emphasize to the students that the discussion with the owner of the vehicle is as important as the evaluation.

### Physics To Go

1. Review and list all the safety features found in today's new cars. As you compile your list, write next to each safety feature one or more of the following designations:

   F: effective in a front-end collision.

   R: effective in a rear-end collision.

   S: effective in a collision where the car is struck on the side.

   T: effective when the car rolls over or turns over onto its roof.

2. Make a list of safety features that could be used for cycling.

3. Make a list of safety features that could be used for in-line skating.

4. Make a list of safety features that could be used for skateboarding.

5. Ask family members or friends if you may evaluate their car. Discuss and explain your evaluation to the car owners. Record your evaluation and their response in your log.

### Stretching Exercises

1. Read a consumer report on car safety. Are any cars on the road particularly unsafe?

2. Collect brochures from various automobile dealers. What new safety features are presented in the brochures? How much of the advertising is devoted to safety?

ANSWERS

## Stretching Exercises

1. There may be many cars on the road that have elements which are unsafe. Have the students evaluate which ones constitute true safety breaches, and which are more along cosmetic lines. For example, electronic fuel filters which may ignite, or dashboards which spontaneously start on fire, or brakes which fail would constitute major safety hazards. Radios that fail or brakes that squeal when stopping, or cars that prematurely rust, are more cosmetic, and don't constitute a safety hazard.

2. Students answers will vary. Students will likely come up with the observation that a great deal of advertising has to do with driver safety.

# Assessment: Participation in Discussion

The following is an **Assessment Rubric**, designed for informal feedback to the students on their discussions related to their personal experiences with safety in a vehicle. While they are discussing this, the students may want to brainstorm on some of the safety features with which they are familiar.



| **Descriptor** | **most of the time** | **some of the time** | **almost never** | **comments** |
|---|---|---|---|---|
| shows interest | | | | |
| stays on task | | | | |
| asks questions related to topic | | | | |
| listens to other students' ideas | | | | |
| shows cooperation in group brainstorming | | | | |
| provides leadership in group activity | | | | |
| demonstrates tolerance of others' viewpoints | | | | |

For use with *Safety*, Chapter 2, Activity 3: Accidents

# ACTIVITY 4
## Life (and Death) before Seat Belts

# Background Information

Newton's First Law of Motion (Inertia) gives us an understanding of constant motion (or rest). Therefore, when an object such as a car is stopping, there is a change in velocity (previously introduced as speed). In other words there is an acceleration. Analyzing a collision involves examining the changes in velocity of a car. The crumple zone refers to the stopping distance as the car is pushed in, while stopping. Cars designed today are built with crumple zones. This is to increase the distance a car takes to stop, therefore reducing the acceleration and ultimately the force being applied to your body.

Increasing the crumple zone will affect the acceleration of the people inside the car, initially using the average velocity ($v_{ave} = d/t$). Average velocity is the sum of the two velocities divided by 2

$v_{ave} = (v_1 + v_2)/2.$

$v_{ave} = (40 \text{ m/s} + 0 \text{ m/s})/2$

$v_{ave} = 20 \text{ m/s}$

Find the time required to stop (crumple zone of the vehicle),

$t = d/ v_{ave}$

$t = 0.50 \text{ s}/20 \text{ m/s}$

$t = 0.025 \text{ s}$

and substitute $t$ into the equation for acceleration

$a = \Delta v/\Delta t$

$a = 40 \text{ m/s} / 0.025 \text{ s}$

$a = 1600 \text{ m/s}^2$

This is clearly enough to rip out the aorta.

Another way to emphasize the impact of this force, is that a 10 m/s$^2$ acceleration is roughly equivalent to holding 1 kg (2.2 pounds) mass (you need 10 N of force to lift 1 kg of mass). Therefore, if we look at the above acceleration, you could picture a 160 kg (352 pound) person being held up by the aorta...ouch that would hurt!

This model should help students realize that the external safety devices we use help, but there is not very much we can do to improve on the safety devices that are built into our own bodies.

Factors which may enter into the discussion, regarding race car drivers traveling at 200 mph, may be the fact that most race car drivers are in very good physical shape, and are generally younger.

One other way to help the students understand that the internal organs undergo acceleration, is how they feel while driving in a car over a hilly road. The feeling often expressed is that of your stomach rising and falling at the crests and valleys.

In the **FYTD** activity, the students will be releasing the carts at different heights on the ramp. If the carts experience an almost frictionless surface, the speed at which the cart hits the barrier can be determined using the height from which the cart is released.

Conservation of Energy Theory (First Law of Thermodynamics) states that energy can be neither created nor destroyed, only transferred from one form to another. Therefore, the energy at the top of the ramp (gravitational energy) $E_p = mgh$ (where $m$ is the mass of the cart and clay, $g$ is gravity (9.81 m/s$^2$) and $h$ is the height from which the cart is released) will be the same as the kinetic energy ($E_k = 1/2\, mv^2$ where $m$ is the mass, $v$ is the velocity) at the bottom.

$E_p = E_k$

$mgh = 1/2mv^2$

masses cancel out

$gh = 1/2v^2$

$v = \sqrt{2gh}$

For example: the cart is released at a height of 0.50 m. Therefore the speed of the cart as it reaches the bottom of the ramp would be

$v = \sqrt{2gh}$

$v = \sqrt{2 \text{ x } 9.81 \text{ m/s}^2 \text{ x } 0.50 \text{ m}}$

$v = 3.1 \text{ m/s}$

## *Active*-ating the Physics InfoMall

The title of this activity leads naturally to a search for "seat belt*". The very first result from this search is from *The Fascination of Physics*, in the Textbook Trove, which says "Automobile accidents involve two collisions. The first occurs when the automobile strikes an object, such as a telephone pole. The pole provides the force needed to change the car's momentum, eventually bringing it to rest. A second collision, which occurs shortly after the first, involves the passengers. If they are not in some way attached to the car, the passengers do not experience the force exerted by the telephone pole. The car may stop, but the passengers continue moving forward at a constant velocity. According to Newton's First Law, their forward motion will continue until they experience a force. Unfortunately, this force is usually exerted by the dashboard or windshield, and serious injuries result." This can be compared to the **For You To Read** passage, which goes a little further by looking at three collisions. You should do this search and read the results.

The second hit from this search is the epilogue to *The Fascination of Physics*, and mentions that "The benefits of not using seat belts are the saving of a few second in buckling and unbuckling, a slight increase in the freedom of movement inside a car, and some psychological or emotional benefits that are difficult to define. The benefits and the belief that the probability of an accident is small convince most people to sit on top of their seat belts." This applies directly to **Question 2** of the **Stretching Exercise**.

A little further down the list from this "seat belt*" search is an article that you might not otherwise find - "The car, the soft drink can, and the brick wall," from *The Physics Teacher*, vol. 13, (1975). This article was written by a co-PI for the InfoMall. It describes an experiment very similar to this activity, but with some variations.

A big concept in this activity is the concept of force. Students' understanding of this concept has been studied extensively. An InfoMall search using "force" AND "misconception*" in only the Articles and Abstracts Attic produced many great references. The first such hit is the article containing the Force Concept Inventory. The second is "Common sense concepts about motion," *American Journal of Physics*, vol. 53, issue 11, 1985 in which it is mentioned that "(a) On the pretest (post-test), 47% (20%) of the students showed, at least once, a belief that under no net force, an object slows down. However, only 1% (0%) maintained that belief across similar tasks. (b) About 66% (54%) of the students held, at least once, the belief that under a constant force an object moves at constant speed. However, only 2% (1%) held that belief consistently." More results are reported in this article.

The third hit in this search is "Physics that textbook writers usually get wrong," in *The Physics Teacher*, vol. 30, issue 7, 1992. This article is good reading for any introductory physics teacher. The list of hits from this search is long. In fact, it had to be limited to just the Articles and Abstracts Attic to prevent the "Too many hits" warning. If you search the rest of the CD-ROM, you will find many other great hits, such as this quote from Chapter 3 of Arons' *A Guide to Introductory Physics Teaching: Elementary Dynamics*: "In the study of physics, the Law of Inertia and the concept of force have, historically, been two of the most formidable stumbling blocks for students, and, as of the present time, more cognitive research has been done in this area than in any other."

Speed is one of the first concepts introduced in introductory physics. It is also one that causes students problems. If you search the InfoMall for "student difficult*" OR "student understand*" you will find several articles that deal with research into how students learn fundamental concepts in physics. Some of these are "Investigation of student understanding of the concept of velocity in one dimension," *American Journal of Physics*, vol. 48, issue 12 (see "Diagnosis and remediation of an alternative conception of velocity using a microcomputer program," *American Journal of Physics*, vol. 53, issue 7 for a discussion of this); "Research and computer-based instruction: Opportunity for interaction," *American Journal of Physics*, vol. 58, issue 5 (if you look at other articles in this volume, you will find "Learning motion concepts using real-time microcomputer-based laboratory tools," *American Journal of Physics*, vol. 58, issue 9); and more.

These last two articles make specific mention of the use of computers in teaching physics. This is a trend that is gaining strength and shows great promise. Look for more such articles in journals that are newer than the CD-ROM. As a starting point, you can always look in the annual indices of the physics journals and look under the names of the authors of the articles mentioned above.

Note that **Physics To Go Step 1** mentions curved motion. This topic was discussed elswhere in this book, where we found that you can give students a feel for curved motion with *The Physics Teacher*, volume 21, issue 3, "People Demos." Curved motion often brings up "centrifugal force," which is discussed in *Physics Education* (in the Articles and Abstracts Attic), issue 3, "Centrifugal force: fact or fiction," by Michael D. Savage and Julian S. Williams. See also Robert P. Bauman, "What is centrifugal force?," *The Physics Teacher*, vol. 18, number 7. Another good reference for centripetal and centrifugal forces is Arnold Arons' book (found in the Book Basement) *A Guide to Introductory Physics Teaching*, Motion in Two Dimensions, sections 4.9 to 4.11. These can all be found with simple searches.

# Planning for the Activity

## Time Requirements

- At least one class period (40 – 50 minutes). Allow extra time for variations on their molded clay figures.

## Materials Needed

### For each group:

- camcorder on tripod
- clay, modeling, 1/2 lb.
- dynamics carts
- starting ramp for lab cart
- VCR having single frame advance mode
- video monitor

## Advance Preparation and Setup

This experiment requires students to crash a loaded cart into a wall, or other suitable barrier. Test out different barriers (walls, desks, bricks, homemade structures) prior to laboratory day.

# Teaching Notes

You might want to show the *Active Physics Transportation* Content Video showing collisions in which a dummy is thrown forward.

The students begin by forming a clay figure of a body with a relatively large head. With the figure seated on the lab cart, allow the cart to be released from various heights or various angles on the ramp. This will simulate the vehicle crashing into a barrier at various speeds. At high speeds, the figure should crash into the barrier head-first. At low speeds, the figure should topple head first, smashing its head into the "dashboard."

Some students may wish to analyze different scenarios. Have the students submit their plans for teacher approval, then allow them to carry out their plans. An example may be to release the cart backwards to demonstrate the rear-end collisions (to be studied later).

Be aware that the ramp needs to be secure on a desk, or the floor as the cart may fall onto the floor or on someone's toes! Remind the students to place down newspaper or some drop cloths while creating their figure.

Students may have the notion that if they were in an accident if they got their arms up in time they would be able to protect themselves. Review with them reaction times, and have them try to bring their hands up to the dashboard level in that time. To further emphasize the danger in this thinking refer to the **Additional Activities A** following this activity in the Teacher's Edition.

# Activity 4 Life (and Death) before Seat Belts

**GOALS**

In this activity you will:

- Understand Newton's First Law of Motion.
- Understand the role of safety belts.
- Identify the three collisions in every accident.

### What Do You Think?

Throughout most of the country, the law requires automobile passengers to wear seat belts.

- **Should wearing a seat belt be a personal choice?**
- **What are two reasons why there should be seat belt laws and two reasons why there should not?**

Record your ideas about these questions in your *Active Physics* log. Be prepared to discuss your responses with your small group and the class.

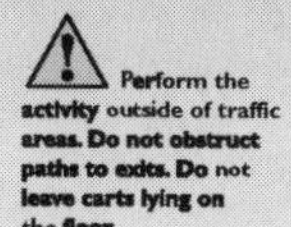

**Perform the activity outside of traffic areas. Do not obstruct paths to exits. Do not leave carts lying on the floor.**

### For You To Do

1. In this activity, you will investigate car crashes where the driver or passenger does not wear seat belts. Your model car is a laboratory cart. Your model passenger is molded from a lump of soft clay. With the "passenger" in place, send the "car" at a low speed into a wall.

## Activity Overview

This activity is an investigation into the inertia of an automobile passenger and will lead directly toward the necessity of a restraint system.

## Student Objectives

### Students will:

- Understand Newton's First Law of Motion.
- Understand the role of safety belts.
- Identify the three collisions in every accident.



ANSWERS FOR THE TEACHER ONLY

## What Do You Think?

Some students will think it should and some will think it not. It is an excellent way to open discussion

Some reasons for wearing might include: safety, decrease the risk of death, decrease the risk of serious injury, lower the total cost of health insurance due to decrease in serious accidents.

Some of the reasons for not wearing a seat belt might include: the right to choose to wear or not, might get trapped in the car, the seat belt will do more damage than the accident.

Discussion should try to center around the effects of the accident rather than a debate on the issue of personal choice. Try to bring in statistics from your local police or automobile association to emphasize the risk of injury or death will always decrease over many different accidents. There are always exceptions to the rule, but the facts are that wearing seat belts will decrease the likelihood of serious injury or death.

 a) Describe, in your log, what happens to the "passenger."

2. Repeat the collision at a high speed. Compare and contrast this collision with the previous one.

a) Compare and contrast requires you to find and record at least one similarity and one difference. A better response includes more similarities and differences.

3. You can conduct a more analytical experiment by having the cart hit the wall at varying speeds. Set up a ramp on which the car can travel. Release the car on the ramp and observe as it crashes into the wall. Repeat the collision for at least two ramp heights.

a) Record the heights of the ramp and describe the results of the collision. Describe the collision by noting the damage to the "passenger."

**Physics Words**

**Newton's First Law of Motion:** an object at rest stays at rest and an object in motion stays in motion unless acted upon by an unbalanced, external force.

**inertia:** the natural tendency of an object to remain at rest or to remain moving with constant speed in a straight line.

**PHYSICS TALK**

**Newton's First Law of Motion**

**Newton's First Law of Motion** (also called the Law of **Inertia**) is one of the foundations of physics. It states:

**An object at rest stays at rest, and an object in motion stays in motion unless acted upon by a net external force.**

There are three distinct parts to Newton's First Law.

Part 1 says that objects at rest stay at rest. This hardly seems surprising.

Part 2 says that objects in motion stay in motion. This may seem strange indeed. After looking at the collisions of this activity, this should seem clearer.

Part 3 says that Parts 1 and 2 are only true when no force is present.



ANSWERS

## For You To Do *(continued)*

2. a) The students should note that at a slower speed, the clay model will fall head-first into the front of the cart, or hit the wall with its head. Difference at a high speed would be that the figure will continue moving at approximately the same speed as the cart was moving before the accident. The difference as to why the figure will not continue at exactly the same speed, or that at higher speeds it is more pronounced, is that there is a certain amount of friction between the figure, and the cart. It should be noted also that the greater the speed, the greater damage there is to the figure.

3. a) The higher the ramp, the faster the speed of the cart down the ramp, and the faster the figure will crash into the wall. Therefore, there should continue to be greater damage as the ramp gets higher and higher.

## FOR YOU TO READ

### Three Collisions in One Accident!

Arthur C. Damask analyzes automobile accidents and deaths for insurance companies and police reports. This is how Professor Damask describes an accident:

Consider the occupants of a conveyance moving at some speed. If the conveyance strikes an object, it will rapidly decelerate to some lower speed or stop entirely; this is called the first collision. But the occupants have been moving at the same speed, and will continue to do so until they are stopped by striking the interior parts of the car (if not ejected); this is the second collision. The brain and body organs have also been moving at the same speed and will continue to do so until they are stopped by colliding with the shell of the body, i.e., the interior of the skull, the thoracic cavity, and the abdominal wall. This is called the third collision.

Newton's First Law of Motion explains the three collisions:

- First collision: the car strikes the pole; the pole exerts the force that brings the car to rest.
- Second collision: when the car stops, the body keeps moving; the structure of the car exerts the force that brings the body to rest.
- Third collision: the body stops, but the heart and brain keep moving; the body wall exerts the force that brings the heart and brain to rest.

Even with all the safety features in our automobiles, some deaths cannot be prevented. In one accident, only a single car was involved, with only the driver inside. The car failed to follow the road around a turn, and it struck a telephone pole. The seat belt and the air bag prevented any serious injuries apart from a few bruises, but the driver died. An autopsy showed that the driver's aorta had burst, at the point where it leaves the heart.

Chapter 2

ANSWERS

## Physics To Go

1. • *You step on the brakes to stop your car.*

   You and the car are moving forward. The brakes apply a force to the tires and stop them from rotating. Newton's law states that an object in motion will remain in motion unless a force acts upon it. In this case, the force is friction between the ground and the tires. You remain in motion since the force that stopped the car did not stop you.

   • *You step on the accelerator to get going.*

   You and the car are stopped. The engine provides a force to turn the wheels, which in turn causes the car to move forward. Inertia will keep you still unless a force acts upon you. This force is provided by the seat back which pushes you forward at the same rate as the car.)

   • *You turn the wheel to go around a curve.*

   You are moving forward with the car. Your force causes the wheels to turn, the friction of the road on the tires produces the centripetal force necessary to turn the vehicle. Inertia causes you to remain moving in a straight line, where the car (doors, seat belt, seat, etc.) produce the centripetal force to allow you to stay in the car, and move in the curved path.

   • *You step on the brakes, and an object in the back of the car comes flying forward.*

   The object was moving with the same velocity as the car. The force which caused the car to stop, was not acting on the object. Inertia of the object kept the object moving in a straight path (until it hits the driver or the windshield).

Safety

### Reflecting on the Activity and the Challenge

In this activity you discovered that an object in motion continues in motion until a force stops it. A car will stop when it hits a pole but the passenger will keep on moving. If the car and passenger have a large speed, then the passenger will continue moving with this large speed. The passenger at the large speed will experience more damage from the fast-moving cart.

Have you ever heard someone say that they can prevent an injury by bracing themselves against the crash? They can't! Restraining devices help provide support. Without a restraining system, the force of impact is either absorbed by the rib, skull, or brain.

Use Newton's First Law of Motion to describe your design. How will your safety system protect passengers from low speed and higher speed collisions?

### Physics To Go

1. Describe how Newton's First Law applies to the following situations:
   - You step on the brakes to stop your car.

   (Sample answer: You and the car are moving forward. The brakes apply a force to the tires and stop them from rotating. Newton's law states that an object in motion will remain in motion unless a force acts upon it. In this case, the force is friction between the ground and the tires. You remain in motion since the force that stopped the car did not stop you.)

   - You step on the accelerator to get going.
   - You turn the wheel to go around a curve. (Hint: You keep moving in a straight line.)
   - You step on the brakes, and an object in the back of the car comes flying forward.

104

Active Physics

2. Give two more examples of how Newton's First Law applies to vehicles or people in motion.
3. According to Newton's First Law, objects in motion will continue in motion unless acted upon by a force. Using Newton's First Law, explain why a cart that rolls down a ramp eventually comes to rest.
4. The skateboard, shown in the picture to the right, strikes the curb. Draw a diagram indicating the direction in which the person moves. Use Newton's First Law to explain the direction of movement.
5. Explain, in your own words, the three collisions during a single crash as described by Professor Damask in **For You To Read**.
6. Use the diagrams below to compare the second and third collisions described by Professor Damask with the impact of a punch during a boxing match.

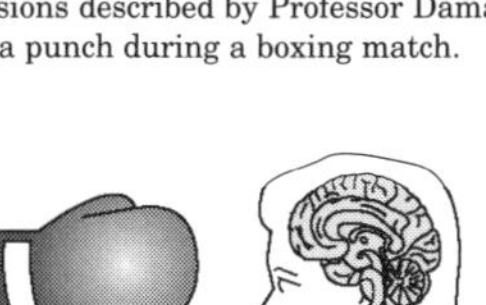

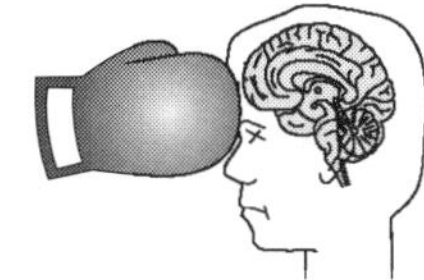

7. When was the law instituted requiring drivers to wear seat belts?

ANSWERS

## Physics To Go
***(continued)***

2. Students' answers will vary. Look for sound physics when describing the situation. Some might include, the sensation of getting "heavier" or "lighter" as an elevator starts up or down, the problem of stopping or rounding a curve on ice, being slammed into the seat of a bus, as it accelerates from the stop.
3. Friction between the cart and the wheels, wheels and ground, and to a very small extent the air friction.
4. Inertia states that the person will continue to move in the same direction as the skateboard, thus indicating the need for helmet, knee pads, and gloves!
5. Answers will vary; first: car hitting tree; second: body hitting car; third: internal organs hitting the internal wall of thoracic cavity, or brain hitting internal skull.
6. First collision: glove hitting head; second collision: head moves backward, and the brain collides with the interior of the skull; third collision: elasticity of the brain attached to the brain stem, causes the brain to move toward the back of the brain colliding with the internal back of the skull.
7. Answers will vary, depending on the state. Forty-nine states and the District of Columbia have mandatory seat-belt laws. In some states, seat-belt laws date back to 1985.

Safety

### Stretching Exercises

1. Determine what opinions people in your community hold about the wearing of seat belts. Compare the opinions of the 60+ years old and 25 to 59 year old groups with that of the 15 to 24 year old group. Survey at least five people in each age group: Group A = 15 to 24 years, Group B = 25 to 59 years, and Group C = 60 years and older. (Survey the same number of individuals in each age group.) Ask each individual to fill out a survey card.

A sample questionnaire is provided below. You may wish to eliminate any question that you feel is not relevant. You are encouraged to develop questions of your own that help you understand what attitudes people in your community hold about wearing seat belts. The answers have been divided into three categories: 1 = agree; 2 = will accept, but do not hold a strong opinion; and 3 = disagree. Try to keep your survey to between five and ten questions.

| Age group: | Date of Survey: | | |
|---|---|---|---|
| Statement | Agree | No strong opinion | Disagree |
| 1. I believe people should be fined for not wearing seat belts. | 1 | 2 | 3 |
| 2. I wouldn't wear a seat belt if I didn't have to. | 1 | 2 | 3 |
| 3. People who don't wear seat belts pose a threat to me when they ride in my car. | 1 | 2 | 3 |
| 4. I believe that seat belts save lives. | 1 | 2 | 3 |
| 5. Seat belts wrinkle my clothes and fit poorly so I don't wear them. | 1 | 2 | 3 |

2. Make a list of reasons why people refuse to wear seat belts. Can you challenge these opinions using what you have learned about Newton's First Law of Motion?

106

## Activity 4 A

# Dropping a Clay Ball to Investigate Inertia

### FOR YOU TO DO

1. This could be a messy activity. Place newspaper or other drop cloth on the desks before building the clay balls. Form a ball using 1.0 kg of clay. Using a balance, measure the exact mass of the ball.

a) Record the mass of the ball in your log.

2. Drop the ball from a variety of heights into a hand wearing a baseball glove. Avoid trying to "cushion" the catch, which will ruin the effect of the force of impact. Look away while catching the ball, so you are better able to describe the force qualitatively.

a) In your log record the distance from which the ball was dropped.

b) Describe how the ball felt as it hit the glove when dropped from different heights.

3. The formula for finding the velocity of an object that is falling is:

$$v = \sqrt{2gh}$$

where $g = 9.81$ m/s$^2$,

$h$ is the height from which the ball is released to the glove.

From this formula you can determine the momentum ($m$ x $\Delta v$) of the object in order to show the increase in the change momentum. Determine the time by averaging the velocities $[(v_2 + v_1)/2]$, then using that $v$ to find the time in $t = d/v$. Now you can determine the force ($F = m\Delta v/\Delta t$) that is being exerted on their hands as the object is dropped from different heights.

a) Determine the force of the ball for the different heights.

Sample Calculation:

Mass of the ball = 1.0 kg; distance from glove to the desk = 1.5 m

- Velocity where $h = 1.5$ m

  $v = \sqrt{2gh}$

  $v = 5.42$ m/s

- Change in Momentum (as the ball goes from the final speed of 5.42 m/s to 0 m/s in the glove)

  $\Delta p = m\Delta v$

  $\Delta p = 5.42$ m/s

- Average velocity

  = (5.42 m/s+ 0 m/s)/2

  = 2.71 m/s

For use with *Safety*, Chapter 2, Activity 4: Life (and Death) before Seat Belts

- Time to stop the ball in 0.10 m

  $t = d/v$

  $t = 0.10$ m/ 2.71 m/s

  $t = 0.037$ s

- Force acting on the glove by the ball

  $F = m\Delta v/\Delta t$

  $F = (1.0$ kg 5.42 m/s)/0.037 s

  $F = 146$ N

Or about 15 kg (31 lb.) of mass (Every 10 N of force is approximately the equivalent of 1 kg of mass.)

b) What was the speed of the clay ball when dropped from 2.0 m?

c) Estimate what the force might be if the clay ball were dropped from the gym roof (approximate height of 10 m).

If the possibility exists, take the clay ball outside and drop it into a cloth held in a way similar to the firefighters net.

a) The force of an accident can be compared to the forces you felt when you were dropping the ball of clay. Relate your experiences of catching the ball, and how they might compare with the forces that the car and the tree exert on each other during their collision.

4. Make clay balls of various sizes (exact masses are not necessary).

5. Place the balls on the ends of chopsticks so that only the end of the chopsticks is showing, as shown.

6. With the bottom of the chopstick in hand, slam your hand down on the desk. Observe what happens.

a) Why did the clay slide down the chopstick?

b) What stopped the clay once it was moving?

c) What could you do to stop the clay from moving down the chopstick when you hit it on the desktop?

For use with *Safety*, Chapter 2, Activity 4: Life (and Death) before Seat Belts

Activity 4 B

# Life [and Death] before Seat Belts Part B: Low-Tech Alternative.

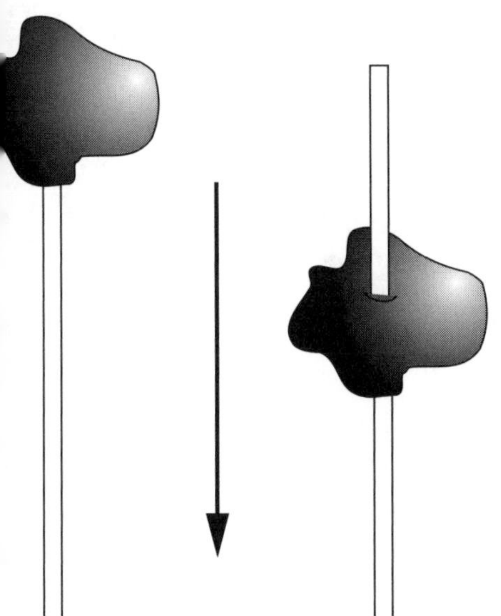

This is a low-tech activity to illustrate Newton's First Law. It is intended to give the students a firsthand look at inertia.



## Materials needed

- quantity of play dough or clay
- chopsticks (or similar sticks)

## Planning for the Activity

- Students will make clay balls of various sizes (exact masses are not necessary).
- Place the balls on the ends of the chopsticks so that only the end of the chopsticks is showing.
- With the bottom of the chopstick in hand, slam your hand down on the desk.
- Observe what happens

### Time Requirements

Approximately 30 minutes to complete this activity.

## Teaching Notes

As the mass of the ball increases, the students will want to think that the acceleration increases. While the ball appears to accelerate more as it increases its mass, the increases in the mass actually means there is an increase of the force (due to gravity) *Fg*.

## Classroom Management Tips

This could be a messy activity. Remind the students to place newspaper or other drop cloth on the desks before building the clay balls.

### Questions:

1. Why did the clay slide down the chopstick?
   A: Inertia of the clay kept the clay moving.

2. What stopped the clay once it was moving?
   A: The friction between the chopstick and the clay?

3. What could you do to stop the clay from moving down the chopstick, when you hit it on the desktop?
   A. Answers will vary. Look for something like seat belts or stops of some kind on the chopstick. Some may say change the friction between the clay and the chopstick (fire it in an oven to harden it), but we are looking for an application to be able to relate it to a moving car.

For use with *Safety*, Chapter 2, Activity 4: Life (and Death) before Seat Belts

# ACTIVITY 5
## Life (and Fewer Deaths) after Seat Belts

# Background Information

Inertia is the tendency for an object to stay in motion. In order to stop that motion, there must be a force applied to that object in order to accelerate the object or change its velocity to 0 m/s ($a = \Delta v/\Delta t = v_2 - v_1/\Delta t$). For example, for an object moving at 3.0 m/s, is brought to 0 m/s in 1.0 s

$$a = \Delta v/\Delta t$$

$$a = v_2 - v_1/\Delta t$$

$$a = (0 \text{ m/s} - 3.0 \text{ m/s}) / 1.0 \text{ s}$$

$$a = -3.0 \text{ m/s}^2$$

Note that this is a negative acceleration, which we often refer to as deceleration. For the purposes of this course, we will refer to slowing down and speeding up as acceleration. There will be references to negative and positive acceleration.

Newton's Second Law of Motion gives a mathematical understanding of forces and acceleration – i.e.: changing velocity.

Symbols
$v$ = velocity (m/s)
$d$ = distance (m)
$t$ = time (s)
$a$ = acceleration (m/s)
$F$ = force (N)
$m$ = mass (kg)
$\Delta$ = change

Newton's Second Law states that the acceleration of an object is proportional to the force in the same direction, and inversely proportional to the mass.

$$a \quad F$$

$$a \quad 1/m$$

Therefore we can state this proportion as $a = F/m$, or as it is commonly known as $F = ma$.

If you have a wooden crate with nothing in it, it is very easy to push (apply a force to) the box. The empty crate is also very easy (or requires a smaller force) to stop. However, increasing the mass in the box, it becomes increasing difficult to push the box. This is caused by the increase in the inertia of the box. As well, it is very difficult (or requires a larger force) to stop the box.

Newton, after showing that forces cause objects to move (accelerate) asked himself where do these forces come from. This led to his third law, where forces come from other objects. Your hand pushing on a desk exerts a force on the desk. You can see, also that there must be a force acting on your hand because you can see the deformation of the "dent". Also, by applying more force, you realize that it hurts. Therefore, the only conclusion can be that the desk must be exerting a force on your hand. Newton's Third Law of Motion states that whenever an objects exerts a force on another object, the second object exerts a force equal in magnitude, but opposite in direction, on the first object. These are dealt with as action-reaction pairs. There can never be one force on an object without an equal but opposite force on the other

Therefore, your inertia will keep you moving in a straight line unless a force stops your motion. The force comes when your body comes in contact with the seat belt. This net force is enough to accelerate you to a stop.

In the previous activity the students examined the acceleration of a vehicle while stopping in a collision. In this activity, the students are manipulating the restraining device to give different pressures, without changing the forces. The physics involved with this activity is the understanding of how the pressure changes as you increase the area of the restraining device. The pressure is the force per unit area of the contact with the body. They will discover that the greater the area of contact, the more the force is spread out over the body.

$$P = F/A.$$

Therefore, even if the force is great (due to a large acceleration), the pressure on the individual will be much less.

## *Active*-ating the Physics InfoMall

This activity has much in common with **Activity 4**. So perform searches that will find something new, and perhaps interesting. For example, see what the InfoMall says about crash dummies. One method is to search for "crash"" AND "dumm*", where we are using AND to mean the logical operation (returns hits that contains both search keywords) and the asterisk (wild character). The first hit is "Forensic physics of vehicle accidents," *Physics Today*, vol. 40, issue 3, (1987). You may wish to look this one up yourself.

Another suggestion is to search for "bed of nails". Try it out.

And if you want more problems for your students, don't forget the Problems Place! You can easily find many, many problems about pressure.

# Planning for the Activity

## Time Requirements

Allow approximately 40 minutes to complete this activity and record the data.

## Materials Needed

### For each group:

- clay, modeling, 1/2 lb.
- copper, bare #22 wire, 1ft.
- dynamics carts
- ribbon, various widths
- starting ramp for lab cart

### For the class:

- *Active Physics Transportation* Content Video (Segment: Crash Dummy)
- VCR & TV monitor

## Advance Preparation and Setup

Prepare as you did for the previous activity. Have several different types of material, as well as large quantities of each available.

# Teaching Notes

Students find that the wire will cut more deeply, and that more, thicker supports will do the least damage. In debriefing, emphasis should still be on the fact that even at high speeds, three-point seat belts may still not save lives. There is lots of discussion that can come from this, especially when referring back to the **Chapter Challenge**. Again, the emphasis will be on decreasing the force that is exerted on the body by spreading out the force which occurs while the vehicle is stopping. The same force will occur if the body is stopped. The idea of having fatter seat belts is that the greater surface area of the belt allows for a smaller pressure.

As with any activity, a review of proper decorum in the classroom should be warranted. Students may use the clay to throw around the room. If the students are to complete the survey (**Question #7** in **Physics To Go**) they need to have permission to survey other students, or parents. This is a good activity to have the students survey more than immediately around the school. Give them the assignment of surveying some members of their community.

Please note that students may still have the misconception that the forces exerted on the seat belts are not that significant.

NOTES

# Activity 5 Life (and Fewer Deaths) after Seat Belts

**GOALS**

In this activity you will:

- Understand the role of safety belts.
- Compare the effectiveness of various wide and narrow belts.
- Relate pressure, force and area.

## What Do You Think?

In a collision, you cannot brace yourself and prevent injuries. Your arms and legs are not strong enough to overcome the inertia during even a minor collision. Instead of thinking about stopping yourself when the car is going 30 mph, think about catching 10 bowling balls hurtling towards you at 30 mph. The two situations are equivalent.

- **Suppose you had to design seat belts for a racecar that can go 200 mph. How would they be different from the ones available on passenger cars?**

Record your ideas about this question in your *Active Physics* log. Be prepared to discuss your responses with your small group and the class.

# Activity Overview

In this activity the students will be investigating the role and requirements of an effective restraint system. They will be simulating different styles of safety belts, and submitting their model to crash simulations as in **Activity 4**.

## Student Objectives

### Students will:

- Understand the role of safety belts.
- Compare the effectiveness of various wide and narrow belts.
- Relate pressure, force and area.

Chapter 2

ANSWERS FOR THE TEACHER ONLY

## What Do You Think?

If you can, go to see the film *Speedway*. The film talks about how the race cars have changed in the past few decades. Because the speeds are greater than they were, they can no longer build the cars like tiny missiles, which do not crush. They have found, as was mentioned above, that the crush zone allows some of the energy of the collision to be absorbed by the car, rather than the driver. Even in well-protected cars, equipped with shoulder belts, air bags, and padded interiors, if the body of the driver is stopped at certain accelerations, the damage to the interior of the body may be fatal.

Most of the answers the students will come up with will be centered on more of what is already in the vehicle. However, the answers should reflect the idea that the faster you go, the greater protection you will need. Also in the discussion, you can emphasize that it is not only more padding, but rather the reduction of the forces, which will be talked about in this activity. In other words, the stopping from high speeds, into a tree, or the front of your dashboard can be equally dangerous with a tight seat belt on, to the interior organs of the body.

## For You To Do

1. In this activity you will test different materials for their suitability for use as seat belts. Your model car is, once again, a laboratory cart; your model passenger is molded from a lump of soft clay. Give your passenger a seat belt by stretching a thin piece of wire across the **front** of the passenger, and attaching it on the sides or **rear** of the car.

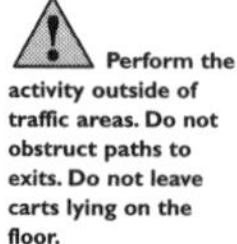

**Perform the activity outside of traffic areas. Do not obstruct paths to exits. Do not leave carts lying on the floor.**

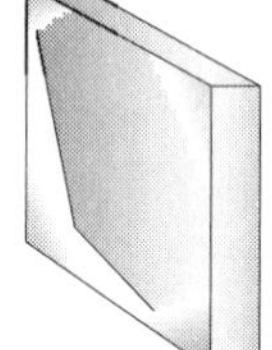

2. Make a collision by sending the car down a ramp. Start with small angles of incline and increase the height of the ramp until you see significant injury to the clay passenger.
   a) In your log, note the height of the ramp at which significant injury occurs.

**Physics Words**

**force:** a push or a pull that is able to accelerate an object; force is measured in newtons; force is a vector quantity.

**pressure:** force per surface area where the force is normal (perpendicular) to the surface; measured in pascals.

3. Use at least two other kinds of seat belts (ribbons, cloth, etc.). Begin by using the same incline of ramp and release height as in **Step 2**.
   a) In your log, record the ramp height at which significant injury occurs to the "passenger" using the other kinds of seat belt material.

4. Crash dummies cost $50,000! Watch the video presentation of a car in a collision, with a dummy in the driver's seat. You may have to observe it more than once to answer the following questions:
   a) In the collision, the car stops abruptly. What happens to the driver?
   b) What parts of the driver's body are in the greatest danger? Explain what you saw in terms of the law of inertia (Newton's First Law of Motion).

ANSWERS

# For You To Do

1. Student activity.

2.a) Students' responses, but as the height of the ramp is increased, there should be an increase in the significant damage.

3.a) Students' responses.

4.a) The driver continues to move in the direction of the car (inertia) until stopped by the seat belt, (or dashboard).

b) The head is probably in the most danger as it is not secured by anything. In terms of the Law of Inertia, as the car stops, the body continues to move. As the body stops, by the seat belt, the head continues to move. This is where damage can occur and usually happens.

## FOR YOU TO READ

### Force and Pressure

When you repeated this experiment accurately each time, the **force** that each belt exerted on the clay was the same each time that the car was started at the same ramp height. Yet different materials have different effects; for example, a wire cuts far more deeply into the clay than a broader material does.

The force that each of the belts exerts on the clay is the same. When a thin wire is used, all the force is concentrated onto a small area. By replacing the wire with a broader material, you spread the force out over a much larger area of contact.

The force per unit area, which is called **pressure**, is much smaller with a ribbon, for example, than with a wire. It is the pressure, not the force, that determines how much damage the seat belt does to the body. A force applied to a single rib might be enough to break a rib. If the same force is spread out over many ribs, the force on each rib can be made too small to do any damage. While the total force does not change, the pressure becomes much smaller.

## PHYSICS TALK

Pressure is the force per unit area:

$$P = \frac{F}{A}$$

where $F$ is force in newtons (N)

$A$ is area in meters squared ($m^2$)

and $P$ is pressure in newtons per meter squared ($N/m^2$).

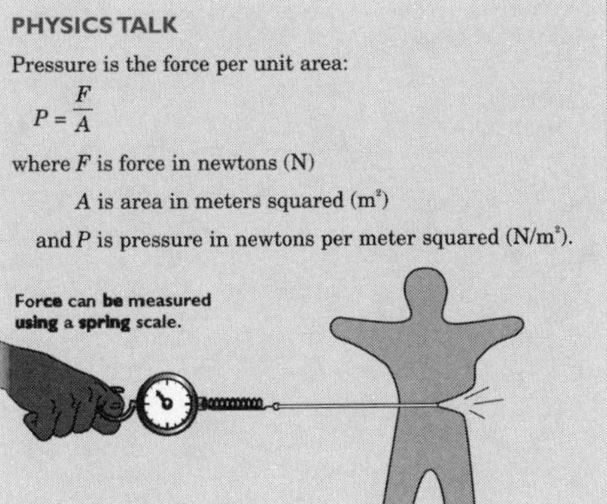

Force can be measured using a spring scale.

### Sample Problem

Two brothers have the same mass and apply a constant force of 450 N while standing in the snow. Brother A is wearing snow shoes that have a base area of 2.0 m². Brother B, without snowshoes, has a base area of 0.1 m². Why does the brother without snowshoes sink into the snow?

***Strategy:*** This problem involves the pressure that is exerted on the snow surface by each brother. You can use the equation for pressure to compare the pressure exerted by each brother.

***Givens:***

$F = 450 \text{ N}$

$A_1 = 2.0 \text{ m}^2$

$A_2 = 0.1 \text{ m}^2$

***Solution:***

| Brother A | Brother B |
|---|---|
| $P = \frac{F}{A}$ | $P = \frac{F}{A}$ |
| $= \frac{450 \text{ N}}{2.0 \text{ m}^2}$ | $= \frac{450 \text{ N}}{0.1 \text{ m}^2}$ |
| $= 225 \text{ N/m}^2$ or $230 \text{ N/m}^2$ | $= 4500 \text{ N/m}^2$ |

The pressure that Brother B exerts on the snow is much greater.

Activity 5 Life (and Fewer Deaths) after Seat Belts

### Reflecting on the Activity and the Challenge

In this activity you gathered data to provide evidence on the effectiveness of seat belts as restraint systems. The material used for the seat belt and the width of the restraint affected the distortion of the clay figure. By applying the force over a greater area, the pressure exerted by the seat belt during the collision can be reduced.

It is important to note that not every safety restraint system will be a seat belt or harness, but that all restraints attempt to reduce the pressure exerted on an object by increasing the area over which a force is applied.

How will your design team account for decreasing pressure by increasing the area of impact? Think about ways that you could test your design prototype for the pressure created during impact. Your presentation of the design will be much more convincing if you have quantitative data to support your claims. Simply stating that a safety system works well is not as convincing as being able to show how it reduces pressure during a collision.

### Physics To Go

1. Use Newton's First Law to describe a collision with the passenger wearing a seat belt during a collision.
2. What is the pressure exerted when a force of 10 N is applied to an object that has an area of:
   a) 1.0 $m^2$?
   b) 0.2 $m^2$?
   c) 15 $m^2$?
   d) 400 $cm^2$?
3. A person who weighs approximately 155 lb. exerts 700 N of force on the ground while standing. If his shoes cover an area of 400 $cm^2$ (0.0400 $m^2$), calculate:
   a) the average pressure his shoes exert on the ground
   b) the pressure he would exert by standing on one foot



ANSWERS

## Physics To Go

1. The car is moving in a given direction. As the car stops, the driver or passenger continues to move forward until the seat belt stops them—a force acts on the body to change the motion from moving relative to the car to stopping with the car. Parts of the body, not directly attached to the seat belt, will continue also to move in the given direction. This is where significant damage to the body, mostly the head, can occur.

2. a) 1.0 $m^2$

   Using formula $P = F/A$,
   $P = 10$ N/ 1.0 $m^2$ therefore
   $P = 10$ N/$m^2$ )

   b) 0.2 $m^2$

   $P = 50$ N/$m^2$

   c) 15 $m^2$

   $P = 0.67$ N/$m^2$

   d) 400 $cm^2$

   $P = 0.025$ N/$cm^2$ or 250 N/$cm^2$

3. a) 17,500 N/$cm^2$

   b) Double the pressure or 35,000 N/$cm^2$

ANSWERS

## Physics To Go

***(continued)***

4. a) Using approximate values, of weight of 500 Newtons, and area of high heels about 1 cm², the $P = F/A$, = 500 N/ 0.0001 m², $P$ = 5,000,000 N/m²

   b) Using approximate values, of weight of 500 N, and area of your hands of about 900 cm², the $P = F/A$, = 500 N/ 0.09 m², $P$ = 5555 N/m²

   c) If your body was about 500 N, and there were 5000 nails, then there is approximately 500 N / 5000 nails, or 0.1 N per nail of pressure. In reality, that is probably less the weight of one nail, and if you were to support one nail on your finger it would not penetrate.

5. The force is identical, but on a smaller surface area. Therefore the pressure exerted on your body is greater. It would be enough to cut your body in two pieces.

6. Students answers will vary. Answers might include people who don't use seat belts may end up having greater injuries, therefore increasing the cost of health care, they may lose control of their vehicle more easily, therefore causing more damage to their own vehicle or others. Have the students explore the concept of social responsibility. What constitutes the need for any law? How does the use or lack of use of safety restraints affect us as a society? How do economics affect the passing of legislation in this area?

7. Have students try to survey more people, and not simply their immediate family.

Safety

4. For comparison purposes, calculate the pressure you exert in the situations described below. Divide your weight in newtons, by the area of your shoes. (To find your weight in newtons multiply your weight in pounds by 4.5 N/lb. You can approximate the area of your shoes by tracing the soles on a sheet of centimeter squared paper.)
   a) How much pressure would you exert if you were standing in high heels?
   b) How much pressure would you exert while standing on your hands?
   c) If a bed of nails contains 5000 nails per square meter, how much force would one nail support if you were to lie on the bed? With this calculation you can now explain how people are able to lie on a bed of nails. It's just physic!
5. Describe why a wire seat belt would not be effective even though the force exerted on you by the wire seat belt is identical to that of a cloth seat belt.
6. Do you think there ought to be seat belt laws? How does not using seat belts affect the society as a whole?
7. Conduct a survey of 10 people. Ask each person what percentage of the time they wear a seat belt while in a car. Be prepared to share your data with the class.

### Stretching Exercises

The pressure exerted on your clay model by a thin wire can be estimated quite easily. Loop the wire around the "passenger," and connect the wire to a force meter.

a) Pull the force meter hard enough to make the wire sink into the model just about as far as it did in the collision.
b) Record the force as shown on the force meter (in newtons).
c) Estimate the frontal area of the wire—its diameter times the length of the wire that contacts the passenger. Record this value in centimeters squared (cm²).
d) Divide the force by the area. This is the pressure in newtons per centimeter squared (N/cm²).

112

NOTES

# Assessment: Activity 5

Place a check mark in the appropriate box. Two check marks will be required for one point.
A sample conversion scale is provided below the chart.
Total marks = 10.

| Descriptor | Yes | No |
|---|---|---|
| **Lab Skills** | | |
| understands the need to control variables | | |
| knows the manipulated variable | | |
| is able to construct reasonable seat belts | | |
| uses more than two types of materials in constructing seat belts | | |
| uses appropriate materials in constructing seat belts | | |
| runs at least three trials with each model of seat belt | | |
| demonstrates or explains why there is a need for several trials | | |
| rebuilds model as necessary | | |
| **Understanding Concepts** | | |
| demonstrates understanding of forces | | |
| demonstrates understanding of pressure | | |
| knows which parts of the body are most vulnerable in a collision | | |
| demonstrates an understanding for seat belt laws | | |
| is able to describe a collision of a person with and without a seat belt | | |
| understands the societal impact of using a seat belt | | |
| understands the societal impact of not using a seat belt | | |
| **Mathematical Skills** | | |
| can calculate the force per unit area (pressure) | | |
| calculates the pressure of the wire | | |
| calculates the pressure of their own weight on their shoes | | |

## Conversion

**20 check marks = 10 points**
**18 check marks = 9 points**
**16 check marks = 8 points, etc.**

For use with *Safety*, Chapter 2, Activity 5: Life (and Fewer Deaths) after Seat Belts

NOTES

# ACTIVITY 6
## Why Air Bags?

# Background Information

Symbols
$v$ = velocity (m/s)
$d$ = distance (m)
$t$ = time (s)
$a$ = acceleration (m/s)
$F$ = force (N)
$m$ = mass (kg)
$\Delta$ = change

Now that we have dealt with forces, how does this have anything to do with accidents? In order to examine accidents, we must first look at momentum. Momentum ($p$) is the product of the mass ($m$) of an object and the velocity ($v$) of that object ($p = mv$). Ask the students which is harder to stop, the 175 pound (80 kg) quarterback running at 4 m/s, or the fullback at 255 pounds (116 kg) running at 4 m/s? Most students will say that the quarterback would be easier to stop. We would say that the momentum of the quarterback is less ($m$ x $v$) than the fullback ($m$ x $v$)

momentum of quarterback

$$p_{qb} = m \times v$$

$$p_{qb} = (80\ \text{kg}) \times (4\ \text{m/s})$$

$$p_{qb} = 320\ \text{kg}\bullet\text{m/s}$$

momentum of fullback

$$p_{fb} = m \times v$$

$$p_{fb} = (116\ \text{kg}) \times (4\ \text{m/s})$$

$$p_{fb} = 464\ \text{kg}\bullet\text{m/s}$$

We can summarize momentum by stating that the greater the momentum of an object, the harder it will be to stop.

According to Newton's First Law – Inertia, an object in motion will remain in motion unless acted upon by an outside, unbalanced force. Therefore, to stop a moving object (the fullback), we need a force. Newton's Second Law

$F = ma$

states that the acceleration (or change in velocity) will be proportional to the force. Since acceleration is the change in velocity divided by the change in time, we can restate Newton's equation as

$$F = m\ \Delta v/\Delta t.$$

Remove $\Delta t$ by multiplying both sides by $\Delta t$ and we now have

$$F\Delta t = m\Delta v.$$

called the impulse (measured in Ns). Impulse is the product of the force applied and the change in time over which the force acted. In other words the force times the time interval is proportional to the change in velocity times the mass.

Therefore, (where $p = mv$)

$$\Delta p = m\Delta v,\ \ [\Delta p = m\ (v_2 - v_1)]$$

or the change in momentum is equal to the mass times the change in velocity.

Substituting $\Delta p$ for $m\Delta v$ in $F\Delta t = m\Delta v$, we get

$$F\Delta t = \Delta p$$

Therefore we get impulse is equal to the change in momentum. As we analyze the equation, we see that in order to stop an object (change its momentum to zero), there must be a force applied to the object over a particular time. Increasing the time decreases the force and decreasing the time increases the force required to change the same momentum. For example: A bowling ball ($m$ = 16 pounds or 7.3 kg) moving at 5.0 m/s. What is the impulse to stop the bowling ball?

$$\text{impulse} = \Delta p = m\Delta v$$

$$\Delta p = 7.3\ \text{kg} \times 5.0\ \text{m/s}$$

$$\Delta p = 36.5\ \text{kg}\bullet\text{m/s}$$

Therefore, what is the force required to bring the bowling ball to rest in 1.0 s?

Impulse = $\Delta p$, therefore $F\Delta t = \Delta p$, therefore

$$F = \Delta p/\Delta t$$

$$F = 36.5\ \text{kg}\bullet\text{m/s} / 1.0\text{s}$$

$$F = 36.5\ \text{kg}\bullet\text{m/s}^2 \text{ or } 36.5\ \text{N}$$

How much force is then needed to bring the ball to rest in 3.0 s?

$$F = 36.5\ \text{kg}\bullet\text{m/s} / 3.0\ \text{s}$$

$$F = 12.2\ \text{N}$$

How much force is then needed to bring the ball to rest in 0.10 s?

$F = 36.5 \text{ kg}\cdot\text{m/s} / 0.10 \text{ s}$

$F = 365 \text{ N}$

You can see as you increase the time needed to stop an object, you decrease the force necessary. As the time decreases, the force increases.

The physics of the air bag is to increase the time required for the object (the driver) to stop, and therefore reduce the force on the driver. The other aspect for which the air bag is designed is that the force, even though decreased, is spread over a larger area. The face slamming into the steering wheel can cause a lot of damage, as compared with the face slamming into an air bag.

## *Active*-ating the Physics InfoMall

Search for "air bag*" and you will not be disappointed. There are several good comments made in various places on the InfoMall, including "Resource letter PE-1: Physics and the environment," *American Journal of Physics*, vol. 42, 267-273 (1974), and "How Things Work" in the Book Basement has a section on air bags (in the 1985 section).

The concept of impulse is introduced in this activity. Searching for "impulse" on the InfoMall provides many great references, including problems your you or your students to work. The textbook *Physics Including Human Applications* has a chapter devoted to Momentum and Impulse.

# Planning for the Activity

## Time Requirements

Allow at least one period for the students to design and run the tests on the air bag. If a video camera is used the students may want to take it home to analyze. If there are not enough video cameras to go around, then rotate through the groups, with the other groups analyzing the data they have collected, while the other students may be watching a video, or redesigning their air bags.

## Materials Needed

### For each group:

- ball, heavy
- clear bag, inflatable
- landing surface materials of 3 hardnesses

## Advance Preparation and Setup

Depending on the material available, you may need to make the equivalent of several different substances to test against the air bag model.

Check proper functioning of the recording device. The VCR must be able to give a clear picture on single frame advance mode. If the students are to use the tape times, they may need additional instructions if they have not used them before. If you are going to use computer software, which digitizes the video, and then allows you to analyze the picture, prior use or training with the software would be an asset, both for the students and the teacher. If there is not enough software, and hardware for all students to use, then a demo would be beneficial to set up for the students in order for them to have a standard to shoot for.

# Teaching Notes

Air bags are becoming more and more common in vehicles. Poll the students as to whether their family cars have air bags. Ask if they are willing to pay the extra money for them. Have them write the answers in their log books.

Remind the students how to make measurements using the videotape system. Encourage experimentation that includes various impact speeds and different stopping surfaces. If tape timers are to be used review the time interval specifications. Measurements of distances are not necessary; only counting the intervals between when the object touches the bag and when the object stops in order to get the stopping time. When using the tape timer, it may be difficult at first to get good results. Be prepared to use a lot of tape in getting accurate results.

There may be video tapes available which show slow motion air bags actually working. These may be available from your local automobile association, or from a car dealership.

Students should work in groups of three or four. Students will be dropping heavy objects onto surfaces which may cause the object to bounce in different and unpredictable directions, and therefore, should take appropriate precautions to prevent injury to damage to property.

Some students may have heard that air bags have caused suffocation once they inflated. This is an incorrect understanding of how the air bag works. The air bag inflates in about 1/32 of a second, and once inflated, it deflates after about one to two seconds. Other fallacies about air bags are that they impair your ability to see, and they inhibit you from exiting the vehicle. (For more information, visit any car dealer and ask for their information on air bags.) While there is some smoke and dust from the $CO_2$ cartridge, and some heat associated with the inflation of the airbag, it will not impair your ability to get out of the vehicle. The effectiveness of the air bag increases in conjunction with the seat and shoulder belts, but is not very effective in side impacts.

Local automotive dealers will have safety videos about air bags and ABS brakes. Many dealers will lend them, or may even give a copy to you.

You may also use the local Automobile Association affiliate in your area to give or lend safety videos. Some will even send instructors to give safety talks.

Some local driving companies (taxi, trucking, courier services) may also have safety supervisors who would be able to come to the classroom and talk about safety.

The *Active Physics Transportation* Content Video has excellent footage showing air bags inflating.

NOTES

# Activity 6 Why Air Bags?

**GOALS**

In this activity you will:

- Model an automobile air bag.
- Relate pressure to force and area.
- Demonstrate that the force of an impact can be reduced by spreading it out over a longer time.

**What Do You Think?**

Air bags do not take the place of seat belts. Air bags are an additional protection. They are intended to be used with seat belts to increase safety.

- **Why are air bags effective?**
- **How does the air bag protect you?**

Record your ideas about these questions in your *Active Physics* log. Be prepared to discuss your responses with your small group and the class.



## Activity Overview

In this lesson the students will study the use of an air bag and how it will spread the force of the impact over both space and time.

## Student Objectives

### Students will:

- Model an automobile air bag.
- Relate pressure to force and area.
- Demonstrate that the force of an impact can be reduced by spreading it out over a longer time.

Answers for the Teacher Only

## What Do You Think?

Students' answers will vary. Some may think of the air bag as cushioning the stop, or making the stop softer. What they are actually saying, is that the air bag will increase the time in which you crash, therefore decreasing the force acting on your body (or more particularly the head).

Again, students' answers will vary. The air bag protects you by increasing the time and the space over which the force is acting on your body. See the **Background Information** for details and the mathematics behind the air bag.

Chapter 2

Safety

### For You To Do

1. You will use a large plastic bag or a partially inflated beach ball as a model for an air bag. Impact is provided by a heavy steel ball, or just a good-sized rock, dropped from a height of a couple of meters.

   Gather the equipment you will need for this activity. Your problem is to find out how long it takes the object to come to rest. What is the total time duration from when the object first touches the air bag until it bounces back?

2. With a camcorder, videotape the object striking the air bag from a given height, such as 1.5 m.

   a) Record the exact height, from which you dropped the object.

3. Play the sequence back, one frame at a time. Count the number of frames during the time the object is moving into the air bag—from the moment it first touches the bag until it comes to rest, before bouncing. Each frame stands for $\frac{1}{30}$ s. (Check your manual.)

   a) In your log, record the number of frames and calculate how long it takes for the object to come to rest.

   If a camcorder is not available, the experiment may be performed, although less effectively, by attaching a ticker-tape timer to the falling object.

   After the object is dropped, with the object still attached, stretch the tape from the release position to the air bag. Mark the dot on the tape that was made just as the object touched the air bag.

   Now push the object into the air bag, about as far as it went just before it bounced. Mark the tape at the dot that was made as the object came to rest. The dots should be close together for a short interval at this point.

   Now count the time that passed between the two marks you made. (You must know how rapidly dots are produced by your timer.)

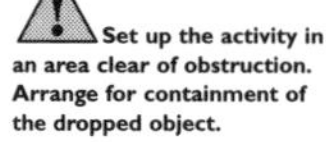

**Set up the activity in an area clear of obstruction. Arrange for containment of the dropped object.**

Active Physics

ANSWERS

## For You To Do

1. Student activity.

2. a) Students' response.

3. a) This may vary, with the camcorder. Most are 1/30 s. Students will be noticing, that the greater the time to stop, the less potential damage will be done.

Activity 6 Why Air Bags?

4. Repeat **Steps 2** and **3**, but this time drop the ball against a hard surface, such as the floor. Keep the height from which the object is dropped constant.
   a) Record how long it takes for the object to come to rest on a floor.
5. Choose two other surfaces and repeat **Steps 2** and **3**.
   a) Record how long it takes the object to come to rest each time.
   b) In your log, list all the surfaces you tested in the order in which you expect the most damage to be done to a falling object, to the least damage.
   c) Is there a relationship between the time it takes for the object to come to rest and the potential damage to the object landing on the surface? Explain this relationship in your log.

**PHYSICS TALK**

**Force and Impulse**

Newton's First Law states that an object in motion will remain in motion unless acted upon by a net external force. In this activity you were able to stop an object with a force. In all cases the object was traveling at the same speed before impact. Stopping the object was done quickly or gradually. The amount of damage is related to the time during which the force stopped the object. The air bag was able to stop the object with little damage by taking a long time. The hard surface stopped the object with more damage by taking a short time.

Physicists have a useful way to describe these observations. An **impulse** is needed to stop an object. That impulse is defined as the product (multiplication) of the force applied and the time that the force is applied. →

**Physics Words**

**impulse:** the product of force and the interval of time during which the force acts; impulse results in a change in momentum.

Chapter 2

ANSWERS

# For You To Do *(continued)*

4. a) Students' response. Again the students should be noticing that the greater the time to stop, the less damage will be done.

5. a) Students' response. The harder the surface, the less time, and potentially the greater damage done.

   b) Students' response. This will vary with the types of materials each group uses.

   c) Students' response. Again, the longer it takes the object to stop, the less damage will be done to the object.

$$\text{Impulse} = F\Delta t$$

where $F$ is force in newtons (N)

$\Delta t$ is the time interval during which the force is applied in seconds (s).

Impulse is calculated in newton seconds (Ns).

An object of a specific mass and a specific speed will need a definite impulse to stop. Any forces acting for enough time can provide that impulse.

If the impulse required to stop is 60 Ns, a force of 60 N acting for 1 s has the required impulse. A force of 10 N acting for 6 s also has the required impulse.

| Force $F$ | Time Interval $\Delta t$ | Impulse $F\Delta t$ |
|---|---|---|
| 60 N | 1 s | 60 Ns |
| 10 N | 6 s | 60 Ns |
| 6000 N | 0.01 s | 60 Ns |

The greater the force and the smaller the time interval, the greater the damage that is done.

### Sample Problem

A person requires an impulse of 1500 Ns to stop. What force must be applied to the person to stop in 0.05 s?

***Strategy:*** You can use the equation for impulse and rearrange the terms to solve for the force required.

$$\text{Impulse} = F\Delta t$$

$$F = \frac{\text{Impulse}}{\Delta t}$$

***Givens:***

Impulse = 1500 Ns

$\Delta t = 0.05$ s

***Solution:***

$$F = \frac{\text{Impulse}}{\Delta t}$$

$$= \frac{1500 \text{ Ns}}{0.05 \text{ s}}$$

$$= 30{,}000 \text{ N}$$

## Reflecting on the Activity and the Challenge

People once believed that the heavier the automobile, the greater the protection it offered passengers. Although a heavy, rigid car may not bend as easily as an automobile with a lighter frame, it doesn't always offer more protection.

In this activity, you found that air bags are able to protect you by extending the time it takes to stop you. Without the air bags, you will hit something and stop in a brief time. This will require a large force, large enough to injure you. With the air bag, the time to stop is longer and the force required is therefore smaller.

Force and impulse must be considered in designing your safety system. Stopping an object gradually reduces damage. The harder a surface, the shorter the stopping distance and the greater the damage. In part this provides a clue to the use of padded dashboards and sun visors in newer cars. Understanding impulse allows designers to reduce damage both to cars and passengers.





Chapter 2

ANSWER

## Physics To Go

1. Any combination of $F$ (e.g., 30 N) multiplied by $\Delta t$ (e.g., 2 s) which give a result of 60 Ns.
2. $\Delta p = F\Delta t$, therefore, $F = \Delta p/\Delta t$,
   a) $F = 1000$ Ns / 0.01 s, $F = 100{,}000$ N;
   b) $F = 1000$ Ns / 0.1 s, $F = 10{,}000$ N;
   c) $F = 1000$ Ns / 1.0 s, $F = 1000$ N;
3. When a car stops, there is an impulse (change in the momentum). This impulse ($F\Delta t$) will be transferred to the body in the car as the body's change in momentum is the same. Therefore, the air bag changes the force being applied to the body by increasing the time that the body is slowing down because the impulse does not change.
4. Hitting a brick wall will cause greater damage, as the time that the car stops is smaller, therefore the force is greater than the car hitting a snow bank.
5. a) The bumper might be mounted on a piston or a spring which will compress when the bumper strikes an object. Even without a spring or piston, the bumper can dent before the rigid frame of the car strikes the object.
   b) The collapsible steering wheel breaks upon impact. Rather than impaling the driver's body on the rigid shaft, the steering column telescopes in, lengthening the time of the impact.
   c) The crush or crumple zones allow the car to compress like an accordion. Previously, the sheet metal parts of the car were welded continuously along the seams, conveying the impact from object to occupants. Now the crumpling of the car increases the time after the front of the car hits until the rest of the car stops.
   d) The padding is not rigid, so as a body hits it, it crushes, lengthening the time for the impulse to be delivered.
6. a) When you catch a hard ball, you use a mitt and you do not hold your arms rigid. The mitt is cushioned and the leather webbing stretches. Your flexed arms ride with the ball. All of this lengthens the time of the impulse, reducing the required force.
   b) When you land on the ground you allow your knees to bend, lengthening the time for your body to come to rest, reducing the required force that will stop your movement.
   c) The bungee cord exerts a force stopping your fall as it stretches. By the time it is fully extended, your speed should be reduced enough so that the jerk of the stop is neither sudden nor severe. The impulse received from the bungee acts over a long period of time.
   d) The net stretches, allowing the victim to stop gradually. The impulse is the same as if the victim had landed on the ground, but much less destruction since the force is less, (due to increased time).
7. Due to the higher speed, the design of a seat belt for the airplane might be similar to the four-point belt that is used by fighter pilots, and racecar drivers.
8. Air bags might be a very effective, but expensive, safety feature in a passenger plane. This is especially so because of the absence of a shoulder strap. One could argue that the seat in front of the passenger is not rigid, and acts as a modified air bag.

### Physics To Go

1. If an impulse of 60 Ns is required to stop an object, list in your log three force and time combinations (other than those given in the **Physics Talk**) that can stop an object.
2. A person weighing 130 lb. (60 kg) traveling at 40 mph (18 m/s) requires an impulse of approximately 1000 Ns to stop. Calculate the force on the person if the time to stop is:
   a) 0.01 s
   b) 0.10 s
   c) 1.00 s

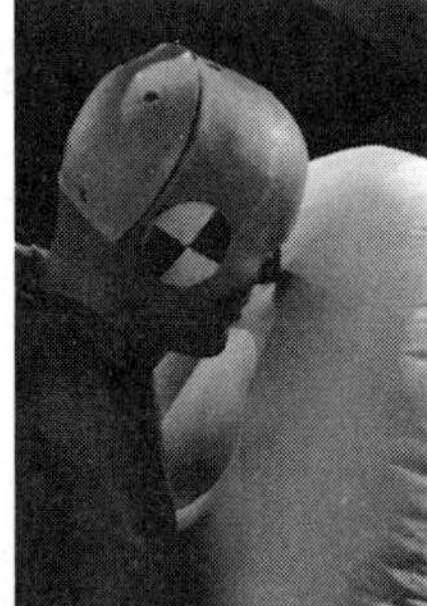

3. Explain in your log why an air bag is effective. Use the terms force, impulse, and time in your response.
4. Explain in your log why a car hitting a brick wall will suffer more damage than a car hitting a snow bank.
5. There are several other safety designs that employ the concept of spreading out the time interval of a force. Describe in your log how the ones listed below perform this task:
   a) the bumper
   b) a collapsible steering wheel
   c) frontal "crush" zones
   d) padding on the dashboard
6. There are many other situations in which the force of an impulse is reduced by spreading it out over a longer time. Explain in your log how each of the actions below effectively reduces the force by increasing the time. Use the terms force, impulse, and time in your response.
   a) catching a hard ball
   b) jumping to the ground from a height
   c) bungee jumping
   d) a fireman's net
7. The speed of airplanes is considerably higher than the speed of automobiles. How might the design of a seat belt for an airplane reflect the fact that a greater impulse is exerted on a plane when it stops?
8. Airplanes have seat belts. Should they also have air bags?

# Assessment: Activity 6

Place a check mark in the appropriate box. Two check marks will be required for one point. A sample conversion scale is provided below the chart.

Maximum = 10

| Descriptor | Yes | No |
|---|---|---|
| **Lab Skills** | | |
| student understands the need to control variables | | |
| student is able to identify the manipulated variable | | |
| student is able to identify the responding variable | | |
| student is able to accurately record the time required to stop | | |
| student uses more than 2 types of materials in constructing air bags | | |
| student uses appropriate materials in constructing air bags | | |
| student runs at least 3 trials with each model of air bag | | |
| student demonstrates or explains why there is a need for several trials | | |
| student tested at least 3 different surfaces | | |
| **Understanding Concepts** | | |
| student demonstrates understanding of forces and effects of air bags | | |
| student demonstrates understanding of pressure | | |
| student understands how the air bag inflates | | |
| student relates inertia with the forces an air bag exerts on the body | | |
| student understands need for seat belt use with the air bag | | |
| student relates increased time of stopping with decreased force on body | | |
| student understands impulse and the air bag | | |
| student understands how impulse is related to damage | | |
| **Mathematical Skills** | | |
| student can calculate impulse given force and different times | | |
| given an impulse and time, student can calculate force | | |

Conversion

19 check marks = 10 points

17+ check marks = 9 points

15+ check marks = 8 points

For use with *Safety*, Chapter 2, Activity 6: Why Air Bags?

# ACTIVITY 7
## Automatic Triggering Devices

# Background Information

Air bags and seat belts do save your life, and prevent serious injuries if used properly. If you pull on your seat belt while you are moving at normal speed, it seems as though there is no way the seat belt could stop you from moving forward. Yet, you have probably slammed on your brakes at some point and the seat belt did in fact tighten, and held you in place. However, the air bag didn't inflate. The safety devices in your car cannot be working all the time (you wouldn't be able to see past the air bag), so you must have a triggering device to make sure that the air bag inflates only when it is needed and when it is it will prevent further damage.

In this activity the students will be using their own imagination, ingenuity, and any materials you are able to bring into the classroom. They will attempt to design a device which will operate like the trigger device in a car.

One such design that is used in cars is a cylinder with a steel ball attached at one end to a magnet, and a open circuit at the other end (See diagram). The ball is held in place by the magnet and only dislodges from the magnet at certain accelerations. When it leaves one end and hits the other end, the circuit is then closed, and the air bag inflates inside the car. This acceleration must be such that it would only happen in a collision, where the speed of the car is greater than about 18 km/h (11 mph).

## *Active*-ating the Physics InfoMall

Again, you may want to find what is known about student difficulties with a concept in this activity. One possible search is "student difficult*" AND "circuit*".

# Planning for the Activity

## Time Requirements

Allow at least 40 minutes for this activity, with the possibility of the students taking their project home to work on. For the presentation of their design, allow approximately 10 minutes for each group to explain their design, and show how it works.

## Materials Needed

### For each group:

- alligator clip leads
- battery, d-cell
- bulb base for miniature screwbase
- bulb, miniature light
- dynamics carts
- single battery holder, d-cell
- starting ramp for lab cart
- switch, spst.

### For the class:

- Collection of components such as tape, rubber bands, string, wire, paper clips, metal foil

# Teaching Notes

Explore the students thoughts on how a triggering device might operate. One type of the device is shown in the **Background Information**. Elicit the conditions under which the device should trigger. The test criterion is purposely vague: that the device must trigger only when the car collides with a large speed and not a slow speed.

Reliability is a key issue. The device that inflates the air bag is a good example of a spin-off from the space program. It is based on a device that demanded 100% reliability: the release mechanism from the lunar launch system.

For a sophisticated class, you may wish to help the class determine the best release points for testing the car by using some kinematics, or through energy conservation. These areas have been explored. A sonic ranger or tape timer may be used in conjunction with your tests.

This activity can lead to increasing amounts of noise and mayhem. Keep students on task with constant monitoring and subtle encouragement while looking at each group. Use groups of four. In groups of three or less, there are generally fewer ideas, and groups greater than five give some an opportunity to blend into the woodwork.

Encourage students to work cooperatively by assigning tasks prior to beginning the activity. The following tasks are designed for groups of four students:

- **Organizer:** helps focus discussion and ensures that all members of the group contribute to the discussion. The organizer ensures that all of the equipment has been gathered and that the group completes all parts of the activity.
- **Recorder:** provides written procedures when required, diagrams where appropriate and records data. The recorder must work closely with the organizer to ensure that all group members contribute.
- **Researcher:** seeks written and electronic information to support the findings of the group. In addition, where appropriate, the researcher will develop and test prototypes. The researcher will also exchange information gathered among different groups.
- **Diverger:** seeks alternative explanations and approaches. The task of the diverger is keep the discussion open. "Are other explanations possible?"

Students will be working with many different electrical devices. Ensure that there are no wires connected to common household circuits. Ensure that the students are only using common dry cell or household batteries.

Some students may think that the triggering device for such safety devices are simple electronic switches. This activity will help them to understand that in order for a switch to be functional for a particular purpose (such as the g-forces that are enough to cause damage to the body, and hence need to have some way to reduce the overall force to the body), that the actual trigger must be linked to the cause of the damage. This is why seat belts are based on a rocker/pendulum switch, which catches the roll that the seat belt is rolled upon, when the forces are high enough—much like the blinds that when pulling down slowly will slowly come down, but to stop them rolling back up a sharp tug is needed to catch the roll.

NOTES

# Activity 7 Automatic Triggering Devices

### GOALS

In this activity you will:

- Design a device that is capable of transmitting a digital electrical signal when it is accelerated in a collision.

### What Do You Think?

An air bag must inflate in a sudden crash, but must not inflate under normal stopping conditions.

- **How does the air bag "know" whether to inflate?**

Record your ideas about this question in your *Active Physics* log. Be prepared to discuss your responses with your small group and the class.

### For You To Do

**Inquiry Investigation**

1. Form engineering teams of three to five students. Meet with your engineering team to design an automatic air bag triggering device using a knife switch, rubber bands, string, wires, and a flashlight bulb. Other materials may also be supplied by you or your teacher.

# Activity Overview

In this activity the students will use imagination and creativity to design their own triggering device. The emphasis on this activity should be that when "inventing" something, there should be no limitations, within reason.

## Student Objectives

### Students will:

- Design a device that is capable of transmitting a digital electrical signal when it is accelerated in a collision.

Answers for the Teacher Only

## What Do You Think?

Students answers will vary. One common way in which the air bag is triggered, is the method described in the **Background Information**.

Chapter 2

Answers

## For You To Do

All the answers for this section involve students observations and their design of the triggering device.

**Be sure to receive your teacher's approval before using any material.**

2. The design parameters are as follows:
   - The device must turn a flashlight bulb on, or turn it off. This will be interpreted as the trigger signal.
   - The device must not trigger if the car is brought to a sudden stop from a slow speed.
   - The device must trigger if the sudden stop is from a high speed.
   - The car containing the device must be released down a ramp. The car will then strike a wall at the bottom of the ramp.
   - The battery and bulb must be attached to the car along with the triggering device. The bulb does not have to remain in the final on or off state, but it must at least flash to show triggering.
3. Follow your teacher's guidelines for using time, space, and materials as you design your triggering device.
4. Demonstrate your design team's trigger for the class.

## FOR YOU TO READ

### Impulse and Changes in Momentum

It takes an unbalanced, opposing force to stop a moving car. **Newton's Second Law** lets you find out how much force is required to stop any car of any mass and any speed.

The overall idea can be shown using a concept map.

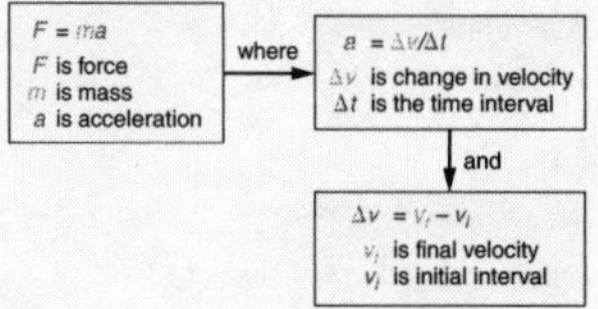


If you know the mass and can determine the acceleration, you can calculate the force using Newton's Second Law:

$$F = ma$$

A moving car has a forward **velocity** of 15 m/s. Stopping the car gives it a final velocity of 0 m/s.

$$\text{The change in velocity} = v_{final} - v_{initial} = 0 - 15 \text{ m/s} = -15 \text{ m/s}.$$

Any change in velocity is defined as **acceleration**.

In this case the change in velocity is –15 m/s. If the change in velocity occurs in 3 s, the acceleration is –15 m/s in 3 s, or –5 m/s every second, or –5 m/s$^2$. You can look at this as an equation:

$$a = \frac{\Delta v}{\Delta t} = \frac{v_f - v_o}{\Delta t} = \frac{0 \text{ m/s}-15 \text{ m/s}}{3\text{s}} = -5 \text{ m/s}^2$$

If the car had stopped in 0.5 s, the change in speed is identical, but the acceleration is now:

$$a = \frac{\Delta v}{\Delta t} = \frac{v_f - v_o}{\Delta t} = \frac{0 \text{ m/s}-15 \text{ m/s}}{0.5 \text{ s}} = -30 \text{ m/s}^2$$

Newton's Second Law informs you that unbalanced outside forces cause all accelerations. The force stopping the car may have been the frictional force of the brakes and tires on the road, or the force of a tree, or the force of another car. Once you know the acceleration, you can calculate the force using Newton's Second Law. If the car has a mass of 1000 kg, the unbalanced force for the acceleration of –5 m/s *every* second would be –5000 N. The force for the larger acceleration of –30 m/s *every* second would be –30,000 N. The negative sign tells you that the unbalanced force was opposite in direction to the velocity.

The change in velocity, acceleration, and force give a complete picture.

There is another, equivalent picture that describes the same collision in terms of **momentum** and impulse.

Any moving car has momentum. Momentum is defined as the mass of the car multiplied by its velocity $p = mv$.

The impulse/momentum equation tells you that the momentum of the car can be changed by applying a force for a given amount of time.

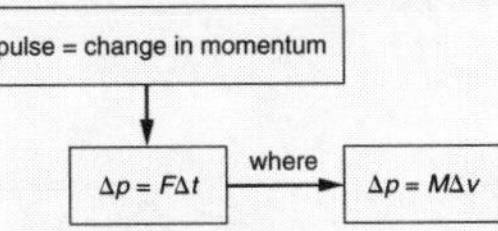


Using the impulse/momentum approach explains something different about a collision. Consider this question, "why do you prefer to land on soft grass rather than on hard concrete?" Soft grass is preferred because the force on your legs is less when you land on soft grass. Let's find out why using Newton's Second Law and then by using the impulse/momentum relation.

→

**Newton's Second Law explanation:** Whether you land on concrete or soft grass, your change in velocity will be identical. Your velocity may decrease from 3 m/s to 0 m/s. On concrete, this change occurs very fast, while on soft grass this change occurs in a longer period of time. Your acceleration on soft grass is smaller because the change in velocity occurred in a longer period of time.

$$a = \frac{\Delta v}{\Delta t}$$

When the change in the period of time gets larger, the denominator of the fraction gets larger and the value of the acceleration gets smaller.

When landing on grass, Newton's Second Law then tells you that the force must be smaller because the acceleration is smaller for an identical mass. $F = ma$. Smaller acceleration on grass requires a smaller force. Smaller forces are easier on your legs and you prefer to land on soft grass.

**Momentum/impulse explanation:** Whether you land on concrete or soft grass, your change in momentum will be identical. Your velocity will decrease from 3 m/s to 0 m/s on either concrete or grass.

$$F\Delta t = \Delta p$$

You can get this change in momentum with a large force over a short time or a small force over a longer time.

If your mass is 50 kg, the amount of your change in momentum may be 150 kg m/s, when you decrease your velocity from 3 m/s to 0 m/s. There are many forces and associated times that can give this change in the value of the momentum.

If you could land on a surface that requires 3 s to stop, it will only require 50 N. A more realistic time of 1 s to stop will require a larger force of 150 N. A hard surface that brings you to a stop in 0.01 s requires a much larger force of 15,000 N.

On concrete, this change in the value of the momentum occurs very fast (a short time) and requires a large force. It hurts. On soft grass this change in the value of the momentum occurs in a longer time and requires a small force: it is less painful and is preferred.

| Change in value of momentum | Force | Change in Time $\Delta t$ | $F\Delta t$ |
|---|---|---|---|
| 150 kg m/s | 50 N | 3 s | 150 kg m/s |
| 150 kg m/s | 75 N | 2 s | 150 kg m/s |
| 150 kg m/s | 150 N | 1 s | 150 kg m/s |
| 150 kg m/s | 1500 N | 0.1 s | 150 kg m/s |
| 150 kg m/s | 15,000 N | 0.01 s | 150 kg m/s |

### Physics To Go

1. How do impulse and Newton's First Law (the Law of Inertia) play a role in your air bag trigger design?
2. Imagine a device where a weight is hung from a string within a metal can. If the weight hits the side of the can, a circuit is completed. How do impulse and the Law of Inertia work in this device?
3. In cars built before 1970, the dashboard was made of hard metal. After 1970, the cars were installed with padded dashboards like you find in cars today. In designing a safe car, it is better to have a passenger hit a cushioned dashboard than a hard metal dashboard.
   a) Explain why the padded dashboard is better using Newton's Second Law.
   b) Explain why the padded dashboard is better using impulse and momentum.
4. Why would you prefer to hit an air bag during a collision than the steering wheel?
   a) Explain why the air bag is better using Newton's Second Law.
   b) Explain why the air bag is better using impulse and momentum.
5. Explain why you bend your knees when you jump to the ground.
6. Catching a fast ball stings your hand. Why does wearing a padded glove help?
7. When a soccer ball hits your chest, you can stiffen your body and the ball will bounce away from you. In contrast, you can "soften" your body and the ball drops to your feet. Explain how the force on the ball is different during each play.

**Physics Words**

**Newton's Second Law of Motion:** if a body is acted upon by an external force, it will accelerate in the direction of the unbalanced force with an acceleration proportional to the force and inversely proportional to the mass.

**velocity:** speed in a given direction; displacement divided by the time interval; velocity is a vector quantity, it has magnitude and direction.

**acceleration:** the change in velocity per unit time.

**momentum:** the product of the mass and the velocity of an object; momentum is a vector quantity.

### Stretching Exercises

1. How does a seat belt "know" to hold you firmly in a crash, but allow you to lean forward or adjust it without locking? Write your response in your log.
2. Go to a local auto repair shop, junk yard, or parts supply store. Ask if they can show you a seat belt locking mechanism. How does it work? Construct a poster to describe what you have learned.



Chapter 2

ANSWERS

## Physics To Go

1. Students' will have individual responses. Impulse is the change in momentum. As the car comes to a stop, the occupants and the triggering device will continue to move in a straight line (Newton's First Law of Motion—Inertia).
2. When in motion at a constant speed the metal pendulum hangs straight down. When the car stops, the pendulum continues to move forward until an impulse, the force of the string and the wall of the can, stop it.
3. a) Newton's Second Law states that the force is proportional to the acceleration. A padded dashboard will require more time for the passenger to stop. More time stopping implies a smaller acceleration. If the acceleration is smaller, the force is smaller ($F = ma$).
   b) $F\Delta t = \Delta mv$
   The change in momentum of the passenger will be identical whether stopped by a hard or soft dashboard. The padded dashboard will require more time to stop. The impulse momentum equation requires that the force be smaller.
4. a) Newton's Second Law states that the force is proportional to the acceleration. An air bag will require more time for the passenger to stop. More time stopping implies a smaller acceleration. If the acceleration is smaller, the force is smaller ($F = ma$).
   b) $F\Delta t = \Delta mv$
   The change in momentum of the passenger will be identical whether stopped by the steering column or the air bag. The air bag will require more time to stop. The impulse momentum equation requires that the force be smaller.
5. Jumping down and stopping requires you to change your momentum. This change in momentum will occur with a large force over a short time or a small force over a large time. You bend your knees in order to increase the time to stop, thereby decreasing the force on your legs.
6. Catching a baseball requires you to change the momentum of the ball. This change in momentum will occur with a large force over a short time or a small force over a large time. You use a padded glove in order to increase the time to stop, thereby decreasing the force on your hands.
7. A soccer ball hitting a hard surface like a stiff body will reverse its velocity and bounce away. A soccer ball hitting a soft surface like a "softened" body, will have an inelastic collision and the two objects will move off together after the collision with a small, almost zero velocity.

## Assessment: Activity 7

The following rubric can be used to assess group work during **Activity** 7: Automatic Triggering Devices.

Each member of the group can evaluate the manner in which the group worked to solve problems in the activity.

**Total = 12 marks**

1. Low level – indicates minimum effort or effectiveness.
2. Average – acceptable standard has been achieved, but that the group could have worked more effectively if better organized.
3. Good – this rating indicates a superior effort. Although improvements might have been made, the group was on task all of the time and completed all parts of the activity.

| **Descriptor** | **Values** | | |
|---|---|---|---|
| 1. The group worked cooperatively to engineer an automatic triggering device. | 1 | 2 | 3 |
| Comments: | | | |
| 2. A plan was established before beginning and all tasks were shared equally. | 1 | 2 | 3 |
| Comments: | | | |
| 3. The group was organized. Materials were collected and the problems were addressed by the entire group. | 1 | 2 | 3 |
| Comments: | | | |
| 4. Data was collected and recorded in an organized fashion in data tables in journals. | 1 | 2 | 3 |
| Comments: | | | |

## Assessment: Scientific and Technological Thinking

Scientific and technological thinking can be assessed using the rubric below. Allow one mark for each check mark.

**Maximum value = 10**

| **Descriptor** | **Yes** | **No** |
|---|---|---|
| 1. The device (either light on or light off) does not trigger at low speeds. | | |
| 2. The device (either light on or light off) does trigger at high speeds. | | |
| 3. The device shows innovation and imagination. | | |
| 4. Controls, such as release height on ramp, are maintained throughout. | | |
| 5. The device is reliable (at least 75%). | | |
| 6. Students recognize the need for several trials. | | |
| 7. Triggering devices are tested and/or modified prior to final demonstration. | | |
| 8. Students identify the role of inertia and Newton's laws in this activity. | | |
| 9. Students can identify variables that would trigger the device. | | |
| 10. Students are able to explain their device to the class. | | |

NOTES

# ACTIVITY 8
## The Rear-End Collision

# Background Information

Whiplash is the mechanism of injury, not an injury itself. However, most people associate whiplash with the injury itself. In this activity, whiplash will be referred to as the mechanism of injury, and the injury will be referred to as whiplash effect.

In a head-on collision, the head will continue to move in a forward direction while the body is strapped in, until the chin hits the chest. As there is a sudden movement, there will be minor soft tissue injury, not dissimilar to a mild sprain, but not as in the rear-end collisions.

In the rear-end collision, the head is snapped backward, and then as the collision stops, snapped forward, in a lashing motion (hence whiplash). However, if there is nothing that will stop the head from moving backward (such as a head rest), the head will keep moving as far as possible, very quickly, and doing tremendous damage to the soft tissue. The whipping back and forth will keep doing the damage, can leave the neck muscles very stretched and sore, in minor cases, or torn and not able to function properly, as well as the possibility of fracture cervical vertebrae in severe cases. Again, this will depend on the severity of the accident.

The physics involved with this accident, is again Newton's First Law – an object (head) in motion will remain in motion unless acted upon by an outside unbalanced force (the elasticity of the muscles in the neck, and the physical motion of the neck). Newton's Second Law, more commonly referred to in the formula $F = ma$, the force involved is proportional to the acceleration of the head. Therefore, if there is a larger vehicle crashing into a smaller vehicle, there will be a larger acceleration. This larger acceleration causes a larger force, which will cause the head to whip backwards with greater force, causing greater injury.

In this activity, car 1 will be released and will collide at the bottom of the ramp with car 2 with the passenger. For this collision we will assume conservation of momentum. Therefore, if the cars are the same mass, there will be conservation of momentum, and the second car will move off with a velocity the same as the first before the collision.

Momentum before the crash will equal momentum after the crash

$$m_1v_1 + m_2 v_2 = m_1v_1' + m_2 v_2'$$

because the first car is stationary before, $m_1v_1 = 0$, and because the second car is stationary after $m_2 v_2' = 0$, then

$$m_2 v_2 = m_1v_1'$$

Now, the passenger in the car, according to Newton's First Law of Motion, will remain stationary unless an outside, unbalanced force acts on it. Therefore, the body as a result of the seat pushing on the back of the individual, will move forward. However, the head remains stationary, and will only go forward after the elasticity of the neck muscles cause the head to snap forward. The result is the whiplash effect and injury.

The extent of the injury will have to do with Newton's Second Law of Motion, more commonly referred to in the formula $F = ma$ ($F$ is the force in newtons, $m$ is the mass in kg, and $a$ is the acceleration in m/s$^2$). The force is proportional to the mass, and is also proportional to the acceleration of the object. In the case of the collision from behind, the greater the change in momentum, the greater the force acting on the car, and hence the body and head. Therefore, being hit from behind by a motorcycle will cause less damage than being hit from behind by a semi.

## *Active*-ating the Physics InfoMall

Search for "whiplash" and you will find a nice discussion in *The Fascination of Physics*, from the Textbook Trove. This discussion includes some nice graphics.

Newton's Second Law is discussed in virtually every physics textbook in existence, not to mention the InfoMall. Depending on the level at which you wish to present this Law, you may wish to examine the conceptual-level texts, the algebra-based texts, or even the calculus-based textbooks on the InfoMall.

If you want more exercises to give to your students, searching the InfoMall is a bad idea - there are too many problems on the CD-ROM. Searching with keywords "force" AND "acceleration" AND "mass" in the Problems Place alone produces "Too many hits." However, you will find more than enough by simply going to the Problems Place and browsing a few of the resources you will find there. For example, *Schaum's 3000 Solved Problems in Physics* has a section on Newton's Laws of Motion. You will surely find enough problems there to keep any student busy for some time!

**For You To Do Step 6** mentions ratios. This is one of those areas in which students are known to have problems. Perform a search using "student difficult*" AND "ratio*" for more information. Alternately, try "misconcept*" AND "ratio*". If any of these searches causes "Too many hits" simply reduce the number of stores you search in, or require that the words occur in the same paragraph.

Try similar searches to find student difficulties with the concept of acceleration. You should find, for example, "Investigation of student understanding of the concept of acceleration in one dimension," *American Journal of Physics*, vol. 49, issue 3.

**Physics To Go, Step 6**, asks students to predict the direction a cork will move. If you were to search the InfoMall to find more about the importance of predictions in learning, you would find that you need to limit your search. For example, a search for "prediction*" AND "inertia" resulted in several hits; the first hit is from *A Guide to Introductory Physics Teaching: Elementary Dynamics*, Arnold B. Arons' Book Basement entry. Here is a quote from that book: "Because of the obvious conceptual importance of the subject matter, the preconceptions students bring with them when starting the study of dynamics, and the difficulties they encounter with the Law of Inertia and the concept of force, have attracted extensive investigation and generated a substantial literature. A sampling of useful papers, giving far more extensive detail than can be incorporated here, is cited in the bibliography [Champagne, Klopfer, and Anderson (1980); Clement (1982); di Sessa (1982); Gunstone, Champagne, and Klopfer (1981); Halloun and Hestenes (1985); McCloskey, Camarazza, and Green (1980); McCloskey (1983); McDermott (1984); Minstrell (1982); Viennot (1979); White (1983), (1984)]." Note that students' preconceptions can have a large effect on how they learn something. It is important that they are forced to consciously acknowledge their preconceptions by making predictions.

# Planning for the Activity

## Time Requirements

Approximately 40 minutes are required to complete the procedure. Longer if the students would like to try to rig up different devices to prevent the injury.

## Materials Needed

### For each group:

- balance platform
- duct tape, roll
- dynamics carts
- spring scale, 0-10 newton range
- starting ramp for lab cart
- stick, meter, 100 cm, hardwood
- tape, masking, 3/4" x 36 yds.
- wood piece, 1" x 2" x 2"
- wood piece, 1" x 3" x 10"
- wood piece, 2" x 4" x 1'

## Advance Preparation and Setup

Show the students your model of the passenger.

# Teaching Notes

This assignment provides an application of the scientific principles of momentum, and Newton's First and Second Laws.

Students will probably be able to articulate the action or anticipated actions. However, it is important that the physics be explained, so as to help them realize that this demonstration can, in fact, be a realistic model.

Give a short explanation as to conservation of momentum. Most students will have a realistic view (from playing billiards, playing marbles, football, or hockey, or other contact sports).

# Activity Overview

The students will be comparing the collision of their model "crash test dummy" with real life accidents, in order to analyze the effect of rear-end collisions on the neck muscles.

# Student Objectives

## Students will:

- Evaluate from simulated collisions, the effect of rear-end collisions on the neck muscles.
- Understand the causes of whiplash injuries.
- Understand Newton's Second Law of Motion.
- Understand the role of safety devices in preventing whiplash injury.

Safety

## Activity 8 The Rear-End Collision

### GOALS

In this activity you will:

- Evaluate from simulated collisions, the effect of rear-end collisions on the neck muscles.
- Understand the causes of whiplash injuries.
- Understand Newton's Second Law of Motion.
- Understand the role of safety devices in preventing whiplash injury.

### What Do You Think?

Whiplash is a serious injury that is caused by a rear-end collision. It is the focus of many lawsuits, loss of ability to work, and discomfort.

- **What is whiplash?**
- **Why is it more prominent in rear-end collisions?**

Record your ideas about these questions in your *Active Physics* log. Be prepared to discuss your responses with your small group and the class.

Active Physics 124

ANSWERS FOR THE TEACHER ONLY

# What Do You Think?

Students will likely say that whiplash is an injury to the neck, when hit from behind. See **Background Information** for explanation of whiplash injury.

Students will likely say that when you go forward, your body will also move forward, so there is less force acting on the neck. However, when hit from behind, the head stays at rest due to inertia, as the body is accelerated forward by the seat. This is true to a certain extent, but the primary reason is due to the lashing effect of the head being whipped back and forth, with nothing to support the head in the backwards direction. In the front the chin will hit the chest, thus stopping the head's forward motion.

### For You To Do

1. You will use two pieces of wood to represent the torso (the trunk of the body) and the head of a passenger. Attach a small piece of wood (about 1" x 2" x 2") to a larger piece of wood (about 1" x 3" x 10") with some duct tape acting like a hinge between the two pieces.
   a) Make a sketch to show your passenger. Label what each part of the model passenger represents.
2. Set up a ramp against a stack of books about 40 cm high, as shown in the diagram below. Place the wooden model passenger at the front of a collision cart positioned about 50 cm from the end of the ramp. Release a second cart from a few centimeters up the ramp.
   a) In your log record what happens to the head and torso of the wooden model.

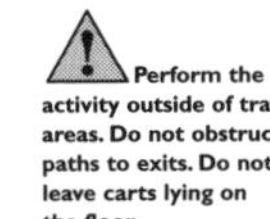

Perform the activity outside of traffic areas. Do not obstruct paths to exits. Do not leave carts lying on the floor.

3. With the first cart still positioned about 50 cm from the end of the ramp, release the second cart from the top of the ramp.
   a) Describe what happens to the head of the model passenger in this collision.
   b) Use Newton's First Law of Motion to explain your observations.

4. The duct tape represents the neck muscles and bones of the vertebral column. How large a force do the neck muscles exert to keep the head from flying off the body, and to return the head to the upright position? To answer this question, begin by estimating the mass of an average head.
   a) Estimate and record in your log the mass of an average human head. The mass would be close to the mass of a filled water container of the same size.

ANSWERS

## For You To Do

1. a) Students' sketches.
2. a) Students' response. They should note that the "head" may snap backward, and the torso may move backward, but without the snapping motion. Because of the low speed, there may not be a lot of movement.
3. a) The head will snap backward, or may even fall off.
   b) An object (the head) at rest will remain at rest unless acted upon by an outside unbalanced force (the first cart crashing into it).
4. a) One way to estimate the mass, is to find the approximate volume of the head, and multiply it by the mass of 1 mL of water. The average head of a human is about 14 pounds (estimate).

5. Mark off a distance about 30-cm long on the lab table or the floor. Obtain a piece of wood and attach it to a spring scale. Pull the wooden mass with the spring scale over the distance you marked.
   a) In your log record the force required to pull the mass and the time it took to cover the distance.
   b) Repeat the step, but vary the time required to pull the mass over the distance. Record the forces and the times in your log.
   c) Use your observations to complete the following statement:

      The shorter the time (that is, the greater the acceleration) the  the force required.

6. The ratio of the mass of the wood to the estimated mass of the head is the same as the ratio of the forces required to pull them.
   a) Use the following ratio to calculate how large a force the neck muscles exert to keep the head from flying off the body, and to return the head to the upright position under different accelerations.

$$\frac{\text{mass of head}}{\text{mass of wood}} = \frac{\text{force to move head}}{\text{force to move wood}}$$

7. Whiplash is a serious injury that can be caused by a rear-end collision. The back of the car seat pushes forward on the torso of the driver and the passengers and their bodies lunge forward. The heads remain still for a very short time. The body moving forward and the head remaining still causes the head to snap backwards. The neck muscles and bones of the vertebral column become damaged. The same muscles must then snap the head back to its place atop the shoulders.
   a) What type of safety devices can reduce the delay between body and head movement to help prevent injury?
   b) What additional devices have been placed in cars to help reduce the impact of rear-end collisions?



ANSWERS

## For You To Do *(continued)*

5. a) Students' response.
   b) Students' response.
   c) Students should note that the shorter the time period, the greater the force needed to accelerate the block of wood.

6. a) Students' response.

7. a) Head rests.
   b) Rear bumpers also have the collapsing bumpers, which absorb some of the energy of the collision, as well as the crumple zone in the rear of the car. Cars also are equipped with more visible brake lights, to enable the driver behind to see the braking vehicle in front.

Activity 8 The Rear-End Collision

## FOR YOU TO READ

### Newton's Second Law of Motion

Newton's First Law of Motion is limited since it only tells you what happens to objects if net force acts upon them. Knowing that objects at rest have a tendency to remain at rest and that objects in motion will continue in motion does not provide enough information to analyze collisions. Newton's Second Law allows you to make predictions about what happens when an external force is applied to an object. If you were to place a collision cart on a level surface, it would not move. However, if you begin to push the cart, it will begin to move.

Newton's Second Law states:

**If a body is acted on by a force, it will accelerate in the direction of the unbalanced force. The acceleration will be larger for smaller masses. The acceleration can be an increase in speed, a decrease in speed, or a change in direction.**

Newton's Second Law of Motion indicates that the change in motion is determined by the force acting on the object, and the mass of the object itself.

### Analyzing the Rear-End Collision

This activity demonstrated the effects of a rear-end collision. Newton's First Law and Newton's Second Law can help explain the "whiplash" injury that passengers suffer during this kind of collision.

Imagine looking at the rear-end collision in slow motion. Think about all that happens.

1. A car is stopped at a red light. This is the car in which the driver is going to be injured with whiplash. The driver is at rest within the car.
2. The stopped car gets hit from the rear.
3. The car begins to move. The back of the seat pushes the driver forward and his torso moves with the car. The driver's head is not supported and stays back where it is.
4. The neck muscles hold the head to the torso as the body moves forward. The muscles then "whip" the head forward. The head keeps moving until it gets ahead of the torso. The head is stopped by the neck muscles. The muscles pull the head back to its usual position. Ouch!

Let's repeat the description of the collision and insert all of the places where Newton's First Law applies. Newton's First Law states that *an object at rest stays at rest and an object in motion stays in motion unless acted upon by an unbalanced, outside force.*

1. A car is stopped at a red light. This is the car in which the driver is going to be injured with whiplash. The driver is at rest within the car. *Newton's First Law: an object (the driver) at rest stays at rest.*

→

2. The stopped car gets hit from the rear.
3. The car begins to move. The back of the seat pushes the driver forward and his torso moves with the car. *Newton's First Law: an object (the driver's torso) at rest stays at rest unless acted upon by an unbalanced, outside force.* The driver's head is not supported and stays back where it is. *Newton's First Law: an object (the driver's head) at rest stays at rest.*
4. The neck muscles hold the head to the body as the body moves forward. The muscles then "whip" the head forward. *Newton's First Law: an object (the head) at rest stays at rest unless acted upon by an unbalanced, outside force.* The head keeps moving until it gets ahead of the torso. *Newton's First Law: an object (the head) in motion stays in motion.* The head is stopped by the neck muscles. *Newton's First Law: an object (the head) in motion stays in motion unless acted upon by an unbalanced, outside force.* The muscles pull the head back to its usual position. *Newton's First Law: an object at rest stays at rest unless acted upon by an unbalanced, outside force.* Ouch!

Let's repeat the description of the collision and insert all of the places where Newton's Second Law applies. Newton's Second Law states that *all accelerations are caused by unbalanced, outside forces,* $F = ma$. An acceleration is any change in speed.

1. A car is stopped at a red light. This is the car in which the driver is going to be injured with whiplash. The driver is at rest within the car.
2. The stopped car gets hit from the rear.
3. The car begins to move. *Newton's Second Law: the car accelerates because of the unbalanced, outside force from the rear;* $F = ma$. The back of the seat pushes the driver forward and his torso moves with the car. *Newton's Second Law: the torso accelerates because of the unbalanced, outside force from the back of the seat;* $F = ma$. The driver's head is not supported and stays back where it is.
4. The neck muscles hold the head to the torso as the body moves forward. The muscles then "whip" the head forward. *Newton's Second Law: the head accelerates because of the unbalanced force of the muscles;* $F = ma$. The head keeps moving until it gets ahead of the torso. The head is stopped by the neck muscles. *Newton's Second Law: the head accelerates (slows down) because of the unbalanced force from the neck muscles;* $F = ma$. The muscles pull the head back to its usual position. *Newton's Second Law: the head accelerates because of the unbalanced force from the rear;* $F = ma$. Ouch!

Newton's Second Law informs you that all accelerations are caused by *unbalanced, outside* forces. It does not say that all forces cause accelerations. An object at rest may have many forces acting upon it. When you hold a book in

Activity 8 The Rear-End Collision

your hand, the book is at rest. There is a force of gravity pulling the book down. There is a force of your hand pushing the book up. These forces are equal and opposite. The "net" force on the book is zero because the two forces balance each other. There is no acceleration because there is no "net" force.

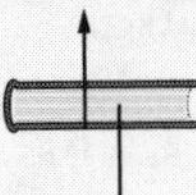

Both forces act through the center of the book. They are shifted a bit in the diagram to emphasize that the upward force of the hand acts on the bottom of the book and the downward force of gravity acts on the middle of the book.

As a car moves down the highway at a constant speed, there are forces acting on the car but there is no acceleration. This indicates that the net force must be zero. The force of the engine on the tires and road moving the car forward must be equal and opposite to the force of the air pushing the car backward. These forces balance each other in this case, where the speed is not changing. There is no net force and there is no acceleration. The car stays in motion at a constant speed. A similar situation occurs when you push a book across a table at constant speed. The push is to the right and the friction is to the left, opposing motion. If the forces are equal in size, there is no net force on the book and the book does not accelerate—it moves with a constant speed.

## Reflecting on the Activity and the Challenge

The vertebral column becomes thinner and the bones become smaller as the column attaches to the skull. The attachment bones are supported by the least amount of muscle. Unfortunately, the smaller bones, with less muscle support, make this area particularly susceptible to injury. One of the greatest dangers following whiplash is the damage to the brainstem. The brainstem is particularly vital to life support because it regulates blood pressure and breathing movements. Consider how your safety device will help prevent whiplash following a collision. What part of the restraining device prevents the movement of the head?

ANSWERS

## Physics To Go

1. The huge forces that are associated with the rear-end collision, in conjunction with little or no support for the neck from behind.
2. Inertia. An object will continue to move in a straight path until an outside force acts on it.
3. The passengers on the bus are not moving. They have a tendency to remain at rest even after the bus begins moving.
4. Because the motorcycle has less mass it will move more quickly if struck by a car.
5. The headrest would be most beneficial if you were in a rear-end collision. The passengers and driver are forced backward during the collision.
6. a) The cork will appear to move in the opposite direction to the push. Emphasize to students, that the cork is not moving, but "trying to stay motionless" (inertia).

   b) Opposite to the original push.

Safety

### Physics To Go

1. Why are neck injuries common after rear-end collisions?
2. Explain why the packages in the back move forward if a truck comes to a quick stop.
3. As a bus accelerates, the passengers on the bus are jolted toward the back of the bus. Indicate what causes the passengers to be pushed backward.
4. Why would the rear-end collision demonstrated by the laboratory experiment be most dangerous for someone driving a motorcycle?
5. Would headrests serve the greatest benefit during a head-on collision or a rear-end collision? Explain your answer.
6. A cork is attached to a string and placed in a jar of water as shown by the diagram to the right. Later, the jar is inverted.

   a) If the glass jar is pushed along the surface of a table, predict the direction in which the cork will move.

   b) If you place your left hand about 50 cm in front of the jar and push it with your right hand until it strikes your left hand, predict the direction in which the cork will move.

**Be sure the outside of the jar is dry so it does not slip out of your hands.**

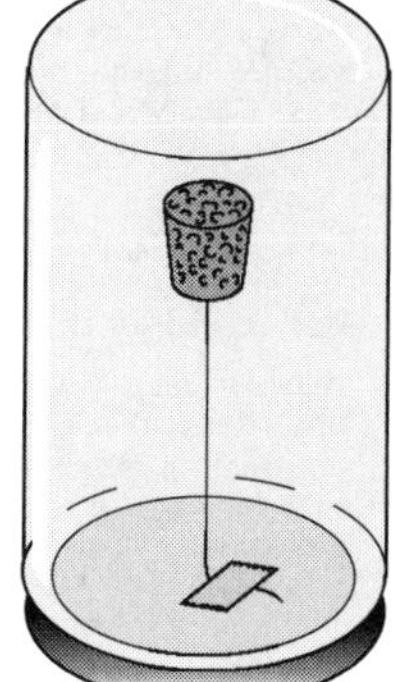

130

# ACTIVITY 9
# Cushioning Collisions (Computer Analysis)

# Background Information

In this activity the students will be putting into practice the concepts that they have learned in the past few activities. Their task is to design a cushioning device which will reduce the damage that can be done in a collision. They should be looking for cushioning material that will reduce the force as much as possible. The change in momentum will be the same with each velocity, but the force that brings the car to a rest should be as small as possible.

One other concept that may help in understanding collisions is the Conservation of Energy.

Collisions can be either elastic (energy is conserved) or inelastic (energy is not conserved). The collisions that we study are collisions (in the laboratory) where there is always a conservation of momentum. While there is conservation of momentum ($p$), there may or may not be conservation of energy. For example, there can be conservation of energy with two billiard balls moving towards each other, and colliding and moving away with the exact opposite speed and direction. This is elastic. However, in the case of most collisions, there is no conservation of energy.

In this activity, there will not be a conservation of energy. Therefore, the students will need to design a cushion which will be able to transform as much kinetic energy (energy of motion) as possible to another substance(s). For example, when you have a large ball and you drop it onto a solid floor (such as concrete), it will bounce back. Therefore, there is energy from kinetic energy falling, transferred to potential in the elastic potential energy of the ball in deformation, and subsequently transferred back into kinetic energy to move it upwards. (Assume an elastic collision, and that the ball will bounce back to its original height.)

However, if you were to drop that same ball into a box of sand, then the kinetic energy is transferred to each particle of sand moving in all directions. There is a conservation of momentum in both cases, but not a conservation of energy. In almost all cases of collisions, in reality, there is not a conservation of energy. Energy is transferred to sound, but mostly to heat due to friction of the parts heating up in the collision, and to the deformation of the car parts.

## *Active*-ating the Physics InfoMall

In addition to the information we found previously (on momentum and graphs, for example), you may wish to examine the Problems Place for even more exercises in momentum conservation. Remember, *Schaum's 3000 Solved Problems in Physics* has the problem and the solution. It can be a source for you, as well as a way to provide your students with solved problems for them to study!



# Planning for the Activity

## Time Requirements

The time for this activity will vary depending on the materials at hand. Allow time to investigate as many different kinds of materials as available, and encourage bringing materials from home. This activity can be introduced (10 minutes) one day and then allow one class (approximately 40 minutes) for the design and measuring. Allow extra time for analyzing the data and graphing the data as necessary.

## Materials Needed

### For each group:

- balance platform
- cushioning materials set, variety of
- MBL or CBL with sonic ranger
- starting ramp for lab car
- VCR & TV monitor
- tape, masking, 3/4" x 36 yds.

## Advance Preparation and Setup

If your materials are limited, use a large-screen monitor to display the data to the class. Prepare a kit for each team consisting of a toy car and a different cushioning material. Each team can prepare one car and run it as a demonstration station.

Try a few sample runs in advance to make sure that the results fall within limits of the available probes. Adjust the height of the ramp accordingly. Demonstrate the operation of the sensors and the software to the class if necessary.

Providing materials produces a certain regularity and predictability to the activity. You may want to present this activity as a challenge a day in advance. The students are sure to bring some unusual cushioning material from home.

# Teaching Notes

Replaying the video of auto crashes is one way to engage students in this activity. Ask students to explain and give examples of the three types of collisions that result from a car crash. Previous activities focused on the secondary collision—the occupants with the interior of the car. The tertiary collision—the internal organs of the occupants is not easily tested. Here the focus is on the primary collision—the vehicle with another vehicle or obstacle. How can the effects of this collision be minimized? Students should be encouraged to think of other systems besides sand canisters that are designed to cushion primary collisions such as crumple zones in cars, energy-absorbing bumpers, standardized height bumpers, break-away signs and light poles, types of plantings near roads, buried guard rail ends, shape of curbs or barriers (so-called "New Jersey barriers" are shaped to rub against the tire to slow the car), etc.

In order to help students with the **WDYT**, demonstrate the operation of the force probe and focus on the impulse graph (force vs. time). Demonstrate how the measurement of the impact force and velocity prior to impact changes when the ramp angle is changed. See if the students understand the operation of the apparatus, make a measurement of the velocity of the toy car before the collision and the impulse during the collision when no cushioning is used. This baseline information (the graphs) should be provided to each lab group or left on the chalkboard during the investigation.

As each group completes the design on the cushioning system, they should test the effectiveness of their system by using the probes at the demonstration work station. Once the students have a print-out of the $F$ vs. $t$ graphs, they can return to their regular work stations to analyze the data and figure out ways of increasing the effectiveness of their cushioning system.

The balance is needed to get the mass of each cushioned car in order to calculate its momentum.

Students can work in groups of four, using similar criteria to the previous assessment criteria for group work. This activity will inevitably lead to noise, but busy noise can still be productive noise. Keep students on task by asking them questions such as: Why are you using that material? What are the independent variables? Why are you keeping the height of the ramp the same for this activity? Why are you changing the material?

Refer to the safety notes in previous activities regarding a secure ramp, and objects falling on students' toes.

Students will need to be guided to realize, again, that the increase in the time of the collision is the most important factor here. The change in momentum will always be the same, as the mass of the object never changes, and the velocity should be constant if they are releasing the car at the same height each time they do the trial.

# Activity 9 Cushioning Collisions (Computer Analysis)

**GOALS**

In this activity you will:

- Apply the concept of impulse in the analysis of automobile collisions.
- Use a computer's motion probe (sonic ranger) to determine the velocity of moving vehicles.
- Use a computer's force probe to determine the force exerted during a collision.
- Compare the momentum of a model vehicle before the collision with the impulse applied during the collision.
- Explore ways of using cushions to increase the time that a force acts during a primary collision.

### What Do You Think?

The use of sand canisters around bridge supports and crush zones in cars are examples of technological systems that are designed to minimize the impact of collisions between a car and a stationary object or another car.

- **How do these technological systems reduce the impact of the primary collision?**

Record your ideas about this question in your *Active Physics* log. Be prepared to discuss your responses with your small group and the class.

### For You To Do

1. In this investigation you will be using a force probe that is attached to a computer to determine the effectiveness of different types of cushions for a toy vehicle. Release a toy car

ANSWERS

## For You to Do

1. Student activity.

# Activity Overview

The students will be analyzing how different types of materials will affect the impact of collisions. They will relate these designs with real-life materials, such as the bumper design on cars. The design of these materials will teach about impulse and momentum.

## Student Objectives

### Students will:

- Apply the concept of impulse in the analysis of automobile collisions.
- Use a computer's motion probe (sonic ranger) to determine the velocity of moving vehicles.
- Use a computer's force probe to determine the force exerted during a collision.
- Compare the momentum of a model vehicle before the collision with the impulse applied during the collision.
- Explore ways of using cushions to increase the time that a force acts during a primary collision.

ANSWERS FOR THE TEACHER ONLY

## What Do You Think?

Looking back to the understanding of impulse and forces, we can say that the impulse is the same in any collision from the same speed—the car traveling on the highway at 55 mph is brought to rest. However, the car which crashes directly into the bridge support experiences tremendously large forces, as the time of the impulse is very small.

The car that crashes into the sand canisters, increases the time of the collision, sometimes many times longer than the first car, which allows the force on the car to be decreased by the same factor. This is similar to the example of the ball falling into the sand.

ANSWERS

## For You To Do
### *(continued)*

2.–3. Data will vary.

4. a) Data will vary.

5. a) Data will vary.

**Perform the activity outside of traffic areas. Do not obstruct paths to exits. Do not leave carts lying on the floor.**

at the top of a ramp and measure the force of impact as it strikes a barrier at the bottom. A sonic ranger can be mounted on the ramp to measure the speed of the toy car prior to the collision. Open the appropriate computer files to prepare the sonic ranger to graph velocity vs. time and the force probe to graph force vs. time.

2. Mount the sonic ranger at the bottom of a ramp and place the force probe against a barrier about 10 cm from the bottom of the ramp, as shown in the diagram. Attach an index card to the back of the car, to obtain better reflection of the sound wave and improve the readings of the sonic ranger.

3. Conduct a few runs of the car against the force probe to ensure that the data collection equipment is working properly.

4. Attach a cushioning material to the front of the car. Conduct a number of runs with the same type of cushioning. Make sure that the car is coasting down the same slope from the same position each time.

a) Make copies of the velocity vs. time and force vs. time graphs that are displayed on the computer.

5. Repeat **Step 4** using other types of cushioning materials.

a) Record your observations in your log.

6. Use the graphical information you obtained in this activity to answer the following:

a) Compare the force vs. time graphs for the cushioned cars with those for the cars without cushioning.

b) Compare the areas under the force vs. time graphs for all of the experimental trials.

c) Compute the momentum of the car (the product of the mass and the velocity) prior to the collision and compare it with the area under the force vs. time graphs.

d) Summarize your comparisons in a chart.

e) How can impulse be used to explain the effectiveness of cushioning systems?

f) Describe the relationship between impulse ($F\Delta t$) and the change in momentum ($m\Delta v$).

**PHYSICS TALK**

**Change in Momentum and Impulse**

Momentum is the product of the mass and the velocity of an object.

$$p = mv$$

where $p$ is the momentum,
$m$ is the mass,
and $v$ is the velocity.

Change in momentum is the product of mass and the change in velocity.

$$\Delta p = m\Delta v$$

Impulse = change in momentum

$$F\Delta t = m\Delta v$$



ANSWERS

## For You To Do
***(continued)***

6. a) Data will vary. Generally, students should notice that the force is greater on the collision without the cushioning.

b) Data will vary. If the speeds before the collisions are the same, then the areas under the graphs will be the same.

c) Mass x velocity should equal the area under the force vs. time graphs. The relevant velocity is the velocity at impact, which should be the highest value recorded by the sonic ranger. The area under the graph should equal the change in momentum of the car, since the final momentum is zero. Student answers may vary due to calculation and measurement errors.

d) Data will vary.

e) Impulse is equal to the change in momentum, and is therefore related to the mass and the velocity of the car. Depending on the cushioning, there will be an increase in the time, impulse remaining constant; thus the force acting on the car and individuals inside will be less.

f) Impulse equals the change in momentum.

### Sample Problem

A vehicle has a mass of 1500 kg. It is traveling at 15.0 m/s. Calculate the change in momentum required to slow the vehicle down to 5.0 m/s.

***Strategy:*** You can use the equation for calculating the change in momentum.

$$\Delta p = m\Delta v$$

Recall that the Δ symbol means "the change in." If you know the final and initial velocities you can write this equation as:

$$\Delta p = m(v_f - v_i)$$

where $v_f$ is the final velocity and

$v_i$ is the initial velocity.

***Givens:***

$m = 1500$ kg

$v_f = 5.0$ m/s

$v_i = 15.0$ m/s

***Solution:***

$$\Delta p = m\ (v_f - v_i)$$

$$= 1500 \text{ kg } (5.0 \text{ m/s} - 15.0 \text{ m/s})$$

$$= 1500 \text{ kg } (-10.0 \text{ m/s})$$

$$= -15{,}000 \text{ kg} \cdot \text{m/s}$$

### Reflecting on the Activity and the Challenge

What you learned in this activity better prepares you to defend the design of your safety system. The principles of momentum and impulse must be used to justify your design. Previously, you discovered objects with greater mass are more difficult to stop than smaller ones. You determined that increasing the velocity of objects also makes them more difficult to stop. Objects that have a greater mass or greater velocity have greater momentum.

Linking the two ideas together allows you to begin examining the relationship between momentum and impulse. For a large momentum change in a short time, a large force is required. A crushed rib cage or broken leg bones often result. The change in the momentum can be defined by the impulse on the object.

What device will you use to increase the stopping time for the **Chapter Challenge** activity? Make sure that you include impulse and change in momentum in your report. Your design features must be supported by the principles of physics.

### Physics To Go

1. Helmets are designed to protect cyclists. How would the designer of helmets make use of the concept of impulse to improve their effectiveness?
2. The Congress of the United States periodically reviews federal legislation that relates to the design of safer cars. For many years, one regulation was that car bumpers must be able to withstand a 5 mph collision. What was the intent of this regulation? The speed was later lowered to 3 mph. Why? Should it be changed again?
3. If a car has a mass of 1200 kg and an initial velocity of 10 m/s (about 20 mph) calculate the change in momentum required to:

   a) bring it to rest

   b) slow it to 5 m/s (approximately 10 mph)

ANSWERS

## Physics To Go

1. Helmets use cushioning materials that lengthen the time of impact and thus decrease the force of impact. Note that helmets are designed to break and should not be used again after an accident. A rigid, unbreakable helmet would be useless.

2. The ability of bumpers to withstand collisions at various speeds is directly related to the ability of the bumpers to minimize the force of the impact via cushioning. The intent of the regulation is to protect the consumer from expensive repairs after every fender-bender or parking lot bump. The industry was able to get the regulation changed by arguing that the initial cost of effective bumpers was too high and the added weight contributed to higher fuel consumption. Arguments for another change can be based on consumer preference, cost of injuries, and the development of new materials and technologies that would reduce initial costs.

3. a) $\Delta p = m\Delta v$;

   $\Delta p = m \times (v_2 - v_1)$;

   $\Delta p = 1200 \text{ kg} \times (0 - 10 \text{ m/s})$;

   $\Delta p = 12{,}000 \text{ kg·m/s}$

   b) $\Delta p = m\Delta v$;

   $\Delta p = m \times (v_2 - v_1)$;

   $\Delta p = 1200 \text{ kg} \times (5 \text{ m/s} - 10 \text{ m/s})$;

   $\Delta p = 6000 \text{ kg·m/s}$

Chapter 2

ANSWERS

## Physics To Go
***(continued)***

4. Impulse $\Delta p = F\Delta t$

   $\Delta p$ = 10,000 N x 1.2 s

   $\Delta p$ = 12,000 kg•m/s

   change in velocity is $\Delta v$ therefore:

   $\Delta p = m\Delta v$

   $\Delta v = \Delta p / m$

   $\Delta v$ = 12,000 k.gm/s / 1200 kg

   $\Delta v$ = 10 m/s

5. $\Delta p = F\Delta t$

   $m\Delta v = F\Delta t$

   $F = m\Delta v/\Delta t$

   $F$ = 1500 kg x 5 m/s / 0.1 s

   $F$ = 75,000 N

6. Increasing the time from 0.1 s to 2.8 seconds decreases the force acting on the car (and the driver).

   $\Delta p = F\Delta t$

   $m\Delta v = F\Delta t$

   $F = m\Delta v/\Delta t$

   $F$ = 1500 kg x 5 m/s / 2.8 s

   $F$ = 2700 N

7. Students' answers will vary.

8. The steering wheel increases the time of contact, and will decrease the force acting on the body. Similar to previous answer.

9. The first graph has a greater force over a shorter time period, and the second graph has a smaller force over a longer period of time. If you measure the area under the graph, you will find the impulse.

4. If the braking force for a car is 10,000 N, calculate the impulse if the brake is applied for 1.2 s. If the car has a mass of 1200 kg, what is the change in velocity of the car over this 1.2 s time interval?
5. A 1500-kg car, traveling at 5.0 m/s after braking, strikes a power pole and comes to a full stop in 0.1 s. Calculate the force exerted by the power pole and brakes required to stop the car.
6. For the car described in **Question 5**, explain why a breakaway pole that brings the car to rest after 2.8 s is safer than the conventional power pole.

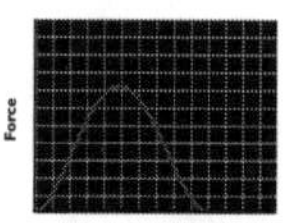


7. Write a short essay relating your explanation for the operation of the cushioning systems to the explanation of the operation of the air bags.
8. Explain why a collapsible steering wheel is able to help prevent injuries during a car crash.

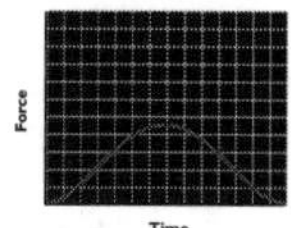


9. Compare and contrast the two force vs. time graphs shown.

### Stretching Exercises

Package an egg in a small container so that the egg will not break upon impact. Your teacher will provide the limitations in the construction of your package. You may be limited to two pieces of paper and some tape. You may be limited to a certain size package or a package of a certain weight. Bring your package to class so that it can be compared in a crash test with the other packages.

(Hint: Place each egg in a plastic bag before packaging to help avoid a messy cleanup.)

# PHYSICS AT WORK

## Mohan Thomas

### DESIGNING AUTOMOBILES THAT SAVE LIVES

Mo is a Senior Project Engineer at General Motors North American Operation's (NAO) Safety Center and his responsibilities include making sure that different General Motors vehicles meet national safety requirements. Several of the design features that Mo has helped to develop have been implemented into vehicles that are now out on the road.

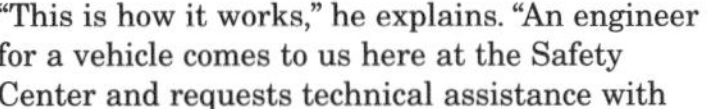

"This is how it works," he explains. "An engineer for a vehicle comes to us here at the Safety Center and requests technical assistance with design features to help them meet the side impact crash regulations required by the government. You have to analyze the physical forces of an event, which involves one car hitting another car on the side and then the door smashing into the driver," he continues. "We'll study the velocity, acceleration, momentum, and inertia in an event, as well as the materials used in the vehicle itself."

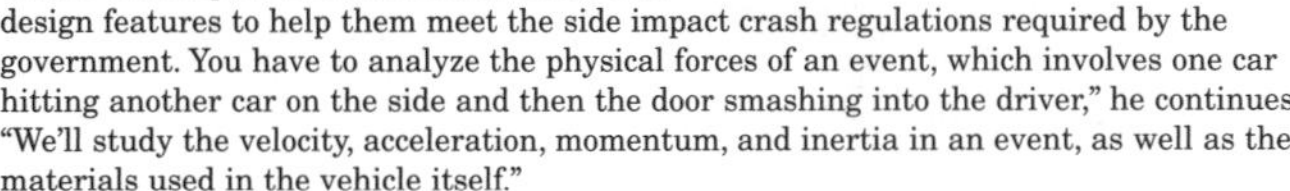

"The initial energy of an impact from one vehicle on another," states Mo, "has to be managed by the vehicle that's getting hit. Our goal is to manage the energy in such a way that the occupant in the vehicle being hit is protected. You take the forces that are coming into the vehicle and you redirect them into areas around the occupant. The framework of the car, therefore, is very important to the design, as well as energy-absorbing materials used in the vehicle."

Mo grew up in Chicago, Illinois, and has always enjoyed math and science, but he was also interested in creative writing. He wanted to combine math and science with creative work and has found that combination in the design work of engineering. "The nice part of being at the Safety Center," states Mo, "is that you know that you are contributing to something meaningful. The bottom line is that the formulas and problems that we are working on are meant to save people's lives."

Chapter 2

# Chapter 2 Assessment

Your design team will develop a safety system for protecting automobile, airplane, bicycle, motorcycle or train passengers. As you study existing safety systems, you and your design team should be listing ideas for improving an existing system or designing a new system for preventing accidents. You may also consider a system that will minimize the harm caused by accidents.

Your final product will be a working model or prototype of a safety system. On the day that you bring the final product to class, the teams will display them around the room while class members informally view them and discuss them with members of the design team. At this time, class members will generate questions about each others' products. The questions will be placed in envelopes provided to each team by the teacher. The teacher will use some of these questions during the oral presentations on the next day. The product will be judged according to the following:

1. The quality of your safety feature enhancement and the working model or prototype.
2. The quality of a 5-minute oral report that should include:
   - **the need for the system**
   - **the method used to develop the working model**
   - **demonstration of the working model**
   - **discussion of the physics concepts involved**
   - **description of the next-generation version of the system**
   - **answers to questions posed by the class**
3. The quality of a written and/or multimedia report including:
   - **the information from the oral report**
   - **documentation of the sources of expert information**
   - **discussion of consumer acceptance and market potential**
   - **discussion of the physics concepts applied in the design of the safety system**

## Criteria

Review the criteria that were agreed to at the beginning of the chapter. If they require modification, come to an agreement with the teacher and the class.

Your project should be judged by you and your design team according to the criteria before you display and share it with your class. Being able to judge the quality of your own work before you submit it is one of the skills that will make you a "treasured employee"!

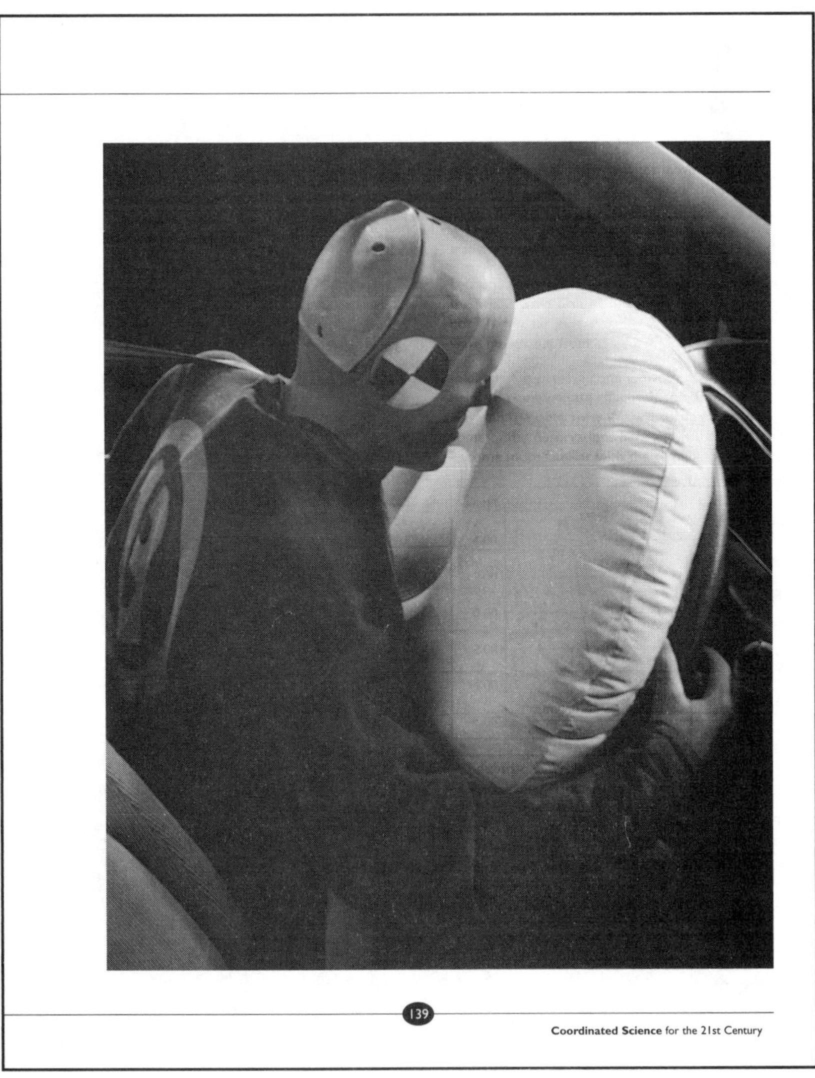

# Physics You Learned

Newton's First Law of Motion (inertia)

Pressure ($N/m^2$)

Pressure = $\frac{\text{force}}{\text{area}}$ $(P = \frac{F}{a})$

Distance vs. time relationships

Time interval = $\text{time}_{(final)} - \text{time}_{(initial)}$ $(\Delta t = t_f - t_i)$

Impulse (N × *time*),

Impulse = force × time interval (Impulse = $F \times \Delta t$)

Stopping distance

Newton's Second Law of Motion, constant acceleration, net force

Force = mass × acceleration

$F = ma$

Momentum = mass × velocity $(p = mv)$

Conservation of momentum

Change in momentum is affected by mass and change in velocity

Change in momentum = mass × change in velocity $(\Delta p = m\,\Delta v)$

Impulse = change in momentum $(F\Delta t = m\Delta v)$

140

Active Physics

NOTES

Chapter 2

# Alternative Chapter Assessment

## Part A: Multiple Choice:

Choose the best answer and place on your answer sheet.

1. Safety belts

a) always protect you from death in an accident

b) seldom protect you from death in an accident

c) increase the chances of you suffering severe injuries or death

d) decrease the chances of you suffering severe injuries or death

2. Seat belts are more effective

a) when used alone

b) when used with shoulder belts

c) when used with air bags

d) when used with shoulder belts and air bags

3. Which of the following are safety features, that have been added to most vehicles in the last ten years?

a) shoulder belts in back seats

b) air bags

c) ABS brakes

d) all of these

4. A seat belt is a safety device used in a vehicle because in a low-speed accident

a) they slow down the vehicle

b) they slow down the occupants in the vehicle

c) they prevent the occupants from flying around inside the vehicle

d) they prevent the occupants from dying

5. In a collision, there is generally more than one collision. Which of the following would describe the collisions of an accident?

a) when the car hits a stationary object, and the occupant hits the dashboard

b) when the car hits a stationary object, the occupant hits the dashboard, and the brain collides with the inside of the skull

c) when the occupant hits the dashboard, and the brain collides with the inside of the skull

d) when the car hits a stationary object, the occupant flies out of the vehicle, and hits the ground and stops

6. Newton's First Law is

a) inertia

b) impulse

c) momentum

d) force

7. In order to have a change in inertia, you must have

a) no force

b) balanced forces

c) a force opposite to the object's motion

d) an accelerating force

8. Momentum is the product of

a) mass and acceleration

b) mass and velocity

c) velocity and acceleration

d) force and velocity

9. The reason for having a wide seat belt as opposed to having a thinner seat belt is

a) it is more comfortable when driving

b) it reduces the force on your body when stopping

c) it increases the impulse on your body when stopping

d) it decreases the pressure of the seat belt on your body while stopping

10. Impulse is the same as

a) change in velocity

b) change in impulse

c) change in momentum

d) change in acceleration

11. By wearing a seat belt, and a shoulder belt, you are potentially reducing the damage to your body because

a) you are decreasing the velocity of the car

b) you are increasing the impulse of the car

c) you are decreasing the change in momentum of the car

d) you are increasing the time of the impulse of the car

12. A car was travelling at 35 m/s. The car hit a tree, and the crush zone in the car was only 0.40 m. What was the acceleration of the driver of the car?

a) 1521 $m/s^2$

b) 3182 $m/s^2$

c) 7 $m/s^2$

d) 14 $m/s^2$

13. The average speed of a car that is slowing down from 25 m/s to 0 m/s in 3.0 s is

a) 25 m/s

b) 12.5 m/s

c) 8.3 m/s

d) 0 m/s

14. The acceleration of an airplane travelling at 65 m/s that stops on a runway 200 m long in 10 s is

a) 6.5 $m/s^2$

b) 650 $m/s^2$

c) 0.325 $m/s^2$

d) 3.1 $m/s^2$

15. The force to stop a vehicle is supplied by

a) the friction caused by the road against the tires

b) the friction caused by the motor slowing down

c) the friction caused by the brake pads against the moving wheels

d) the inertia of the car

16. The bungee cord works to reduce the force on your body by

a) decreasing the time you are falling

b) increasing the time you are falling

c) increasing the time the elastic part of the cord is slowing you down

d) decreasing the time the elastic part of the cord is slowing you down

17. What is the impulse required to bring a 1000-kg vehicle moving at 25 m/s to a stop in 5.0 s?

a) 25,000 kgm/s

b) 125,000 kgm/s

c) 40 kgm/s

d) 5000 kgm/s

18. What is the momentum of a 75-kg tennis player moving at 6.5 m/s?

a) 488 kgm/s

b) 736 kgm/s

c) 11.5 kgm/s

d) 81.5 kgm/s

Use the following information to answer questions 19 - 21.

A cyclist (mass of cyclist and bicycle is 90 kg) is travelling along a road at 58 km/h (16 m/s).

19. What force is necessary to bring him to rest in 10 s?

a) 56 N

b) 144 N

c) 522 N

d) 14,400 N

20. Where does the force necessary to bring him to a stop come from?
a) no force is necessary, as he will glide to a stop
b) the friction between the road and the tires
c) the friction between the wheels and the brakes
d) friction from wind resistance and the road

21. If time were to change, which time would most likely do the most damage to the cyclist and his bike?
a) 1 s
b) 5 s
c) 15 s
d) 20 s

22. Which of the following would be very important when designing an automatic triggering device?
a) the device must trigger 100 % of the time
b) the device must trigger at high speeds when brought to a slow stop
c) the device must trigger at low speeds when brought to a sudden stop
d) none of these

23. The sand or water canisters you see around bridge supports are similar to air bags in that they
a) are large and protect something hard (cement or dashboard)
b) reduce the impulse of the collision
c) increase the force by decreasing the time of the collision
d) decrease the force by increasing the time of the collision

## Part B: Matching

Match the words or phrases on the left with the matching word or phrase on the right.

| | |
|---|---|
| __ 1. average speed | A. increase the time of collision |
| __ 2. acceleration | B. final velocity + initial velocity /2 |
| __ 3. impulse | C. change in momentum |
| __ 4. momentum | D. part of the car to reduce injuries in collisions |
| __ 5. decrease force | E. reduce energy and therefore velocity in bungee cord |
| __ 6. friction | F. change velocity divided by change in time |
| __ 7. crush zone | G. product of mass and velocity |
| __ 8. inertia | H. Newton's First Law of Motion |

## Part C: Written Response

1. Describe why seat belts are an important safety device. Be sure to include how physics plays a role.

   A: Answers will vary. Look for seat belts keep the occupant in the car; seat belts slow the occupant down in a more controlled fashion; they spread out the force over a large area of the body, rather than concentrate it in one spot (head on dashboard); they decrease the extent of injury, thus keeping the cost of health care lower.

2. Describe what type of material would make the best kind of seat belt. Include in your answer such qualities as comfort, safety, pressure, cost.

   A: Answers will vary. The ideal type of material, would be inexpensive, have high tensile strength, washable, would decrease the force per unit area (pressure) to the body to an acceptable level, easy to install, readily available materials, be soft to touch so as to be more comfortable, be relatively light.

3. Describe the role of inertia in the three collisions that occur in an accident.

   A: Inertia is the tendency for an object to stay in motion, therefore, when a vehicle stops abruptly (in a collision), the first collision is when the car hits the wall, and the force of the wall causes the vehicle to come to a rest. Because the passenger is in motion, the second collision comes when the force of the vehicle, at rest now, exerts the force against the passenger causing it to come to a rest. The third collision is the force exerted by the inside of the skull, now at rest, against the moving brain.

4. Describe how a parachute works in helping prevent injury to the body. (Be sure to include impulse and force into your answer.)

   A: A parachute will slow down your descent, where the velocity is much decreased. This slower velocity gives a lower momentum ($m$ x $v$), which therefore lowers the impulse of the parachutist landing on the ground. $F\Delta t = m\Delta v$.

5. Explain why a helmet is designed to break upon impact.

   A: A helmet is designed to increase the time of the collision, therefore decreasing the force of the impact. If the helmet was rigid, then the total force of the collision would be transferred to the head, and serious damage would result.

6. Your baseball coach has been reminding you to stay in contact with the ball and follow through when you hit. Explain why, in terms of the physics you learned about in this chapter, this is important.

   A: The longer ($\Delta t$) you can keep in contact ($F$) with the ball the greater the impulse ($F\Delta t$). Therefore, the greater the impulse, the greater the change in momentum ($F\Delta t = m\Delta v$). (Mass stays constant).

7. Explain how an air bag works. Include the terms pressure, force, impulse, change in momentum.

   A: Answers will vary. An air bag essentially changes the force that is applied to the head during a collision because the time of the collision is larger, and the impulse or change in momentum is the same. Because the air bag has a large surface area, the force is spread out over a large area, causing the pressure on the head to be smaller.

8. You have recently dislocated your shoulder, but it doesn't hurt that badly and you are sitting in the seat next to the emergency exit on the plane. Your friend says that you should change seats with him so that he is next to the seat. Explain why your friend thinks it is a good idea.

   A: Because your need to be able to operate the emergency exit, efficiently, and with a dislocated shoulder, even though it doesn't hurt, may hinder or slow your actions and response time in an emergency.

# Alternative Chapter Assessment Answers

## Part A: Multiple Choice:

1. d
2. d
3. d
4. c
5. b
6. a
7. d
8. b
9. d
10. c
11. d
12. c
13. b
14. a
15. a
16. c
17. a
18. a
19. b
20. b
21. a
22. a
23. d

## Part B: Matching:

1. B
2. F
3. C
4. G
5. A
6. E
7. D
8. H.

NOTES

# Chapter 3

# LET US ENTERTAIN YOU

## Chapter 3- Let Us Entertain You

# National Science Education Standards

### Chapter Challenge

The use of sound and light in the entertainment industry provides the scenario for this chapter. Students are challenged to design a sound and light show that demonstrates the physics principles they learned, yet is low budget. They are limited to using only sounds that come from human voices or homemade instruments and light from conventional household lamps.

### Chapter Summary

To gain understanding of science principles necessary to meet this challenge, students work collaboratively on activities to learn about wave motion, sound waves, light rays, and how mirrors and lenses change the direction of light rays and result in formation of images. They learn to use the iterative process of engineering design, refining designs based on the physics they learn. These experiences engage students in the following content from the National Science Education Standards.

# Content Standards

### Unifying Concepts

- Evidence, models and explanations
- Constancy, change, and measurement

### Science as Inquiry

- Identify questions and concepts that guide scientific investigations
- Use technology and mathematics to improve investigations
- Design and conduct scientific investigations

### Science and Technology

- Abilities of technological design
- Identify a problem or design an opportunity
- Propose designs and choose between alternate solutions

### Physical Science

- Structure and properties of matter
- Motions and forces
- Interactions of energy and matter

# Key Physics Concepts and Skills

| Activity Summaries | Physics Principles |
|---|---|
| **Activity 1: Making Waves**<br>Students begin the chapter by making waves with a Slinky®, observing pulses, periodic, standing, and compressional waves. From these observations, then measurements, they establish the relationships among wavelength, frequency, and speed of the wave. | • Wave motion<br>• Periodic, standing, and compressional waves<br>• Wavelength – Frequency, speed |
| **Activity 2: Sounds in Strings**<br>To connect waves to sound, students observe the vibration of a plucked string and compare how vibration and pitch vary when the tension of the string changes. They then explore the affect on vibrations and pitch when the length of the string is changed. Reading explains the physics concepts in the observed phenomena. | • Sound waves<br>• Wave motion<br>• Tension and pitch<br>• Frequency |
| **Activity 3: Sounds from Vibrating Air**<br>Drinking straws and test tubes filled with water are used to model instruments that use columns of vibrating air to produce sounds. The relationship of pitch to length of the column of air provides another look at frequency and wavelength, helping students understand how sound is produced by compressional and standing waves. | • Sound waves<br>• Wave motion<br>• Compressional waves<br>• Frequency and wavelength |
| **Activity 4: Reflected Light**<br>In this activity, students begin looking at how light can be incorporated into the chapter challenge. They explore the result of changing the angle at which light rays are aimed at a mirror and learn to predict and control where images will be visible. | • Light rays<br>• Reflection of light<br>• Real images—focus, focal length |
| **Activity 5: Curved Mirrors**<br>Shining a light beam on concave and convex mirrors increases student understanding of the variables that are involved in creating an image. They apply what they have learned to predict the path of a light beam reflected off a mirror. | • Reflection of light<br>• Real images—focus, focal length<br>• Controlling variables |
| **Activity 6: Refraction of Light**<br>In this activity, a block of gelatin allows students to explore what happens when light goes from air into another substance. They observe and measure the angle of incidence and the angle of refraction as they learn about Snell's Law and how to mathematically predict where the beam of light can be observed. | • Refraction of light<br>• Snell's Law |
| **Activity 7: Effect of Lenses on Light**<br>Shining a light through different lenses enables students to observe how focal length and the size of the image changes as the light source moves closer to, then farther away from a lens. They then consider how the variables in this phenomenon can enhance their sound and light show for the chapter challenge. | • Refraction of light<br>• Lenses and image formation<br>• Focus, focal length |
| **Activity 8: Color**<br>This final activity adds to the study of light with observations of shadows. By carefully tracing the light ray and noting the areas without any light and the areas of gray light, students begin to learn about diffusion of light. They extend their investigations to include the effect of shining different colored lights on objects. | • Light and shadows<br>• White light<br>• Color addition |

# GETTING STARTED WITH EQUIPMENT NEEDED TO CONDUCT THE ACTIVITIES.

### Items needed—not supplied in Material Kits

Preparing the equipment needed for each activity in this chapter is an important procedure. There are some items, however, needed for the chapter that are not supplied in the It's About Time material kitpackage. Many of these items may already be in your school and would be an unnecessary expense to duplicate. Please read carefully the list of items to the right which are not found in the supplied kits and locate them before beginning activities.

**Items needed—not supplied by It's About Time:**

- **Copies of Blackline Master of Sine Waves**
- **Copies of Blackline Master of Wavelength Tables**
- **Copies Of Blackline Master Pitch vs. Tube Length**
- **Ripple Tank**
- **Pencil**
- **Water, about 300 cc in a container**
- **Overhead Projector**

# Equipment List For Chapter 3 (Serves a Classroom of 30 Students)

| PART | ITEM | QTY | ACTIVITY |
|---|---|---|---|
| 22-2000-C1 | 40-W Bulb and Socket, Ceramic w/Switch | 6 | 6, 8, 9 |
| AC-2279-C1 | Acrylic Block | 6 | 7 |
| AS-6105-C1 | Acetate Sheet, 8 cm Square | 6 | 8 |
| BS-1608-C1 | AA Battery | 6 | 5 |
| BS-5903-C1 | Bulb, Clear 100 Watt | 6 | 8 |
| BU-0004-C1 | Bulb, Set/4, Red | 6 | 9 |
| BU-0003-C1 | Bulb, Set/4, Blue | 6 | 9 |
| BU-0002-C1 | Bulb, Set/4, Green | 6 | 9 |
| BU-0001-C1 | Bulb, Set/4, White | 6 | 9 |
| CS-3246-C1 | C-Clamp | 6 | 2 |
| CS-6111-C1 | Cardboard Sheet, 14x14 Square | 6 | 5, 6, 7 |
| CS-5011-C1 | Unlined Index Cards | 600 | 1, 6, 7 |
| FF-7704-C1 | File Folder | 36 | 9 |
| FL-0030-C1 | Fishing Line Spool | 6 | 2 |
| GS-5016-C1 | Goggles, Safety | 6 | 2 |
| LH-6534-S-C1 | Laser Pointer | 6 | 5, 6, 7 |
| LS-0015-C1 | 1.5 Volt Miniature Light Bulb | 6 | 5 |
| LS-1001-C1 | Lens, Convex Focal Point 10 cm | 6 | 8 |
| LS-4119-C1 | Stand for Clip-on Lamp | 18 | 9 |
| LS-6600-C1 | Light Source for Optical Bench Apparatus | 6 | 5 |
| LS-9000-C1 | Lamp, Clip-On w/Shade | 18 | 9 |
| MH-2621-C1 | Mass Hanger | 6 | 2 |
| MS-1006-C1 | Convex/Concave Mirror Surface | 6 | 6 |
| MS-1425-C1 | Marker, Felt-Tip | 6 | 8 |
| MS-2114-C1 | Mirror, Plane, 3x4x1/2 | 12 | 5 |
| CM-0002 | Cosmetic Mirror | 6 | 4 |
| OH-1404-C1 | Optical Bench Apparatus | 6 | 5 |
| PH-3070-C1 | Pulley on Mount | 6 | 2 |
| PM-0521-C1 | Protractor | 6 | 5, 7 |
| PR-0001-C1 | Right Angle Prism | 6 | 5, 6, 7 |
| RH-6131-C1 | Rod, Glass | 6 | 5, 6, 7 |
| RS-2826-C1 | Ruler, Metric | 6 | 3, 5, 6, 7, 8 |
| SH-3639-C1 | Heavy Duty Slinky® | 6 | 1 |
| SM-1676-C1 | Stick, Meter, 100 cm, Hardwood | 6 | 1, 2, 6, 8 |
| SS-1281-C1 | Steel Scissors | 6 | 1, 3, 8, 9 |
| SS-6113-C1 | Drinking Straws | 60 | 3 |
| SS-6133-C1 | Support for Mirrors | 12 | 5 |
| SS-6135-C1 | Screen, White | 24 | 9 |
| SS-7778-C1 | Stopwatch | 6 | 1 |
| TS-0337-C1 | Test Tube | 30 | 3 |
| TS-2662-C1 | Tape, Masking 3/4x60 yds. | 6 | 1, 6 |
| WH-0500-C1 | Hooked Mass Weight, 500 Grams | 12 | 2 |
| WH-9012-C1 | Slotted Weight Set | 6 | 2 |
| WS-6136-C1 | Wood Block to Raise Light Source | 18 | 7 |
| ZZ-0066-C1 | 66 qt. Clear Plastic Storage Container | 6 | |
| | | | |
| THINGS NEEDED NOT SUPPLIED | | | |
| | Ripple Tank | 1 | 1 |
| | Pencil | 6 | 2 |
| | Key or other small metal object | 6 | 2 |
| | Water, about 300 cc in a container | 6 | 3 |
| | Various Wind Instruments | 1 | 3 |
| | Sound probe and software | 1 | 3 |
| | Small amount of play sand | 6 | 4 |
| | Oscilloscope | 1 | 4 |
| | Overhead Projector | 1 | 8 |
| | File cards one with an arrow shaped opening | 6 | 8 |

## Organizer for Materials Available in Teacher's Edition

| Activity in Student Text | Additional Material | Alternative / Optional Activities |
|---|---|---|
| ACTIVITY 1: Making Waves, p. 144 | Amplitude and Wavelength Measurements, p. 287<br>Sine Waves for Wave Model, p. 288 | Activity 1 A: Exploring Water Waves p. 286 |
| ACTIVITY 2: Sounds in Strings, p. 158 | | |
| ACTIVITY 3: Sounds from Vibrating Air, p. 164 | Pitch vs. Tube Length, p. 309 | Activity 3 A: Wind Instruments and Resonant Frequencies, p. 308 |
| ACTIVITY 4: Reflected Light, p. 172 | Angles of Incidence and Reflection, p. 323 | Activity 4 A: Low-Tech Reflection and Mirror Symmetry, p. 322 |
| ACTIVITY 5: Curved Mirrors, p. 180 | Bulb and Image Distances, p. 338<br>Graph Paper, p. 339 | Activity 5 A: Low-Tech Reflection: Curved Mirror, p. 337 |
| ACTIVITY 6: Refraction of Light, p. 189 | Sines of Angles of Incidence and Reflection, p. 349 | |
| ACTIVITY 7: Effects of Lenses on Light, p. 195 | Determining Distances of Objects and Images and Appearances, p. 362 | |
| ACTIVITY 8: Color, p. 204 | | |

## Scenario

Most entertainment today comes from the communication of sound and light signals. You look forward to television shows, movies, and rock concerts. The sound signals that entertain you come from voices or musical instruments. Light signals make the images you see on TV or in the movies. Specially designed light patterns add to the effect of a rock concert.

## Challenge

You have been made part of a committee to design a two- to four-minute sound and light show to entertain other students your age. But unlike the professionals, you have neither the funds nor the technology available to rock stars or MTV™ productions. All the sounds you use must come from musical instruments or sound makers that you build yourself, or from human voices. Some of these sounds may be prerecorded and then played back during your show. If your teacher has a laser and is willing to allow you to use it, you may do so. All other light must come from conventional household lamps.

142

## Chapter and Challenge Overview

Chapter 3 challenges groups of students to produce a two–to four–minute light show. In addition, each student must write a report that lists the physics concepts used in the show, gives an example of each, and explains why each concept is important. The shows are evaluated for creativity, and the reports are evaluated for the understanding of the concepts.

Sound and light are the basis of much of the entertainment industry. Beyond fireworks and light shows, sound is transmitted by radio, sound and light are transmitted by television, and the videotape and CD store the information for bringing us the sound and light on demand. A sound and light show is an excellent opportunity for your students to show what they have learned and to have some fun at the same time.

Students begin the unit in **Activity 1** by making pulses and standing waves on the Slinky®. They determine the relationship among speed, frequency, and wavelength. Also, they build a model to help understand wave motion. Next, students begin the study of musical instruments. In **Activity 2** they investigate the sounds made by a vibrating string. They observe the effect of changes in string length and tension on the pitch the string produces. Students continue the study of musical instruments by investigating the resonance of pipes in **Activity 3**. They observe how the pitch produced depends on the length of the pipe and whether or not one end is closed. Through reading, they then relate these observations to the patterns of standing waves in a tube.

Next, the students begin the investigation of mirrors. In **Activity 4**, they investigate reflection from a plane mirror. They measure the angles of incidence and refraction for reflected light beams, locate the reflected image, and observe the reversals in reflections of letters of the alphabet. Finally, they investigate multiple reflections. **Activity 5** introduces curved mirrors. In this activity, students investigate concave and convex mirrors. Students begin by shining parallel light beams at curved mirrors to find the focal length and the location of the focus. With a light bulb, students look for real and virtual images and investigate image distance vs. object distance.

To prepare for the study of lenses, students investigate refraction in **Activity 6** by shining a beam of light into a rectangular acrylic block. The students measure the angles of incidence and refraction. They also observe total internal reflection and measure the critical angle. Having studied refraction, students then explore convex lenses in **Activity 7**. They observe the real image made by a convex lens and how its size and position change as the object distance changes. To prepare for the light show, the students project the image of a slide onto a wall. In **Activity 8** the students investigate additive color mixing. They observe shadows of objects illuminated by combinations of colored lights. For each such combination they make a drawing to record the pattern of the shadow colors.

# Criteria

Work with your classmates to agree on the relative importance of the following assessment criteria. Each item in the list has a point value given after it, but your class must decide what kind of grading system to use.

1. The variety and number of physics concepts used to produce the light and sound effects: 30 points

   four or more concepts: 30 points

   three concepts: 25 points

   two concepts: 20 points

   one concept: 10 points

2. Your understanding of the physics concepts: 40 points

   Following your production, you will be asked to:

   a) Name the physics concepts that you used. 10 points
   b) Explain each concept. 10 points
   c) Give an example of something that each concept explains or an example of how each concept is used. 10 points
   d) Explain why each concept is important. 10 points

   As a class, you will have to decide if your answers will be in an oral report or a written report.

3. Entertainment value: 30 points

   Your class will need to decide on a way to assign points for creativity. Note that an entertaining and interesting show need not be loud or bright.

You will have a chance later in the chapter to again discuss these criteria. At that time, you may have more information on the concepts and how you might produce your show. You may want to then propose changes in the criteria and the point values.

## Assessment Rubric: Sound and Light Show

| | |
|---|---|
| **Meets the standard of excellence.**<br>**5** | • Four or more physics concepts are appropriately used to produce the sound and light show.<br>• All sounds produced have excellent quality and come from more than one musical instrument or sound maker designed and constructed by the student.<br>• Additional research, beyond basic concepts presented in the chapter, is evident.<br>• Physics concepts are integrated in a creative manner into the show.<br>• The show is entertaining and interesting. |
| **Approaches the standard of excellence.**<br>**4** | • Three physics concepts are appropriately used to produce the sound and light show.<br>• All sounds produced have good quality and come from more than one musical instrument or sound maker designed and constructed by the student.<br>• Physics concepts are integrated in a creative manner into the show.<br>• The show is entertaining. |
| **Meets an acceptable standard.**<br>**3** | • Two physics concepts are appropriately used to produce the sound and light show.<br>• Sounds produced have a reasonable quality and come from a musical instrument or sound maker designed and constructed by the student.<br>• The presentation shows creativity.<br>• The show has some entertainment value. |
| **Below acceptable standard and requires remedial help.**<br>**2** | • One physics concept is used to produce the sound and light show.<br>• An attempt has been made to design and construct a musical instrument. Quality of sound produced is extremely poor.<br>• The presentation lacks creativity.<br>• The show has little entertainment value. |
| **Basic level that requires remedial help or demonstrates a lack of effort.**<br>**1** | • No physics concepts are used in the production of the sound and light show.<br>• No attempt has been made to design and construct a musical instrument.<br>• The presentation lacks creativity.<br>• The show has no entertainment value. |

For use with *Let Us Entertain You*, Chapter 3

# Assessment Rubric: Written Report

| | |
|---|---|
| **Meets the standard of excellence.** **5** | • Four physics concepts are named and clearly explained.<br>• The importance of each concept is explained.<br>• An example is provided for each concept.<br>• Scientific vocabulary and symbols for units are used consistently and precisely.<br>• Sentence structure is consistently controlled. Spelling, punctuation, and grammar are consistently used in an effective manner.<br>• Where appropriate, data is organized into tables or presented by graphs or diagrams. |
| **Approaches the standard of excellence.** **4** | • Three physics concepts are named and clearly explained.<br>• The importance of each concept is explained.<br>• An example is provided for each concept.<br>• Scientific vocabulary and symbols for units are used appropriately.<br>• Sentence structure is consistently controlled. Spelling, punctuation, and grammar are consistently used in an effective manner.<br>• Some data is organized into tables or presented by graphs or diagrams. |
| **Meets an acceptable standard.** **3** | • Two physics concepts are named and clearly explained.<br>• The importance of each concept is explained.<br>• An example is provided for each concept.<br>• Use of scientific vocabulary and symbols for units is evident.<br>• Sentence structure is generally controlled. Spelling, punctuation, and grammar do not impede the meaning.<br>• Very limited presentation of data by diagrams, tables, or graphs. |
| **Below acceptable standard and requires remedial help.** **2** | • One physics concept is named and explained.<br>• An attempt is made to explain the importance of the concept.<br>• Limited use of scientific vocabulary and symbols for units. Usage is not always consistent or precise.<br>• Sentence structure is poorly controlled. Spelling, punctuation, and grammar impedes the meaning.<br>• No presentation of data by diagrams, tables, or graphs. |
| **Basic level that requires remedial help or demonstrates a lack of effort.** **1** | • No physics concepts are named and explained or physics concepts named are explained incorrectly.<br>• No attention to use of scientific vocabulary or scientific symbols for units of measurement.<br>• Sentence structure is poorly controlled. Spelling, punctuation, and grammar impedes the meaning. |

For use with *Let Us Entertain You*, Chapter 3

# What is in the Physics InfoMall for Chapter 3?

While teaching a physics course, it is often nice to have a source of information other than only the textbook used in the course. One source of information for not just physics, but how to teach physics, is the Physics InfoMall CD-ROM. This duo-platform CD-ROM is a huge database containing 19 textbooks, thousands of articles from physics journals, thousands of homework-style problems (many with solutions available to you), numerous demonstration and laboratory ideas, catalogs, and books covering related topics. In addition to this, you can also find historical information, and more.

The organization of the database is similar to a shopping mall; there are various stores that contain related types of information. The Textbook Trove has textbooks, the Book Basement has related material, the Article & Abstracts Attic has thousands of journal articles, etc. You can browse the stores, just looking to see if there is something interesting you have not seen before, or you can ask for specific information. The database is not interactive; it is a collection of text and graphics, with hyperlinks and a powerful search engine. The power of the search engine in nearly as important as the volume of material on the CD-ROM.

If you are already familiar with using the search engine, you may wish to skip this paragraph and the next. The search engine for the InfoMall enables the user to locate almost any word or passage included in the many articles, textbooks, etc. If you use a Macintosh (remember—the CD-ROM works on both PC and Mac), you can search the entire database (that is, all stores) at once. This discussion was prepared using a Mac. If you use a PC, there will be minor differences but the database is identical. The same information can be found on either platform. There are two categories for searches: Simple and Compound. Simple searches look for a single word or phrase. Compound searches can locate multiple words or phrases that may not occur together. The Compound search also provides search options that are not available in the Simple search. Since a Compound search can always be done for a single word, it provides the same function as a Simple search. All searches in this discussion will be done using the Compound search.

The search option can be found under the "Functions" menu. When you select "Compound Search", you will get a window that has all the options you need for just about any search you are likely to want. Among these choices are database selection (which store or stores you want searched), category (perhaps only certain articles), and search words. You can enter any words or words you want the engine to search for. These words can be combined with the logical operators AND, OR, and NOT. That is, you can find passages that contain, for example, "phone" AND "Bell" but NOT "monopoly". In addition, you can use an asterisk * as a wild character. That is, search word "fun*" will search for any word beginning with "fun", such as "fun", "funny", "funding", "fundamental", etc. Wild characters allow great freedom in searching, but as this example shows, you need to be careful. And finally, if your search is too broad, you will get "Too Many Hits" and you will have to restrict your search parameters to something that will not be found so many times on the CD-ROM. Such restriction techniques include searching in fewer stores, using fewer wild characters, or searching for less common words followed by a search of only the search hits. With a little practice, you should have very little trouble finding the information you desire.

When using the Physics InfoMall CD-ROM, it is easy to get in the "search" mode, where every time you want something, you engage the search engine. Don't forget that that InfoMall is also great for simply browsing. While searching for items, it is good to keep an open mind. When looking up an article, for example, there might be another article in the same journal that you can use later. You can locate it later by browsing, or place a Bookmark or Note there, so you can quickly come back to it when convenient. (Notes and Bookmarks are not used in this discussion, but you may find them useful. Consult the User's Guides for more information. Incidentally, the User's Guides are on the CD-ROM in the Utility Closet.)

# ACTIVITY 1
## Making Waves

# Background Information

Waves are a collective phenomena. The neighboring parts of the wave medium—the matter the wave moves through—interact and propogate a disturbance from one place to another. For spring waves, each part of the spring influences the neighboring parts through the force of the stretched spring. A disturbance in one part of the spring creates forces on neighboring parts, and the disturbance moves. The disturbance is a stretching of the spring from its equilibrium position, which is a straight spring. In **Activity 1**, the students disturb this equilibrium by whipping one end of spring back and forth to make a pulse, as shown in the drawing on page 145 of the Student Edition.

In the pulse, the spring coils are stretched. This stretching exerts forces on these coils and pulls them back towards the equilibrium position. The stretching also puts forces on the nearby parts of the spring. These forces are reaction forces. When part of the spring is pulled one way, these reaction forces pull adjacent parts the other way. That is how the pulse moves.

Notice that the spring does not have to be motionless to be in the equilibrium position. Look again at the diagram of the standing wave on page 146 of the Student Edition. For all standing waves, the whole spring goes through the equilibrium position twice in each wave cycle. As suggested above, the spring is moving at this time, so all of the energy in the spring is kinetic as the spring moves through the equilibrium position.

The most important wave properties are amplitude, frequency, wavelength, and speed. The first example discussed in teacher notes for **WDYT** is how a tsunami grows. For this, amplitude depends on wave speed. When two waves pass through each other, the two amplitudes simply add at every point, as shown. This is the principle of superposition.

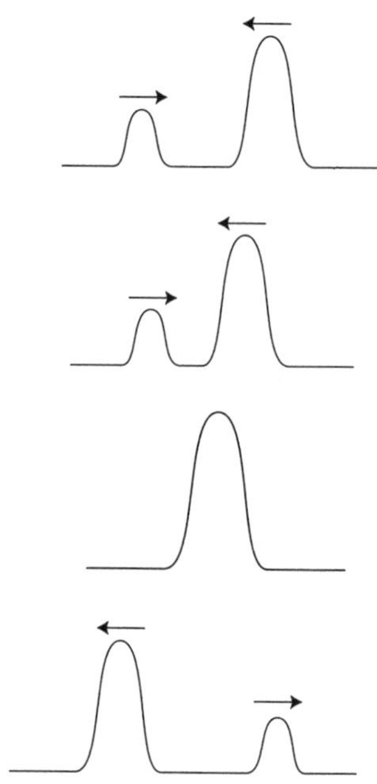

If you took a snapshot of the wave, with a meter stick right above the crests, you could measure the wavelength (the distance between adjacent crests). If you made a movie of the motion, with a clock in the frame, you could measure the frequency of the waves by timing one complete cycle. The period is simply the inverse of the frequency. Notice how the units invert as well. One hertz is a cycle per second. It is the inverse of the period, which is measured in seconds (cycles are dimensionless).

Suppose you are watching water waves washing over a rock. If you watch a particular crest for a time of one wave period, it moves one wavelength. The frequency gives the number of wave periods in one second. Thus the speed is simply the wavelength times the frequency.

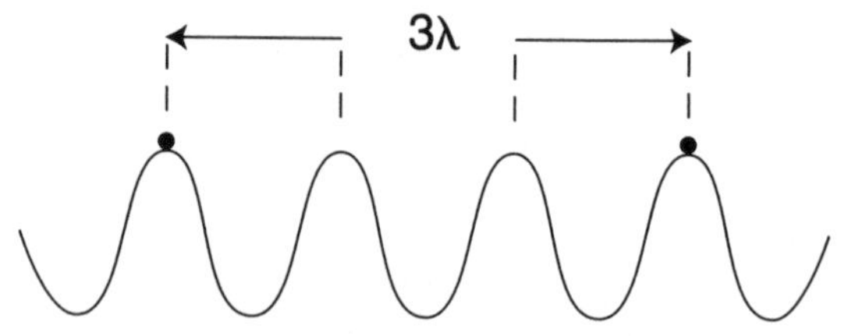


In 1s, a point on this wave moves 3λ.
Wave speed = $\lambda f$

A wave moves through a medium. The spring is the medium for Slinky waves, water is the medium for water waves, and air is the medium for sound waves. But there is one wave that does not require a medium. That is the electromagnetic wave.

Electromagnetic waves move freely through space, which is an excellent vacuum. The Sun gives off light and radio waves. Various distant objects in the universe give off every kind of electromagnetic radiation, and all these waves travel through the emptiness of space to reach the Earth.

In a standing wave, the spring vibrates but no waves are seen to move along its length. Standing waves occur only at certain frequencies, which correspond to simple patterns of motion. In the lowest-frequency standing wave on a spring, the spring bows out in the middle, with the ends, of course, fixed in place. If the frequency is doubled, there is a node, a point of no motion, in the middle of the spring. (See the drawings on page 146.)

In a standing wave, a small amount of oscillation with just the right frequency can excite a huge oscillation in the vibrating system, a phenomena called resonance. Moving the frequency away from a resonant frequency produces disorganized motion with a very small amplitude.

## *Active*-ating the Physics InfoMall

As you might imagine, there is plenty of information on "waves" on the InfoMall. Let's take a hint from the **What Do You Think?** question, and search for tsunamis. Since this word may be spelled poorly (perhaps by various authors on the InfoMall?), we can save some time and trouble by searching for "tsunam*". Note that by doing this, some spelling errors are avoided, plus we get the benefit of a search for all forms of the word (how many words begin with these six letters?). While we might not expect much from such a search, we are greatly rewarded for our efforts.

The first hit is from The Fascination of Physics, Chapter 14, Making Waves, found in the Textbook Trove. This has great information, graphics, and even the necessary wave vocabulary. All this is live text, meaning you can copy and paste your findings right into your own handouts. This includes the graphics! Much of the information you may want related to **Activity 1** is right in this single search hit. There are nice drawings of standing waves, discussion of wave speed, interference, and much more.

Other hits include a rather inexpensive experiment with a "tsunami": "Tsunamis, sealing wax, and string:" Citation: F. L. Curzon B. Neilson Physics Department, University of British Columbia, Vancouver, British Columbia V6T 1W5, Canada *Am. J. Phys.*, Vol. 44, No. 11, November 1976 Pages 1073 - 1076: A ground glass screen on the side of a wave tank has been used to record the profile left by the peak of a tsunami propagating along a wave tank with a sloping base. Using traditional sealing wax-and-string technology and old-fashioned but cheap data processing, we have carried out a $12 verification of tsunami growth. The features of the experiment are presented in enough detail to enable anyone with the necessary $20 (allowing for inflation) to investigate interesting anomalies which personal circumstances prevented us from pursuing.

You are encouraged to try this search for yourself. These two items were not the only hits from the search. Of course, you might want other types of information. Suppose you want something on "wave speed." Search with these keywords, and you find that our earlier hit is again first. Look a little further down the list and you can find some information on teaching about waves in *Teaching Physics: A Guide For The Non-Specialist*, found in the Book Basement: "Teaching of the subject of waves usually starts with an exploration of wave behaviour on springs. There are two types of springs commonly in use for this, a tightly coiled fairly massive spring, about 2 m in length, and the Slinky. The latter is very suitable for demonstration but not for use by pupils as it easily ties itself in knots. Pupils can perform set exercises with these springs to explore the factors determining wave speed, what happens to waves that cross and what happens to a wave on reflection. Essentially this is an exercise in observation which pupils enjoy but find difficult to draw the correct conclusions." You should look this up to find what difficulties students have with this topic.

Of course, waves are discussed in nearly all physics textbooks. This means that many of the words you may want to search for occur so many times that you may get too many search hits to handle effectively. This is where browsing may help (of course, you can always limit searches to single stores, for example). Simply go to the InfoMall Entrance and click on the Textbook Trove. Select any text you like. Note that there are different levels of texts, from conceptual to those that use calculus. Choose the one that best suits your students. Then simply look over the table of contents of the book you select, find the appropriate chapter, and read away! You can easily find the same information that is contained in **Physics Talk**, along with more examples of how to use the equations.

For activities, you may want to search the Demo & Lab Shop. Or use a rather crude search: "wave demo*". This produces a few hits, and some are good, but you can probably (and should try to) find better searches.

As an example of other ways you may wish to use the InfoMall, see **Physics To Go Step 3**. Students are asked to consider using a photograph to take measurements. Search the InfoMall for "measurement*" AND "photograph*". You will find several references that discuss the benefits of doing exactly that — using photographs to make measurements. Check these out.

# Planning for the Activity

## Time Requirements

two class periods

## Materials Needed

### For the class:

- ripple tank

### For each group:

- masking tape, 3/4" × 60 yds
- meter stick, 100 cm, hardwood
- Slinky®, heavy duty
- steel scissors
- stopwatch
- unlined index cards

NOTES

NOTES

# Activity Overview

To help understand basic wave properties, students make pulses on a Slinky and measure the speed of the pulse for different wave amplitudes. They create standing waves on the Slinky and determine the relationship among speed, frequency, and wavelength. Students make both transverse and compressional waves on the Slinky and contrast the motion of the two kinds of waves. Finally, they build a model to help relate basic wave properties to observations of wave motion.

## Student Objectives

### Students will:

- Observe the motion of a pulse.
- Measure the speed of a wave.
- Observe standing waves.
- Investigate the relationship among wave speed, wavelength, and frequency.
- Make a model of wave motion.

ANSWERS FOR THE TEACHER ONLY

## What Do You Think?

A tsunami begins when an earthquake shifts the ocean bottom and creates a small step in the height of the water surface. This step is usually only a few tens of centimeters high. This small disturbance in the surface spreads out in circular waves from the point above the earthquake, but the amplitude remains small as long as the water depth does not change.

However, when the wave reaches more shallow water, the wave speed drops, and the back of the wave overruns the front, increasing the amplitude just as if the waves were piling up on top of each other. This is exactly the same effect, but on a much larger scale, that creates breaking waves at ocean beaches.

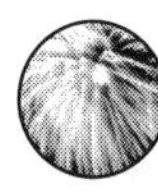

Let Us Entertain You

# Activity 1 Making Waves

**GOALS**

In this activity you will:

- Observe the motion of a pulse.
- Measure the speed of a wave.
- Observe standing waves.
- Investigate the relationship among wave speed, wavelength, and frequency.
- Make a model of wave motion.

### What Do You Think?

On December 26, 2004, one of the largest tsunamis (tidal waves) hit many countries in the Indian Ocean in Southeast Asia, causing massive damage. Some of the waves reached almost 35 feet in height.

- **How does water move to make a wave?**
- **How does a wave travel?**

Record your ideas about these questions in your *Active Physics* log. Be prepared to discuss your responses with your small group and with your class.

144

Active Physics

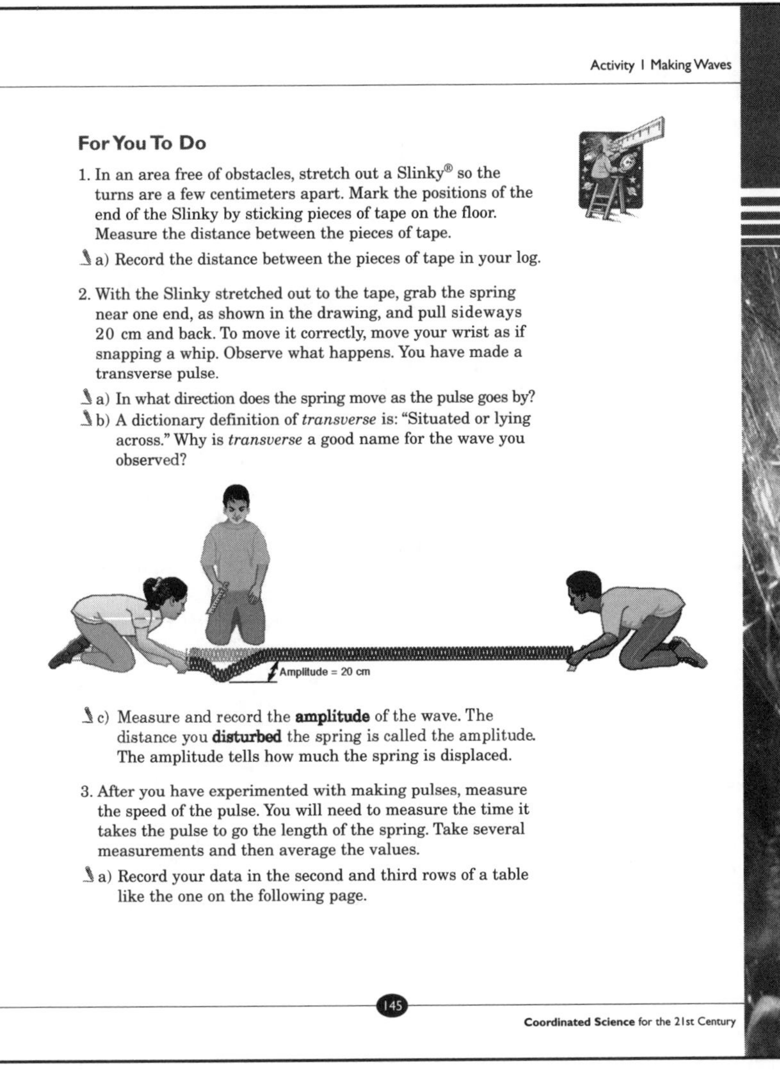



**For You To Do**

1. In an area free of obstacles, stretch out a Slinky® so the turns are a few centimeters apart. Mark the positions of the end of the Slinky by sticking pieces of tape on the floor. Measure the distance between the pieces of tape.
   a) Record the distance between the pieces of tape in your log.
2. With the Slinky stretched out to the tape, grab the spring near one end, as shown in the drawing, and pull sideways 20 cm and back. To move it correctly, move your wrist as if snapping a whip. Observe what happens. You have made a transverse pulse.
   a) In what direction does the spring move as the pulse goes by?
   b) A dictionary definition of *transverse* is: "Situated or lying across." Why is *transverse* a good name for the wave you observed?



   c) Measure and record the **amplitude** of the wave. The distance you **disturbed** the spring is called the amplitude. The amplitude tells how much the spring is displaced.
3. After you have experimented with making pulses, measure the speed of the pulse. You will need to measure the time it takes the pulse to go the length of the spring. Take several measurements and then average the values.
   a) Record your data in the second and third rows of a table like the one on the following page.



ANSWERS

## For You To Do

1. Answers will vary.
2. a) Sideways.
   b) The motion of the Slinky itself is at right angles to the motion of the pulse. Like a bridge built across a river, the Slinky motion is "across" the Slinky itself.
   c) Answers will vary.
3. The students' data for the speed of the Slinky waves will depend on the Slinky material (metal or plastic) and on how often they stretch the spring. Sample data is shown below.

3. a) Sample data

Length of Slinky = 2.5 m

| time (s) | average time (s) | speed (m/s) |
|---|---|---|
| .47 | | |
| .44 | .47 | 5.3 |
| .5 | | |

Also shown is data for a standing wave. Here the wavelength is twice the length of the Slinky.

| wavelength (m) | frequency (cycle/s) | wavelength x frequency (m/s) |
|---|---|---|
| 5 | 1 | 5 |
| 2.5 | 2 | 5 |
| 1.7 | 3 | 5 |

ANSWERS

## For You To Do
*(continued)*

4. a) Answers will vary.

   b) The speed is independent of the pulse size.

5. a) The motion will probably be disorganized.

6. Now the motion is organized. The spring looks like a sine wave.

7. a) Many wavelengths are possible.

| Amplitude | Time for pulse to travel from one end to the other | Average time | Speed = $\frac{\text{length of spring}}{\text{average time}}$ |
|---|---|---|---|
| | | | |
| | | | |
| | | | |
| | | | |

4. Measure the speed of the pulses for two other amplitudes, one larger and one smaller than the value used in **Step 3**.
   a) Record the results in the table in your log.
   b) How does the speed of the pulse depend on the amplitude?

5. Now make waves! Swing one end back and forth over and over again along the floor. The result is called a periodic wave.
   a) Describe the appearance of the periodic wave you created.

6. To make these waves look very simple, change the way you swing the end until you see large waves that do not move along the spring. You will also see points where the spring does not move at all. These waves are called standing waves.

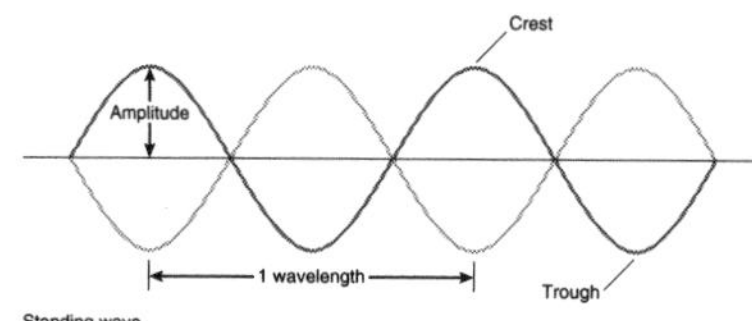

Standing wave

7. The distance from one crest (peak) of a wave to the next is called the wavelength. Notice that you can find the wavelength by looking at the points where the spring does not move. The wavelength is twice the distance between these points. Measure the wavelength of your standing wave.
   a) Record the wavelength of your standing wave in your log.

8. You can also measure the wave frequency. The frequency is the number of times the wave moves up and down each second. Measure the frequency of your standing wave. (Hint: Watch the hands of the person shaking the spring. Time a certain number of back-and-forth motions. The frequency is the number of back-and-forth motions of the hand in one second.)

a) Record the wave frequency in your log. The unit of frequency is the hertz (Hz).

9. Make several different standing waves by changing the wave frequency. Try to make each standing wave shown in the drawings (at right). Measure the wavelength. Measure the frequency.

a) Record both in a table like the one below.

| Wavelength (m/cycle) | Frequency (cycles/s or Hz) | Speed (m/s) wavelength × frequency |
|---|---|---|
| | | |
| | | |
| | | |
| | | |

Wavelength = twice Slinky length

Wavelength = Slinky length

Wavelength = 2/3 Slinky length

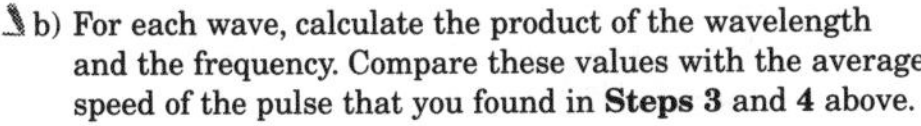

b) For each wave, calculate the product of the wavelength and the frequency. Compare these values with the average speed of the pulse that you found in **Steps 3** and **4** above.

10. All the waves you have made so far are transverse waves. A different kind of wave is the compressional (or longitudinal) wave. Have the members of your group stretch out the Slinky between the pieces of tape and hold the ends firmly. To make a compressional wave, squeeze part of the spring and let it go. Measure the speed of the compressional wave and compare it with the speed of the transverse wave.

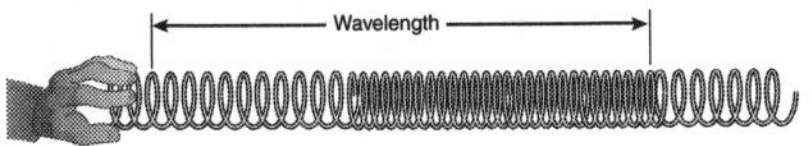


a) Record your results in a table partly like the one after **Step 3**.

b) In what direction does the Slinky move as the wave goes by?

Answers

## For You To Do

***(continued)***

8. a) The frequency measured will depend on how the spring is shaken.

9. a) Wavelength and frequency are inversely proportional. If one doubles the other is cut in half.

   b) The speeds probably agree to within ten or twenty percent.

10. a) Student data

    b) Looking along the string it is alternately stretched and compressed. There is no sideways motion. Back and forth along the direction the Slinky is stretched.

Chapter 3

ANSWERS

## For You To Do
*(continued)*

10. c) The wave consists of compressions and extensions of the spring. The wave runs lengthwise along the spring.

11.-12. Student activity,

13. If you stand in water and watch how the water moves as a wave passes by, you would see the same up-and-down motion.

14. a) As in **Step 13**, you will see a part of the wave going up and down.

**Physics Words**

**periodic wave:** a repetitive series of pulses; a wave train in which the particles of the medium undergo periodic motion (after a set amount of time the medium returns to its starting point and begins to repeat its motion).

**crest:** the highest point of displacement of a wave.

**trough:** the lowest point on a wave.

**amplitude:** the maximum displacement of a particle as a wave passes; the height of a wave crest; it is related to a wave's energy.

c) A dictionary definition of *compressional* is: "*a*. The act or process of compressing. *b*. The state of being compressed." A dictionary definition of *longitudinal* is: "Placed or running lengthwise." Explain why *compressional* or *longitudinal wave* is a suitable name for this type of wave.

11. To help you understand waves better, construct a wave viewer by cutting a slit in a file card and labeling it as shown.

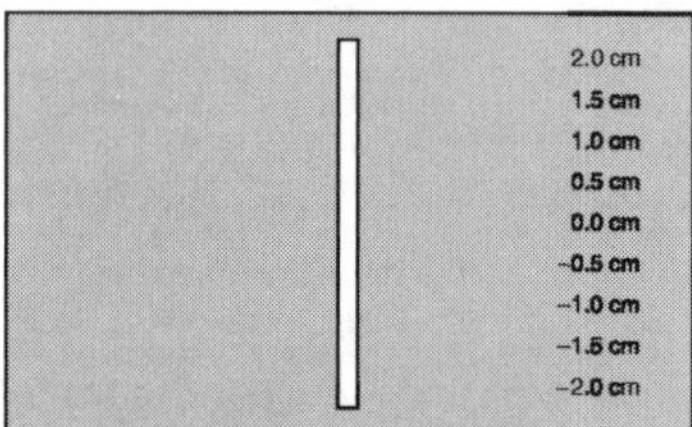


12. Make a drawing of a transverse wave on a strip of adding machine tape. Place this strip under the wave viewer so you can see one part of the wave through the slit.

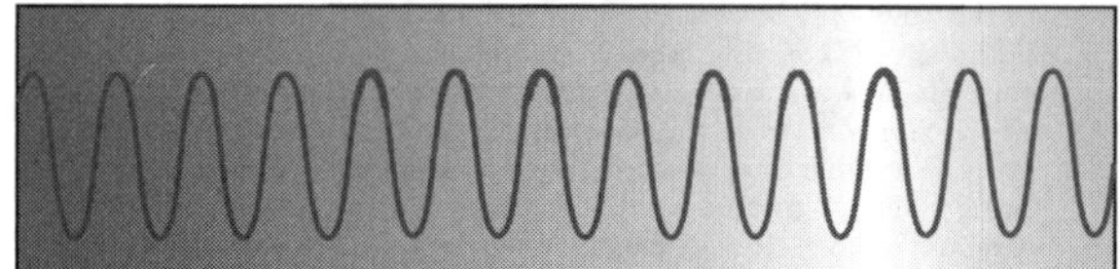

13. With the slit over the tape, pull the tape so that the wave moves. You will see a part of the wave (through the slit) going up and down.

14. Draw waves with different wavelengths on other pieces of adding machine tape. Put these under the slit and pull the adding machine tape at the same speed.

a) Describe what you see.

## FOR YOU TO READ

### Wave Vocabulary

In this activity, you were able to send energy from one end of the Slinky to the other. You used chemical energy in your muscles to create mechanical energy in your arms that you then imparted to the Slinky. The Slinky had energy. A card at the other end of the Slinky would have moved once the wave arrived there. The ability to move the card is an indication that energy is present. The total energy is transferred but it is always conserved.

Of course, you could have used that same mechanical energy in your arm to throw a ball across the room. That would also have transferred the energy from one side of the room to the other. It would have also moved the card.

There is a difference between the Slinky transferring the energy as a wave and the ball transferring the energy. The Slinky wave transferred the energy, but the Slinky basically stayed in the same place. If the part of the Slinky close to one end were painted red, the red part of the Slinky would not move across the room. The Slinky wave moves, but the parts of the Slinky remain in the same place as the wave passes by. A wave can be defined as a transfer of energy with no net transfer of mass.

Leonardo da Vinci stated that "the wave flees the place of creation, while the water does not." The water moves up and down, but the wave moves out from its center.

In discussing waves, a common vocabulary helps to communicate effectively. You observed waves in the lab activity. We will summarize some of the observations here and you can become more familiar with the terminology.

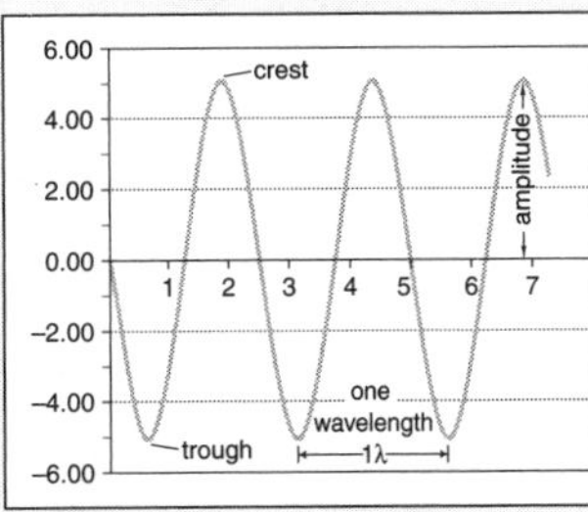


A **periodic wave** is a repetitive series of pulses. In the periodic wave shown in the diagram above, the highest point is called the **crest.** The lowest point is called the **trough**. The maximum disturbance, the **amplitude**, is 5.00 cm. Notice that this is the height of the crest or the height of the trough. It is *not* the distance from the crest to the trough.

→

Chapter 3

The **wavelength** of a periodic wave is the distance between two consecutive points in phase. The distance between two crests is one wavelength or 1 λ. (The Greek letter lambda is used to signify wavelength.) The wavelength of the wave in the diagram is 2.5 cm.

The amplitude of a periodic wave is the maximum disturbance. A large amplitude corresponds to a large energy. In sound, the large amplitude is a loud sound. In light, the large amplitude is a bright light. In Slinkies, the large amplitude is a large disturbance.

The wavelength of the wave in the diagram is 2.5 cm. It is the distance between two crests or the distance between two troughs.

The **frequency** is the number of vibrations occurring per unit time. A frequency of 10 waves per second may also be referred to as 10 vibrations per second, 10 cycles per second, 10 per second, 10 $s^{-1}$, 10 Hz (hertz). The human ear can hear very low sounds (20 Hz) or very high sounds (20,000 Hz). You can't tell the frequency by examining the wave in the diagram. The "snapshot" of the wave is at an instant of time. To find the frequency, you have to know how many crests pass by a point in a given time.

The **period**, $T$, of a wave is the time it takes to complete one cycle. It is the time required for one crest to pass a given point. The period and the frequency are related to one another. If three waves pass a point every second, the frequency is three waves per second. The period would be the time for one wave to pass the point, which equals $\frac{1}{3}$ s. If 10 waves pass a point every second, the frequency is 10 waves per second. The period would be the time for one wave to pass the point, which equals $\frac{1}{10}$ second. Mathematically, this relationship can be represented as:

$$T = \frac{1}{f} \text{ or } f = \frac{1}{T}$$

Points in a periodic wave can be "in phase" if they have the same displacement and are moving in the same direction. All crests of the wave shown below are "in phase."

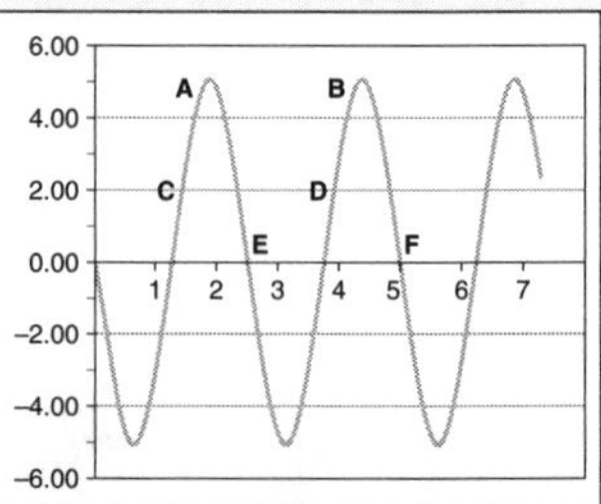


In the wave shown, the following pairs of points are in phase A and B, C and D, E and F.

A **node** is a spot on a standing wave where the medium is motionless. There are places along the medium that do not move as the standing wave moves up and down. The locations of these nodes do not change as the standing wave vibrates. A **transverse wave** is a wave in which the motion of the medium is perpendicular to the motion of the wave. A **longitudinal wave** is a wave in which the motion of the medium is parallel to the direction of the motion of the wave.

**PHYSICS TALK**

**Calculating the Speed of Waves**

You can find the speed of a wave by measuring the distance the crest moves during a certain change in time.

$$\text{speed} = \frac{\text{change in distance}}{\text{change in time}}$$

In mathematical language:

$$v = \frac{\Delta d}{\Delta t}$$

where $v$ = speed

$d$ = distance

$t$ = time

Suppose the distance the crest moves is 2 m in 0.2 s. The speed can be calculated as follows:

$$v = \frac{\Delta d}{\Delta t}$$

$$= \frac{2 \text{ m}}{0.2 \text{ s}}$$

$$= 10 \text{ m/s}$$

The distance from one crest of a wave to the next is the wavelength. The number of crests that go by in one second is the frequency. Imagine you saw five crests go by in one second. You measure the wavelength to be 2 m. The frequency is 5 crests/second, so the speed is $(5 \times 2) = 10$ m/s. Thus, the speed can also be found by multiplying the wavelength and the frequency.

speed = frequency × wavelength

In mathematical language:

$$v = f\lambda$$

where $v$ = speed

$f$ = frequency

$\lambda$ = wavelength

**Physics Words**

**wavelength:** the distance between two identical points in consecutive cycles of a wave.

**frequency:** the number of waves produced per unit time; the frequency is the reciprocal of the amount of time it takes for a single wavelength to pass a point.

**period:** the time required to complete one cycle of a wave.

**node:** a point on a standing wave where the medium is motionless.

**transverse pulse or wave:** a pulse or wave in which the motion of the medium is perpendicular to the motion of the wave.

**longitudinal pulse or wave:** a pulse or wave in which the motion of the medium is parallel to the direction of the motion of the wave.

Standing waves happen anywhere that the length of the Slinky and the wavelength have a particular mathematical relationship. The length of the Slinky must equal $\frac{1}{2}$ wavelength, 1 wavelength, $1\frac{1}{2}$ wavelengths, 2 wavelengths, etc. Mathematically, this can be stated as:

$$L = \frac{n\lambda}{2}$$

where $L$ is the length of the Slinky,
$\lambda$ is the wavelength
$n$ is a number (1, 2, 3...)



**Sample Problem 1**

You and your partner sit on the floor and stretch out a Slinky to a length of 3.5 m. You shake the Slinky so that it forms one loop between the two of you. Your partner times 10 vibrations and finds that it takes 24.0 s for the Slinky to make these vibrations.

a) How much of a wave have you generated and what is the wavelength of this wave?

***Strategy:*** Draw a sketch of the wave you have made and you will notice that it looks like one-half of a total wave. It is! This is the maximum wavelength that you can produce on this length of Slinky. You can use the equation that shows the relationship between the length of the Slinky and the wavelength.

***Givens:***

$L = 3.5$ m

$n = 1$

Activity 1 Making Waves

***Solution:***

$$L = \frac{n\lambda}{2}$$

Rearrange the equation to solve for λ.

$$\lambda = \frac{2L}{n}$$

$$= \frac{2\,(3.5\text{ m})}{1}$$

$$= 7.0\text{ m}$$

b) What is the period of vibration of the wave?

***Strategy:*** The period is the amount of time for one vibration. You have the amount of time for 10 vibrations.

***Solution:***

$$T = \frac{\text{time for 10 vibrations}}{10} = \frac{24.0\text{ s}}{10} = 2.4\text{ s}$$

c) Calculate the wave frequency.

***Strategy:*** The frequency represents the number of vibrations per second. It is the reciprocal of the period.

***Given:***

$$T = 2.4\text{ s}$$

***Solution:***

$$f = \frac{\text{number of vibrations}}{\text{time}} \text{ or } f = \frac{1}{T}$$

$$= \frac{1}{2.4\text{ s}}$$

$$= 0.42 \text{ vibrations per second}$$

$$= 0.42\text{ s}^{-1} \text{ or } 0.42\text{ Hz}$$

→



Chapter 3

d) Determine the speed of the wave you have generated on the Slinky.

***Strategy:*** The speed of the wave may be found by multiplying the frequency times the wavelength.

***Givens:***

$f = 0.42$ Hz

$\lambda = 7.0$ m

***Solution:***

$$\begin{aligned} v &= f\lambda \\ &= 0.42 \text{ Hz} \times 7.0 \text{ m} \\ &= 29 \text{ m/s} \end{aligned}$$

Remember that Hz may also be written as 1/s so the unit of speed is m/s.

**Sample Problem 2**

You stretch out a Slinky to a length of 4.0 m, and your partner generates a pulse that takes 1.2 s to go from one end of the Slinky to the other. What is the speed of the wave on the Slinky?

***Strategy:*** Use your kinematics equation to determine the speed.

***Givens:***

$d = 4.0$ m

$t = 1.2$ s

***Solution:***

$$\begin{aligned} v &= \frac{d}{t} \\ &= \frac{4.0 \text{ m}}{1.2 \text{ s}} \\ &= 3.3 \text{ m/s} \end{aligned}$$

Activity 1 Making Waves

### Reflecting on the Activity and the Challenge

Slinky waves are easy to observe. You have created transverse and compressional Slinky waves and have measured their speed, wavelength, and frequency. For the **Chapter Challenge**, you may want to create musical instruments. You will receive more guidance in doing this in the next activities. Your instruments will probably not be made of Slinkies. You may, however, use strings that behave just like Slinkies. When you have to explain how your instrument works, you can relate its production of sound in terms of the Slinky waves that you observed in this activity.

### Physics To Go

1. a) Four characteristics of waves are amplitude, wavelength, frequency, and speed. For each characteristic, tell how you measured it when you worked with the Slinky.
   b) For each characteristic, give the units you used in your measurement.
   c) Which wave characteristics are related to each other? Tell how they are related.
2. a) Suppose you shake a long Slinky slowly back and forth. Then you shake it rapidly. Describe how the waves change when you shake the Slinky more rapidly.
   b) What wave properties change?
   c) What wave properties do not change?
3. Suppose you took a photograph of a wave on a Slinky. How can you measure wavelength by looking at the photograph?
4. Suppose you mount a video camera on a tripod and aim the camera at one point on a Slinky. You also place a clock next to the Slinky, so the video camera records the time. When you look at the video of a wave going by on the Slinky, how could you measure the frequency?
5. a) What are the units of wavelength?
   b) What are the units of frequency?
   c) What are the units of speed?
   d) Tell how you find the wave speed from the frequency and the wavelength.

155

Coordinated Science for the 21st Century

ANSWERS

## Physics To Go

1. a) *amplitude*: measure the distance of a point on the Slinky from its rest position.

   *wavelength*: make a standing wave; measure the distance between the points of the Slinky that do not move; twice this distance is the wavelength.

   *frequency*: measure how many times you shook the Slinky in ten seconds and divide by ten.

   *speed:* make a pulse; measure distance traveled and time elapsed; divide time into distance to find the speed.

   b) amplitude and wavelength: meters
   frequency: 1/s
   speed: m/s

   c) Speed is frequency times wavelength.

ANSWERS

## Physics To Go
***(continued)***

2. a) When you shake the Slinky more rapidly, you can see more crests and the crests are closer together. If you are making standing waves, the more rapidly you shake the Slinky, the more complete waves you will see.

   b) The wavelength and frequency change.

   c) The speed does not change.

3. If there is a meter stick in the photograph, use its image to measure the distance between one crest and the next.

4. Measure the time for a point on the Slinky to go through one complete cycle. You may have to measure for ten cycles and divide the time by ten.

5. a) m

   b) 1/s

   c) m/s

   d) speed = wavelength × frequency

Chapter 3

## ANSWERS

# Physics To Go *(continued)*

5. e) m/s = m × 1/s

6. a) A standing wave is a repeating back-and-forth motion that does not move from one place to another. For transverse waves on a Slinky, the parts of the spring move from side-to-side, but nothing moves along the spring (nothing moves like the pulse did).

   b) See diagram on page 146.

   c) See diagram on page 146. At places where the wave amplitude is zero, there is no wave motion. At other places, the wave amplitude increases, reaches a maximum, then decreases, goes through zero, changes sign, reaches a maximum, goes back through zero, etc.

   d) Find the length of one complete wave of the spring. Or, figure out what fraction of a wave you see on the spring and from that compute the wavelength.

7. a) In a compressional wave, the back-and-forth motion is in the same direction that the disturbance moves (or in the opposite direction). In a transverse wave, the back-and-forth motion is perpendicular to the direction the disturbance moves.

   b) In transverse waves, the Slinky moves back-and-forth perpendicular to its length. In compressional waves, the Slinky moves back-and-forth along its length.

8. a) You shook the spring at a higher frequency.

   b) You shook the spring at a lower frequency.

e) Using your answer to **Part (d)**, show how the units of speed are related to the units of wavelength and frequency.

6. a) What is a standing wave?
   b) Draw a standing wave.
   c) Add labels to your drawing to show how the Slinky moves.
   d) Tell how to find the wavelength by observing a standing wave.

7. a) Explain the difference between transverse waves and compressional waves.
   b) Slinky waves can be either transverse or compressional. Describe how the Slinky moves in each case.

8. a) When you made standing waves, how did you shake the spring (change the frequency) to make the wavelength shorter?
   b) When you made standing waves, how did you shake the spring (change the frequency) to make the wavelength longer?

9. Use the wave viewer and adding machine tape to investigate what happens if the speed of the wave increases. Pull the tape at different speeds and report your results.

10. A Slinky is stretched out to 5.0 m in length between you and your partner. By shaking the Slinky at different frequencies, you are able to produce waves with one loop, two loops, three loops, four loops, and even five loops.
    a) What are the wavelengths of each of the wave patterns you have produced?
    b) How will the frequencies of the wave patterns be related to each other?

11. A tightrope walker stands in the middle of a high wire that is stretched 10 m between the two platforms at the ends of the wire. He is bouncing up and down, creating a standing wave with a single loop and a period of 2.0 s.

## ANSWERS

# Physics To Go *(continued)*

9. If the speed of the wave seen with the viewer increases, the frequency increases but the wavelength remains the same. If these were water waves and you photographed them, the wavelength would remain the same no matter what the speed of the wave (you could measure the wavelength from a still photo). But if you made a video, you would see more crests go by if the wave speed increased. That is consistent with the relationship speed = wavelength × frequency.

10. The number of loops and the wavelength are related by the following equation:

$$\frac{n\lambda}{2} = L$$

then $\lambda = \frac{2L}{n}$

a) What is the wavelength of the wave he is producing?
b) What is the frequency of this wave?
c) What is the speed of the wave?

12. A clothesline is stretched 9 m between two trees. Clothes are hung on the line as shown in the diagram. When a particular standing wave is created in the line, the clothes remain stationary.

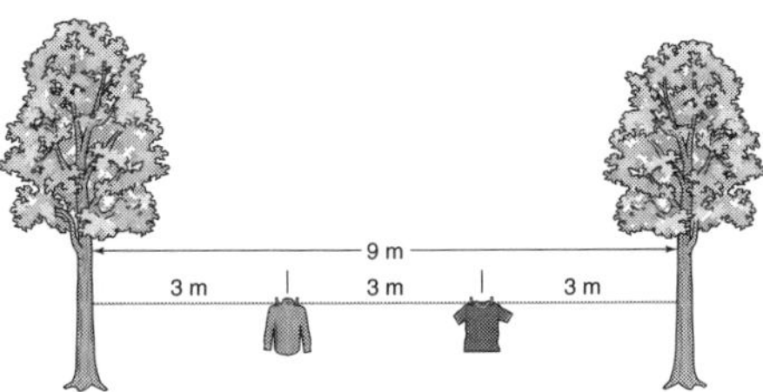


a) What is the term for the positions occupied by the clothes?
b) What is the wavelength of this standing wave?
c) What additional wavelengths could exist in the line such that the clothes remain stationary?

13. During the Slinky lab, your partner generates a wave pulse that takes 2.64 s to go back and forth along the Slinky. The Slinky stretches 4.5 m along the floor. What is the speed of the wave pulse on the Slinky?

14. A drum corps can be heard practicing at a distance of 1.6 km from the field. What is the time delay between the sound the drummer hears ($d = 0$ m) and the sound heard by an individual 1.6 km away? (Assume the speed of sound in air to be 340.0 m/s.)

ANSWERS

## Physics To Go *(continued)*

a) for one loop: $\lambda = \frac{2(5.0\text{m})}{1}$

$= 10\text{m}$

two loops: $\lambda = 5.0\text{m}$

three loops: $\lambda = 3.3\text{m}$

four loops: $\lambda = 52.5\text{m}$

five loops: $\lambda = 2.0\text{m}$

b) As the wavelength decreases, the frequency at which you will need to shake the Slinky increases.

11. First find the wavelength, then the frequency, and finally put them together to find the wave speed.

ANSWERS

## Physics To Go *(continued)*

a) $\lambda = \frac{2L}{n}$

$= \frac{2(10\text{m})}{1}$

$= 20\text{m}$

b) $f = \frac{1}{T}$

$= \frac{1}{2.0 \text{ s}}$

$= 0.5 \text{ s}^{-1}$ of $0.5$ Hz

c) $v = f\lambda$

$= 0.5 \text{ Hz} \times 20\text{m}$

$= 10$ m/s

12. a) The clothes are located at the nodes.

b) $\lambda = 6$ m

c) Wavelengths in which nodes will occur at the same locations will allow the clothes to remain stationary. These would include (but not limited to): l = 0.75 m, 2 m and 3 m.

13. You can solve this problem using the basic kinematics equation:

$d = vt$

so $v = \frac{d}{t}$

$= \frac{4.5 \text{ m}}{2.64 \text{ s}}$

$= 1.7$ m/s

14. $v = d/t$

$t = d/v$

$= (1600 \text{ m}) / (340 \text{ m/s})$

$= 4.7$ s

Chapter 3

## Activity 1 A

# Exploring Water Waves

### FOR YOU TO DO

If you have ripple tanks, set them up and let students explore ripples and also water waves with a single frequency and wavelength.

For use with *Let Us Entertain You*, Chapter 3, Activity 1: Making Waves

## Amplitude and Wavelength Measurements

| Amplitude | Time for pulse to travel from one end to another | Average time | Speed= $\frac{\text{length of spring}}{\text{average time}}$ |
|---|---|---|---|
| | | | |
| | | | |
| | | | |
| | | | |

| Wavelength (m/cycles) | Frequency (cycles/s or Hz) | Speed (m/s) wavelength x frequency |
|---|---|---|
| | | |
| | | |
| | | |
| | | |



For use with *Let Us Entertain You*, Chapter 3, Activity 1: Making Waves

# Sine Waves for Wave Model

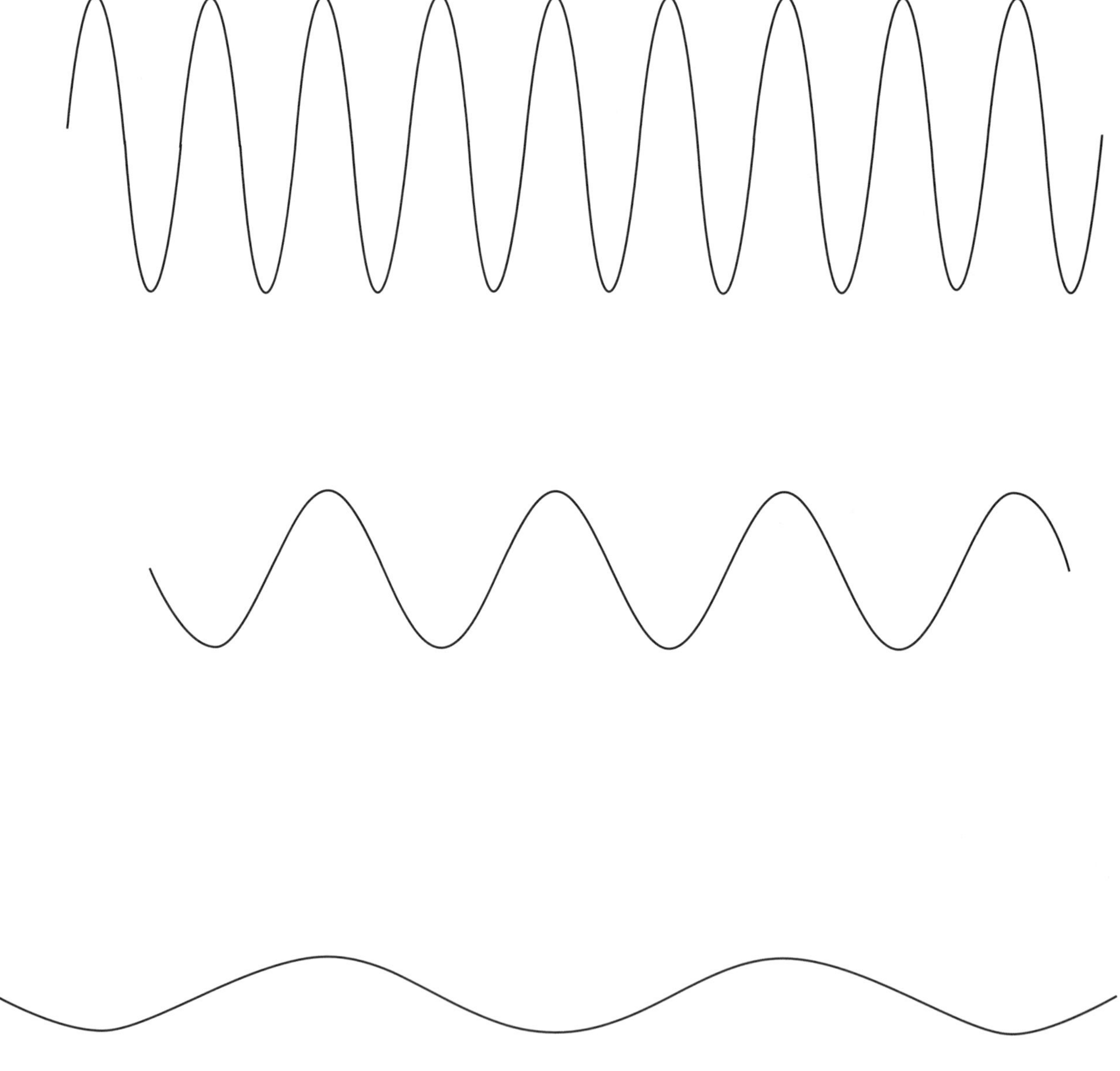

For use with *Let Us Entertain You*, Chapter 3, Activity 1: Making Waves

NOTES

# ACTIVITY 2
## Sounds in Strings

# Background Information

A tensioned string is a vibrating system. When you pull back the string and release it, the string tension accelerates the string back towards its equilibrium position, as the drawing shows.

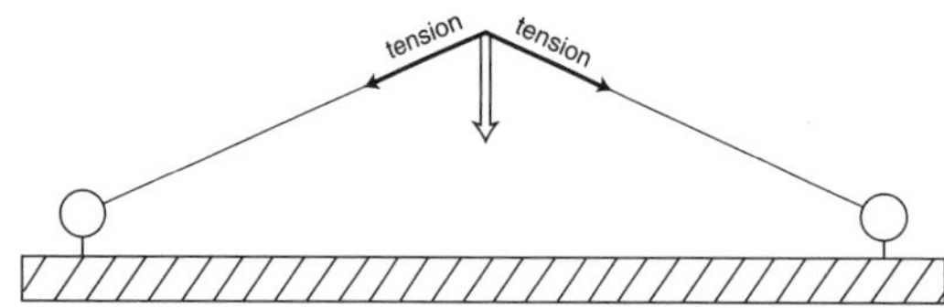


The variables that affect the frequency of a vibrating string are the string tension, the string length, and the string's mass per unit length. When the string is pulled back, the restoring force is equal to the sum of the two components of the tension that are perpendicular to the undisturbed string. The larger the tension, the larger the restoring force, the more rapidly the string will accelerate back towards the equilibrium position. The higher the tension, the higher the frequency, and, when the sound is heard, the higher the pitch.

Suppose you plucked a long and a short string. The drawing shows what happens.

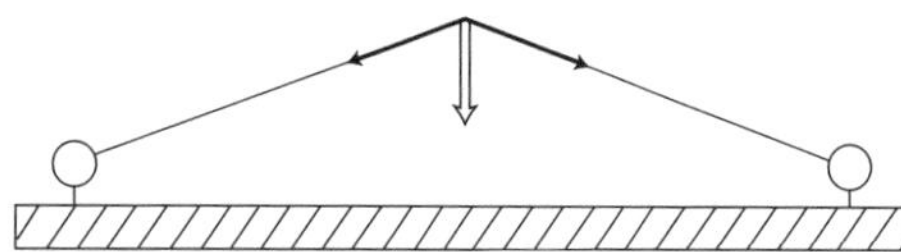

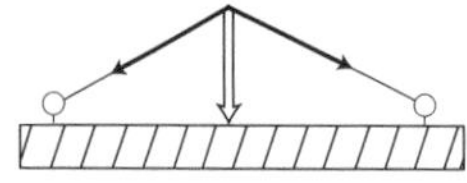

The component of the tension that is restoring the string to the equilibrium position is much higher for the shorter string. For the same amplitude, the shorter string experiences a larger restoring force. Consequently, its frequency will be correspondingly higher, as will the pitch it produces.

The third variable is the mass per unit length of the string (although students do not investigate it in this activity). Take a look at piano strings or guitar strings. The bass strings of a piano are much more heavily wound than the midrange or treble strings. In a nylon-string guitar, the base strings are wound with metal wire, and the lower the string, the heavier the wire. The larger the mass per unit length of the string, the less it responds to the tension that pulls it back towards the equilibrium position. The larger the mass per unit length, the lower the frequency.

Standing waves form on the string in the same way as on the spring.

When the string in a musical instrument vibrates, it moves in a combination, or superposition, of many standing wave patterns. A guitarist can suppress certain of these standing waves to emphasize the remaining ones by making "chimes." The technique is to simultaneously pluck the string and lightly tap it at the right spot for the particular chime. If the guitarist taps the string right in the middle, that suppresses the fundamental and all other odd harmonics with large amplitudes in the center of the string (the fundamental is the first harmonic).

NOTE: In this activity students find the wavelength of the fundamental from the length of the string. Since the activity does not introduce harmonics, the wavelength is called "the wavelength of the sound."

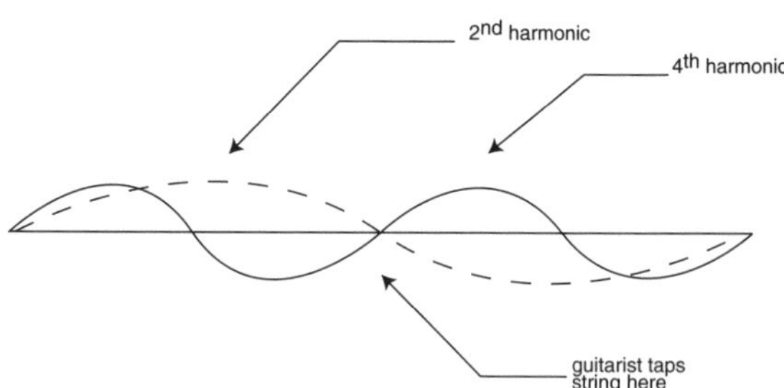


## *Active*-ating the Physics InfoMall

Here we get a great chance to really test the search engine. Sounds in strings? Stringed instruments? What should we search for? After all, are they "string" or "stringed" instruments? Why bother with deciding? Search for "string* instrument*". That's right—two wild characters in one search term. Is the InfoMall up to this task? You bet! Try it.

One of the hits is from the Sources of Musical Sounds chapter of *Modern College Physics*, found in the Textbook Trove. There is wonderful discussion as well as some nice graphics. One of the sections in this chapter is on stringed instruments. A discussion of wind instruments is also in this chapter (it might be useful later).

**Stretching Exercises, Step 2**, mentions a frequency meter. Search for key words "frequency meter" and you will find journal articles on using a frequency meter, plus a reference to the Vernier catalog on how to order the frequency meter software.

Also in the **Stretching Exercises**, students are asked to create a graph. You may wish to search for "student difficult*" AND "graph*". This produces several wonderful hits, including "Student difficulties with graphical representations of negative values of velocity," in *The Physics Teacher*, vol. 27, issue 4 on the InfoMall. Also found in that search is "Student difficulties in connecting graphs and physics: Example from kinematics," in the *Physics American Journal*, vol. 55, issue 6. Don't let these titles make you think that graphs are a bad thing! It is good to be aware that students do not always understand graphs, a powerful tool in the study of physics. Note that many of the hits you will find from this search involve the use of computers. Perhaps you may wish to have your students create their graphs on computers.

# Planning for the Activity

## Time Requirements

one and a half class periods

## Materials Needed

### For each group:

- c-clamp
- fishing line spool
- goggles, safety
- hooked mass weight, 500 gm
- key or other small metal object
- mass hanger
- meter stick, 100 cm, hardwood
- pencil
- pulley on mount
- slotted weight set

## Advance Preparation and Setup

Set up the experiment in advance to find out how long the string must be. Use nylon monofilament rated at least 15 lb. test. Cut the string in advance, and be sure the string is long enough to drape over the pulley and tie to the mass hanger.

# Teaching Notes

You may have to help students use the drawing for **FYTD, Step 5**, of the string fretted in different places. Encourage them to order the lengths they pluck from longest to shortest and look for a pattern in the pitches they observe. To facilitate ordering the pitches, encourage the students to compare the strings in pairs. After recording each result, they can readily order the pitches and corresponding string lengths.

Encourage the students to work quietly so they can observe the pitch of the vibrating string.

Students should wear safety goggles while working with the tensioned string.

Students may relate amplitude and frequency, so they would interpret louder sounds as lower in pitch. This prior conception could hinder students from recognizing that amplitude and frequency are independent variables. To help, set up a frequency generator, amplifier, and speaker. Then vary the frequency and amplitude to demonstrate that all possible combinations of pitch and volume can exist.

# Activity Overview

Students begin the study of musical instruments by investigating two variables—tension and length—that determine the sounds made by a vibrating string. They hang a weight on the end of the string to set the tension. With the tension fixed, they vary the length of the string and observe the pitch produced. Then with the length fixed, students vary the tension and again observe the pitch. They make general statements about the effect of changes in string length and tension.

## Student Objectives

### Students will:

- Observe the effect of string length and tension upon pitch produced.
- Control the variables of tension and length.
- Summarize experimental results.
- Calculate wavelength of a standing wave.
- Organize data in a table.

ANSWERS FOR THE TEACHER ONLY

## What Do You Think?

Both guitarists and violinists make different sounds in a similar way. The player pushes the string against the neck of the instrument at different positions to play different notes. One significant difference is that the guitar has frets, thin metal strips that run across the neck right under the strings. The guitarist pushes the string down against the fret, whereas the violinist must find the correct position to push the string down to make a particular note.

In addition, each string is tuned to a different note, so each string has a different range. Guitarists and violinists tune their instruments by adjusting the tension in the strings. Guitarists turn knobs on the tuning machine, a set of gears that increase torque through mechanical advantage. The violinist turns pegs, which have larger knobs than those on the guitar but which are connected directly to the strings.

Activity 2 Sounds in Strings

GOALS

In this activity you will:

- Observe the effect of string length and tension upon pitch produced.
- Control the variables of tension and length.
- Summarize experimental results.
- Calculate wavelength of a standing wave.
- Organize data in a table.

What Do You Think?

When the ancient Greeks made stringed musical instruments, they discovered that cutting the length of the string by half or two-thirds produced other pleasing sounds.

- **How do guitarists or violinists today make different sounds?**

Record your ideas about this question in your *Active Physics* log. Be prepared to discuss your responses with your small group and with your class.

For You To Do

1. Carefully mount a pulley over one end of a table. Securely clamp one end of a string to the other end of the table.
2. Tie the other end of the string around a mass hanger. Lay the string over the pulley. Place a pencil under the

Active Physics  158

ANSWERS

## For You To Do

1. Be sure the masses hang beyond the pulley but all are up off the floor.
2. If the pencil moves when the sting is plucked, tape or clamp the pencil in place.

string near the clamp, so the string can vibrate without hitting the table, as shown in the drawing.

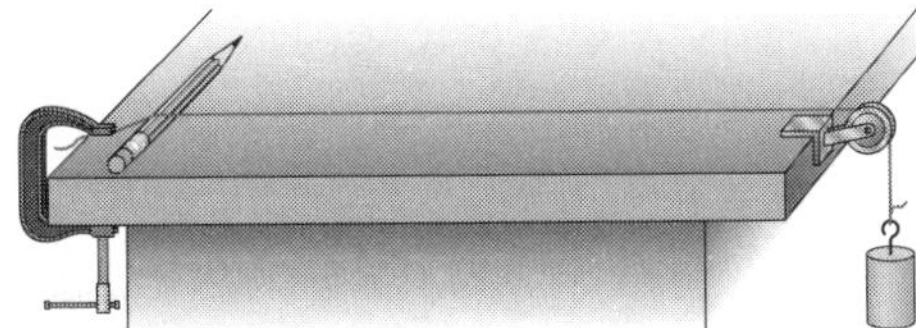

3. Hang one 500-g mass on the mass hanger. Pluck the string, listen to the sound, and observe the string vibrate.

a) Record your observations in your log in a table similar to the following:

| Length of vibrating string | Load on mass hanger | Pitch (high, medium, low) |
|---|---|---|
| | | |
| | | |
| | | |
| | | |

**Make sure the area under the hanging mass is clear (no feet, legs). Also monitor the string for fraying.**

4. Use a key or some other small metal object. Press this object down on the string right in the middle, to hold the string firmly against the table. Pluck each half of the string.

a) Record the result in your table.

5. To change the string length, press down with the key at the different places shown in the diagrams on the next page. Pluck each part of the string.

a) Record the results in your table.



Chapter 3

ANSWERS

## For You To Do *(continued)*

3. a) If the strings are the same material and length, the sounds will have the same pitch.

4. a) The frequency will be higher.

5. a) The shorter the length, the higher the pitch.

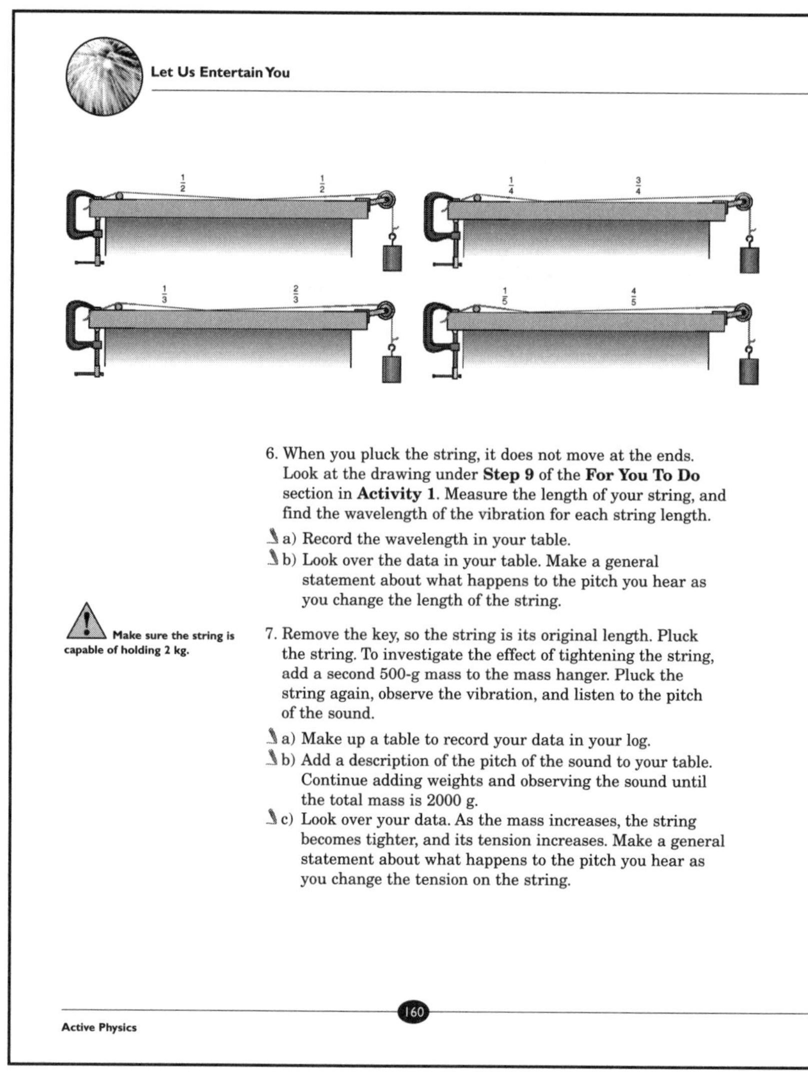
Let Us Entertain You



6. When you pluck the string, it does not move at the ends. Look at the drawing under **Step 9** of the **For You To Do** section in **Activity 1**. Measure the length of your string, and find the wavelength of the vibration for each string length.
   a) Record the wavelength in your table.
   b) Look over the data in your table. Make a general statement about what happens to the pitch you hear as you change the length of the string.

Make sure the string is capable of holding 2 kg.

7. Remove the key, so the string is its original length. Pluck the string. To investigate the effect of tightening the string, add a second 500-g mass to the mass hanger. Pluck the string again, observe the vibration, and listen to the pitch of the sound.
   a) Make up a table to record your data in your log.
   b) Add a description of the pitch of the sound to your table. Continue adding weights and observing the sound until the total mass is 2000 g.
   c) Look over your data. As the mass increases, the string becomes tighter, and its tension increases. Make a general statement about what happens to the pitch you hear as you change the tension on the string.

Active Physics 160

ANSWERS

## For You To Do *(continued)*

6. a) Assume the tone produced is the fundamental which is shown in the top drawing on p. 147. The wavelength is twice the length of the part of the string that is vibrating.
   b) As the string is made shorter, the pitch it makes goes up.

7. a) See the table on p. 159. Here the variables are the load on the mass hanger and pitch.
   b) The results are qualitative. (Have the students compare the pitch produced by each load.)
   c) As the mass on the mass hanger goes up, the pitch the string makes goes up.

Activity 2 Sounds in Strings

### FOR YOU TO READ

#### Changing the Pitch

Sound comes from vibration. You observed the vibration of the string as it produced sound. You investigated two of the variables that affect the sound of a vibrating string.

When you pushed the vibrating string down against the table, the length of the string that was vibrating became shorter. Shortening the string increased the **pitch** (resulted in a higher pitch). In the same way, a guitarist or violinist pushes the string against the instrument to shorten the length that vibrates and increases the pitch.

When you hung weights on the end of the string, that increased the pitch too. These weights tightened the string, so they created more tension in it. As the string tension increased, the pitch of the sound also increased. In tuning a guitar or violin, the performer changes the string tension by turning a peg attached to one end of a string. As the peg pulls the string tighter, the pitch goes up.

Combining these two results into one expression, you can say that increasing the tension or decreasing the length of the string will increase the pitch.

The string producing the pitch is actually setting up a standing wave between its endpoints. The length of the string determines the wavelength of this standing wave. Twice the distance between the endpoints is the wavelength of the sound. The pitch that you hear is related to the frequency of the wave. The higher the pitch, the higher the frequency. The speed of the wave is equal to its frequency multiplied by its wavelength.

$$v = f\lambda$$

where $v$ = speed

$f$ = frequency

$\lambda$ = wavelength

If the speed of a wave is constant, a decrease in the wavelength will result in an increase in the frequency or a higher pitch. A shortened string produces a higher pitch.

## Reflecting on the Activity and the Challenge

Part of the **Chapter Challenge** is to create a sound show. In this activity you investigated the relationship of pitch to length of the string and tension of the string: the shorter the string, the higher the pitch; the greater the tension, the higher the pitch. You also learned that the string is setting up a standing wave between its two ends, just like the standing wave that you created in the Slinky in **Activity 1**. That's the physics of stringed instruments! If you wanted to create a stringed or multi-string instrument for your show, you would now know how to adjust the length and tension to produce the notes you want. If you were to make such a stringed instrument, you could explain how you change the pitch by referring to the results of this activity.

**Physics Words**

**pitch:** the quality of a sound dependent primarily on the frequency of the sound waves produced by its source.

ANSWERS

## Physics To Go

1. a) By changing the weight hung on the end of the string (or, in an instrument, by turning the knobs on the tuning machine).

   b) Increasing the tension increases the pitch of the sound.

2. a) By pressing the string down against the tabletop.

   b) Decreasing the length of the string increases the pitch of the sound.

3. The right combination of decreasing the length and decreasing the tension (or increasing both) can keep the pitch the same.

4. In general, the pitch would change. (See answer to **Question 3.**)

5. a) For the guitar, by pressing the string down on the neck of the guitar; for the piano, by striking a different string.

   b) For the guitar, by changing the string tension; for the piano, also by changing the string tension.

6. a) To change the tension of the strings.

   b) To make it easier to press the string against the neck at the right place, so the note has the right pitch.

   c) No.

   d) Because the player must judge where to press the string down so each note has the correct pitch.

7. a)-c) Student activity.

Let Us Entertain You

### Physics To Go

1. a) Explain how you can change the tension of a vibrating string.
   b) Tell how changing the tension changes the pitch.

2. a) Explain how you can change the length of a vibrating string.
   b) Tell how changing the length changes the sound produced by the string.

3. How would you change both the tension and the length and keep the pitch the same?

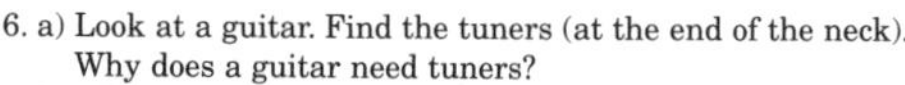

4. Suppose you changed both the length and the tension of the string at the same time. What would happen to the sound?

5. a) For the guitar and the piano, tell how a performer plays different notes.
   b) For the guitar and the piano, tell how a performer (or tuner) changes the pitch of the strings to tune the instrument.

6. a) Look at a guitar. Find the tuners (at the end of the neck). Why does a guitar need tuners?
   b) What is the purpose of the frets on a guitar?
   c) Does a violin or a cello have frets?
   d) Why do a violinist and a cellist require more accuracy in playing than a guitarist?

7. a) Using what you have learned in this activity, design a simple two-stringed instrument.
   b) Include references to wavelength, frequency, pitch, and standing waves in your description.
   c) Use the vocabulary of wavelength, frequency, and standing waves from **Activity 1** to describe how the instrument works.

162

Active Physics

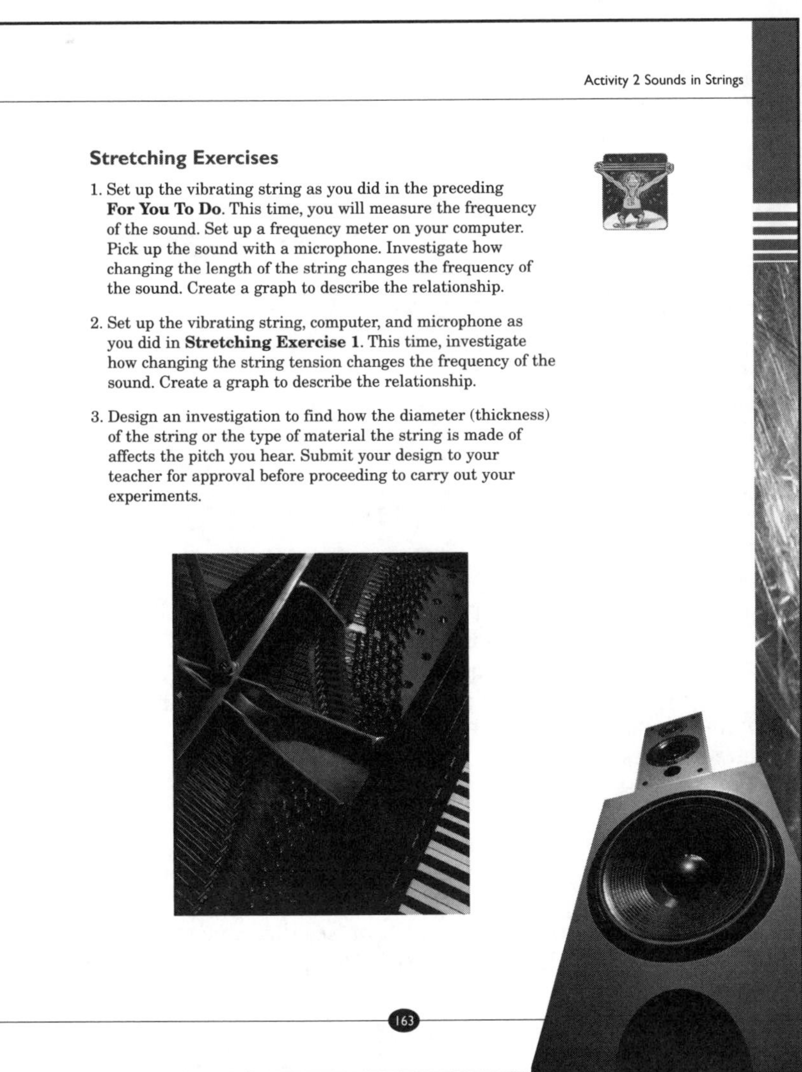


ANSWERS

## Stretching Exercises

See **Program Resources** for packages of computer sound probe and software.

# ACTIVITY 3
## Sounds from Vibrating Air

# Background Information

The phenomena of generating sound by blowing into a tube is an example of resonance. In resonance, a system that has a natural frequency is excited by an oscillation of the same frequency. A familiar example is pushing a child on a swing. The swing is a pendulum, with a single natural frequency. When the child is pushed so that the push is in the same direction as the child's velocity, energy feeds into the system rapidly and the child swings higher and higher (the system here is the child on the swing). If the push is opposed to the child's velocity, energy drains out of the system rapidly.

Sound, too, can be pictured as a regular series of pushes and pulls. The sound wave is compressional, so the wave is a pattern of regular pressure changes.

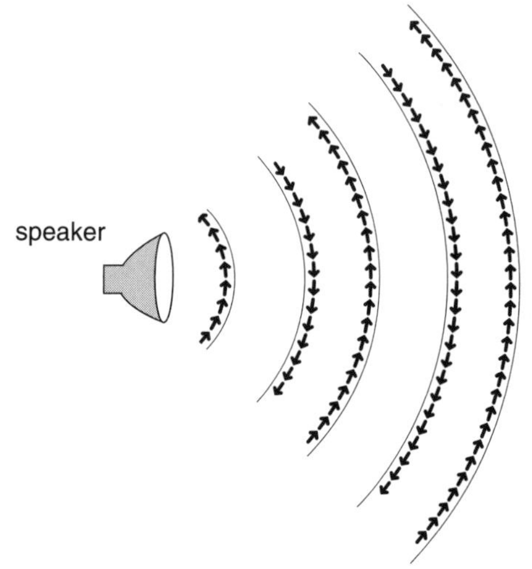


One important difference between blowing into a tube and pushing a child on a swing is that the push is made at just the right time to get the child swinging with large amplitude. The push has a single frequency. By contast, the sound of blowing into a tube is noise—a random spectrum of unrelated frequencies. But when that spectrum contains a frequency that makes the tube resonate, then the sound level for that frequency in the tube builds up and up. Listening to a sea shell is the same effect.

Now imagine a tube, with length "*L*," that is closed at one end. A sound wave reaches the open end of the tube. The wave moves to the end of the tube, reflects, and returns to the open end, which requires a time of $2L/v$, as shown.

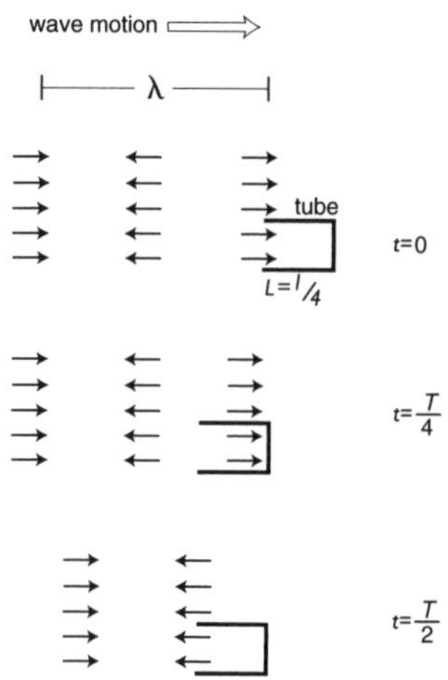


When the wave returns to the open end, the wave will be pushing the air at this end out of the tube. For resonance to occur, the sound wave exciting the tube must have gone through a half a cycle since it entered the tube, which requires a time of half the period *T*.

$2L/v = T/2 = 1/(2f)$

Substituting $v = \lambda \times f$ gives

$L = \lambda/4$

This equation expresses the same result as the drawings on page 166 in the Student Edition.

## *Active*-ating the Physics InfoMall

As we found in the previous activity, there is a passage on wind instruments in the Sources of Musical Sounds chapter of Modern College Physics, found in the Textbook Trove.

Of course, you may want to perform a new search for this activity. One idea is to search for "flute". The very first hit is "Acoustics of the flute," from *Physics Today*, vol. 21, issue 11, 1968. The second hit is "Observations on the acoustical characteristics of the English flute," from the *American Journal of Physics*, vol. 27, issue 1, 1959. The is also a hit for "The physics of the organ flue pipe," in the *American Journal of Physics*, vol. 21, issue 5, 1953.

Also in this search is a chapter from *Household Physics*, from the Textbook Trove. There is an entire chapter on Musical Sounds and Musical Instruments. This chapter includes absolutely wonderful discussions of the topics we have in this activity as well as the previous activities. The graphics are good, too. There are discussions of vibrating air columns (with pipes open or closed at one end), pitch, the human voice and vocal cords, and more.

If you want more problems than you find in **Physics To Go**, simply visit the Problems Place. All six sources in the Problems Place have sections on Sound or Waves. This includes *Schaum's 3000 Solved Problems in Physics*, so don't hesitate to use this store for problems or examples for your students.

# Planning for the Activity

## Time Requirements

one class period

## Materials Needed

### For the class:

- sound probe and software
- various wind instruments

### For each group:

- drinking straws (10)
- ruler, metric
- steel scissors
- test tubes (5)
- water, about 300 cc, in a container

### For each student:

- 1 copy of the Blackline Master, Pitch vs. Tube Length

## Advance Preparation and Setup

Make one copy of the Blackline Master for each student. Be sure that the each group has enough water to partly fill the test tubes as explained in the student book.

# Teaching Notes

Encourage the students to make direct comparisons by blowing into first one tube and then another. By making systematic comparisons of pairs of tube, they can order the tubes according to pitch.

Keep the student groups as well separated as possible.

Students are reasonably successful in associating vibration with sound when they can directly observe the vibrating object, such as a guitar string. A vibrating column of air is much more difficult to conceptualize, as is the idea that vibrations can move out from the column towards a listener. If you have both a guitar and a recorder in the room, have students play a note on each. First have the class trace the vibration from the guitar string to the students' ears. Then do the same with the recorder.

# Activity Overview

Students continue the study of musical instruments by investigating the resonance of pipes. They blow into different lengths of drinking straw and observe the tone produced and also observe how the tone changes if they cover one end of the straw. They perform a similar experiment with a test tube partially filled with water. Through reading they then relate these observations to the patterns of standing waves in a tube.

## Student Objectives

### Students will:

- Identify resonance in different kinds of tubes.
- Observe how resonance pitch changes with length of tube.
- Observe the effect of closing one end of the tube.
- Summarize experimental results.
- Relate pitch observations to drawings of standing waves.
- Organize observations to find a pattern.

Let Us Entertain You

# Activity 3 Sounds from Vibrating Air

**GOALS**

In this activity you will:

- Identify resonance in different kinds of tubes.
- Observe how resonance pitch changes with length of tube.
- Observe the effect of closing one end of the tube.
- Summarize experimental results.
- Relate pitch observations to drawings of standing waves.
- Organize observations to find a pattern.

**What Do You Think?**

The longest organ pipes are about 11 m long. A flute, about 0.5 m long, makes musical sound in the same way.

- **How do a flute and organ pipes produce sound?**

Record your ideas about this question in your *Active Physics* log. Be prepared to discuss your responses with your small group and with your class.

Active Physics 164

Answers for the Teacher Only

## What Do You Think?

A flute and an organ produce sound from a resonating air column. The flutist's lips direct a stream of air into the mouthpiece to set up the vibration of the air inside the instrument. The strength, direction of the air stream, and length of the air column determine the harmonic content of the vibration and thereby the octave of the musical note. In the organ, compressed air is blown into each tube to create the resonance.

Each pipe in the organ has a different length and produces a different note, with longer pipes producing lower notes. The flutist produces different notes by pressing keys to expose openings in the instrument that, in effect, change its length.

### For You To Do

1. Carefully cut a drinking straw in half. Cut one of the halves into two quarters. Cut one of the quarters into two eighths. Pass one part of the straw out to one member of your group.
2. Gently blow into the top of the piece of straw.
   a) Describe what you hear.
   b) Listen as the members of your group blow into their straw pieces one at a time. Describe what you hear.
   c) Write a general statement about how changing the length of the straw changes the pitch you hear.

3. Now cover the bottom of your straw piece and blow into it again. Uncover the bottom and blow again.
   a) Compare the sound the straw makes when the bottom is covered and then uncovered.
   b) Listen as the members of your group blow into their straw pieces, with the bottom covered and then uncovered. Write a general statement about how changing the length of the straw changes the pitch you hear when one end is covered.
4. Obtain a set of four test tubes. Leave one empty. Fill the next halfway with water. Fill the next three-quarters of the way. Fill the last one seven-eighths of the way.

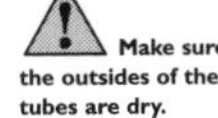

Make sure the outsides of the tubes are dry.

5. Give each test tube to one member of your group. Blow across your test tube.
   a) Describe what you hear.
   b) Listen as the members of your group blow, one at a time, across their test tubes. Record what you hear.
   c) What pattern do you find in your observations?
   d) Compare the results of blowing across the straws with blowing across the test tubes. How are the results consistent?

ANSWERS

## For You To Do

1. Student activity.

2. a)-b) A faint tone is produced.

   c) The shorter the straw, the higher the pitch.

3. a) When the bottom of the straw is covered, the pitch is much lower than when both ends are open.

   b) When one end of the straw is covered, the shorter the straw, the higher the pitch.

4. Student activity.

5. a) A faint tone is produced.

   b)-c) The shorter the air space in the tube, the higher the pitch.

   d) The shorter the length, the higher the pitch.

**Physics Words**

**diffraction:** the ability of a sound wave to spread out as it emerges from an opening or moves beyond an obstruction.

## PHYSICS TALK

### Vibrating Columns of Air

The sound you heard when you blew into the straw and test tube was produced by a standing wave. If both ends of the straw are open, the air at both ends moves back and forth. The above drawing shows the movement of the air as a standing wave.

Tube is open at both ends.
1/2 wavelength fits in straw.

When you covered the other end of the straw, you prevented the air from moving at the covered end. This drawing shows the movement of the air as a standing wave.

Tube is closed at one end.
1/4 wavelength fits in straw.

The velocity of a wave is equal to the frequency multiplied by the wavelength. Therefore,

$$\text{frequency} = \frac{\text{wave speed}}{\text{wavelength}}$$

Using mathematical symbols,

$$f = \frac{v}{\lambda}$$

As the wavelength increases, the frequency decreases. The wavelength in the open straw is half the wavelength in the straw closed at one end. This equation predicts that the frequency of the standing wave in the open straw is twice the frequency of the standing wave in the straw closed at one end.

## FOR YOU TO READ

### Compressing Air to Make Sound

Sound is a compression wave. The molecules of air bunch up or spread apart as the sound wave passes by.

At the end where the tube is closed, the air cannot go back and forth, because its motion is blocked by the end of the tube. That's why the wave's amplitude goes to zero at the closed end. At the open end, the amplitude is as large as it can possibly be. This back-and-forth motion of air at the open end makes a sound wave that moves from the tube to your ear.

In the compressional Slinky wave, the coils of the Slinky bunched up in a similar fashion when the Slinky wave passed by.

### Wave Diffraction

As the sound wave leaves the test tube in this activity, it spreads out. In the same way, when you speak to a friend, the sound waves leave your mouth and spread out. You can speak to a group of friends because the sound leaves your mouth and moves out to the front and to the sides.

This ability of the sound wave to spread out as it emerges from an opening is called **diffraction**. The smaller the opening, the more spreading of the sound. The spreading of the wave as it emerges from two holes can be shown with a diagram.

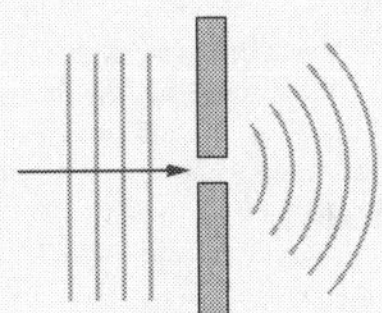

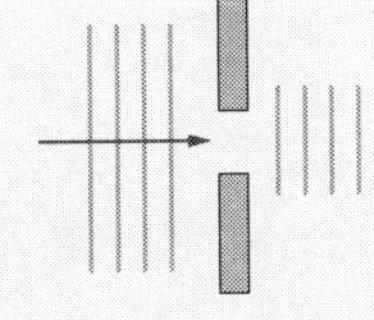

The wave on the top is going through a small opening (in comparison to its wavelength) and diffracts a great deal. The wave on the bottom is going through a large opening (in comparison to its wavelength) and shows little diffraction.

Cheerleaders use a megaphone to limit the diffraction. With a megaphone, the mouth opening becomes larger. The sound wave spreads out less, and the cheering crowd in front of the cheerleader hears a louder sound.

→



Chapter 3

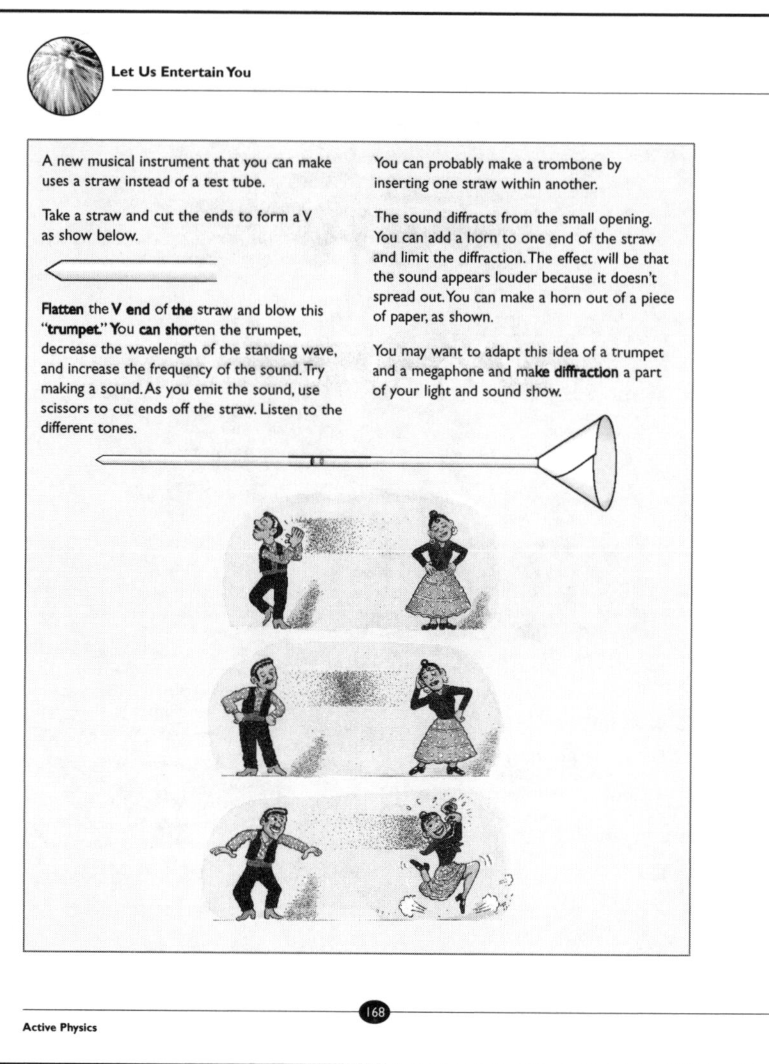



A new musical instrument that you can make uses a straw instead of a test tube.

Take a straw and cut the ends to form a V as show below.

**Flatten** the **V end** of **the** straw and blow this "**trumpet.**" You **can shor**ten the trumpet, decrease the wavelength of the standing wave, and increase the frequency of the sound. Try making a sound. As you emit the sound, use scissors to cut ends off the straw. Listen to the different tones.

You can probably make a trombone by inserting one straw within another.

The sound diffracts from the small opening. You can add a horn to one end of the straw and limit the diffraction. The effect will be that the sound appears louder because it doesn't spread out. You can make a horn out of a piece of paper, as shown.

You may want to adapt this idea of a trumpet and a megaphone and ma**ke diffraction** a part of your light and sound show.

## Reflecting on the Activity and the Challenge

In this activity you have observed the sounds produced by different kinds of pipes. If the pipe is cut to a shorter length, the pitch of the sound increases. Also, when the pipe is open at both ends, the pitch is much higher than if the pipe were open at only one end. You have seen how simple drawings of standing waves in these tubes help you find the wavelength of the sound. If the tube is closed at one end, the air has zero displacement at that end. If the tube is open at one end, the air has maximum displacement there.

For your sound show, you may decide to create some "wind" instruments using test tubes or straws, or other materials approved by your teacher. When it comes time to explain how these work, you can refer to this activity to get the physics right.

## Physics To Go

1. a) You can produce a sound by plucking a string or by blowing into a pipe. How are these two ways of producing sound similar?
   b) How are these two ways different?

2. a) For each piece of straw your group used, make a full-sized drawing to show the standing wave inside. Show both the straw closed at one end and open at both ends.
   b) Next to each drawing of the standing waves, make a drawing, at the same scale, of one full wavelength. You may need to tape together several pieces of paper for this drawing.
   c) Frequency times the wavelength is the wave speed. The speed is the same for all frequencies. From your answer to **Part (b)**, what can you predict about the frequencies of the standing waves in the straw pieces?
   d) How well do your predictions from **Part (c)** agree with your observations in this activity?



### Answers

## Physics To Go

1. a) The string is vibrating, and the air in the pipe is also vibrating.
   b) The string vibration is transverse, and the air vibration is compressional.

2. a)-b) See drawings on page 166.
   c) If you double the length of the straw, you cut the frequency it produces in half. When both ends are open, the frequency will be twice as high as when one end is closed.

d) There should be good agreement.

3. a) 11 m

b) See drawings on page 166.

c) 4 × 11 m = 44 m

d) The speed of sound is independent of the wavelength. But the product of wavelength and frequency equals speed, so this product is a constant. Thus, a longer wavelength means a lower frequency.

4. a) $\lambda = 4 \times 3 \text{ m} = 12 \text{ m}$

b) $f = v/\lambda = 340/12 \text{ Hz} = 28 \text{ Hz}$

c) $2 \times 3 \text{ m} = 6 \text{ m}$

d) $f = 340/6 \text{ Hz} = 57 \text{ Hz}$

5. The frequency is three times higher.

6. a) Waves spreading into a region behind an obstruction is called diffraction.

3. a) What is the length, in meters, of the longest organ pipe?
   b) Assume this pipe is closed at one end. Draw the standing wave pattern.
   c) For this pipe, how long is the wavelength of this standing wave?
   d) Why does a long wavelength indicate that the frequency will be low? Give a reason for your answer.

4. a) Suppose you are listening to the sound of an organ pipe that is closed at one end. The pipe is 3 m long. What is the wavelength of the sound in the pipe?
   b) The speed of sound in air is about 340 m/s. What is the frequency of the sound wave?
   c) Now suppose you are listening to the sound of an organ pipe that is open at both ends. As before, the pipe is 3 m long. What is the wavelength of the sound in the pipe?
   d) What is the frequency of the sound wave?

5. Suppose you listen to the sound of an organ pipe that is closed at one end. This pipe is only 1 m long. How does its frequency compare with the frequency you found in **Question 4, Part (b)**?

6. Waves can spread into a region behind an obstruction.
   a) What is this wave phenomenon called?
   b) Draw a diagram to illustrate this phenomenon.

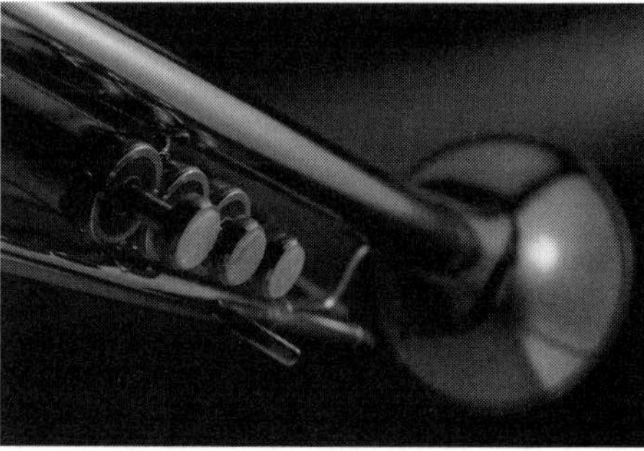

### Stretching Exercises

1. If you have a good musical ear, add water to eight test tubes to make a scale. Play a simple piece for the class.
2. Obtain a 2- to 3- meter-long piece of a 7- to 10-centimeter-diameter plastic pipe, like that used to filter water in small swimming pools. In an area free of obstructions, twirl the pipe overhead. What can you say about how the sound is formed? Place some small bits of paper on a stool. Twirl the pipe and keep one end right over the stool. What happens to the paper? What does that tell you about the air flowing through the pipe? Try to play a simple tune by changing the speed of the pipe as you twirl it.
3. Carefully cut new straw pieces, as you did in **For You To Do, Step 1**. This time, you will measure the frequency of the sound. Set up a frequency meter on your computer. Place the microphone near an open end of the straw.
   As before, each person blows into only one piece of straw. Make the sound and record the frequency. Now cover the end of the straw and predict what frequency you will measure. Make the measurement and compare it with your prediction. Repeat the measurements for all of the lengths of straw. Record your results, and tell what patterns you find.

ANSWERS

## Stretching Exercises

Try the lengths for a major scale:

11 cm
9.8 cm
8.8 cm
8.3 cm
7.4 cm
6.6 cm
5.9 cm
5.5 cm

Movement of the air in the pipe produced a standing wave. If paper bits are sucked into the pipe, that shows the direction of the air flow.

## Activity 3 A

# Wind Instruments and Resonant Frequencies

### FOR YOU TO DO

1. Wind Instruments: Have students bring wind instruments to class. Discuss whether each instrument is open at one end or both ends. Explain how the performer plays different notes. Also discuss how to tune the instrument.

2. Measuring Resonant Frequencies of Tubes: With sound probe and software, measure the resonant frequency of tubes of various lengths. Make the measurements for tubes open at both ends and closed at one end. Compare your results with the predictions of the standing wave diagrams.

For use with *Let Us Entertain You*, Chapter 3, Activity 3: Sounds from Vibrating Air

# Pitch vs. Tube Length

| Length of test tube or straw | End closed or open | Length of air column | Pitch |
|---|---|---|---|
| | | | |
| | | | |
| | | | |
| | | | |
| | | | |
| | | | |
| | | | |
| | | | |
| | | | |
| | | | |
| | | | |
| | | | |
| | | | |
| | | | |
| | | | |
| | | | |
| | | | |
| | | | |
| | | | |
| | | | |
| | | | |
| | | | |
| | | | |
| | | | |
| | | | |
| | | | |
| | | | |
| | | | |
| | | | |
| | | | |
| | | | |
| | | | |



For use with *Let Us Entertain You*, Chapter 3, Activity 3: Sounds from Vibrating Air

# ACTIVITY 4
## Reflected Light

# Background Information

When light strikes an ordinary object, the light is reflected in all directions. This reflected light can be represented by light rays that fan out from each point on the object.

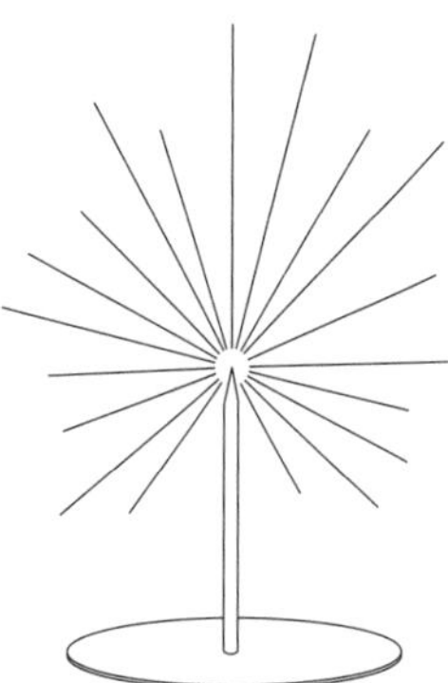

When the object is placed before a mirror, the light rays that fan out from the object can be extended to the mirror. As shown in the drawing on page 174 of the Student Edition, the angles of incidence and reflection for these rays are equal. These angles are straightforward to measure, but there is a minor complication to interpreting the results. Most lab mirrors are rear-surface mirrors. That is, the reflecting coating is on the rear of the mirror, so the light must first pass through glass before being reflected, and then must pass through glass again before again entering the air. The light is refracted at each air-glass interface, as shown in the

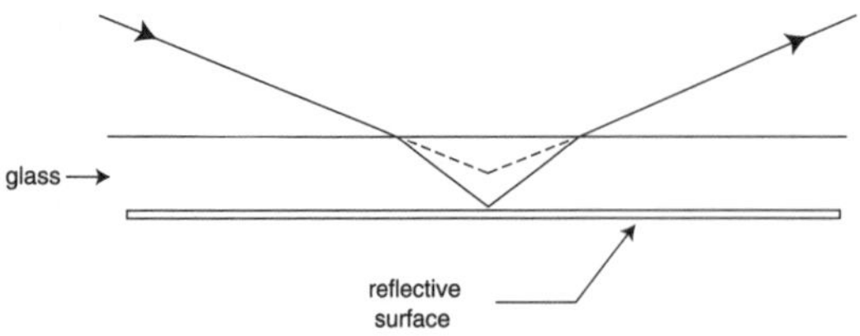


drawing. The net effect is that the light acts as if it had been reflected from a surface located about in the middle of the lab mirror (where the dotted lines meet in the drawing). Notice that the angle of incidence is still equal to the angle of reflection, by symmetry. However, when the reflected ray is extended behind the mirror to locate the image, the extension will be shifted away from the point where the incident ray hits the mirror. The size of the shift increases with the angle of incidence (and the shift is zero for a light beam coming in along the normal). The result is that the image is slightly blurred.

Extending the paths of the reflected rays back behind the mirror produces the location of the image.

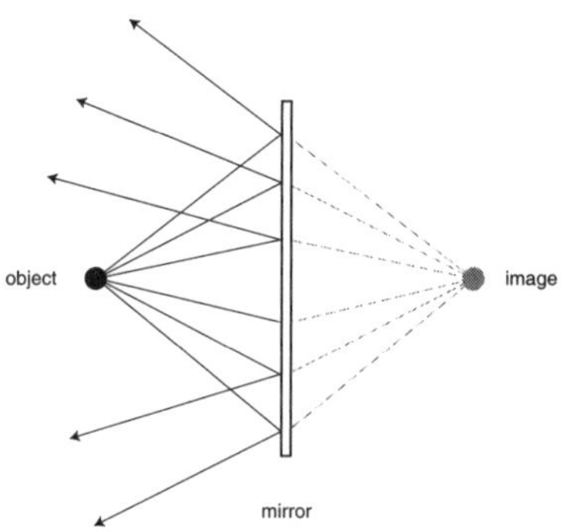


Note that the image is as far behind the mirror as the object is in front. Also, the image is located on the line that goes through the object and is perpendicular to the mirror surface. If you walk towards a mirror, your reflection seems to walk towards you. If you reach out to shake hands with your reflection, a reflected hand reaches towards you . . . but it's a left hand. What happened? Move your hand to the left, and the reflection moves to the left. Move your hand to the right, and the reflection moves to the right. But move your hand towards the mirror, and the reflection moves towards you (in the opposite direction). This is the source of the inversion of your hand. What is inverted in the mirror is not right to left, and not up to down, but rather towards the mirror and away from the mirror. The reflection of an asymmetrical letter like a "R" shows what happens. Note that the reflection of the left hand is a right hand.

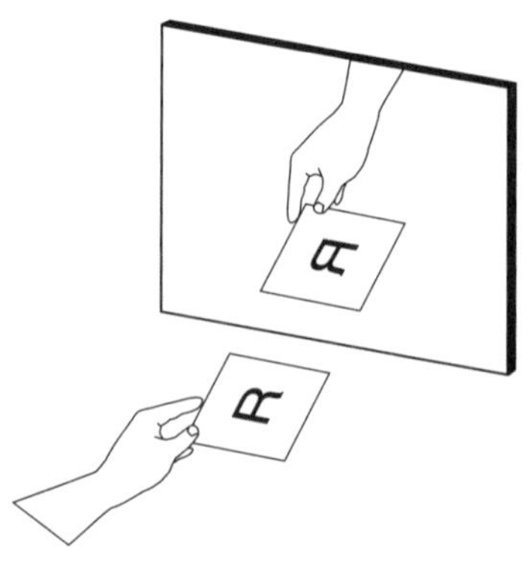

## *Active*-ating the Physics InfoMall

This is the first of several activities that involve light and optics. A good strategy is to find what problems students may have with this area of physics. One way to find out is to search for "student difficult*" AND "optics". You may not want to limit these terms to "the same paragraph" for the first attempt. Indeed, there are several nice articles found in this search. (Note that most textbooks do not directly address the issue of student difficulties or misconceptions. Articles, being more recent and flexible in most cases, do treat these topics.)

The first hit on the list is "An investigation of student understanding of the real image formed by a converging lens or concave mirror," in the *American Journal of Physics*, vol. 55, issue 2, 1987. Note that this article is actually on the CD-ROM for your use. But, since this activity is concerned with plane mirrors, we may wish to keep this article for later. Just a little lower on the list is "Student difficulties in understanding image formation by a plane mirror," from *The Physics Teacher*, vol. 24, issue 8, 1986. This is something you should read.

A related search is "misconcept*" and "optics". This provides a larger number of hits, but includes both articles above. One new hit is "Misinterpretation of theories of light," from the *American Journal of Physics*.

Suppose the question about the mirror on the moon gets you curious. Search for "mirror" AND "moon" to find some information.

What if you want to find additional activities or demonstrations for your class? You can search, or go straight to the Demo & Lab Shop. You can choose any of the books you find there; by chance, the first book found while preparing this document was Physics Lab Experiments and Computer Aids, which has an activity similar to the *Coordinated Science for the 21st Century* activity, with graphics that might be helpful. In addition, there are many more ideas you may wish to try yourself.

In **For You To Do Step 11**, students are asked to place two mirrors at right angles. To see what might happen, search for "mirror*" AND "right angle*", but restrict the search to " "within paragraph." Not only will your question be answered, but we find several articles that might be useful for other things. For example, "Optics: It's Elementary," from Teachers Treasures, has some great ideas. "Some reflections on plane mirrors and images," from *The Physics Teacher*, vol. 29, issue 7, 1991, is also worth finding. Repeat this search for some great sources.

# Planning for the Activity

## Time Requirements

one and a half class periods

## Materials Needed

### For the class:

- oscilloscope

### For each group:

- AA battery
- cardboard sheet, 14" × 14" square
- glass rod
- laser pointer or ray bar
- miniature light bulb, 1.5 volt
- optical bench apparatus
- light source for optical bench apparatus
- plane mirrors, 3 × 4 × 1/2 (2)
- makeup mirror, convex/concave surface
- protractor
- right angle prism
- ruler, metric
- small amount of play sand
- supports for mirrors (2)

### For each student:

- 1 copy of the Blackline Master:
  Angles of Incidence and Angles of Reflection

## Advance Preparation and Setup

Make a copy, for each student, of the Blackline Master Angles of Incidence and Angles of Reflection. You can support the mirrors on blocks of wood or tape them to boxes. Also, you can place small mirrors in slit rubber stoppers. You may also want to insert the glass rod in a rubber stopper and hold the stopper in a test tube clamp, mounted on a ring stand. Be sure that each group has two mirrors that can meet edge-to-edge to make the multiple reflections most effectively. Mirrors in frames cannot meet in this way. Test each light source in advance. Be sure that all mirrors are clean. Darken the room as much as possible.

# Teaching Notes

When the students measure the angle of incidence vs. the angle of reflection, encourage them to take data throughout the range of the angle of incidence (zero to 90 degrees). Every 15 degrees is a good interval. Encourage the students to sight carefully as they place the dots on the paper to mark the position of the beam. Point out that they must be directly over the beam when they place their dots. You might mention the effect of parallax—an object seems to shift position when seen from different places. To help students visualize the position of the image, have them hold a meter stick or ruler perpendicular to the mirror, as shown.

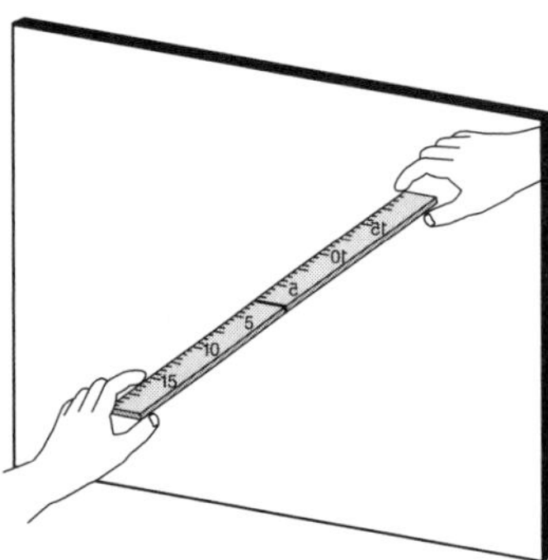

If the students use lasers as the light source, each group must keep its laser's beam under control at all times. Have each group set up a backstop behind its mirrors to block the beam. A large piece of cardboard makes a good backstop.

An easy way to attach the glass rod to the laser is to drill a hole through a flat rubber eraser slightly smaller than the barrel of the pen laser. Push the pen laser through the hole until the end of the laser just comes through. Then tape the glass rod to the rummer eraser so that it is positioned in the laser beam.

SAFETY PRECAUTION: Warn the students not to look into the laser beam. Also emphasize that the laser should only be turned on when it is resting on the tabletop.

Many secondary students have not yet understood the path of light from a light source to an object to their eye. Some may still believe that vision is something that emanates from the eyes, so the reflection of light from the object is not important. It may help to darken the room and illuminate an object with a flashlight. Have the students trace the path of light (or even walk along the path) from the flashlight to their eyes. Ask about the fact that many students, in different locations in the classroom, can see the object, and what that must mean in terms of the reflected light (that it's reflected in many directions).

There should be good agreement (within a few degrees) between the values of the angles of incidence and angles of refraction, especially if the mirrors are made of polished metal. Remember that if the students are using plastic or glass mirrors, these are usually rear-surfaced. In effect, the reflection happens in the middle of the mirror, but the angle of reflection is still equal to the angle of incidence. The main effect will be to shift the extension of the beam used to locate the image.

NOTES

# Activity Overview

Students begin the study of mirrors by investigating reflection from a plane mirror. To find the direction of a reflected light ray, they measure the angles of incidence and refraction. Then they locate the reflected image and observe the reversals in reflections of letters of the alphabet. Finally, students investigate multiple reflections from two mirrors and the resulting symmetrical patterns.

## Student Objectives

### Students will:

- Identify the normal of a mirror.
- Measure angles of incidence and reflection.
- Observe the relationship between the angle of incidence and the angle of reflection.
- Observe changes in the reflections of letters.
- Identify patterns in multiple reflections.

## Activity 4 Reflected Light

**GOALS**

In this activity you will:

- Identify the normal of a mirror.
- Measure angles of incidence and reflection.
- Observe the relationship between the angle of incidence and the angle of reflection.
- Observe changes in the reflections of letters.
- Identify patterns in multiple reflections.

### What Do You Think?

Astronauts placed a mirror on the Moon in 1969 so that a light beam sent from Earth could be reflected back to Earth. By timing the return of the beam, scientists found the distance between the Earth and the Moon. They measured this distance to within 30 cm.

- **How are you able to see yourself in a mirror?**
- **If you want to see more of yourself, what can you do?**

Record your ideas about these questions in your *Active Physics* log. Be prepared to discuss your responses with your small group and with your class.

### For You To Do

1. Place a piece of paper on your desk. Carefully aim the laser pointer, or the light from a ray box, so the light beam moves horizontally, as shown on the opposite page.

ANSWERS FOR THE TEACHER ONLY

## What Do You Think?

You can see yourself in a mirror by observing light reflected from you that the mirror reflects back into your eyes. The mirror creates a virtual image of you the same distance behind the mirror that you are in front of it. But you can see only that part of your body that faces the mirror, since light from the back of your head moves off in another direction and does not hit the mirror.

To see the back of your head requires a second mirror, which is why hair stylists hold up a second mirror to show their clients their new haircut. If you covered the walls of a room with mirrors, you could see many different reflections of the back of your head.

ANSWERS

## For You To Do

1. Student activity.

2. Place a glass rod in the light beam so that the beam spreads up and down. Shine the beam on the piece of paper to be sure the beam passes through the glass rod.

3. Carefully stand the plane mirror on your desk in the middle of the piece of paper. Draw a line on the paper along the

**Do not use mirrors with chipped edges. Make sure the ends of the glass rod are polished.**

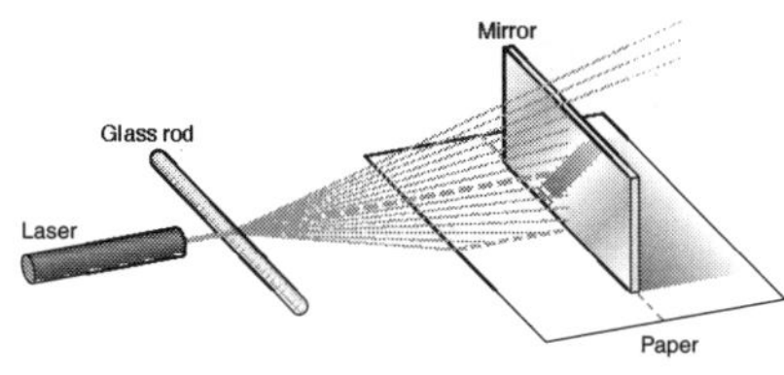


**Never look directly at a laser beam or shine a laser beam into someone's eyes. Always work above the plane of the beam and beware of reflections from shiny surfaces.**

front edge of the mirror. Now remove the mirror and draw a dotted line perpendicular to the first line, as shown. This dotted line is called the **normal**.

Physics Words

normal: at right angles or perpendicular to.

4. Aim the light source so the beam approaches the mirror along the normal. Be sure the glass rod is in place to spread out the beam.

 a) What happens to the light after it hits the mirror?

5. Make the light hit the mirror at a different angle.

 a) What happens now?

 b) On the paper, mark three or more dots **under** the beam to show the direction of the **beam as** it travels to the mirror. The line you traced shows the incident ray. Also make dots to show the light going away from the mirror. **This** line shows the reflected ray. Label this pair of rays to show they go **together**.

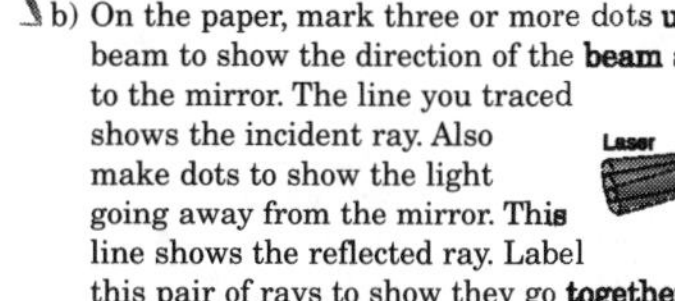

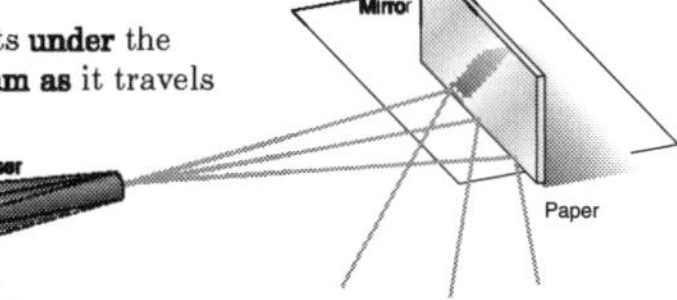


6. Turn the light source so it starts from the same point but strikes the mirror at different angles. For each angle, mark dots on the paper to show the direction of the incident and reflected rays. Also, label each pair of rays.

Chapter 3

ANSWERS

## For You To Do *(continued)*

2. Find some way to support the glass rod, such as with the optional stopper, test-tube holder and ringstand (or the rod can be taped to a stack of books).

3. Student activity. Support the mirror, using the mirror support in the kit.

4. a) It reflects back along the normal.

5. a) The beam reflects but not along the normal.

   b) Encourage students to look straight down to avoid parallax.

6. Student activity.

7. Most lab mirrors have the reflecting surface on the back. In addition, the light bends as it enters and leaves the glass part of the mirror. In your drawing, the rays may not meet at the mirror surface. Extend the rays until they do meet.
   a) Measure these angles for one pair of your rays.

8. Turn off the light source and remove the paper. Look at one pair of rays. The diagram shows a top view of the mirror, the normal, and an incident and reflected ray. Notice the angle of incidence and the angle of reflection in the drawing. Using a protractor, measure these angles for one pair of rays.
   a) Record your data in a table.
   b) Measure and record the angles of incidence and reflection for all of your pairs of rays.

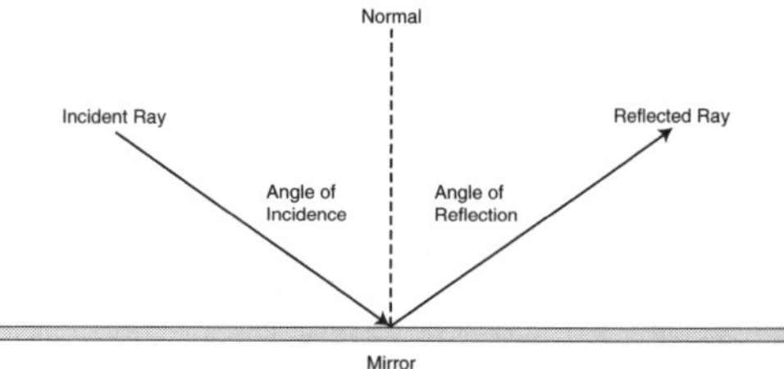


   c) What is the relationship between the angles of incidence and reflection?
   d) Look at the reflected rays in your drawing. Extend each ray back behind the mirror. What do you notice when you have extended all the rays? The position where the rays meet is the location of the image of the light source. All of the light rays leave one point in front of the mirror. The reflected rays all seem to emerge from one point behind the mirror. Wherever you observed the reflection, you would see the source at this point behind the mirror.
   e) Tape a copy of your diagram in your log.

9. Hold the light source, or any object, near the mirror and look at the reflection. Now hold the object far away and again look at the reflection.

ANSWERS

## For You To Do *(continued)*

7. See drawing in **Backround Information**. The angles should be equal.

8. a)-c) The angle of incidence is (approximately) equal to the angle of reflection.

   d) The extended rays intersect in the same place (or nearly so).

   e) See drawing of extended rays in **Background Information**.

a) How is the position of the reflection related to the position of the object?

10. Set up a mirror on another piece of paper, and draw the normal on the paper. Write your name in block capital letters along the normal (a line perpendicular to the mirror). Observe the reflection of your name in the mirror.

a) How can you explain the reflection you see?

b) Which letters in the reflection are closest to the mirror? Which are farthest away?

c) In your log, make a sketch of your name and its reflection.

11. Carefully stand up two mirrors so they meet at a right angle. Be sure they touch each other, as shown in the drawing.

12. Place an object in front of the mirrors.

a) How many images do you see?

b) Slowly change the angle between the mirrors. Make a general statement about how the number of images you see changes as the angle between the mirrors changes.

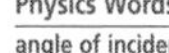

**Physics Words**

**angle of incidence:** the angle a ray of light makes with the normal to the surface at the point of incidence.

**angle of reflection:** the angle a reflected ray makes with the normal to the surface at the point of reflection.

**ray:** the path followed by a very thin beam of light.

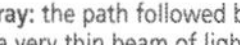

**FOR YOU TO READ**

**Images in a Plane Mirror**

An object like the tip of a nose reflects light in all directions. That is why everybody in a room can see the tip of a nose. Light reflects off a mirror in such a way that the **angle of incidence** is equal to the **angle of reflection.** You can look at the light leaving the tip of a nose and hitting a mirror to see how an image is produced and where it is located. Each **ray** of light leaves the nose at a different angle. Once it hits the mirror, the angle of incidence must equal the angle of reflection. There are now a set of rays diverging from the mirror. If you assume that the light always travels in straight lines, you can extend these rays behind the mirror and find where they "seem" to emerge from. That is the location of the image.

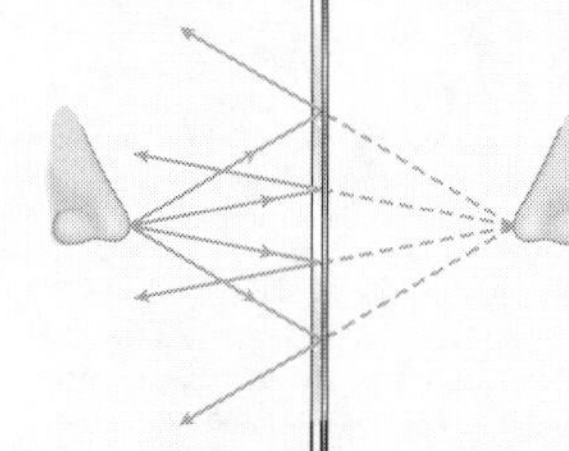

The mirror does such a good job of reflecting that it looks as if there is a tip of a nose (and all other parts of the face) behind the mirror. If you measure the distance of the image behind the mirror, you will find that it is equal to the distance of the nose (object) in front of the mirror. This can also be proved using geometry.

→

Chapter 3

ANSWERS

# For You To Do *(continued)*

9. a) The position of the reflection is behind the mirror. It is the same distance behind the mirror that the object was in front of the mirror.

10. a)-b) The letters that are closest to the mirror make reflections that look closest to the mirror. The letters that are farthest from the mirror make reflections that are farthest from the mirror. Moreover, within a letter, the part of letter closer to the mirror makes a reflection closer to the mirror than the rest of the letter.

c) Student diagrams will differ.

11. Student activity.

12. a) Three (one is behind each mirror, as before; the third image is behind the corner where the mirrors meet).

b) As the angle between the mirrors reduces, the number of reflections increases.

## Diffraction of Light

As you begin to study the reflection of light rays, it is worthwhile to recognize that light is a wave and has properties similar to sound waves.

In studying sound waves, you learned that sound waves are compressional or longitudinal. The disturbance is parallel to the direction of motion of the wave. In sound waves, the compression of the air is left and right as the wave travels to the right. You saw a similar compressional wave using the compressed Slinky.

Light waves are transverse waves. They are similar to the transverse waves of the Slinky. In a transverse wave, the disturbance is perpendicular to the direction of the wave. In the Slinky, the disturbance was up and down as the wave traveled to the right. In light, the fields (the disturbance) are perpendicular to the direction of motion of the waves.

You also read that sound waves diffract—they spread out as they emerge from small openings. You can find out if light waves spread out as they emerge from a small opening. Try this: Take a piece of aluminum foil. Pierce the foil with a pin to create a succession of holes, one smaller than the next. Shine the laser beam through each hole and observe its appearance on a distant wall. You will be able to observe the diffraction of light.

## Sample Problem

Light is incident upon the surface of a mirror at an angle of 40°.

a) Sketch the reflection of the ray.

***Strategy:*** The angles of incidence and reflection are always measured from the normal. The Law of Reflection states that the angle of incidence is equal to the angle of reflection. Since the angle of incidence is equal to 40°, the angle of reflection is also 40°.

***Given:***

$\theta_i = 40°$

***Solution:***

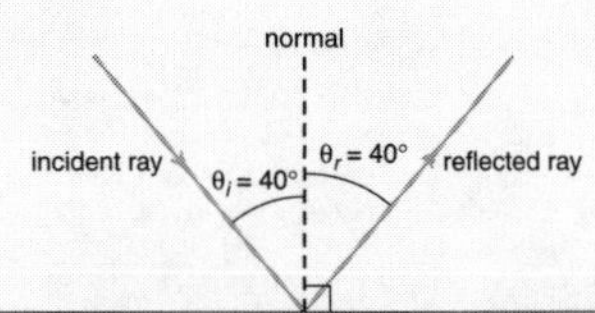


b) At what angle, as measured from the surface of the mirror, did the beam strike the mirror?

***Strategy:*** The angle of incidence is measured from the normal. The question is asking for the complementary angle.

***Solution:***

$\theta_i = \theta_r = 40°$

$90° - 40° = 50°$

The angle between the light beam and the mirror is 50°.

### Reflecting on the Activity and the Challenge

In this activity you aimed light rays at mirrors and observed the reflections. From the experiment you discovered that the angle of incidence is equal to the angle of reflection. Therefore, you can now predict the path of a reflected light beam. You also experimented with reflections from two mirrors. When you observed the reflection in two mirrors, you found many images that made interesting patterns.

This activity has given you experience with many interesting effects that you can use in your sound and light show. For instance, you may want to show the audience a reflection in one mirror or two mirrors placed at angles. You can probably create a kaleidoscope. You will also be able to explain the physics concept you use in terms of reflected light.

### Physics To Go

1. How is the way light reflects from a mirror similar to the way a tennis ball bounces off a wall?
2. a) What is the normal to a plane mirror?
   b) When a light beam reflects from a plane mirror, how do you measure the angle of incidence?
   c) How do you measure the angle of reflection?
   d) What is the relationship between the angle of incidence and the angle of reflection?
3. Make a top-view drawing to show the relationships among the normal, the angle of incidence, and the angle of reflection.
4. a) Suppose you are experimenting with a mirror mounted vertically on a table, like the one you used in this activity. Make a top-view drawing, with a heavy line to represent the mirror and a dotted line to represent the normal.
   b) Show light beams that make angles of incidence of 0°, 30°, 45°, and 60° to the normal.
   c) For each of the above beams, draw the reflected ray. Add a label if necessary to show where the rays are.



ANSWERS

## Physics To Go

1. For the tennis ball and for light, the angle of incidence equals the angle of reflection (except for the effect of the spin of the ball).
2. a) The normal is a line perpendicular to the mirror surface.
   b) The angle of incidence is the angle between the incident beam and the normal. You measure this angle by tracing the path of the incident beam and then drawing the normal where the beam hits the mirror. Use a protractor to measure the angle.
   c) The angle of reflection is the angle between the reflected beam and the normal. See answer to **Part (b)**.
   d) They are equal.
3. See the drawing on page 174.
4. a) See the drawing on page 174.
   b)-c) See below.

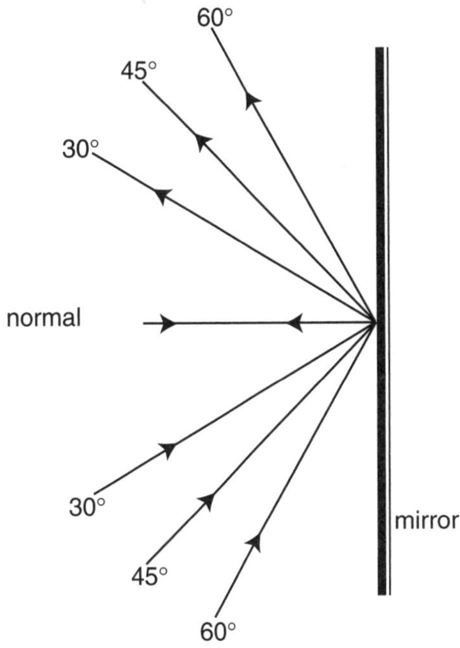


Chapter 3

ANSWERS

## Physics To Go

***(continued)***

5. a) Student activity.

   b) The reflection moves towards the mirror and towards you.

   c) The reflection moves away from the mirror and away from you.

   d) The reflection is the same distance behind the mirror as the object is in front of the mirror. If you move your hand away from the mirror, the distance to the mirror increases, so the reflection is further from the mirror, too.

6. a) The letters A, I, T, M, O, W, U, and V.

   b) MOM, TOT, WOW.

   c) K, N, and S.

   d) See drawing below.

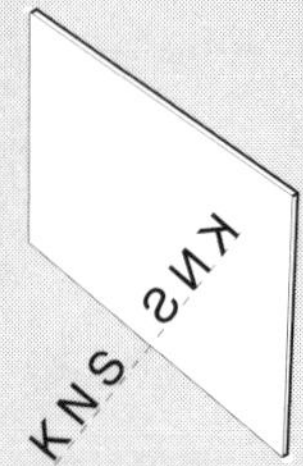


7. It is usually seen in a rear-view mirror in which it would read normally.

8. Placement of the original light ray is random. But once the ray has struck the mirror, angles of incidence to the normal and angles of reflection from the normal must be carefully measured.

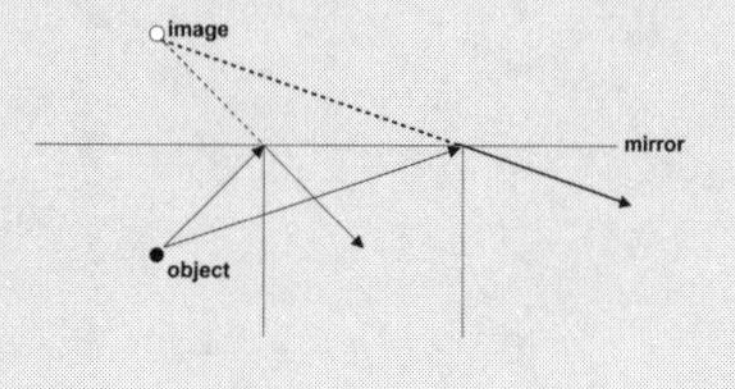


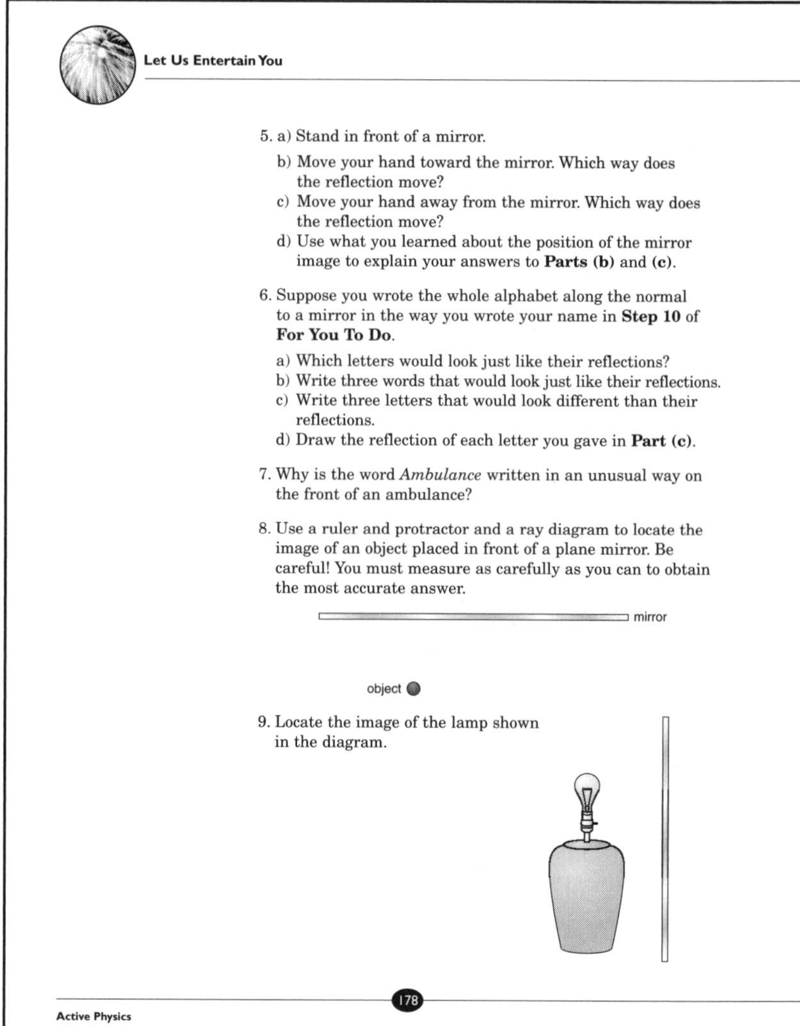

Let Us Entertain You

5. a) Stand in front of a mirror.
   b) Move your hand toward the mirror. Which way does the reflection move?
   c) Move your hand away from the mirror. Which way does the reflection move?
   d) Use what you learned about the position of the mirror image to explain your answers to **Parts (b)** and **(c)**.

6. Suppose you wrote the whole alphabet along the normal to a mirror in the way you wrote your name in **Step 10** of **For You To Do**.
   a) Which letters would look just like their reflections?
   b) Write three words that would look just like their reflections.
   c) Write three letters that would look different than their reflections.
   d) Draw the reflection of each letter you gave in **Part (c)**.

7. Why is the word *Ambulance* written in an unusual way on the front of an ambulance?

8. Use a ruler and protractor and a ray diagram to locate the image of an object placed in front of a plane mirror. Be careful! You must measure as carefully as you can to obtain the most accurate answer.

mirror

object

9. Locate the image of the lamp shown in the diagram.

178

Active Physics

ANSWERS

## Physics To Go *(continued)*

9.

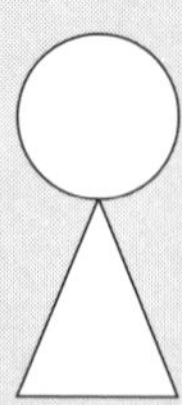

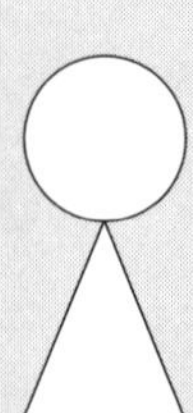

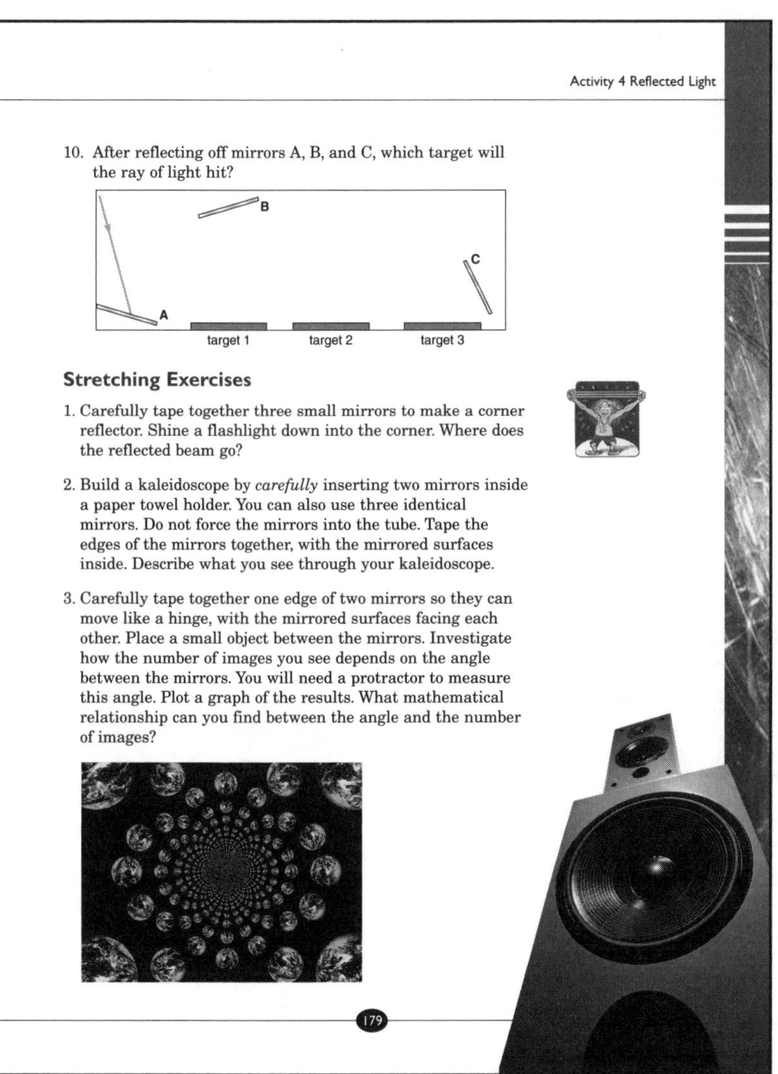

Activity 4 Reflected Light

10. After reflecting off mirrors A, B, and C, which target will the ray of light hit?

B
C
A
target 1
target 2
target 3

**Stretching Exercises**

1. Carefully tape together three small mirrors to make a corner reflector. Shine a flashlight down into the corner. Where does the reflected beam go?
2. Build a kaleidoscope by *carefully* inserting two mirrors inside a paper towel holder. You can also use three identical mirrors. Do not force the mirrors into the tube. Tape the edges of the mirrors together, with the mirrored surfaces inside. Describe what you see through your kaleidoscope.
3. Carefully tape together one edge of two mirrors so they can move like a hinge, with the mirrored surfaces facing each other. Place a small object between the mirrors. Investigate how the number of images you see depends on the angle between the mirrors. You will need a protractor to measure this angle. Plot a graph of the results. What mathematical relationship can you find between the angle and the number of images?

179

ANSWERS

## Stretching Exercises

1. The beam reflects back along the direction it entered the corner mirror.
2. Student activity.
3. Number of images = (360/angle); however when the formula dictates a whole number of images, some of the images overlap and can only be seen as the angle between the mirrors is further decreased.

Chapter 3

ANSWERS

## Physics To Go *(continued)*

10. Target 3

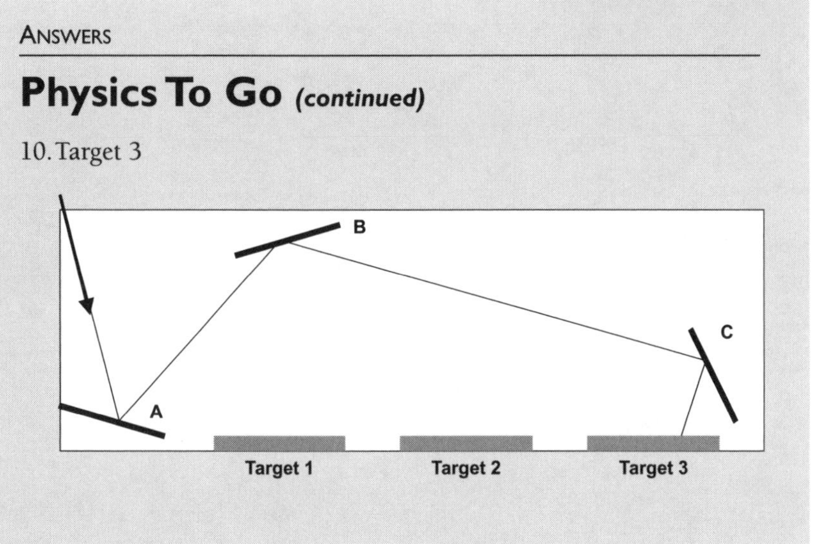

# Activity 4 A

# Low-Tech Reflection and Mirror Symmetry

## FOR YOU TO DO

1. Low-Tech Reflection: You can do the Angle of Incidence/Angle of Reflection experiment with a flashlight. The results are less accurate, of course, than they would be with a laser beam, but students will still get the idea. You might give the students string and have them stretch it along the normal, from the flashlight to the mirror, and from the mirror to the reflection. Then they can measure the angles between the strings.

2. Mirror Symmetry: Give each group a simple design. Have them make a drawing to predict how the reflection of the design will look. Then they hold the design up to a mirror to observe the reflection. Arrows on the design point to the location where the mirror will be. When the students have made and discussed their predictions, give them the mirrors.

For use with *Let Us Entertain You*, Chapter 3, Activity 4: Reflected Light

| Table of Angles of Incidence | Angles of Reflection |
| --- | --- |
| | |
| | |
| | |
| | |
| | |
| | |
| | |
| | |
| | |
| | |
| | |
| | |
| | |
| | |
| | |
| | |

For use with Let Us Entertain You, Chapter 3, Activity 4: Reflected Light



| Table of Angles of Incidence | Angles of Reflection |
| --- | --- |
| | |
| | |
| | |
| | |
| | |
| | |
| | |
| | |
| | |
| | |
| | |
| | |
| | |
| | |
| | |
| | |

For use with *Let Us Entertain You*, Chapter 3, Activity 4: Reflected Light

# ACTIVITY 5
## Curved Mirrors

# Background Information

A good starting point for thinking about the curved mirror is the plane mirror. The plane mirror makes an erect, virtual image.

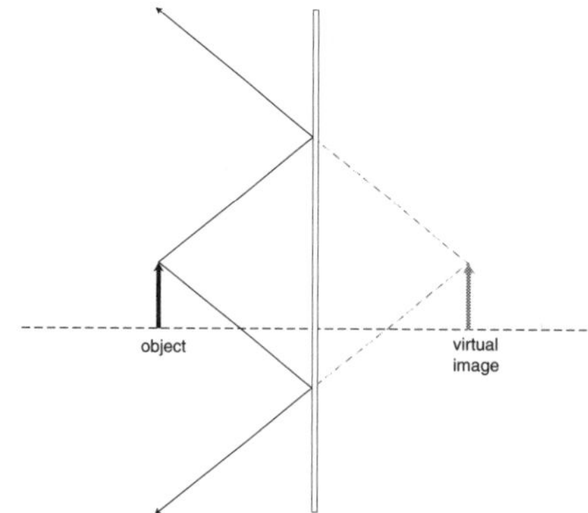


Now imagine the mirror is bowed out to make a convex mirror. How does the image change? The light rays show that the image is still behind the mirror, so the image is still virtual.

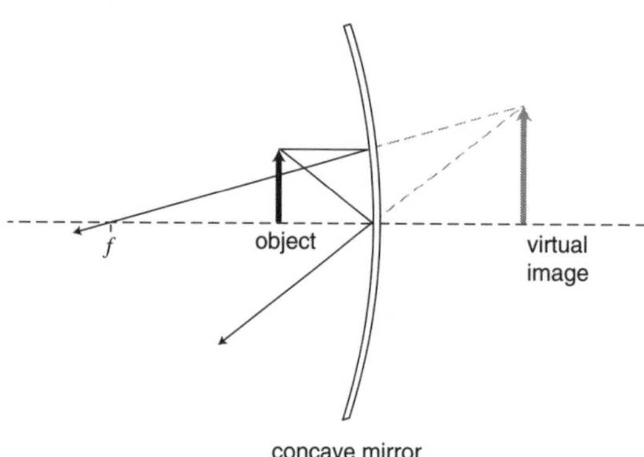


Moreover, the image is reduced, since parallel rays are reflected out away from the mirror. Also, the image is closer to the mirror than is the object. The focus is labeled. The more strongly the mirror is curved, the shorter the focal length and the more the virtual image is reduced compared to the object.

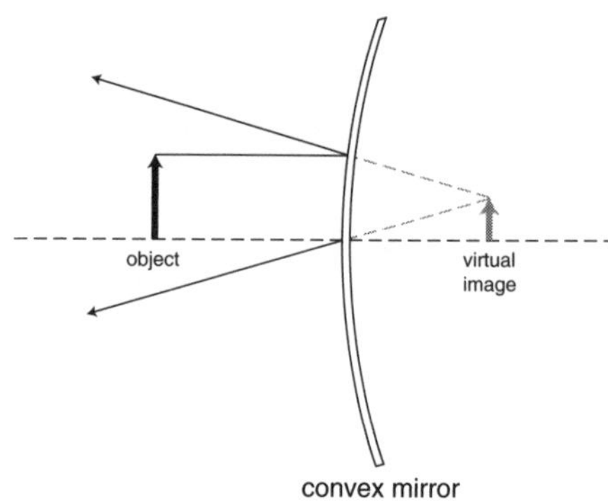


Now imagine bowing the plane mirror inward, to make a concave mirror.

Since parallel rays now bend in towards the mirror axis, objects now appear magnified when the object is close to the mirror. Only when the object is within a focal length of the mirror is the object virtual. For such object positions, the image is further from the mirror than is the object. If the object is outside the focal length, the reflected rays cross, so a real image is formed, as shown in the drawings in **Physics Talk** on page 184 in the Student Edition. This image can be seen on a card. It can, of course, also be seen by looking at the mirror, rather than on a card, and this is the basis of the famous "floating coin" illusion made by two parabolic mirrors fastened together. The concave mirror is exactly analogous to the convex lens, which is a magnifier when the object is up close and which produces a real image otherwise. The lens equation also describes the location of the image and object for mirrors:

$1/f = 1/D_o + 1/D_i$

Image formation in curved mirrors, including the mathematical description of the equation above, are exactly analagous to image formation in lenses, the subject of **Activity** 7.

## *Active*-ating the Physics InfoMall

This activity is all about images formed by curved mirrors. Research has been conducted to see how students understand this phenomenon, as reported in "An investigation of student understanding of the real image formed by a converging lens or concave mirror," from the *American Journal of Physics*, vol. 55, issue 2, 1987, mentioned above. This article may prepare you for problems that might arise with your students.

The first search to perform here might be "curved mirror*". This does produce a few good hits. But if you want to find more, or better, sources of information, you may want to be more specific: search for concave or convex mirrors.

One item you may find while searching is "Physics Review Text" from the Regent's Physics Review in the Problems Place. In Opt. Unit 4 – Geometric Optics you can find a wonderful set of instructions for drawing ray diagrams. Again, these can be copied right into your own documents. As the InfoMall goes, the graphics in this section are great! You may also want to look in the Optical Elements section of *Physics Including Human Applications*, found in the Textbook Trove. Of course, these are just a fraction of the things you can find on the InfoMall.

**Step 11** of **For You To Do** uses an equation. If you want to see what the InfoMall has for this, go back to the InfoMall Entrance (the colorful screen with the directory map). In the lower right-hand corner you will see a button with equations on it. Clicking on this takes you to the equations dictionary. You will be only a few clicks away from this equation, along with examples.

# Planning for the Activity

## Time Requirements

two class periods

## Materials Needed

### For each group:

- 40-W bulb and socket, ceramic w/switch
- cardboard sheet, 14" × 14" square (for a backstop to absorb a laser beam)
- convex/concave mirror surface
- glass rod
- laser pointer
- masking tape, 3/4" × 60 yds.
- paper, plain
- right angle prism
- ruler, metric
- unlined index cards (100)
- stick, meter, 100 cm hardwood

### For each student:

- 1 copy of Blackline Master: Object and Image Distances
- 1 copy of Blackline Master: Graph Paper
- Second copy of Blackline Master: Graph Paper (if **PTG #5** is assigned)

## Advance Preparation and Setup

Make copies of the Object and Image Distances and Graph Paper, for each student. If you assign **Question 5** from **Physics To Go**, make an additional copy of the Graph Paper for each student. If possible, find a large-diameter mailing tube to make the forms for the mirrors used in the ray-tracing. Cut a section about 5 cm high. Give each group a complete section and also one cut in half (to be the form for the concave lens). If it is necessary to use a flashlight as a light source, you will need a very large-diameter form and a large piece of mylar to trace the parallel rays, because the flashlight beam will spread out so much. When the students investigate real images, use a large, well-shaped concave mirror so the image will be bright and clear. A cosmetics mirror with a diameter of about 15 cm works well. If your students use lasers, collect large pieces of cardboard to use as backstops.

# Teaching Notes

If you have neither a laser nor a light box, and you must use a flashlight for a light source, you can still encourage the students to get the general idea of how parallel light beams are reflected from curved mirrors (towards the mirror axis or away, as discussed in **Background Information**). The students can still see images made by a light bulb in the second half of the activity. When the students observe images, give them the best mirrors you have available.

Once the students have observed the images, you may want to remind them about **Steps 1** to **7**, when they observed reflections of individual light beams. Encourage them to use their observations to predict which mirrors make real images.

You may want to set up a light bulb in the front of the room where it is visible to all groups. This bulb can serve as the distant object, to minimize how the students must move about the lab. If your students use lasers, emphasize the importance of keeping their laser beam within their work area.

If students use a laser, have each group prop up a large piece of cardboard to absorb the beam. If the students use a 110-V bulb as a light source, caution them to handle the bulb carefully and to leave the bulb in its socket at all times. Be sure there are no exposed connections to the socket. If the socket is ceramic and has screw-terminal posts underneath, screw the socket to a piece of wood so the terminals are inaccessible.

Many students believe that the virtual image is located inside the mirror. Many others believe that the image of an object moves if the observer moves. To help students, you can have them set the mirror on a table and place a small object nearby. Have them observe the reflection of the object from different positions and report what they see. You might also have them repeat this observation with a plane mirror.

This data was taken with a cosmetics mirror.

| $D_o$ (m) | $D_i$ (m) |
|---|---|
| 26.5 | 0.37 |
| 5.13 | 0.45 |
| 2.74 | 0.52 |
| 1.37 | 0.52 |
| 0.51 | 1.30 |

In the above data, note how in going from 5.13 m to 26.5 m, more than a factor of five in object distance, the image moved in less than fifteen percent. The focal length of the mirror is very close to .37 m. Also notice how the image distance increases rapidly as the object approaches the focus of the mirror.

NOTES

# Activity Overview

Having worked with plane mirrors in **Activity 4**, students now investigate curved mirrors. As in **Activity 4**, students begin by shining parallel light beams at the curved mirrors to record the pattern of reflection. From these patterns students find the focal length and the location of the focus. With a small light bulb, students look for real and virtual images. For the concave mirror, students draw a graph of the image distance vs the object distance. To use a quantitative model of reflection, students examine an equation that relates image distance, object distance, and focal length.

## Student Objectives

### Students will:

- Identify the focus and focal length of a curved mirror.
- Observe virtual images in a convex mirror.
- Observe real and virtual images in a concave mirror.
- Measure and graph image distance vs object distance for a convex mirror.
- Summarize observations in a sentence.

ANSWERS FOR THE TEACHER ONLY

## What Do You Think?

The reflection in a flat mirror has the same proportions as the object. To explain this observation, recall that the reflected image is the same distance behind the mirror as the object is in front, and on a line from the object perpendicular to the mirror. A displacement of the object produces an equal displacement of the image, so the proportions are preserved. But in a curved mirror, this proportionality is lost.

The ray diagrams on pages 181 and 182 show how curved mirrors reflect a parallel beam of light. Imagine standing close to these mirrors and looking at the reflection of the distant object. Since the reflection is seen along the extension of the reflected light rays, the concave mirror makes the object look larger (in the direction of the curvature) and the convex mirror makes the object look smaller.

Let Us Entertain You

# Activity 5 Curved Mirrors

**GOALS**

In this activity you will:

- Identify the focus and focal length of a curved mirror.
- Observe virtual images in a convex mirror.
- Observe real and virtual images in a concave mirror.
- Measure and graph image distance versus object distance for a convex mirror.
- Summarize observations in a sentence.

## What Do You Think?

The curved mirror of the Palomar telescope is five meters across. Mirrors with varying curvatures are used in amusement parks as fun-house mirrors. Store mirrors and car side-view mirrors are also curved.

- **How is what you see in curved mirrors different from what you see in ordinary flat mirrors?**

Record your ideas about this question in your *Active Physics* log. Be prepared to discuss your responses with your small group and with your class.

## For You To Do

1. Carefully aim a laser pointer, or the light from a ray box, so the light beam moves horizontally, as you did in the previous activity. Place a glass rod in the light beam so that the beam spreads up and down.

Active Physics

180

ANSWERS

## For You To Do

1. Support the glass rod in the same way, as in **Activity 4**.

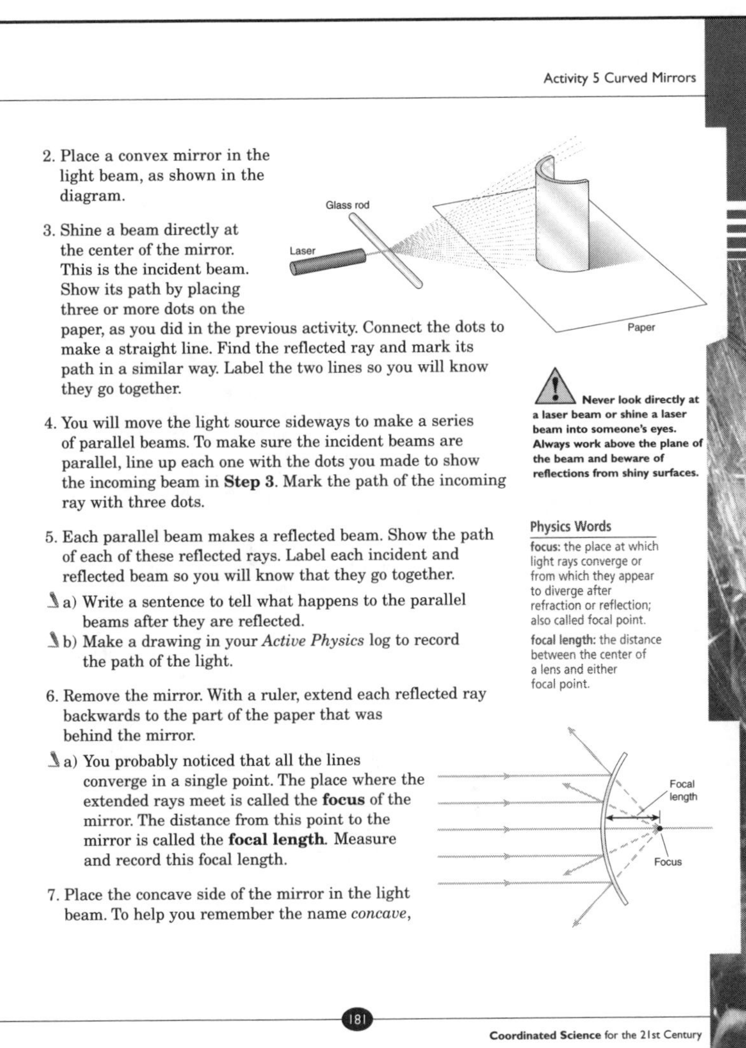



2. Place a convex mirror in the light beam, as shown in the diagram.



3. Shine a beam directly at the center of the mirror. This is the incident beam. Show its path by placing three or more dots on the paper, as you did in the previous activity. Connect the dots to make a straight line. Find the reflected ray and mark its path in a similar way. Label the two lines so you will know they go together.

4. You will move the light source sideways to make a series of parallel beams. To make sure the incident beams are parallel, line up each one with the dots you made to show the incoming beam in **Step 3**. Mark the path of the incoming ray with three dots.

5. Each parallel beam makes a reflected beam. Show the path of each of these reflected rays. Label each incident and reflected beam so you will know that they go together.
   a) Write a sentence to tell what happens to the parallel beams after they are reflected.
   b) Make a drawing in your *Active Physics* log to record the path of the light.

6. Remove the mirror. With a ruler, extend each reflected ray backwards to the part of the paper that was behind the mirror.
   a) You probably noticed that all the lines converge in a single point. The place where the extended rays meet is called the **focus** of the mirror. The distance from this point to the mirror is called the **focal length**. Measure and record this focal length.



7. Place the concave side of the mirror in the light beam. To help you remember the name *concave*,

**Never look directly at a laser beam or shine a laser beam into someone's eyes. Always work above the plane of the beam and beware of reflections from shiny surfaces.**

**Physics Words**

**focus:** the place at which light rays converge or from which they appear to diverge after refraction or reflection; also called focal point.

**focal length:** the distance between the center of a lens and either focal point.



## Answers

# For You To Do *(continued)*

2. Pull the mylar tight to make a smooth curved surface.

3. Student activity.

4. It may be easier to pull the paper sideways to move the mirror instead of the light source.

5. a) After reflecting from the convex mirror, the parallel beams spread apart.

   b) See the drawing on page 181.

6. a) The rays may not meet at the same place, but the students should get the idea.

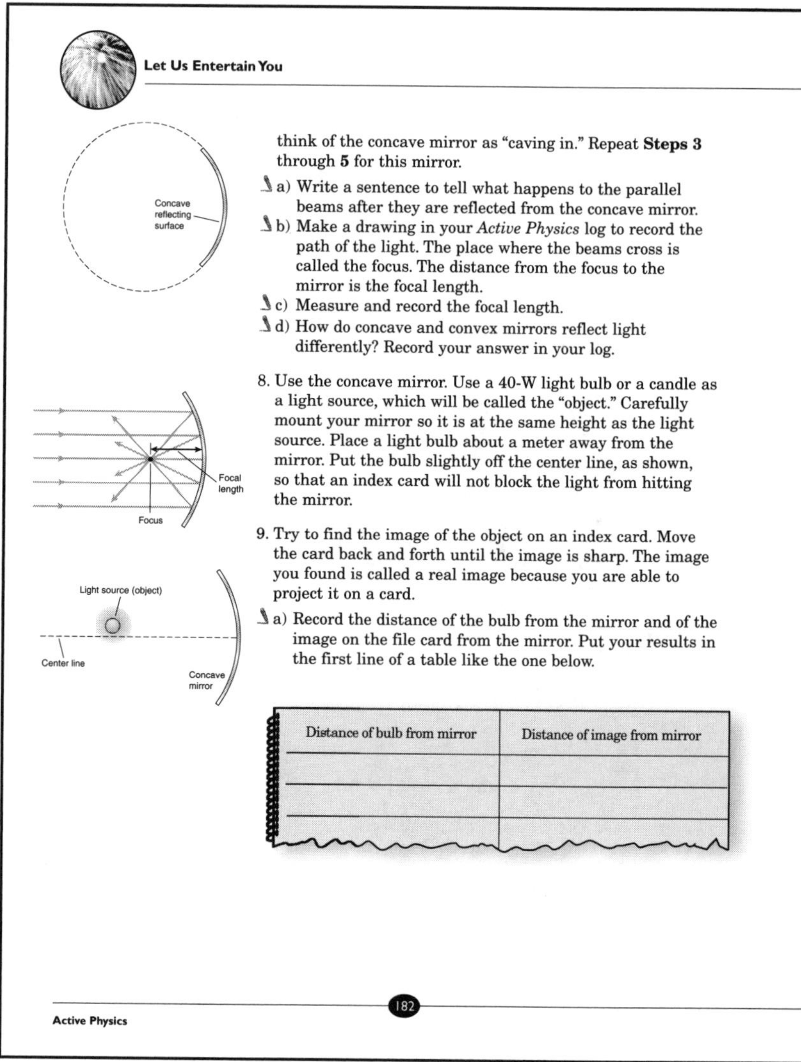





think of the concave mirror as "caving in." Repeat **Steps 3** through **5** for this mirror.

a) Write a sentence to tell what happens to the parallel beams after they are reflected from the concave mirror.

b) Make a drawing in your *Active Physics* log to record the path of the light. The place where the beams cross is called the focus. The distance from the focus to the mirror is the focal length.

c) Measure and record the focal length.

d) How do concave and convex mirrors reflect light differently? Record your answer in your log.



8. Use the concave mirror. Use a 40-W light bulb or a candle as a light source, which will be called the "object." Carefully mount your mirror so it is at the same height as the light source. Place a light bulb about a meter away from the mirror. Put the bulb slightly off the center line, as shown, so that an index card will not block the light from hitting the mirror.



9. Try to find the image of the object on an index card. Move the card back and forth until the image is sharp. The image you found is called a real image because you are able to project it on a card.

a) Record the distance of the bulb from the mirror and of the image on the file card from the mirror. Put your results in the first line of a table like the one below.

| Distance of bulb from mirror | Distance of image from mirror |
|---|---|
| | |
| | |
| | |



ANSWERS

## For You To Do *(continued)*

7. a) After reflecting from the concave mirror, the parallel beams come together.

   b) See the drawing on page 182.

   c) If the students used the same form for both the convex and concave mirror the focal lengths should be approximately the same.

   d) A convex mirror spreads out a light beam. A concave mirror brings a light beam to a focus.

8.-9. a) Students may need help in finding the image.

10. Carefully move the mirror closer to the object. Find the sharp image, as before, by moving the index card back and forth.
   a) Record the image and object distances in your table.
   b) Repeat the measurement for at least six object locations.
   c) Draw a graph of the image distance (*y*-axis) versus the object distance (*x*-axis).
   d) Write a sentence that describes the relationship between the image distance and the object distance.

11. A mathematical relation that describes concave mirrors is

$$\frac{1}{f} = \frac{1}{D_o} + \frac{1}{D_i}$$

where

$f$ is the focal length of that particular mirror

$D_o$ is the object distance

$D_i$ is the image distance

You have measured $D_o$ and $D_i$. Calculate $\frac{1}{D_o}$ and $\frac{1}{D_i}$. Find their sum for each pair of data.

   a) Record your calculations in your log.
   b) Are your sums approximately equal? If so, you have mathematically found the value of $\frac{1}{f}$ for the mirror you used.

12. A convex mirror cannot form a real image that can be projected onto a screen. It can form an image behind the mirror, like a plane mirror.
   a) Record in your log descriptions of the image in a convex mirror when the mirror is held close and when the mirror is held far from the object.

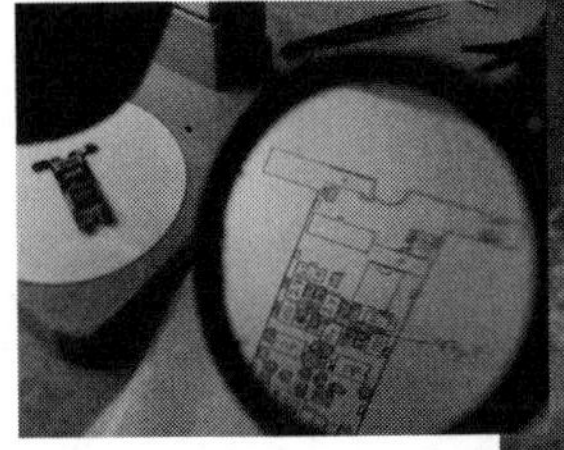





Chapter 3

ANSWERS

## For You To Do *(continued)*

10. a) See sample data.
   b) See sample data.
   c) See graph in the margin of page 184.
   d) As the object distance decreases, the image distance increases (and vice versa).

11. a)-b) The sums should be equal to within 10% to 15%.

12. a) The students should see a reduced image in the convex mirror at any distance. If the object is very close, the image looks similar to the image in the plane mirror.

### Sample Data

| $D_o$ (m) | $D_i$ (m) |
|---|---|
| 26.5 | 0.37 |
| 5.13 | 0.45 |
| 2.74 | 0.52 |
| 1.37 | 0.52 |
| 0.51 | 1.30 |

**Physics Words**

**real image:** an image that will project on a screen or on the film of a camera; the rays of light actually pass through the image.

## PHYSICS TALK

### Making Real Images

To find how a concave mirror makes a **real image**, you can view a few rays of light. Each ray of light obeys the relation you found for plane mirrors (angle of incidence = angle of reflection). In this case, you choose two easily drawn rays.

Look at the drawing. It shows rays coming into a concave mirror from a point on a light bulb. One ray comes in parallel to the dotted line, which is the axis of the mirror. This ray reflects through the focus. The other ray hits the center of the mirror. This ray reflects and makes the same angle with the mirror axis going out as it did coming in. Where these rays meet is the image of the top of the light bulb.

The next drawing shows the same mirror, but with the object much further from the mirror. Notice how the image in this second drawing is much smaller and much closer to the focus.

As you have seen, the position of the object and image are described by the equation below.

$$\frac{1}{f} = \frac{1}{D_o} + \frac{1}{D_i}$$

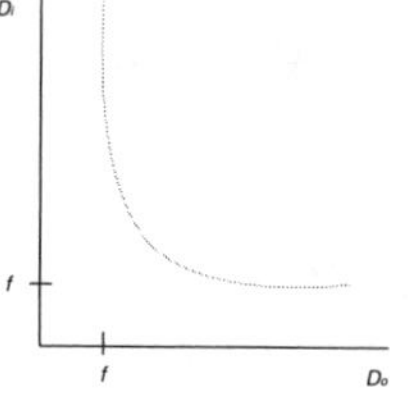


Look at the graph of this equation at left. Notice that as the object distance decreases, the image distance becomes very large. As the object distance increases, the image distance moves towards the focal length ($f$). Also notice that neither the object distance nor the image distance can be less than the focal length.

### Reflecting on the Activity and the Challenge

You have observed how rays of light are reflected by a curved mirror. You have seen that a concave mirror can make an upside-down real image (an image on a screen). You have also seen that the image and object distances are described by a simple mathematical relationship. In addition, you have seen that there is no real image in a convex mirror, and the image is always smaller than the object.

You may want to use a curved mirror in your sound and light show. You may want to project an image on a screen or produce a reflection that the audience can see in the mirror. What you have learned will help you explain how these images are made.

Since the image changes with distance, you may try to find a way to have a moving object so that the image will automatically move and change size. A ball suspended by a string in front of a mirror may produce an interesting effect. You may also wish to combine convex and concave mirrors so that some parts of the object are larger and others are smaller. Convex and concave mirrors could be shaped to make some kind of fun-house mirror.

Remember that your light show will be judged partly on creativity and partly on the application of physics principles. This activity has provided you with some useful principles that can help with both criteria.

### Physics To Go

1. a) Make a drawing of parallel laser beams aimed at a convex mirror.
   b) Draw lines to show how the beams reflect from the mirror.
2. a) Make a drawing of parallel laser beams aimed at a concave mirror.
   b) Draw lines to show how the beams reflect from the mirror.
3. a) Look at the back of a spoon. What do you see?
   b) Look at the inside of a spoon. What do you see?





ANSWERS

## Physics To Go

1. a)-b) See drawing on page 181.
2. a)-b) See drawing on page 182.
3. a) You see a right-side-up image of yourself and most of the room you are in. If you hold your finger very close to the spoon, you see a right-side-up, reduced image.
   b) You see an upside-down image of yourself. If you hold your finger very close to the spoon, you see a right-side-up, enlarged image.

ANSWERS

## Physics To Go

***(continued)***

4. a) It should be concave, so it will magnify.

   b. One side is concave and the other is convex. The concave side provides magnification. The convex side provides a wider view.

   c. This curved mirror is convex. It provides a wide-angle view, as does the outside of the spoon, so you can see cars out to the side, in what is called the blind spot. But the image is smaller, so cars appear further behind than they actually are.

   d. A dentist uses a curved mirror to get a magnified view of your teeth.

5. a)

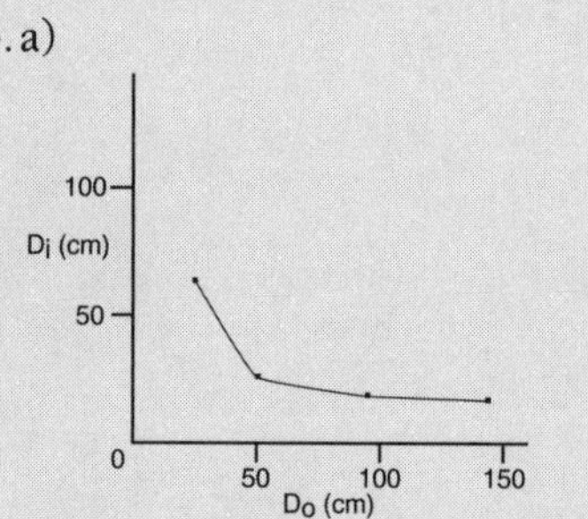


   b) As the object distance decreases, the image distance increases.

   c) The greatest object distance (142 cm) is already so large that the image (at 14 cm) is very close to the focus. Doubling the object distance would bring the image even closer to the focus, but the change might be impossible for the students to detect.

   d) The smallest object distance is 15 cm, and the focal length of the mirror is about 14 cm (see **Part (c)** above). If the smallest object distance were halved, the object would be within one focal length of the mirror, so there would be no real image.

6. As the ball swings towards the mirror, the image swings towards the mirror, too.

7. a) The answer depends on whether or not the ball swings through the focus of the mirror. If it does, then while swinging in, the image goes from upside-down to right-side-up and magnified. At the focus, the image fills the mirror.

   b) You could have several balls swinging in concave mirrors with different focal lengths. You could ask a member of the audience to move his or her head like the ball and then report what he or she sees.

8. The image gradually moves away from the mirror and becomes larger.

4. a) If you were designing a shaving mirror, would you make it concave or convex? Explain your answer.
   b) Why do some makeup mirrors have two sides? What do the different sides do? How does each side produce its own special view?
   c) How does a curved side mirror on a car produce a useful view? How can this view sometimes be dangerous?
   d) Why does a dentist use a curved mirror?

| $D_i$ (cm) | $D_o$ (cm) |
|---|---|
| 549 | 15 |
| 56 | 25 |
| 20 | 50 |
| 18 | 91 |
| 14 | 142 |

5. a) A student found the real image of a light bulb in a concave mirror. The student moved the light bulb to different positions. At each position, the student measured the position of the image and the light bulb. The results are shown in the table on the left. Draw a graph of this data.
   b) Make a general statement to summarize how the image distance changes as the object distance changes.
   c) If the object were twice as far away as the greatest object distance in the data, estimate where the image would be.
   d) If the object were only half as far from the mirror as the smallest object distance in the data, estimate what would happen to the image.

6. A ball is hung on a string in front of a flat mirror. The ball swings toward the mirror and back. How would the image of the ball in the mirror change as the ball swings back and forth?

7. a) A ball is hung on a string in front of a concave mirror. The ball swings toward the mirror and back. How would the image of the ball in the mirror change as the ball swings back and forth?
   b) How could you use this swinging ball in your light show?

8. Outdoors at night, you use a large concave mirror to make an image on a card of distant auto headlights. You make the image on a card. What happens to the image as the car gradually comes closer?

9. The diagram shows a light ray $R$ parallel to the principal axis of a spherical concave (converging) mirror. Point $F$ is the focal point and $C$ is the center of curvature. Draw the reflected light ray.

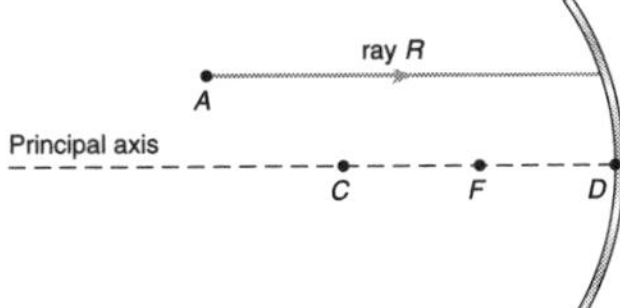


10. The diagram shows a curved mirror surface and a light bulb and its image. In relation to the focal point of the mirror, where is the light bulb (object) most likely located?

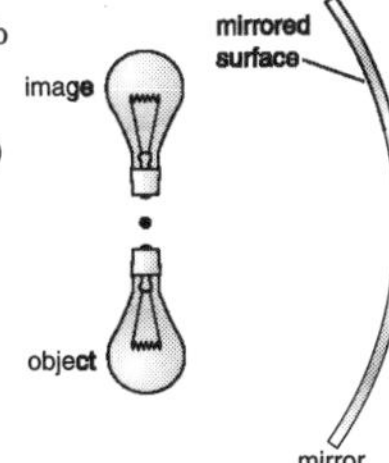


11. A candle is located beyond the center of curvature, $C$, of a concave spherical mirror having a principal focus, $F$, as shown in the diagram. Sketch the image of the candle.

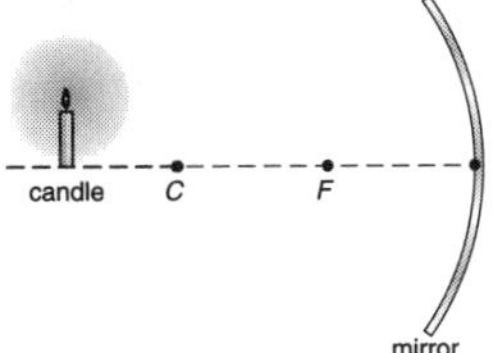




ANSWERS

## Physics To Go

***(continued)***

9.

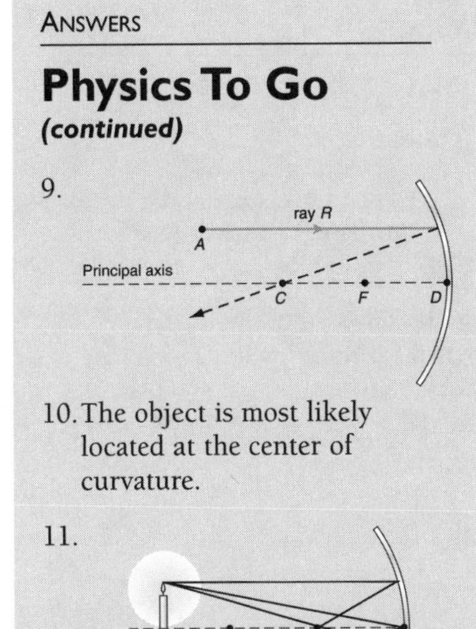


10. The object is most likely located at the center of curvature.

11.

ANSWERS

## Physics To Go
***(continued)***

12. a) The ray that is parallel to the principal axis will pass through *F*.

b) The size of the image produced will get smaller.

12. The diagram shows four rays of light from object *AB* incident on a spherical mirror with a focal length of 0.04 m. Point *F* is the principal focus of the mirror, point *C* is the center of curvature, and point *O* is located on the principal axis.

a) Which ray of light will pass through *F* after it is reflected from the mirror?

b) As object *AB* is moved from its position toward the left, what will happen to the size of the image produced?

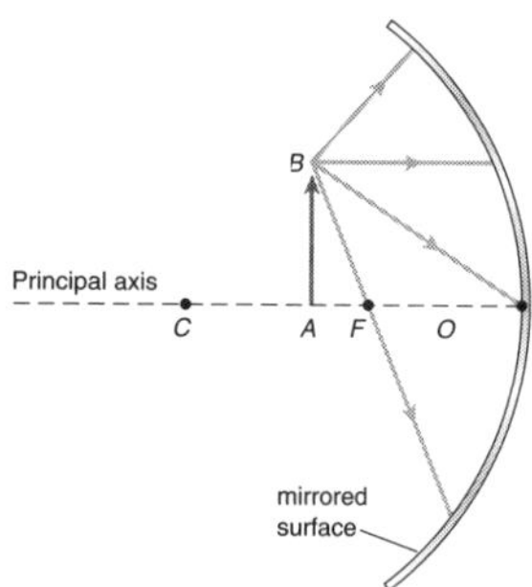

## Activity 5 A

# Low-Tech Reflection: Curved Mirror

### FOR YOU TO DO

If you do not have lasers or light boxes available, give the students penlights to use as the light source. Build long-focal-length mirrors by gently curving cardboard and taping mylar to the cardboard surface. Investigate the reflection of parallel flashlight beams. See if a concave mirror makes the beams cross.



For use with *Let Us Entertain You*, Chapter 3, Activity 5: Curved Mirrors

## Bulb and Image Distances

| Distance of bulb from mirror | Distance of image from mirror |
| --- | --- |
| | |
| | |
| | |
| | |
| | |
| | |
| | |
| | |
| | |
| | |
| | |
| | |

For use with *Let Us Entertain You*, Chapter 3, Activity 5: Curved Mirrors

## Bulb and Image Distances

| Distance of bulb from mirror | Distance of image from mirror |
| --- | --- |
| | |
| | |
| | |
| | |
| | |
| | |
| | |
| | |
| | |
| | |
| | |
| | |

For use with *Let Us Entertain You*, Chapter 3, Activity 5: Curved Mirrors

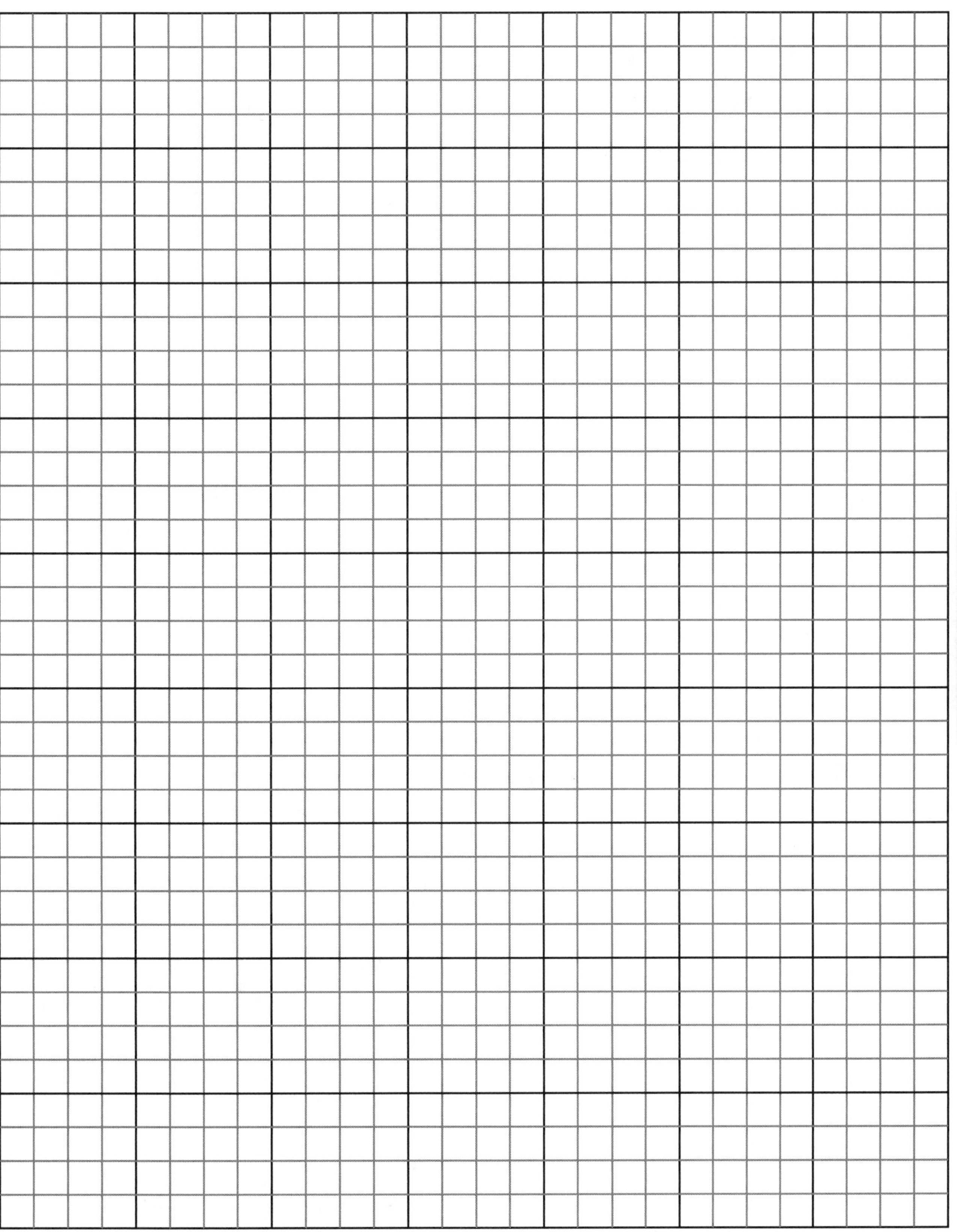



For use with *Let Us Entertain You*, Chapter 3, Activity 5: Curved Mirrors

# ACTIVITY 6
## Refraction of Light

# Background Information

Refraction is the basis of the operation of lenses. Light rays bend as they enter and leave the convex lens, as shown in the drawing.

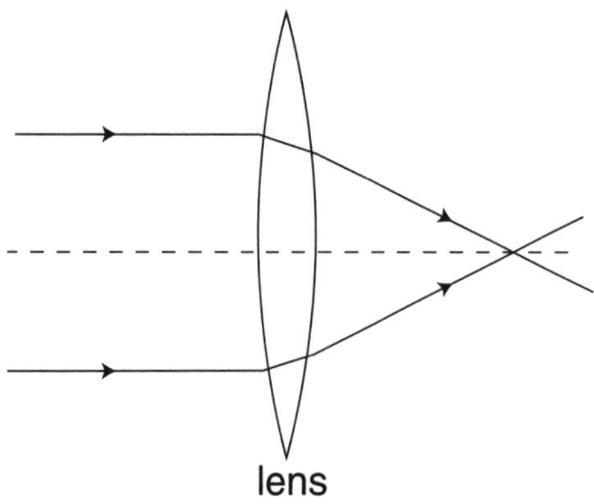


The angles of incidence ($i$) and refraction ($R$) specify the path of the light as it moves from one substance into another, as shown in the drawing.

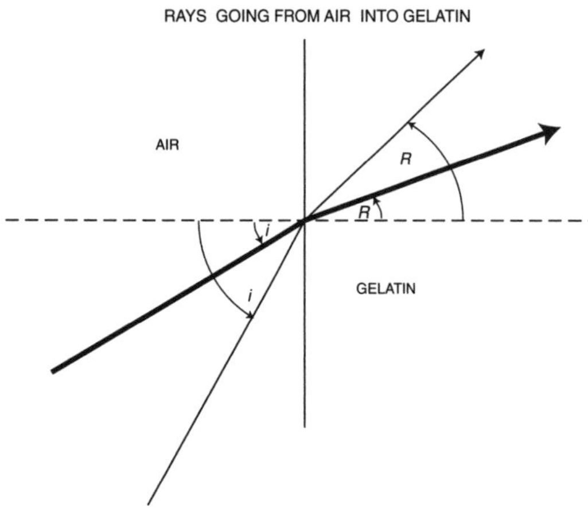


When light moves from air into gelatin, the equation relating the angles $i$ and $R$ can be written as

$$\sin(i) = n(\text{gelatin}) \times \sin(R)$$

The quantity $n$(gelatin) is the index of refraction of gelatin. (The value of the index of refraction of air is 1.000, so it is not shown in the equation above.) Going from air into gelatin, light bends towards the normal. Consequently, the ratio $\sin(i)/\sin(R)$ is greater than one (except for the special case of $i = 0°$).

Going from gelatin into air, light bends away from the normal. The angle of refraction is greater than the angle of incidence

The critical angle occurs as light leaves the medium with the higher index of refraction (in this activity, moving from gelatin into air). Look at the diagram.

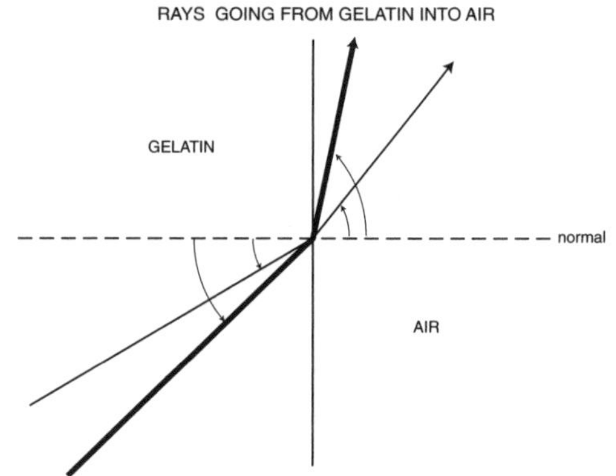


It shows how the angle of refraction changes as the angle of incidence, in the gelatin, increases from zero.

Eventually the angle of refraction reaches 90°. The corresponding angle of incidence is called the critical angle. What happens when the angle of incidence exceeds the critical angle? The light cannot be refracted, because the angle of refraction would be greater than 90°. It turns out that all the light is reflected back into the medium with the higher index of refraction. This effect is called total internal reflection. Although there appears to be an abrupt change at the critical angle, internal reflection begins considerably before reaching the critical angle and becoming total. Likewise, the intensity of the refracted beam diminishes progressively as the critical angle is approached, and this effect can make the critical angle challenging to measure.

Total internal reflection is the basis of the operation of the light pipe. A light beam enters a small diameter, flexible glass rod. When the beam hits the wall of the pipe, total internal reflection occurs and the beam reflects entirely within the pipe. In this way, light signals can carry information through a system of bundled glass rods called fiber optic cables.

## *Active*-ating the Physics InfoMall

Many of the searches done previously will provide great information for refraction, since most optics topics are closely related. For example, "An investigation of student understanding of the real image formed by a converging lens or concave mirror," from the *American Journal of Physics*, vol. 55, issue 2, 1987, will work here, as well as in **Activity** 7. Additional searches you should try would include "critical angle" and "internal reflection." Although quite simple, these are effective.

In **For You To Read**, the index of refraction (another good search term?) is described as a ratio. In his book, *A Guide to Introductory Physics Teaching*, (in the Book Basement) Arons says "One of the most severe and widely prevalent gaps in cognitive development of students at secondary and early college levels is the failure to have mastered reasoning involving ratios." The discussion continues, and you are urged to read this section (it appears early in the "Underpinnings" chapter). Arons also addresses this issue in "Student patterns of thinking and reasoning, part one of three parts," from *The Physics Teacher*, vol. 21, issue 9 (1983).

Again, Snell's Law is a great search item, and the equation can again be found in the *Equations Dictionary* (although this one requires a few more clicks, as it is buried relatively deep). Don't forget to search for more information on fiber optics and prisms while you are looking on the InfoMall.

# Planning for the Activity

## Time Requirements

one and a half class periods

## Materials Needed

### For each group:

- acrylic block
- cardboard sheet, 14" × 14" square
- laser pointer
- cosmetic mirror
- glass rod
- protractor
- right angle prism
- ruler, metric
- unlined index cards (100)
- wooden blocks to raise light source (3)

### For each student:

- 1 copy of the Blackline Master, Angles of Incidence and Refraction

## Advance Preparation and Setup

Make one copy of the Blackline Master for each student. There are many kinds of gelatin you can use. Let the gelatin harden in a rectangular tray and cut out the blocks in advance. Take care to cut the blocks as precisely as possible. You can run hot water on the bottom of the tray to soften the bottom of the gelatin so the blocks will come out easily. Keep them wrapped and refrigerated until the lab so no microorganisms will grow on their surface. Be sure you have supports for the gelatin and the light source so they are at the proper height to permit light beams to go through the gelatin block. If your students are using lasers, find large pieces of cardboard for laser backstops. If you have neither a laser nor a raybox, each group can use a candle and a narrow (2 mm to 3 mm) slit in a piece of cardboard. With this arrangement the students can at least get semiquantitative data and observe total internal reflection.

# Teaching Notes

Suggest that the students use a file card to follow the beam after it hits the glass rod. If the students say that they have difficulty seeing the refracted beam as they approach the critical angle, point out that less and less of the beam is refracted and more and more is reflected (as they approach the critical angle). You can introduce **Steps** 7 and 8 by pointing out that **Steps 1** through **6** produced relatively small angles of incidence inside the gelatin block (as the light approached the gelatin-air surface). This occurs because the light is bent towards the normal upon entering the block, and the angle of refraction on the front surface is equal to the angle of incidence on the rear surface. However, in **Step 8**, the angle of incidence inside the block is the complement of the angle of refraction in the block, so a small angle of refraction entering the block makes a large angle of incidence leaving.

If the students use lasers, be sure they understand the importance of keeping their laser beam within their group's work area (by building some sort of backstop).

If your class makes observations with laser beams, have each group build backstops of cardboard to absorb the beam. Also, explain to the students that they must not eat the gelatin. Keep the gelatin blocks covered and refrigerated before use. If your light source is a candle, have the students remove any loose papers from the area around the candle.

The concept that reflected light fans out from each point on an object is extremely difficult for students. Consequently, when asked to explain examples of refraction, they typically make a broad statement like "The water bent the light," rather than considering the path of light reflecting from the object in the water. For this reason it is particularly effective to have them trace the path of the laser beam. Asking a qualitative question, like "How does the laser beam bend when it goes from air into the gelatin," can help students organize their observations and contrast what happens in going from air to block with what happens going the other way.

Air to Acrylic Block

| angle of incidence | angle of refraction |
|---|---|
| 15 degrees | 10 degrees |
| 35 degrees | 22 degrees |
| 42 degrees | 28 degrees |

For the largest angle of incidence

$$1 \times \sin 42 \text{ degrees} = n_{gelatin} \times \sin 28 \text{ degrees}$$

$$n_{gelatin} = 1.4$$

# Activity 6 Refraction of Light

**GOALS**

In this activity you will:

- Observe refraction.
- Measure angles of incidence and refraction.
- Measure the critical angle.
- Observe total internal reflection.

**What Do You Think?**

The Hope Diamond is valued at about 100 million dollars. A piece of cut glass of about the same size is worth only a few dollars.

- **How can a jeweler tell the difference between a diamond and cut glass?**

Record your ideas about this question in your *Active Physics* log. Be prepared to discuss your responses with your small group and with your class.

**For You To Do**

1. Place an acrylic block on a piece of white paper on your desk.

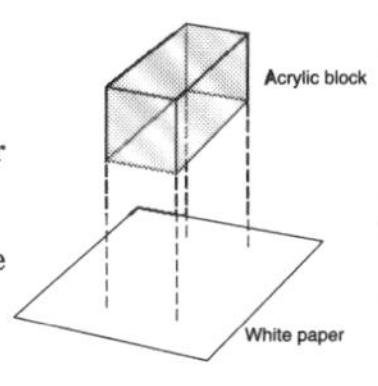


2. Carefully aim a laser pointer, or the light from a ray box, so the light beam moves horizontally, as you

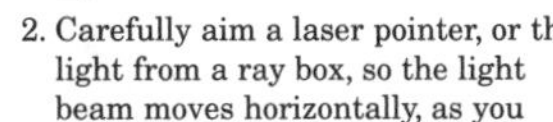



# Activity Overview

To prepare for the study of lenses, students investigate refraction. By shining a beam of light into a rectangular block of clear gelatin, they observe refraction as the light enters and leaves the gelatin. The students measure the angles of incidence and refraction and tabulate the results. They also observe total internal reflection and measure the critical angle.

# Student Objectives

## Students will:

- Observe refraction.
- Measure angles of incidence and refraction.
- Measure the critical angle.
- Observe total internal reflection.

Answers for the Teacher Only

# What Do You Think?

A jeweler with a trained eye can identify a diamond with a careful look. A diamond reflects more light than cut glass, because of the diamond's greater index of refraction. The diamond also has a range of color not found in glass and also particular small imperfections.

Despite these visual differences, many jewelers rely on an electrical device that applies a small amount of heat to an unknown stone and identifies a diamond from the resulting temperature change.

ANSWERS

## For You To Do

1. Be sure the front and back of the acrylic block are paralled.

2.-3. The students do not need to mark the path of the beam inside the block They find the angles inside by connecting the points where the light enters and leaves.

4. The students can measure every ten or fifteen degrees.

5. Remind the students to look straight down when marking the path of the beam to avoid parallax.

6. a) Remember that angles are measured from the normal.

b) Each ratio is >1 since $\measuredangle i > \measuredangle R$

c) This ratio is (approximately) a constant.

7. Be sure the side of the block is perpendicular to the front.

**Never look directly at a laser beam or shine a laser beam into someone's eyes. Always work above the plane of the beam and beware of reflections from shiny surfaces.**

did in previous activities. Place a glass rod in the light beam so that the beam spreads up and down.

3. Shine the laser pointer or light from the ray box through the acrylic block. Be sure the beam leaves the acrylic block on the side opposite the side the beam enters. Mark the path of each beam. You may wish to use a series of dots as you did before. Label each path on both sides of the acrylic block so you will know that they go together.

Normal
Angle of Incidence
1
2
Angle of Refraction
Normal

4. The angle of incidence is the angle between the incident laser beam and the normal, as shown in the diagram. Choose two other angles of incidence and again mark the path of the light, as you did in **Step 3**. As before, label each pair of paths.

5. Trace the outline of the acrylic block on the paper and remove the acrylic block. Connect the paths you traced to show the light beam entering the acrylic block, traveling through the acrylic block, and emerging from the acrylic block. Draw a perpendicular line at the point where a ray enters or leaves the acrylic block. Label this line the normal.

6. Measure the angles of incidence (the angle in the air) and refraction (the angle in the acrylic block).

 a) Record your measurements in tables like the one shown.

| Angle of incidence | Angle of refraction | Sine of angle of incidence | Sine of angle of refraction | Sin $\angle i$ / Sin $\angle R$ |
|---|---|---|---|---|
| | | | | |
| | | | | |

b) Use a calculator to complete the chart by finding the sines of the angles (sin button on calculator).

c) Is the value of $\frac{\sin \angle i}{\sin \angle R}$ a constant? This value is called the index of refraction for the acrylic block.

7. Set up the acrylic block on a clean sheet of white paper. This time, as shown in the drawing (next page), aim the beam so it leaves the acrylic block on the side, rather than at the back.

**Physics Words**

**critical angle:** the angle of incidence for which a light ray passing from one medium to another has an angle of refraction of 90°.

**index of refraction:** a property of a medium that is related to the speed of light through it; it is calculated by dividing the speed of light in vacuum by the speed of light in the medium.

**Snell's Law:** describes the relationship between the index of refraction and the ratio of the sine of the angle of incidence and the sine of the angle of refraction.

Activity 6 Refraction of Light

8. Make the first angle of incidence (angle 1) as small as possible, so the second angle of incidence (angle 2) will be as large as possible. Adjust angle 1 so that the beam leaves the acrylic block parallel to the side of the acrylic block, as shown. Measure the value of angle 2.
   a) Record the value of angle 2. It is called the **critical angle**.
   b) What happens to the beam if you make angle 2 greater than the critical angle?
   c) What you observed in **(b)** is called "total internal reflection." What is reflected totally, and where?

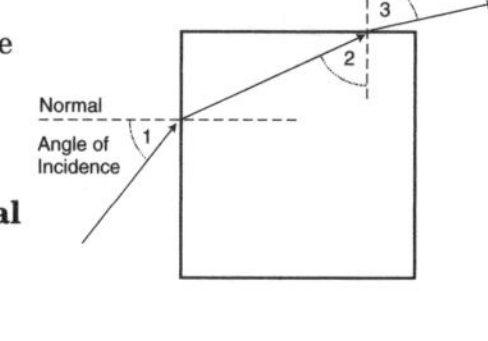


9. It is possible to bend a long, rectangular acrylic block so the light enters the narrow end of the acrylic block, reflects off one side of the acrylic block, then reflects off the other and back again to finally emerge from the other narrow end. Try to bend an acrylic block rectangle so that the light is reflected as described.

### FOR YOU TO READ

**Snell's Law**

Light refracts (bends) when it goes from air into another substance. This is true whether the other substance is gelatin, glass, water, or diamond. The amount of bending is dependent on the material that the light enters. Each material has a specific **index of refraction,** *n*. This index of refraction is a property of the material and is one way in which a diamond (very high index of refraction—lots of bending) can be distinguished from glass (lower index of refraction—less bending). The index of refraction is a ratio of the sine of the angle of incidence and the sine of the angle of refraction.

Index of refraction: $n = \dfrac{\sin \angle i}{\sin \angle R}$

This equation is referred to as **Snell's Law.**

As light enters a substance from air, the light bends toward the normal. When light leaves a substance and enters the air, it bends away from the normal. If the light is entering the air from a substance, the angle in that substance may be such that the angle of refraction is 90°. In this special case, the angle in the substance is called the critical angle. If the angle in the substance is greater than this critical angle, then the light does not enter the air but reflects back into the substance as if the surface were a perfect mirror. This is the basis for light fibers where laser light reflects off the inner walls of glass and travels down the fiber, regardless of the bend in the fiber.

ANSWERS

## For You To Do *(continued)*

8. a) Note that as the critical angle is approached, the refracted intensity diminishes and the reflected intensiy grows.
   b) If angle 2 is greater than the critical angle, the refracted ray disappears.
   c) The incident ray (inside the acrylic block) is totally reflected, with an angle of reflection equal to angle 2.
9. This is a model of a light pipe. When the light hits the walls, there is total internal reflection.

Chapter 3

ANSWERS

## Physics To Go

1. The light is bent towards the normal, so the angle of incidence is larger (except when the angle of incidence is zero).
2. a) See drawing on page 191.

   b) Yes
3. See drawing on page 191.
4. See drawing in **Background Information**.
5. a) 90°

   b) The angle of incidence equals 90°. The angle of refraction equals 42°, the critical angle.

### Reflecting on the Activity and the Challenge

The bending of light as it goes from air into a substance or from a substance into air is called refraction. It is mathematically expressed by Snell's Law. When light enters the substance at an angle, it bends towards the normal. When light leaves the substance at an angle, it bends away from the normal. As you create your light show for the **Chapter Challenge**, you may find creative uses of refraction. You may decide to have light bending in such a way that it spells out a letter or word or creates a picture. You may wish to have the light travel from air into glass to change its direction. You may have it bend by different amounts by replacing one material with another. Regardless of how you use refraction effects, you can now explain the physics principles behind them.

### Physics To Go

1. A light ray goes from the air into an acrylic block. In general, which is larger, the angle of incidence or the angle of refraction?
2. a) Make a sketch of a ray of light as it enters a piece of acrylic block and is refracted.

   b) Now turn the ray around so it goes backward. What was the angle of refraction is now the angle of incidence. Does the turned-around ray follow the path of the original ray?
3. A light ray enters an acrylic block from the air. Make a diagram to show the angle of incidence, the angle of refraction, and the normal at the edge of the acrylic block.
4. Light rays enter an acrylic block from the air. Make drawings to show rays with angles of incidence of 30° and 60°. For each incident ray, sketch the refracted ray that passes through the acrylic block.
5. a) Light is passing from the air into an acrylic block. What is the maximum possible angle of incidence that will permit light to pass into the acrylic block?

   b) Make a sketch to show your answer for **Part (a)**. Include the refracted ray (inside the acrylic block) in your sketch.

6. a) A ray of light is already inside an acrylic block and is heading out. What is the name of the maximum possible angle of incidence that will permit the light to pass out of the acrylic block?
   b) If you make the angle of incidence in **Part (a)** greater than this special angle, what happens to the light?
   c) Make a sketch to show your answer for **Part (b)**. Be sure to show what happened to the light.
7. a) Make a drawing of a light ray that enters the front side of a rectangular piece of acrylic block and leaves through the back side.
   b) What is the relationship between the direction of the ray that enters the acrylic block and the direction of the ray that leaves the acrylic block?
   c) Use geometry and your answer to **Question 2 (b)**, to prove your answer to **Question 7 (b)**.
8. You have seen the colored bands that a prism or cut glass or water produce from sunlight. Light that you see as different colors has different wavelengths. Since refraction makes these bands, what can you say about the way light of different wavelengths refracts?

ANSWERS

## Physics To Go
***(continued)***

6. a) The critical angle.
   b) All the light is reflected inside the block.
   c) See drawing on page 191. (Look at the incident ray inside the acrylic block.) With a slightly larger value for angle 2, the refracted ray in air will have an angle of refraction of 90°.
7. a) See drawing on page 191.
   b) These rays are parallel.
   c) By alternate interior angles, inside the gelatin the angle of refraction is equal to the angle of incidence at the rear of the block. Since the light paths can be reversed, the angle of refraction in the air equals the angle of incidence in the air. (It may help to rotate the diagram at the back side of the block by 180° and compare it with the diagram at the front.)
8. The amount of refraction depends on the wavelength. Equivantly, the angle of incidences is a function of the wavelength.

Chapter 3

### Stretching Exercises

1. Cover the acrylic block with a red filter. Shine a red laser beam into the acrylic block, as you did in **For You To Do, Steps 1** through **3**. What happens? How can you explain what happens?

2. Find some $\frac{1}{2}$" diameter clear tubing, about 2 m long. Plug one end. Pour clear gelatin in the other end, through a funnel, before the gelatin has had time to set. Arrange the tubing into an interesting shape and let the gelatin set. You may wish to mount your tube on a support or a sturdy piece of cardboard, which can be covered with interesting reflective material. Fasten one end of the tube so laser light can easily shine straight into it. When the gelatin has set, turn on the laser. What do you see? This phenomena is called total internal reflection.

3. Place a penny in the bottom of a dish or glass. Position your eye so you can just see the penny over the rim of the glass. Predict what will happen when you fill the glass with water. Then try it and see what happens. How can you explain the results?

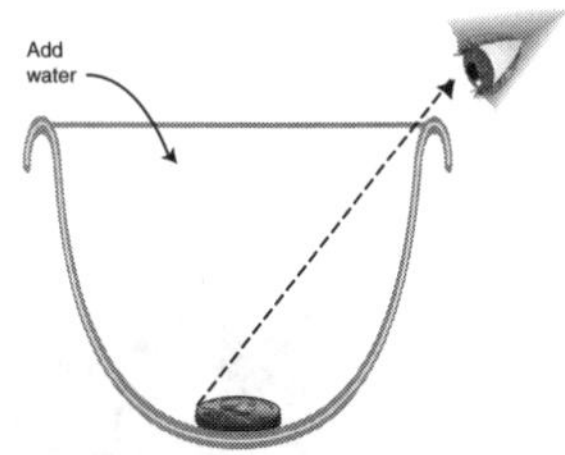


4. Place an empty, clear drinking glass over a piece of a newspaper. When you look through the side of the glass near the bottom, you can see the printing on the newspaper. What do you think will happen if you fill the glass with water? Try it and see. How can you explain the result? Does it help to hold your fingers over the back of the glass?

ANSWERS

## Stretching Exercises

1. The red laser light passes through the acrylic block, which absorbs light of other colors.

2. The laser beam will pass through the tubing both times it reaches the outside at the gelatin, total internal reflection occurs, so this is a model of a light pipe.

3. When the light leaves the water, the light bends away from the normal (the vertical). If you find the ray from the water that enters your eye, extend this ray backward to find where you will see the coin.

4. The print disappears. Suggest that students draw diagrams to explain what happens.

| Angle of incidence | Angle of refraction | Sine of angle of incidence | Sine of angle of refraction | Sin< *i*<br>Sin< *R* |
|---|---|---|---|---|
| | | | | |
| | | | | |
| | | | | |
| | | | | |
| | | | | |
| | | | | |
| | | | | |
| | | | | |
| | | | | |
| | | | | |
| | | | | |
| | | | | |

For use with *Let Us Entertain You*, Chapter 3, Activity 6: Refraction of Light



| Angle of incidence | Angle of refraction | Sine of angle of incidence | Sine of angle of refraction | Sin< *i*<br>Sin< *R* |
|---|---|---|---|---|
| | | | | |
| | | | | |
| | | | | |
| | | | | |
| | | | | |
| | | | | |
| | | | | |
| | | | | |
| | | | | |
| | | | | |
| | | | | |
| | | | | |

For use with *Let Us Entertain You*, Chapter 3, Activity 6: Refraction of Light

# ACTIVITY 7
## Effect of Lenses on Light

# Background Information

A light bulb sends out light in all directions. If a few of these rays pass through a pinhole, and the rest are blocked, a dim image of the bulb can be seen on a screen. Opening up the pinhole makes the image brighter but also less sharp. Only if the light is refracted can the image be both bright and sharp. A convex lens bends light to make such an image. Look at the lower ray diagram. It shows two of the countless rays that fan out from each point on the object.

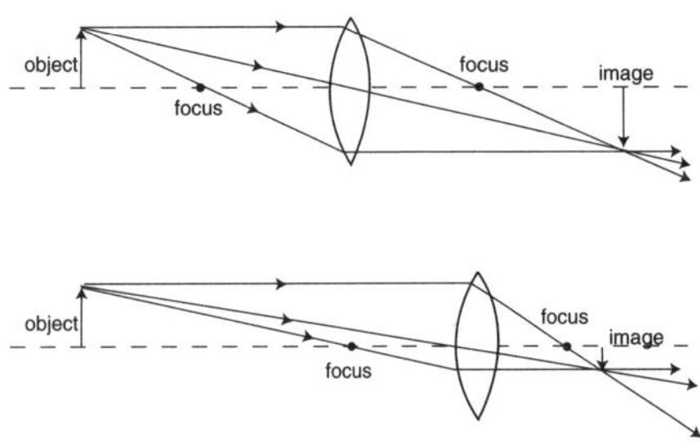


One ray enters the lens parallel to its axis, so this ray is refracted through the focus, the position of the image when the object is extremely distant. The other ray goes through the center of the lens. If the lens is thin, and the object is small, this ray maintains its original direction, since the front and back of the lens are parallel right at the center. Where these two rays meet is the image of the arrowhead. The image is upside down. Also, the image is smaller than the object and is located between $2f$ and $f$ ($f$ is the focal length) from the lens. The object is beyond $2f$. Notice that the object and image could be interchanged, without changing the ray diagram. When the object is close to the lens, the image is far away and larger than the object. When the image is close to the lens, the object is far away and larger than the image. The object and image can be equidistant from the lens, when each is located at $2f$. (See the upper ray diagram.) In this symmetrical arrangement, the object is the same size as the image.

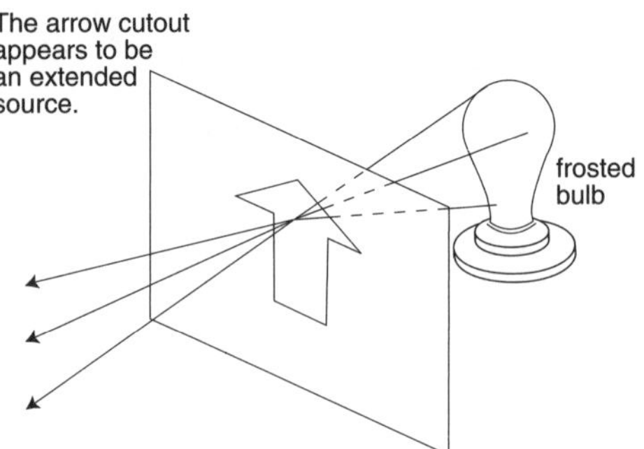


Projecting a slide is challenging to analyze. Think about **Steps** 7 and 8, where a student holds a card with an arrow-shaped cutout between the bulb and the lens. The lens projects the arrow on the screen.

Since the light bulb is an extended source, light rays from the bulb cross at each point in the cutout. In effect, the cutout arrow acts like an extended light source too. This extended source has a sharp edge right at the card, so a sharp arrow can be projected on the screen. See below for notes on changing the size of the projected image.

## *Active*-ating the Physics InfoMall

Search using keyword "Convex lens*" and you will get a short list. Don't forget; using an asterisk in your search words tells the InfoMall to look for words that are similar ("lens*" could be "lens", "lenses", or "lensless"). Some of these include rays diagrams, labs, or homework problems. You may wish to check these out if you do not know what to expect from a convex lens. You may also want to read "Lenses, pinholes, screens, and the eye," from *The Physics Teacher*, vol. 29, issue 4 (1991), which also discusses student difficulties; and don't forget "An investigation of student understanding of the real image formed by a converging lens or concave mirror," from the *American Journal of Physics,* vol. 55, issue 2, 1987.

Another search to perform here might be "converging lens*". Indeed. that is the search that produced the two articles just mentioned. From the same search, we find that "The Camera" (found in the Pamphlet Parlor) is high on the list of search hits. We also find that the textbook *Modern College Physics* has an entire chapter on lenses.

Repeat these searches for concave lenses, also called diverging lenses.

# Planning for the Activity

## Time Requirements

one and a half class periods

## Materials Needed

### For the class:

- overhead projector (several, if possible)

### For each group:

- 40-W bulb and socket, ceramic w/switch
- acetate sheet, about 8 cm square
- bulb, clear 100-W
- convex lens with focal length about 10 cm
- felt-tip marker
- ruler, metric
- steel scissors
- stick, meter 100 cm, hardwood

### For each student:

- 1 copy of the Blackline Master: Table of Object and Image Distances

## Advance Preparation and Setup

Make a copy of the Table of Object and Image Distance following this activity in the Teacher's Edition. Darken the room as much as possible so students can see the images easily. If you use 110-V light bulbs, be sure there are no exposed wires. Use the kind of sockets that have the wiring connections inside a plastic or ceramic cover. If the socket is ceramic, then the connections are probably exposed underneath. Screw each ceramic socket down to a piece of wood, so the wiring is safely out of reach.

# Teaching Notes

Most students need help when they first make a real image with a lens. They often have difficulty getting started because they are unable to line up the lens, the file card, and the object (with the lens facing the object). To help them, you can refer them to the drawing in the book on page 196. Projecting the slide (**Step 9**) can be challenging. Many students make an image of the light bulb as seen through the arrow, but it is in fact the bright arrow that is the image. To enlarge the size of the image, they must bring the lens and arrow closer together. You may need to help them move both lens and arrow to find the position that gives a sharp image.

Keep the groups as well separated as possible. Even with the room dark, stray light from one group's bulb can easily interfere with other group's images.

If the students use a 110-V light bulb for the light source, be sure the students cannot touch any exposed wiring. If you use ceramic sockets with exposed screw terminals underneath, screw down the socket to a board so the terminals are inaccessible. See **Advance Preparation and Setup**. If the students use a candle, be sure no flammable materials are nearby.

Light is a challenging topic for students. Students do not easily understand the fanning out of reflected light nor, for that matter, even the idea that light travels from one place to another yet is not material. You can help your students by asking them to explain how they hold the lens when they make an image (facing the object) and how that orientation produces an image (the lens focuses the light that hits its surface, and the light, which fans out from the object, hits all of the surface).

Focusing Parallel Beams: If you have lasers available, set up the lens and laser so you can move the laser back and forth in front of the lens, with the beam parallel to the lens axis. Observe what happens to the beam after it passes through the lens. Measure the focal length by making an image of a distant object and compare the value with what you found from the laser beam.

NOTES

# Activity 7 Effect of Lenses on Light

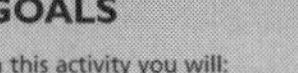

**GOALS**

In this activity you will:

- Observe real images.
- Project a slide.
- Relate image size and position.

**What Do You Think?**

Engineers have created special lenses that can photograph movie scenes lit only by candlelight.

- **How is a lens able to project movies, take photographs, or help people with vision problems?**

Record your ideas about this question in your *Active Physics* log. Be prepared to discuss your responses with your small group and with your class.

**For You To Do**

1. Look at the lens your teacher has given you.

   a) Make a side-view drawing of this lens in your log. This is a *convex* lens.

2. Point the lens at a window or at something distant

ANSWERS

## For You To Do

1. a) See lens drawing in **Physics To Go, Question 3**, in the Student Edition.

## Activity Overview

Having investigated refraction in the last activity, students now move ahead to convex lenses. They observe the real image made by a convex lens and how its size and position change as the object distance changes. To prepare for the light show, the students project the image of a slide onto a wall and vary the size of the projected image.

## Student Objectives

### Students will:

- Observe real images.
- Project a slide.
- Relate image size and position.

ANSWERS FOR THE TEACHER ONLY

## What Do You Think?

The lens does all these things by bending light, a process called refraction. Without refraction, cameras could take only pinhole images, as though the lens let in light through only a tiny opening. The images would be very dim. But by bending light, the lens can direct all the light that strikes the lens onto the proper place in the image, so the image is far brighter than a pinhole image.

To make an image, a lens must be convex. If you make a ray diagram of a light beam passing through a convex lens, you will see that it bends towards the lens axis as it enters the glass and then again towards the axis as it leaves the glass. The result is that the ray can be part of an image, which is located near the axis.

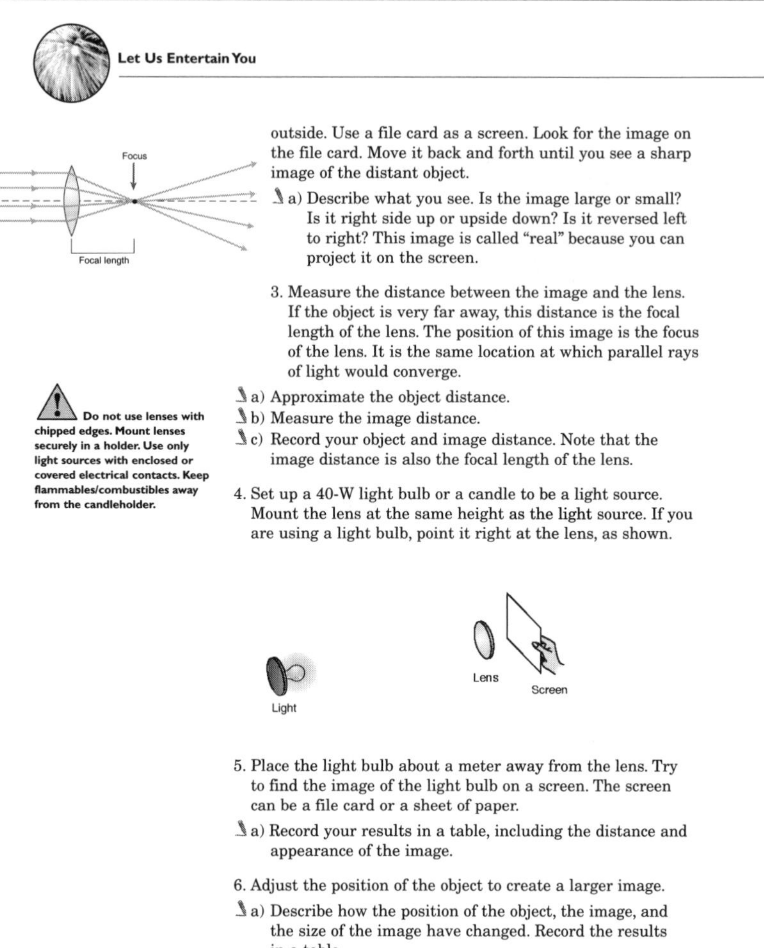

Let Us Entertain You



outside. Use a file card as a screen. Look for the image on the file card. Move it back and forth until you see a sharp image of the distant object.

a) Describe what you see. Is the image large or small? Is it right side up or upside down? Is it reversed left to right? This image is called "real" because you can project it on the screen.

3. Measure the distance between the image and the lens. If the object is very far away, this distance is the focal length of the lens. The position of this image is the focus of the lens. It is the same location at which parallel rays of light would converge.

a) Approximate the object distance.

b) Measure the image distance.

c) Record your object and image distance. Note that the image distance is also the focal length of the lens.

**Do not use lenses with chipped edges. Mount lenses securely in a holder. Use only light sources with enclosed or covered electrical contacts. Keep flammables/combustibles away from the candleholder.**

4. Set up a 40-W light bulb or a candle to be a light source. Mount the lens at the same height as the light source. If you are using a light bulb, point it right at the lens, as shown.



5. Place the light bulb about a meter away from the lens. Try to find the image of the light bulb on a screen. The screen can be a file card or a sheet of paper.

a) Record your results in a table, including the distance and appearance of the image.

6. Adjust the position of the object to create a larger image.

a) Describe how the position of the object, the image, and the size of the image have changed. Record the results in a table.

196

Active Physics

ANSWERS

# For You To Do *(continued)*

2. a) The image is upside down. If the object is outside, the image is quite small. It is inverted and reversed left-to-right.

3.-5. If the object is far away, the image is at the focus (close to the lens). As the object moves toward the lens, the image moves away.

6. a) If the image is larger, the object is closer to the lens. The image is further from the lens.

Activity 7 Effect of Lenses on Light

7. Create an object by carefully cutting a hole in the shape of an arrow in an index card. Have someone in your group hold the card close to the light bulb.
   a) Can you see the object on the screen? Describe what you see.
   b) Have the person holding the object move it around between the light bulb and the convex lens. What happens?
8. Project the object onto the wall. Can you make what you project larger or smaller?
   a) In your log indicate what you did to change the size of the image.

9. Create a slide by **drawing** with a marking pen on clear acetate. Try **placing** a 100-W light bulb and the slide in different positions.
   a) Describe how you can project a real, enlarged image of your slide onto a screen or wall.
   b) How can you use the lens to change the size of the image?
   c) Explore the effect of different lenses. In your log, record how you think this effect might be part of your light and sound show.

**Caution:** **Lamps get very hot. Be careful not to touch the bulb or housing surrounding the bulb.**





ANSWERS

## For You To Do *(continued)*

7. a) Yes, you see a bright arrow.
   b) By moving the screen back and forth to the right place, the edge of the object can be seen sharply.
8. a) Yes, you can change the size of what you see on the wall by changing the distance between the lens and the object.
9. a) Line up the bulb, slide, lens, and wall. Move the lens toward and away from the slide until the image is sharp.
   b) If you move the lens closer to the slide, and then refocus by moving lens and slide together until the image is sharp, the size of the image will be larger.
   c) In your show you can project slides on the wall and change the size of the projected image.

## FOR YOU TO READ

### Lens Ray Diagrams

You are probably more familiar with images produced by lenses than you are with images from curved mirrors. The lens is responsible for images of slides, overhead projectors, cameras, microscopes, and binoculars.

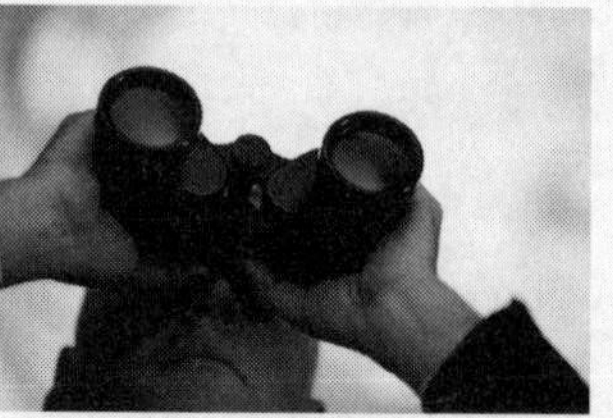

Light bends as it enters glass and bends again when it leaves the glass. The **convex converging lens** is constructed so that all parallel rays of light will bend in such a way that they meet at a location past the lens. This place is the focal point.

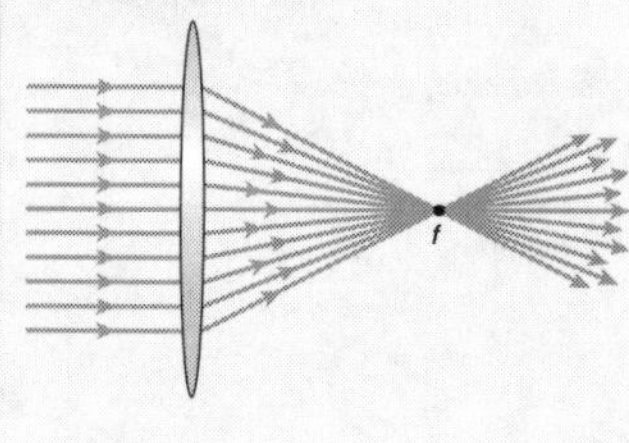


If an object is illuminated, it reflects light in all directions. If these rays of light pass through a lens, an image is formed.

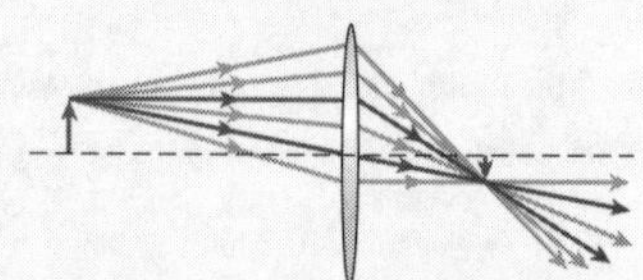

Although all of the light rays from the object help to form the image, you can locate an image by looking at two easy rays to draw—the ray that is parallel to the principal axis and travels through the focal point and the ray that travels through the center of the lens undeflected. (These rays are in red in the diagram.)

You can use this technique to see how images that are larger (movie projector), smaller (camera), and the same size (copy machine) as the object can be created with the same lens.

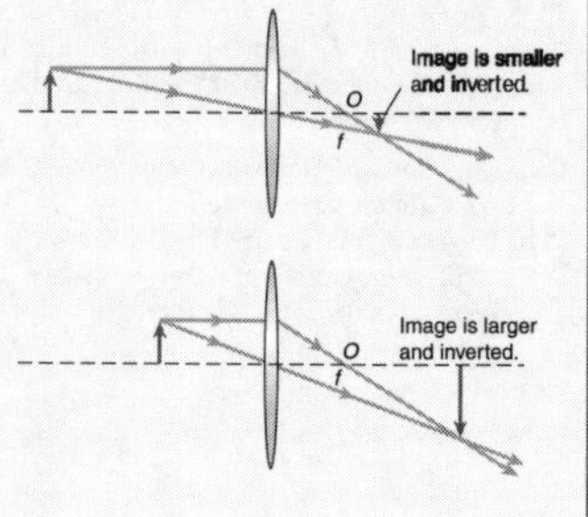

Activity 7 Effect of Lenses on Light

If the object is close to the lens (an object distance smaller than the focal distance), then an image is not formed. However, if you were to view the rays emerging, they would appear to have come from a place on the same side of the lens as the object. To view this image, you put your eye on the side of the lens opposite the object and peer through it—it's a magnifying glass!

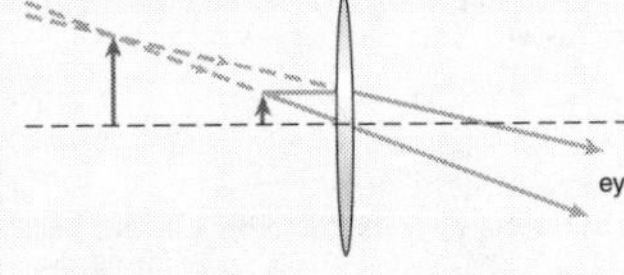


### Sample Problem

The diagram shows a lens and an object.

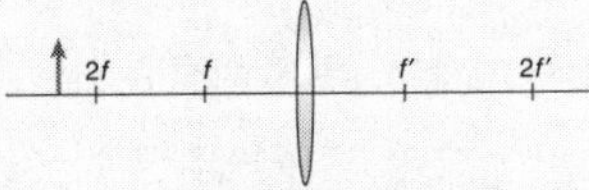


a) Using a ray diagram, locate the image of the object shown.

b) Describe the image completely.

***Strategy:*** Choose a location on the object to be the origin of the rays. A simple choice would be the tip of the arrow. At least two rays must be drawn to locate the image.

***Givens:***

See the diagram.

***Solution:***

a)

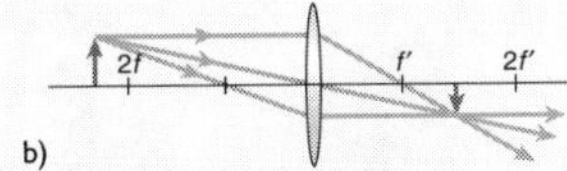


b) The image is real, reduced, and inverted.

As the object moves closer to the lens, its size will increase. At $d_o = f$ there will be no image and at $d_o < f$ the image will be virtual and upright.

**Physics Words**

**converging lens:** parallel beams of light passing through the lens are brought to a real point or focus (if the outside index of refraction is less than that of the lens material); also called a convex lens.

Chapter 3

ANSWERS

## Physics To Go

1.a) The focus is the position of the image of a very distant object.

b) The object is very far away (in fact, it's at infinity).

c) The focal length is the distance between the focus and the lens.

d) See **For You To Do, Steps 2** and **3**.

2. a) Yes.

b) Upside-down.

c) Yes. Light spreads out as it moves from the object to the lens (to see this, imagine the object is a tiny light). If the light were not bent, it would continue spreading after it went through the lens, so there could not be an image.

d) Move the screen back from the lens.

3.a) Move the object away from the lens (more than twice the focal length away).

b) Bring the object close to the lens (closer than twice the focal length), but not less than the focal length for a real image.

4.a) The image of the closer light is blurred.

b) The image of the more distant light is blurred.

Let Us Entertain You

### Reflecting on the Activity and the Challenge

You have explored how convex lenses make real images. You have found these images on a screen by moving a card back and forth until the image was sharp and clear, so you know that they occur at a particular place. Bringing the object near the lens moves the image away from the lens and enlarges the image, but if the object is too close to the lens, there is no real image. These images are also reversed left to right and are upside down. You may be able to use this kind of image in your sound and light show. You have also projected images of slides on a wall. You may be able to add interest by moving the lens and screen to change the size of these images.

### Physics To Go

1. a) What is the focus of a lens?
   b) If the image of an object is at the focus on a lens, where is the object located?
   c) What is the focal length of a lens?
   d) How can you measure the focal length of a lens?

2. a) You set up a lens and screen to make an image of a distant light. Is the image in color?
   b) Is the image right side up or upside down?
   c) Did the lens bend light to make this image? How can you tell?
   d) A distant light source begins moving toward a lens. What must you do to keep the image sharp?

3. a) You make an image of a light bulb. What can you do to make the image smaller than the light bulb?
   b) What can you do to make the image larger than the light bulb?

4. a) You have two lights, a lens, and a screen, as shown on opposite page. One light is at a great distance from the lens. The other light is much closer. If you see a sharp image of the distant light, describe the image of the closer light.
   b) If you see a sharp image of the closer light, describe the image of the more distant light.

200

Active Physics

c) Could you see a sharp image of both lights at the same time? Explain how you found your answer.

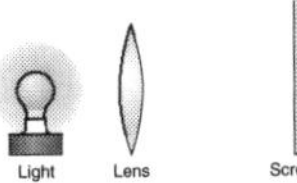


5. Research how a camera works. Find out where the image is located. Also find out how the lens changes so that you can photograph a distant landscape and also photograph people close up.

6. Using a ray diagram, locate the image formed by the lens below.

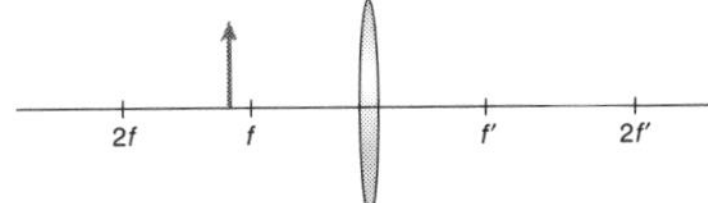


7. An object 1.5 cm tall is placed 5.0 cm in front of a converging lens of focal length 8.0 cm.
   a) Determine the location of the image.
   b) Completely describe the image.

8. A relative wants to show you slides from her wedding in 1972. She brings out her slide projector and screen.
   a) If she puts the screen 2.8 m from the projector and the lens has a focal length of 10.0 cm, how far from the lens will the slide be so that her pictures are in focus?
   b) If each slide is 3.0 cm tall, how big will the image be on the screen?

## Answers

## Physics To Go *(continued)*

4. c) Not if one object is very distant and the other is very close. But if you find a sharp image and then move the object a small distance, the image may remain sharp. That means there is a range of object distances that will produce a sharp image. (In photography, this range of object distances is called depth of field.)

5. The image is located on the film. Part or all of the lens moves in and out to change the focus. Notes: As the aperture—the opening—of the lens decreases (this corresponds to increasing the *f*-number), the range of sharp focus increases. For some cameras, the aperture is kept small to maintain a large range of object distances that produce a reasonably sharp image. For these cameras, the focus of the lens is fixed.

6.

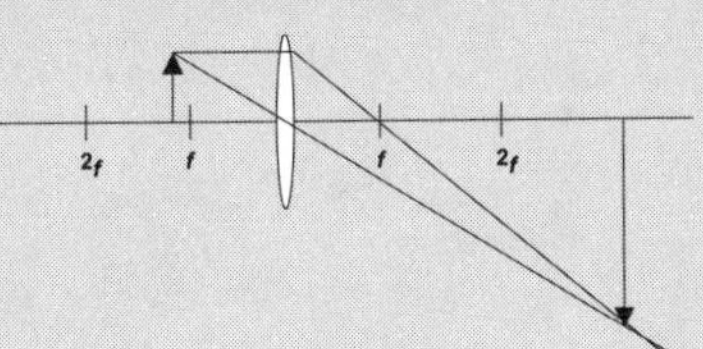


## Answers

## Physics To Go *(continued)*

7. a) $\frac{1}{f} = \frac{1}{d_o} + \frac{1}{d_i}$

$$\frac{1}{d_i} = \frac{1}{f} + \frac{1}{d_o}$$

$$d_i = \left(\frac{1}{f} - \frac{1}{d_o}\right)^{-1} =$$

$$\left(\frac{1}{8cm} - \frac{1}{5cm}\right)^{-1} =$$

$$(-0.075cm^{-1})^{-1} = -13.3cm$$

b) The image is virtual, upright, and enlarged.

8. Apply the lens equation for **Part (a)**. For **Part (b)**, as with mirrors, there is a proportion between the heights (sizes) of the image and object and their respective distances from the lens.

a) $1/f = 1/d_o + 1/d_i$

Rearrange:

$1/d_o = 1/f - 1/d_i$
$= 1/0.1\text{ m} - 1/2.8\text{ m}$
$= 10\text{ m}^{-1} - 0.357\text{ m}^{-1}$
$= 9.643\text{ m}^{-1}$
$d_o = 0.103\text{ m}$

b) $h_i/h_o = d_i/d_o$

Rearrange:

$h_i = h_o d_i/d_o$
$= (3.0\text{ cm})$
$(2.8\text{ m})/(0.103\text{ m})$
$= 81.5\text{ cm}$

You can change the 3.0 cm to meters before using the equation, but the distance units will divide out and the unit of your answer will be whatever you had as the unit for your height. It could be centimeters, inches, etc. As long as the students carry the units through the problem, they should have no trouble identifying the units of the answer.

Chapter 3

ANSWERS

## Physics To Go

*(continued)*

9. The image is real, reduced, and inverted.

10. The image will be located at $2f$.

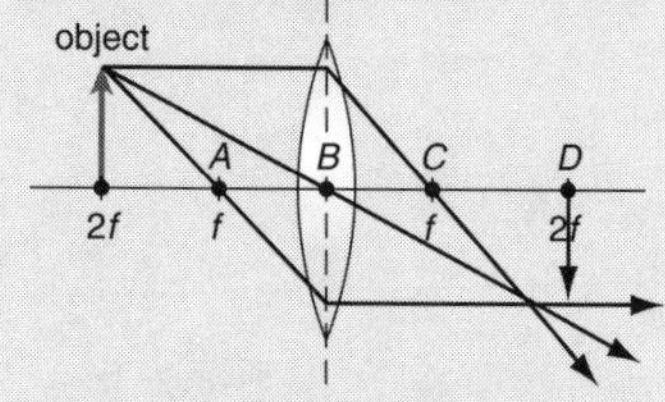


11. As the object moves closer to the lens, its size will increase.

9. The diagram shows an object 0.030 m high placed at point *X*, 0.60 m from the center of the lens. An image is formed at point *Y*, 0.30 m from the center of the lens. Completely describe the image.

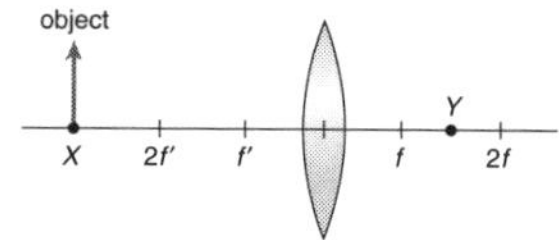


10. The diagram represents an object placed two focal lengths from a con**verging** lens. At which point will the image be located?

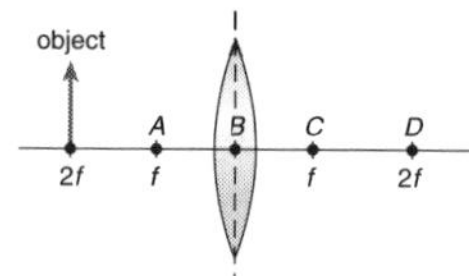


11. The diagram shows a lens with an object located at position *A*. Describe what will happen to the image formed as the object is moved from position *A* to position *B*.

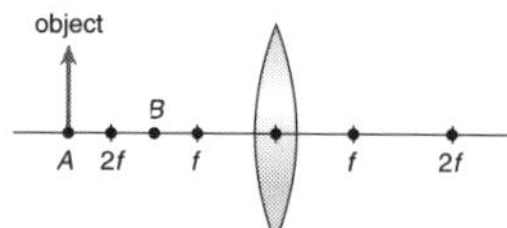

### Stretching Exercises

1. To investigate how the image position depends on the object position, find a convex lens, a white card, and a light source. Find the image of the light source, and measure the image and object distance from the lens. Make these measurements for as wide a range of object distances as you can. In addition, make an image of an object outside, such as a tree. Estimate the distance to the tree. The image of a distant object, like the tree, is located very near the focus of the lens. Draw a graph of the results. Compare the graph with the equation

$$\frac{1}{f} = \frac{1}{D_o} + \frac{1}{D_i}$$

2. Find a camera with a shutter that you can keep open (with a bulb or time setting). Place a piece of waxed paper or a piece of a plastic bag behind the lens, where the film would be if you took a picture. Find the image and compare it to the images you made in this activity. Focus the lens for objects at different distances. Investigate how well the object and image location fit the lens equation $\frac{1}{f} = \frac{1}{D_o} + \frac{1}{D_i}$.

   Remember that the focal length of the lens is typically printed on the lens.

3. Research how the concept of "depth of field" is important in photography. Report to the class on what you learn.

ANSWERS

## Stretching Exercises

1. To compare the graph with the equation, students can calculate $(1/D_o + 1/D_i)$ for each pair of images and object distances. If the values of $(1/D_o + 1/D_i)$ are approximately equal, then the equation shown describes their data (and they can calculate $f$).
2. The real image will lie on the wax paper.
3. Depth of fluid is the range of object distances that will produce a sharp image. It depends on the lens' focal length aperture ($f$ - number) and the object distance.

| Object Distance | Image Distance | Appearance of Image |
|---|---|---|
| | | |
| | | |
| | | |
| | | |
| | | |
| | | |
| | | |
| | | |
| | | |
| | | |
| | | |
| | | |

For use with *Let Us Entertain You*, Chapter 3, Activity 7: Effect of Lenses on Light

| Object Distance | Image Distance | Appearance of Image |
|---|---|---|
| | | |
| | | |
| | | |
| | | |
| | | |
| | | |
| | | |
| | | |
| | | |
| | | |
| | | |
| | | |

For use with *Let Us Entertain You*, Chapter 3, Activity 7: Effect of Lenses on Light

NOTES

# ACTIVITY 8
## Color

# Background Information

Shadows are produced when one object blocks some of the light that is hitting another object. Outside on a sunny day, you can see a sharp shadow of your foot on the ground. You can also see shadows of leaves on the ground, but notice how sharp—or rather how fuzzy—is the border of these shadows. You can see the same effect in a dark room by turning on a single frosted bulb. Because the bulb and the sun are extended light sources, they make shadows with a penumbra region, a fuzzy border around the shadow illuminated by some, but not all, of the light source.

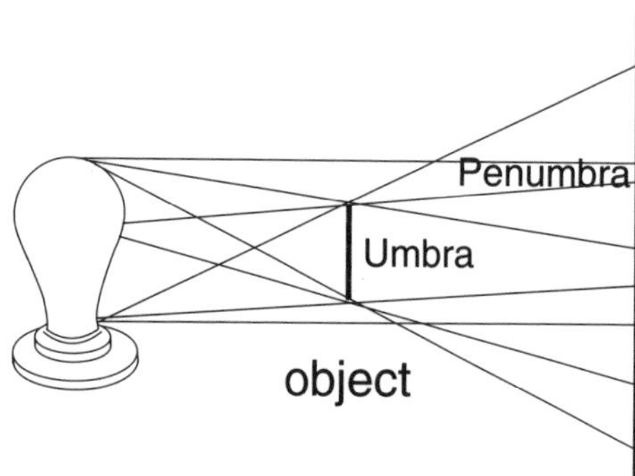


The size of this fuzzy border is directly proportional to the distance from the object to the screen. The top of a flagpole, or even better a tall building, casts a fuzzy shadow indeed.

Color is produced by the eye and the brain. The corresponding physical property of light is the wavelength. Visible wavelengths range from about 0.4 × 10[-6] m to 0.7 × 10[-6] m. Light with a wavelength in this range stimulates the eye and brain to produce a spectral color, as shown by the sprectral colors of sunlight formed by a prism (the index of refraction of the prism glass depends on the wavelength of light). These spectral colors are also produced by combinations of lights with different wavelengths. Red light plus green light produces light we see as yellow. Sometimes these color combinations contain white, and the result is an unsaturated color. Such a combination would be a red, green, and blue light, with a second blue light added, which would be seen as a pale blue. In a saturated color, there is no white. Saturated colors are often called "strong" or "intense." As different colored lights are progressively turned on to illuminate a screen, more and more light falls on the screen, which looks brighter and brighter. In the drawing of the three overlapping disks on page 389, the central area, illuminated by light of all three colors, would be the brightest area and would be white.

Overlapping colored filters gives quite different results. As more and more filters are overlapped, the light transmitted is less and less, so the stack of filters looks dark. Typically a filter passes a range of wavelengths, which correspond to a range of colors, so two overlapped filters pass only those wavelengths that both pass in common.

## *Active*-ating the Physics InfoMall

Many of the searches to try here are probably obvious now. A warning may be helpful; if you search for "color" you will need to limit the search to only a few stores at a time, as the Articles & Abstracts Attic alone provides over 2000 hits. You should search for "pigments" to find the differences between light and pigment mixing. There are many useful searches you could perform here. Don't limit yourself; use your imagination. In other words, be colorful.

# Planning for the Activity

## Time Requirements

two class periods

## Materials Needed

### For the class:

- overhead projector

### For each group:

- 40-W bulb and socket, ceramic w/switch
- bulbs: white, red, blue, and green (4 each)
- file folder
- lamp, clip-on w/shade
- stand for clip-on lamp
- steel scissors
- white screen, about 2' × 3'

## Advance Preparation and Setup

Make a copy of the two Blackline Masters for each student. If possible, buy inexpensive "cans," which can be aimed and which have a slot to hold a filter (get these at a theatrical supply company). Set up the three colored lights so the beams all overlap on the screen.

Be sure all sockets have the wiring safely covered. If you use ceramic sockets with screw terminals underneath, screw the sockets down to a board so these terminals will be inaccessible.

# Teaching Notes

The geometry of the shadows can be challenging. It may help to point a meter stick from a light to the puppet and to ask where that light will reach the screen. Also, with all three lights on, you can ask the students what will happen if you block the lights one at a time (for example, by covering one with a book or clipboard). Encourage the students to compare their observations for the colors they observe with their expectations. You might ask how the unexpected colors might become part of the student's light show.

If you have only one set of colored lights, at any give time you will have to keep the students busy and productive on other parts of the activity. If you have prisms available, you might have each group do a simple prism activity at this time.

SAFETY PRECAUTION: Be extremely careful with 110-V bulbs. Explain to the students that they must not touch the bulbs. Be sure all sockets have the wiring safely covered. If you use ceramic sockets with screw terminals underneath, screw the sockets down to a board so these terminals will be inaccessible.

Most students have little experience with combining colored lights. They probably will be surprised that red light plus green light looks yellow. It may help to explain that what is compicated here is not the physics but rather the functioning of the eye and brain. In fact, the way we commonly refer to light as "red" or "green" may create the preconception that the color is somehow embedded in the light, rather than being a sensory response to the physical characteristic of wavelength.

# Activity Overview

To investigate color addition, students make shadows of colored bulbs. First they observe shadows of an object illuminated by one bulb. They make a drawing to show the size of the shadow and the relationship of size to the positions of bulb, object, and screen. Then they progress to combinations of colored lights, and for each combination they make a drawing to record the pattern of the shadow colors.

## Student Objectives

### Students will:

- Analyze shadow patterns.
- Explain the size of shadows.
- Predict pattern of colored shadows.
- Observe combinations of colored lights.

Let Us Entertain You

## Activity 8 Color

**GOALS**

In this activity you will:

- Analyze shadow patterns.
- Explain the size of shadows.
- Predict pattern of colored shadows.
- Observe combinations of colored lights.

**What Do You Think?**

When a painter mixes red and green paint, the result is a dull brown. But when a lighting designer in a theater shines a red and a green light on an actress, the actress's skin looks bright yellow.

- **How could these two results be so different?**
- **How are the colors you see produced?**

Record your ideas about these questions in your *Active Physics* log. Be prepared to discuss your responses with your small group and with your class.

**For You To Do**

1.Carefully cut out a cardboard puppet that you will use to make shadows.

Active Physics 204

ANSWERS FOR THE TEACHER ONLY

## What Do You Think?

The answer to these questions lies in the field of both physiology and physics. The human eye and brain respond to the wavelength of light by producing the sensation of color, and each wavelength produces a distinct color, as seen in the spectrum of sunlight formed by a prism. (Recall that the index of refraction of the prism is a function of the wavelength of light.)

The nature of the light-sensitive cells in the eye, together with the structure of the brain, determine in a complex way the color we see. Mixing paints produces color subtraction, because only light that can be transmitted by both paints is seen (the other colors are subtracted out). Overlapping lights in a theater produce color addition, since more light is reflected. The eye and brain do the rest to produce the colors we see.

ANSWERS

## For You To Do

1. The puppet can have any shape.

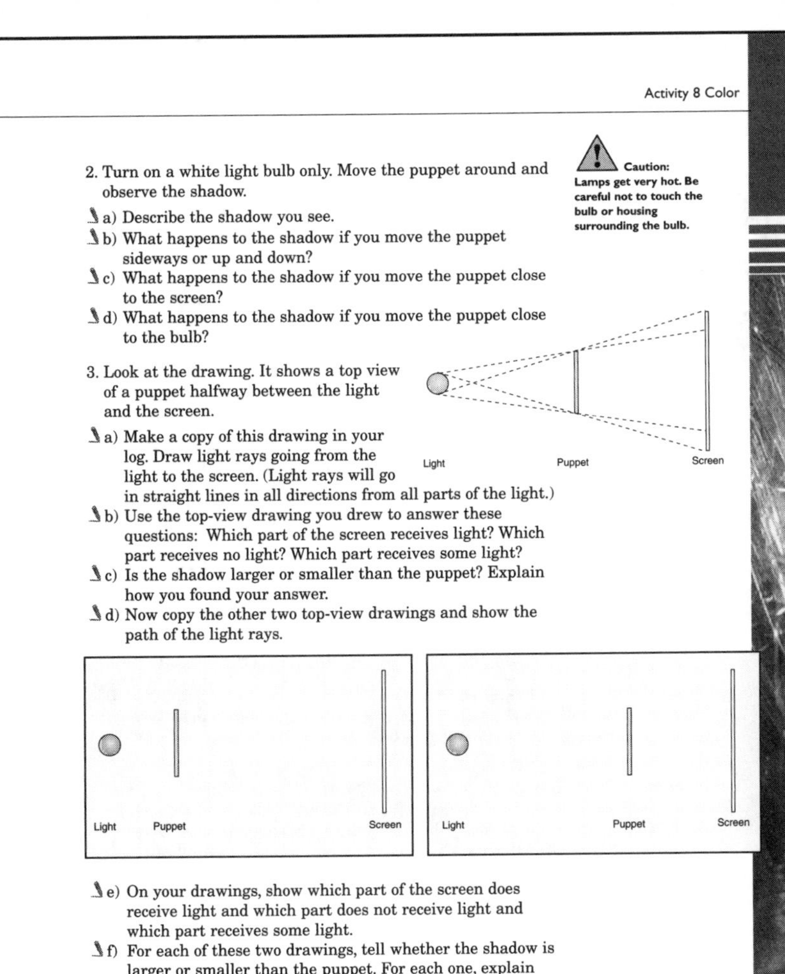


2. Turn on a white light bulb only. Move the puppet around and observe the shadow.
   - a) Describe the shadow you see.
   - b) What happens to the shadow if you move the puppet sideways or up and down?
   - c) What happens to the shadow if you move the puppet close to the screen?
   - d) What happens to the shadow if you move the puppet close to the bulb?

**Caution:** Lamps get very hot. Be careful not to touch the bulb or housing surrounding the bulb.

3. Look at the drawing. It shows a top view of a puppet halfway between the light and the screen.
   - a) Make a copy of this drawing in your log. Draw light rays going from the light to the screen. (Light rays will go in straight lines in all directions from all parts of the light.)
   - b) Use the top-view drawing you drew to answer these questions: Which part of the screen receives light? Which part receives no light? Which part receives some light?
   - c) Is the shadow larger or smaller than the puppet? Explain how you found your answer.
   - d) Now copy the other two top-view drawings and show the path of the light rays.



   - e) On your drawings, show which part of the screen does receive light and which part does not receive light and which part receives some light.
   - f) For each of these two drawings, tell whether the shadow is larger or smaller than the puppet. For each one, explain how you found your answer.



ANSWERS

## For You To Do
### *(continued)*

2\. a) The sharpness of the edge of the shadow depends on the distance from the puppet to the screen.

b) The shadow moves in the same direction.

c) The shadow becomes sharp and is the same size as the puppet.

d) The shadow becomes quite large, and the edge of the shadow will become fuzzy if the light bulb is frosted.

3\. a) Notice that you have to draw lines from only the top and bottom of the bulb.

b) The part of the screen between the inner dashed lines receives no light. The part of the screen between the inner and outer dashed lines receives some light (since it is possible to draw a line from some part of the bulb to this part of the screen). No part of the screen receives light from all of the bulb.

c) The shadow is larger than the puppet, since light from the bulb continues to spread out after it passes the puppet.

d) In the drawing on the left, the shadow is larger than the puppet. Imagine lines from the light source to the ends of the puppet. Extend these lines to the screen. Since these lines spread way out as they go towards the screen, the shadow will be much larger than the puppet. For the drawing on the right, the shadow will be only slightly larger than the puppet.

e) Draw dashed lines from the top and bottom of the bulb, as in the drawing in **Step 3**.

f) See answer to **Step d)** above.

ANSWERS

## For You To Do
***(continued)***

4. a) Where both lights strike the screen, you will see yellow (but it may appear white).

   b) Where the red light is blocked but the green is not, you will see a green shadow. Where the green light is blocked but the red is not, you will see a red shadow.

   c)-d) See diagram below. Also, note that where light from both bulbs hits the screen, you see yellow. Moving toward the center of the screen, the yellow gradually changes to red (above) and to green (below).

5.- 6. See answer to 4. above.

7. Where all three lights strike the screen, you will see white. Where one light is blocked, you will see a shadow with the color of the combination of the other lights. See **Steps 4** through **6** for these colors. Where two lights are blocked, you will see a shadow with the color of the unblocked light.

Let Us Entertain You

4. Turn off the white bulb. Turn on red and green bulbs. They should be aimed directly at the center of the screen.
   a) What color do you see on the screen?
   b) Predict what color the shadows will be if you bring your puppet between the bulbs and the screen. Record your prediction, and give a reason for it.
   c) Make a top-view drawing to show the path of the light rays from the red and green bulbs.
   d) On your drawing, label the color you will see on each part of the screen.

5. Turn off the green bulb and turn on a blue one. Repeat what you did in **Step 4**, but with the blue and red bulbs lit.

6. Turn off the red bulb and turn on the green one. Repeat what you did in **Step 4**, but with the blue and green bulbs lit.

7. Turn on the red bulb so all three—red, blue, and green—are lit. Repeat what you did in **Steps 5** and **6**.

### Reflecting on the Activity and the Challenge

Different colored lights can combine to make white light. When an object blocks all light, it creates a dark shadow. Since some light comes from all parts of the bulb, there are places where the shadow is black (no light) and places where the shadow is gray (some light reaches this area). An object illuminated by different colored lights can create shadows that prevent certain colors from reaching the wall and allowing other colors to pass by.

In your light show creation, you may choose to use the ideas of colored shadows to show how lights can be added to produce interesting combinations of colors. By moving the object or the lights during the show, you may be able to produce some interesting effects. Lighting design is used in all theater productions. It requires a knowledge and understanding of how lights work, as well as an aesthetic sense of what creates an enjoyable display.

206

Active Physics

4. c - d)

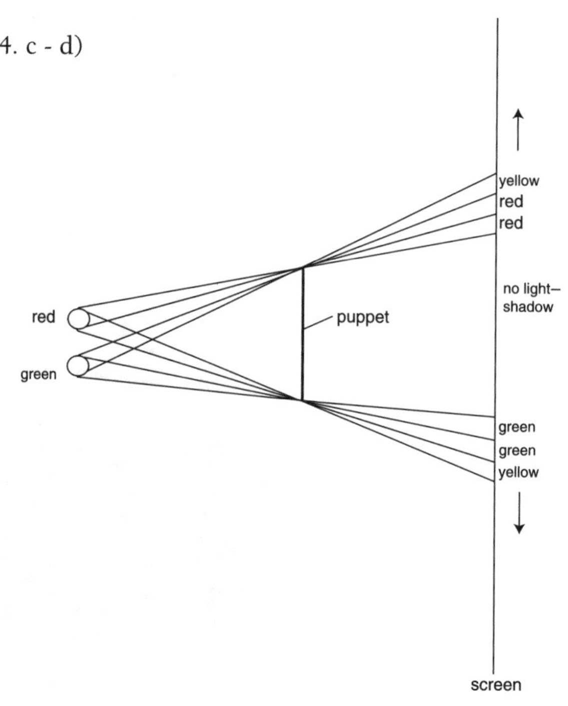

**Physics To Go**

1. Show how a shadow is created.
2. How can moving the light, the object, and the screen all produce the effect of enlarging the shadow?
3. Explain why a gray halo surrounds a dark shadow made by a light bulb and an object.
4. a) Why is your shadow different at different times of the day?
   b) What is the position of the Sun when your shadow is the longest? The shortest?
5. Why is the gray halo about your shadow so thin when you are illuminated by the Sun?
6. a) Suppose you shine a red light on a screen in a dark room. The result is a disk of red light. Now you turn on a green light and a blue light. The three disks of light overlap as shown. Copy the diagram into your journal. Label the color you will see in each part of the diagram.

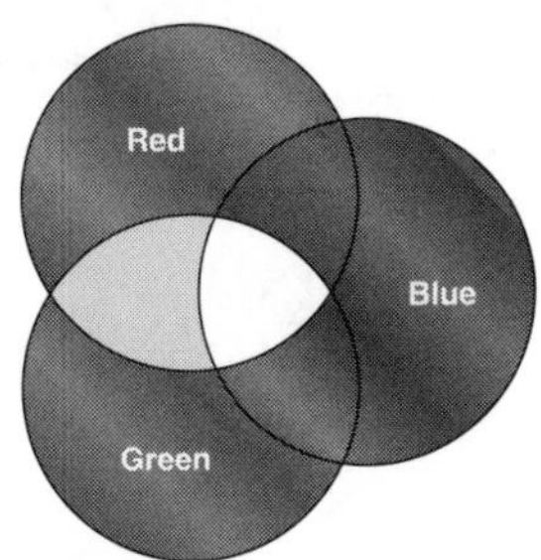


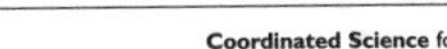

   b) Add the labels "bright," "brighter," and "brightest" to describe what you would see in each part of your diagram.



ANSWERS

## Physics To Go

1. When an object blocks light from hitting part of a screen, the result is a shadow.
2. The shadow becomes larger when light can spread out more after it passes the object. Light can spread out more when the screen is moved away from the light, the object is moved towards the light, or the light is moved towards the object.
3. The halo is a region of intermediate brightness at the edge of the shadow. The effect is to make the shadow edge appear fuzzy. The gray halo occurs when the light bulb is frosted (an "extended source"). Since the light comes from different places on the bulb, the area at the edge of the shadow is illuminated by light from only part of the bulb.

ANSWERS

## Physics To Go *(continued)*

4. a) The length of the shadow is determined by the altitude of the Sun (the angle between the horizon and the Sun). Since the light rays of sunlight are approximately parallel, the orientation of your body and also the orientation of the ground determine the length of the shadow.
   b) Your shadow is longest at sunrise and sunset. It is shortest at noon.
5. Because the Sun, although quite large, is so very far away.
6. a) Where all three lights overlap, in the center, you see white. Where red and green overlap, you see yellow. Where red and blue overlap, you see purple. Where blue and green overlap, you see blue-green.
   b) The central region is the brightest. Three regions with two overlapping lights (yellow, blue-green, and purple) are the next brightest. The red, blue, and green regions are the least bright.

ANSWERS

## Physics To Go
***(continued)***

7. a)-b) See drawing on p. 205. The rest of the screen is red.

c) See drawings on previous pages.

8. Answers will vary.

7. a) Make a drawing of an object in red light. The object casts a shadow on the screen. Label the color of the shadow and the rest of the screen.
   b) Repeat **Part (a)** for an object in green light.
   c) Now make a copy of your drawing for **Part (b)**. Add a red light, as in **Part (a)**. Label the color of all the shadows.

8. List some imaginative ways that you can add colors to your light show.

### Stretching Exercises

1. With the room completely dark, shine a red light on various colored objects. Compare the way they look in red light with the way they look in ordinary room light.
2. View 3-D pictures with red and blue glasses. Explain how each eye sees a different picture.
3. Shine a white light and a red light on a small object in front of a screen. What colors are the shadows? How is this surprising?
4. Prepare a large drawing of the American flag but with blue-green in the place of red, and yellow in the place of blue. Stare at the drawing for 30 seconds and then look at a white surface.

ANSWERS

## Stretching Exercises

1. A green object illuminated in red light will look black (since a green object reflects only green light). A red object will still appear red.
2. The inks in the 3-D pictures are matched to the filters in the glasses. The eye looking through the red glasses sees the blue lines, which appear very dark. The red lines blend into the background, since both appear red. The other eye sees only the red lines, which also appear very dark. The red and blue images are slightly displaced to produce the 3-D effect.
3. It would seem that the shadow of the red light should be white. But often this shadow is seen as blue-green. The screen is a pale red, but the eye and brain often take this for white. If that happens, then subtracting red from this color (in the shadow of the red light) produces what the brain takes to be white minus red, or blue-green, the complement of red (remember that red plus blue-green gives white). It is a remarkable illusion.
4. Staring at the strangely-colored flag tires the color receptor cells in the retina, so they cease responding in that color. They continue to respond in the complementary color, however, so the result is that the brain perceives the complementary colors when viewing a white surface.

# PHYSICS AT WORK

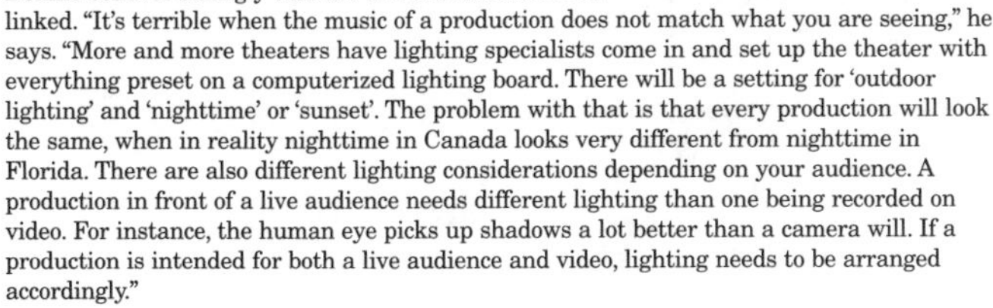

## Alicja and Dennis Phipps

Alicja Phipps has always been interested in electronics. As a child she wanted to be a television repair person. She now works with her husband, Dennis Phipps, in their company, Light & Sound Entertainment, which designs original content programming in a variety of areas—from rock concerts to the Olympics.

Light & Sound Entertainment got its name because Dennis believes strongly that the two are and should be linked. "It's terrible when the music of a production does not match what you are seeing," he says. "More and more theaters have lighting specialists come in and set up the theater with everything preset on a computerized lighting board. There will be a setting for 'outdoor lighting' and 'nighttime' or 'sunset'. The problem with that is that every production will look the same, when in reality nighttime in Canada looks very different from nighttime in Florida. There are also different lighting considerations depending on your audience. A production in front of a live audience needs different lighting than one being recorded on video. For instance, the human eye picks up shadows a lot better than a camera will. If a production is intended for both a live audience and video, lighting needs to be arranged accordingly."

Dennis continues, "The sound of a production is only as good as its setup, and nothing can replace the actual setting. The Red Rocks Theater in Colorado is terrific, for example, because stone has a very high reverberation rate, which is great for guitars. A huge wooden room like Carnegie Hall also provides a unique sound. However, these spaces and materials are not readily available." The hardest projects, Dennis says, are those in which you cannot control the elements. "Sound elements include the size and shape of the space, reverberation, feedback, and temperature."

"New media has its own set of challenges," explains Alicja who oversees the conversion of live events into various other formats, such as CD-ROMs, virtual reality, and Web sites. "We have to think about how much information (sound and image) we will be able to fit on a disc or on to a Web page and how long it will take to load. If it takes too long, no one will ever see or hear what we've done."

"We enjoy the creative process of every production," claim Alicja and Dennis. "Each one is a unique challenge."



Chapter 3

## Chapter 3 Assessment

With what you learned about sound and light in this chapter, you are now ready to dazzle the world. However, you have neither the funds nor the technology available to professionals. All sounds you use to capture the interest of the class must come from musical instruments that you build yourself, or from human voices. Some of these sounds may be prerecorded and then played in your show. If your teacher has a laser and is willing to allow you to use it, you may do so. All other light must come from conventional household lamps. Gather with your committee to design a two- to four-minute sound and light show to entertain other students your age.

Review the criteria by which you decided that your show will be evaluated. The following suggestions were provided at the beginning of the chapter:

1. The variety and number of physics concepts used to produce the light and sound effects.
2. Your understanding of the physics concepts:
   a) Name the physics concepts that you used.
   b) Explain each concept.
   c) Give an example of something that each concept explains or an example of how each concept is used.
   d) Explain why each concept is important.
3. Entertainment value

At this time you may wish to propose changes in the criteria. Also decide as a class if you wish to modify or keep the point value you established at the beginning of the chapter.

Enjoy the sound and light productions!

## Physics You Learned

Compressional and transverse waves

Wave speed = wavelength × frequency

Standing waves

Pitch and frequency

Sound production in pipes and vibrating strings

Controlling frequency of sounds produced electronically

Angle of incidence and angle of reflection

Location of image in plane and curved mirrors

$\frac{1}{f} = \frac{1}{D_o} + \frac{1}{D_i}$ in curved mirrors

Real images

Angle of incidence and angle of refraction

Lenses and image formation

$\frac{1}{f} = \frac{1}{D_o} + \frac{1}{D_i}$ in lenses

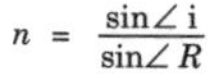

Color addition

$n = \frac{\sin\angle i}{\sin\angle R}$

NOTES

Chapter 3

# Alternative Chapter Assessment

1. You shake a Slinky back and forth. No wave moves along the spring. At a few places, the spring does not move at all. You have made a:

   a) pulse

   b) frequency

   c) standing wave

   d) reflection

2. In a wave, the number of cycles per second is called the:

   a) wavelength

   b) amplitude

   c) speed

   d) frequency

3. As you decrease the tension on a string, the pitch it makes:

   a) is unchanged

   b) increases

   c) decreases

4. As you decrease the length of a tube, the pitch it makes:

   a) is unchanged

   b) increases

   c) decreases

5. You built an electronic circuit that made sound. When you changed the value of a capacitor in this circuit, the sound:

   a) became louder

   b) became softer

   c) stopped

   d) changed its pitch

6. If you walk towards a mirror, your reflection:

   a) doesn't move

   b) moves away from the mirror

   c) moves towards the mirror

7. The image in a convex mirror:
   a) is real
   b) is virtual
   c) can be real or virtual

8. At the critical angle, there is:
   a) total internal reflection
   b) no reflection
   c) partial reflection

9. You make an image on a card with a convex lens. As the object moves away from the lens, the image:
   a) moves away from the lens
   b) stays in the same place
   c) moves closer to the lens

10. One light shines on a screen. If you shine a second light on the screen, the screen looks:
    a) yellow
    b) dimmer
    c) brighter

11. You make a spring wave with a wavelength of 1.5 m. You shake the spring with a frequency of 2 Hz. The speed of the wave is _________________.

12. The distance from one wave crest to the next is called the ______________.

13. As a wave goes by on a spring, the spring moves from its rest position. The distance from the rest position is called the ______________.

14. A concave mirror makes an image of a distant light at one point. The location of the image is called the ________________ of the mirror.

15. The bending of light when it goes from air into gelatin is called ________________.

16. Explain how to measure the focal length of a lens.

17. a) Explain how to measure the *frequency* of a wave.

    b) If you have the frequency of a wave, how could you find the *period*?

18. Explain the difference between a *transverse* wave and a *compressional* wave on a Slinky. You can use drawings to give your answer.

19. a) A beam of light hits a flat mirror. Make a drawing to show the angle of incidence, the angle of reflection, and the normal.

    b) What is the relationship between the angle of incidence and the angle of reflection?

20. a) You are building an organ with pipes that are 3 m, 1.5 m, and .5 m long. Order these tubes according to how high a pitch they will make.

    b) Label the tube with the highest pitch and the tube with the lowest pitch.

21. a) A beam of light goes from air into a piece of gelatin. Make a drawing to show the angle of incidence, the angle of refraction, and the normal.

    b) On your drawing, show which way the beam of light bends.

22. a) You shine a red light and a blue light on a screen (see drawing). The puppet casts shadows on the screen. Make a drawing of the lights, the puppets, and the screen to show the different shadows.

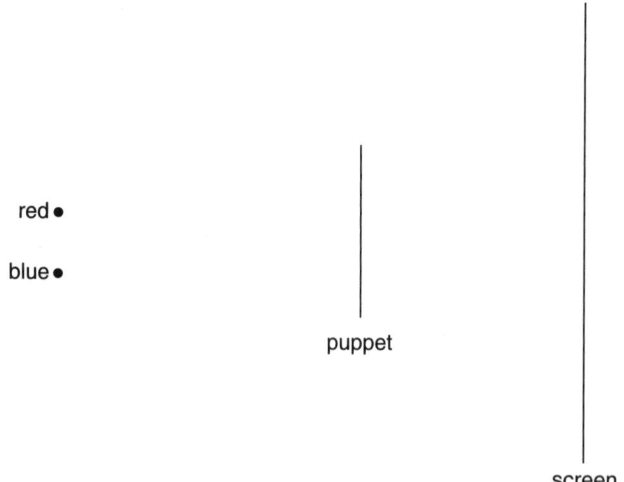


    b) Label the drawing to show the color of the screen and the color of each shadow.

    c) Are the shadows larger or smaller than the puppet?

    d) Explain your answer to part (c).

# Alternative Chapter Assessment Answers

1. c
2. d
3. c
4. b
5. d
6. c
7. b
8. a
9. c
10. c
11. 3 m/s
12. wavelength
13. amplitude or displacement
14. focus
15. refraction
16. Make an image of a distant object. Measure the distance from the image to the lens.
17. a) Count how many cycles of the wave there are in a certain time. Divide the time into the number of cycles.

    b) Invert the frequency (divide it into 1).
18. The difference is in the direction of the back-and-forth motion.

In a transverse wave, the Slinky goes back and forth perpendicular to the direction the wave moves. In a compressional wave, the Slinky goes back and forth in the same direction the wave moves. See drawings on p. 145 and 147.

19. a) See the drawing on page 174.

    b) the angles are equal
20. a) from longest to shortest.

    b) The shortest makes the highest pitch; the longest makes the lowest pitch.
21. a) See angles 1 and 2 in the drawing on page 190.

    b) The light bends towards the normal (going into the gelatin).

22. a) and b)

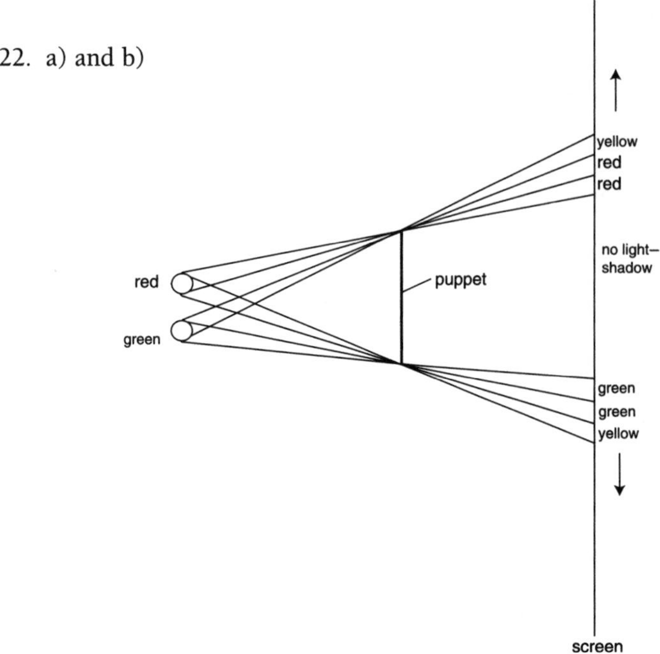


c) and d) The shadows are larger, because the light is small, and the light rays spread out to reach the puppet. After passing the puppet, they keep on spreading.

# Chapter 4

# TOYS FOR UNDERSTANDING

## Chapter 4 – Toys for Understanding

# National Science Education Standards

### Chapter Challenge

Homes for Everyone (HFE) identified a need to educate children about electricity and how it is generated. HFE wants to approach this with toys. Students are challenged to prepare a kit, with materials and instructions, that children use to build a toy with a motor and/or generator. This toy will serve as a tool to illustrate how the electric motors in home appliances work or how electricity can be produced from an energy source such as wind, moving water, or some external force.

### Chapter Summary

To gain understanding of the science concepts of energy conversions necessary to meet this challenge, students are engaged in activities to learn about electricity and magnetism. These experiences engage students in the following content from the National Science Education Standards.

## Content Standards

### Unifying Concepts

- Evidence, models and explanations
- Form and function

### Science as Inquiry

- Identify questions and concepts that guide scientific investigations
- Use technology and mathematics to improve investigations
- Formulate and revise scientific explanations and models using logic and evidence

### Science and Technology

- Abilities of technological design
- Understanding about science and technology

### History and Nature of Science

- Historical perspectives

### Science in Personal and Social Perspectives

- Natural resources
- Science and technology in local, national, and global challenges

### Physical Science

- Motion and force
- Conservation of energy and increase in disorder
- Interactions of energy and matter

# Key Physics Concepts and Skills

| Activity Summaries | Physics Principles |
|---|---|
| **Activity 1: The Electricity and Magnetism Connection**<br>Students investigate the relationship between electricity and magnetism by testing the effect of a magnetic field on current-bearing wire and on a compass. | • Electricity<br>• Magnetism<br>• Magnetic fields |
| **Activity 2: Electromagnets**<br>Using the hand generator to construct an electromagnet is the first step in this continued investigation of the relationship between electricity and magnetism. Students test the strength and find polarity of electromagnets made with different core materials. | • Electromagnets<br>• Solenoids |
| **Activity 3: Detect and Induce Currents**<br>Students construct a galvanometer as they learn that a compass can detect the presence of a magnetic field. They use the galvanometer to create a current similar to the process used by Faraday and Henry, manually alternating the motion of a magnet. | • Galvanometers<br>• Induced currents |
| **Activity 4: AC & DC Currents**<br>The use of human energy to produce electricity is replaced in this activity by a rotating coil motor. While using this motor, students test and describe the voltage in this induced current. They learn the difference between how AC and DC currents are generated. Students also learn to sketch output wave forms. | • Energy conversion<br>• AC and DC currents<br>• Electrical waves |
| **Activity 5: Building an Electric Motor**<br>Students construct, operate, and explain the workings of a DC motor. This enables them to measure and express the efficiency of energy transfers. They also read to learn more about the discoveries that led to the generators and motors we use today to obtain useable power and electricity. | • Electricity<br>• Magnetism<br>• Energy transfer |
| **Activity 6: Building a Motor/Generator Toy**<br>In this final activity, students apply what they have learned about the workings of an electromagnetic motor and how both AC and DC currents are generated. They must use given materials to design, construct, and demonstrate the physics of a motor or generator. | • Energy conversion<br>• Electricity<br>• Magnetism<br>• AC and DC currents<br>• Energy conversion<br>• Energy flow and power<br>• Electrical efficiency |

# Equipment List For Chapter 4 (Serves a Classroom of 30 Students)

| PART | ITEM | QTY | ACTIVITY |
|---|---|---|---|
| AC-6002-H3 | Alligator Clip Leads | 12 | 3, 5, 6 |
| BS-1596-H3 | Battery, D-Cell | 6 | 5, 6 |
| CP-6081-H3 | Compass | 6 | 1, 2 |
| CS-6047-H3 | Cup, Styrofoam® | 36 | 5 |
| GH-0001-H3 | Galvanometer, 0-500 ma. | 6 | 4 |
| GH-0002-H3 | Tangent Galvanometer | 6 | 1 |
| GH-7704-H3 | DC Generator, Hand-Operated W/Wire Leads | 6 | 2, 3, 5 |
| LS-6959-H3 | Miniature Bulb, Light | 18 | 3 |
| LS-6960-H3 | Bulb Base for Miniature Screwbase | 6 | 3 |
| MS-1886-H3 | Magnet,Ceramic Ring,Refrigerator | 12 | 5 |
| MS-0090-H3 | Magnet, Small Bar | 6 | 6 |
| MS-0354-H3 | Magnet, Bar, Large | 6 | 3 |
| NS-7613-H3 | Nail, 16 d. | 6 | 2 |
| PA-7614-H3 | Clips, Paper, Small | 60 | 2 |
| PS-7620-H3 | Safety Pins, Large | 12 | 5 |
| RB-6108-H3 | Rubber Band, #64 | 6 | 5 |
| SS-6078-H3 | Sandpaper, Fine Square | 36 | 2, 3, 5 |
| SS-6113-H3 | Drinking Straw, Each | 60 | 2 |
| TS-2662-H3 | Tape, Masking, 3/4 × 60 yds. | 6 | 3, 5 |
| TS-5000-H3 | Test Tube, Plastic, 2×10 cm | 6 | 5 |
| TS-7601-H3 | Tube, Cardboard, 1×4 | 6 | 3 |
| WS-1148-H3 | #24 Magnet Wire-60 Foot Rolls | 360 ft | 2, 3, 5 |
| WS-7612-H3 | Wire, Insulated Copper For Tangent Galvanometer 18 Inch Sections | 108 in | 1 |
| ZZ-0066-H3 | 66 Quart Clear Plastic Storage Container | 6 | |
| | | | |
| TEACHER DEMONSTRATION ITEMS | | | |
| AC-7600-H3 | AC/DC Demonstration Generator | 1 | 4 |
| BS-7602-H3 | DC Source, High Current | 1 | 1 |
| CT-0006-H3 | 5' Copper Tube For Cow Magnet | 1 | 4 |
| HS-7617-H3 | 1 3/4" Buret Clamp (Use To Support Copper Tube) | 1 | 4 |
| MS-0061-H3 | Magnet, Cow | 1 | 4 |
| | | | |
| THINGS NEEDED NOT SUPPLIED | | | |
| | All materials needed have been supplied | | |

## Organizer for Materials Available in Teacher's Edition

| Activity in Student Text | Additional Material | Alternative / Optional Activities |
|---|---|---|
| ACTIVITY 1:<br>The Electricity and Magnetism Connection<br>p. 214 | | |
| ACTIVITY 2:<br>Electromagnets<br>p. 219 | | |
| ACTIVITY 3:<br>Detect and Induce Currents<br>p. 223 | | Activity 3 A:<br>Twin Coil Swings<br>p. 414 |
| ACTIVITY 4:<br>AC & DC Currents<br>p. 228 | Current Induced by<br>AC and DC Generators<br>pp. 426-427 | Activity 4 A:<br>Falling Magnet<br>p. 425 |
| ACTIVITY 5:<br>Building an Electric Motor<br>p. 235 | | |
| ACTIVITY 6:<br>Building a Motor/Generator Toy<br>p. 241 | | |

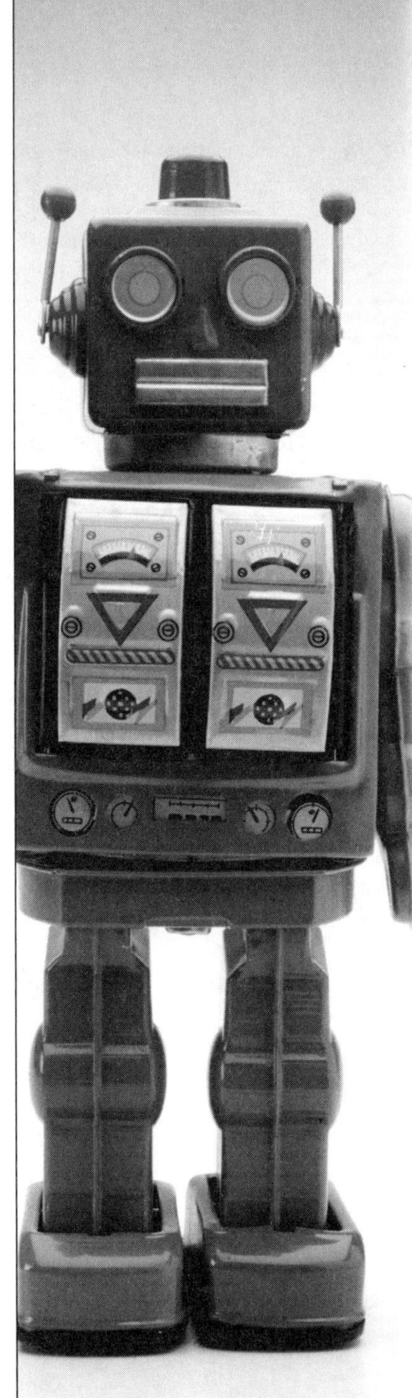

# Scenario

In this *Active Physics* chapter, you will try to help educate children through the use of toys. With your input, the Homes for Everyone (HFE) organization has developed an appliance package that will allow families living in the "universal dwelling" to enjoy a healthy and comfortable lifestyle. The HFE organization would now like to teach the children living in these homes, and elsewhere, more about electricity and the generation of electricity. They hope that this may encourage interest in children to use electricity wisely, as well as encourage development of alternative sources for electrical energy by future generations.

The HFE organization will work with a toy company to provide kits and instructions for children to make toy electric motors and generators. These toys should illustrate how electric motors and generators work and capture the interest of the children.

In an effort to help others, people often make changes or introduce new products without considering the personal and cultural impact on those whom they are trying to assist. If you ever become involved in a self-help community group, such as HFE, it would be important for you to work together with the people you are assisting to assess their needs, both personal and cultural. Although that is not possible given your limited time in class, you should recognize the need for collaborative teamwork in evaluating the impact of any new product on an established community.

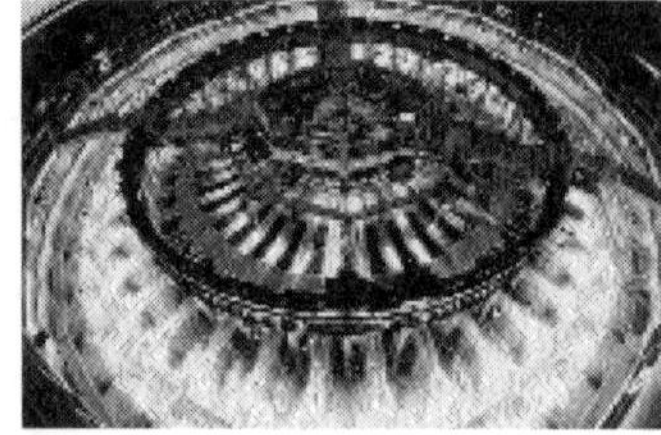

212

## Chapter and Challenge Overview

In this chapter, the students are applying some of the principles they have learned about electricity and applying it to the building of a toy. Although the toy is designed for children, it is important for the students to focus on the learning that is taking place. A generator can be a motor, and vice versa. The students need to realize that the chapter is designed to get them to apply a particular concept (electromagnetism) to a real life situation. It is hoped that the students will gain a greater respect for physics and be able to apply their knowledge in developing their overall appreciation of science.

A specific design for a DC motor which could also serve as a generator is not presented until the final activity of this chapter. Earlier, perhaps right away, you may wish to have students search for alternate designs—references on science projects would be good sources. Also, you may wish to make a rule on whether motor/generator kits available from hobby stores or toy kits will be allowed; greater benefit may result from having students build the device from "junk."

An understanding of the principles of an electric motor or generator must be brought forward to the students. However, quantitative analysis is not necessary at this stage. Some aspects of the quantitative forces may be used as an alternative project for enrichment.

## Challenge

Your task is to prepare a kit of materials and instructions that the toy company will manufacture. Children will use these kits to make a motor or generator, or a combination electric motor/generator. It will serve both as a toy and to illustrate how the electric motors in home appliances work or how electricity can be produced from an energy source such as wind, moving water, a falling weight, or some other external source.

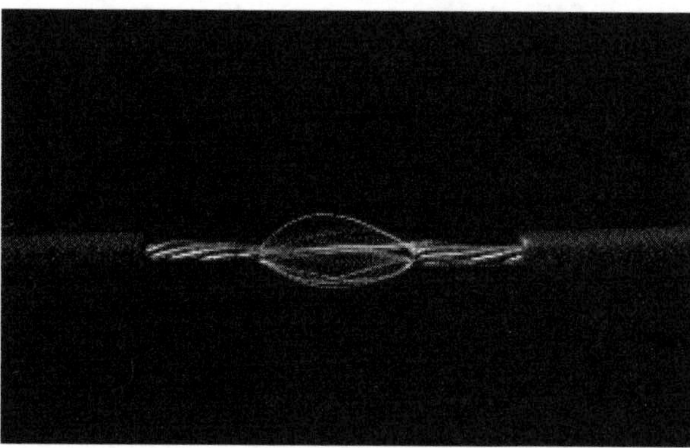

## Criteria

Your work will be judged by the following criteria:

- **(30%) The motor/generator is made from inexpensive, common materials, and the working parts are exposed but with due consideration for safety.**
- **(40%) The instructions for the children clearly explain how to assemble and operate the motor/generator device, and explain how and why it works in terms of basic principles of physics.**
    - **(30%) If used as a motor, the device will operate using a maximum of four 1.5-V (volt) batteries (D cells), and will power a toy (such as a car, boat, crane, etc.) that will be fascinating to children.**

    **OR**

    - **(30%) If used as a generator, the device will demonstrate the production of electricity from an energy source such as wind, moving water, a falling weight, or some other external source and be fascinating to children.**

213

# Assessment Rubric for Challenge

The following is a possible guideline for evaluation of the challenge.

Place a check mark in the appropriate box. If you would rather, you could mark it holistically, where all the statements taken together earn, for example, a mark of 30%.

You may change the criteria, by adding or deleting any of the points, determined by a class decision.

| Descriptor | Yes | No |
|---|---|---|
| **Construction of motor/generator (30%)** | | |
| made from simple, inexpensive materials | | |
| easy to assemble (or already assembled) | | |
| working parts are exposed | | |
| safe device | | |
| creative and imaginative | | |
| **Instructions (40%)** | | |
| well-written | | |
| clear and concise | | |
| explanation of how it works is clear | | |
| simple explanation of the physics involved | | |
| includes safe operating instructions | | |
| clearly stated as motor or generator (or both) | | |
| provides evidence of testing with children | | |
| clear understanding that there is no "creation " of electricity, but transformation | | |
| **A) Motor – toy (30%)** | | |
| will operate on maximum of four 1.5-v batteries | | |
| will operate a toy | | |
| is fascinating to children | | |
| is durable in the hands of children | | |
| toy shows character and imagination | | |
| **OR B ) Generator (30%)** | | |
| clear understanding as to difference between motor and generator | | |
| demonstrates the production of electricity from an energy source | | |
| is fascinating to children | | |
| is durable in the hands of children | | |
| generator shows character and imagination in the transformation of energy | | |

For use with *Toys For Understanding*, Chapter 4

NOTES

## What is in the Physics InfoMall CDROM for Chapter 4?

Although this chapter is called *Toys for Understanding*, this chapter is not really about toys, but about motors and generators. And we will want to use the Physics InfoMall CD-ROM to gather information about these.

After entering the InfoMall, click on "Functions" from the menu bar and select "Compound Search". This will bring up a window which will allow you to specify the search you wish to do. (Caution: this information is generated on a Macintosh, and may vary slightly from what you will find on a PC. The same CD-ROM works on both platforms, and the database is the same. The interface is a little different.) Click the "Search under databases" button and select all databases, (it is sometimes faster to de-select the Keyword Kiosk, since it often provides information you may not want) then click "Apply". Enter "dc generator" in the uppermost box followed by clicking "OK". The search will take less than a minute, and will return all occurrences of the words "dc generator". The first hit on the list of search results is pretty good. You should check them out — they are on the InfoMall in their entirety. And the text is all live, meaning you can print it, copy it, and even paste it into your own handouts. This first hit even has reasonably good graphics

Because this entire chapter deals with electricity, it is a good idea to find out what problems students have with learning about electricity and electrical circuits. Perform a search using "student" AND "difficult*" AND "electric*" on the entire CD-ROM. This search initially provides "Too Many Hits" so it must be limited. One method is to limit it with Terms Must Appear in the Same Paragraph. This finds several references on the difficulties students have with concepts related to electricity. Knowing these difficulties will allow the teacher to expect them, and have ideas on how to overcome them. A similar search can be done using "misconcept*" in place of "difficult*". The results are not identical, so both searches should be tried.

Another idea for a search comes from realizing that the word "student" in the previous searches is almost redundant; we should not expect to find "difficulties" or "misconceptions" in the same paragraph with "electricity" very often unless we are talking about students. In addition, getting rid of the word "student" also gets rid of the "Too Many Hits" problem, so we can search entire articles again. If you perform the searches without "student" you will find many additional references, including a passage from the "Current Electricity" chapter of *A Guide to Introductory Physics Teaching* found in the Book Basement: "Research has been showing that the most basic concepts underpinning simple direct current (d.c.) circuits offer very serious difficulties to many students and that certain misconceptions are widely prevalent [Arons (1982); Cohen, Eylon, and Ganiel (1983); Fredette and Clement (1981)]. As in the case of static electricity, the learning problems are aggravated by the remoteness of the underlying phenomena from direct sense perception. The observable effects are not easily linked to abstractions such as "electrical charge," "current," and "energy." Since students are aware that batteries "run down" and that one "uses" household electricity, they believe that "something is used up" in electric circuits, and, to many of them, the most reasonable thing to be "used up" is "electricity" itself.

Another hit from this search is "LAB MANUAL: Recipes for Science: Misconceptions." You should check this out.

We searched above for "dc generator." What if we search now for "motor"? You might expect many, many hits. And you might expect that many of those hits are useless. Try it anyway. The very first hit is from *Household Physics* from the Textbook Trove, and is a great source of information, including a discussion of the theory of electric motors. We should not be surprised that this book shows up on our search, since this *Coordinated Science for the 21st Century* book has a chapter about the home. Look through the rest of the hits. Some even have nice graphics and homework problems you can use.

# ACTIVITY 1
## The Electricity and Magnetism Connection

# Background Information

The physics phenomenon involved in **Activity 1** is the magnetic field surrounding a long, straight wire carrying a direct current.

**Magnetic field near a current-carrying straight wire.** Before providing **Background Information** about the magnetic field near a current-carrying straight wire, two definitions are offered which apply to magnetic fields in general:

**Definition**: The direction of a magnetic field at a particular location is the direction that the north-seeking pole of a magnetic compass would point if the compass were placed at the location.

**Definition**: The strength of a magnetic field, in Tesla, at a particular location is the force, in newtons, that a long, straight wire carrying a current of one ampere would experience per meter of wire length when the wire is placed at the location so that the length of the wire is oriented perpendicular to the direction of the magnetic field.

**1 Tesla = 1 newton/(ampere)(meter) = 1N/(amp)m**

Magnetic field strength is a vector quantity since it has both direction and magnitude.

As will be indicated by the compass in **For You To Do, Activity 1**, the magnetic field near a long, straight wire carrying a direct current has a circular shape which is concentric on the wire as shown below:

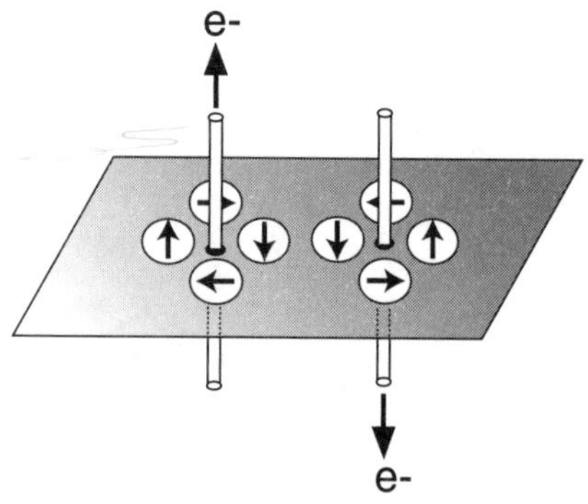


The direction of the magnetic field depends on the direction of the current flow; if the flow of electrons is reversed, so is the direction of the magnetic field around the wire.

SAFETY PRECAUTION: Some physics textbooks use the flow of positive charges (generally called conventional current) as the direction of current flow; this text uses the flow of [negative] electrons which travel around a circuit from the negative terminal of the energy source to the positive terminal.

A convenient rule for determining the direction of the magnetic field surrounding a straight wire in relation to the direction of the current in the wire is shown below as a left-hand rule:

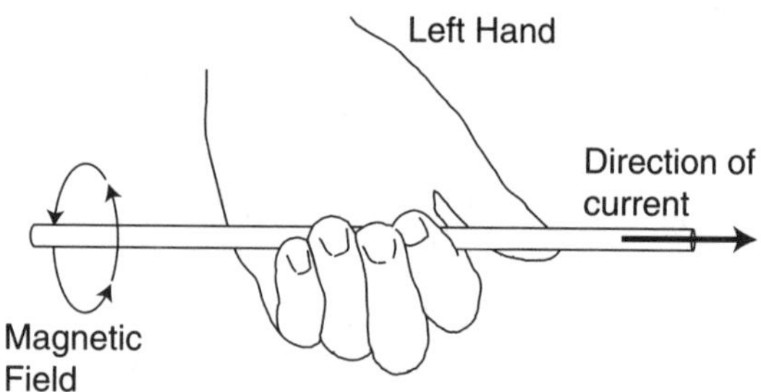


**Left-hand rule for the direction of the magnetic field surrounding a straight current-carrying wire.** Use the left hand to grasp the wire with the extended thumb pointing in the direction of the current in the wire. The fingers point in the direction in which the magnetic field surrounds the wire.

**The strength of a magnetic field near a straight current-carrying wire.** The strength of the magnetic field near a straight wire varies directly with the amount of current flowing in the wire and inversely with the distance from the wire. The equation is:

$$B = \mu_0 I/2\pi d$$

where $B$ is the field strength in Tesla, $\mu_0$ is the permeability of air or vacuum which has the value $4\pi \times 10^{-7}$ newton/(amp)$^2$, $I$ is the current in amps, and $d$ is the distance from the wire in meters.

## *Active*-ating the Physics InfoMall

One of the topics in this activity is the magnetic field. While students often have difficulties learning about fields, this is one area that was not studied widely in time for inclusion on the Physics InfoMall. Some studies have been done more recently, so there is some information to be found, but little of it is on the CD-ROM. Still, you may wish to try a few searches to see what information does exist in this database.

A good search for this activity (with results that we will want in later activities) is "current" AND "compass". You should limit this search to Within

Paragraph. Some of the results are great, including this passage, this time taken from the Electromagnetism chapter of *A Guide to Introductory Physics Teaching:* found in the Book Basement: "Very few students will conduct a meaningful investigation without guidance, however. The homework assignment should guide them Socratically into: (1) investigating the compass deflection both above and below the current-carrying wire; (2) investigating the effect of reversing the connection to the battery terminals; (3) ascertaining the pattern of the effect all the way around the wire—not just above and below; (4) qualitatively noting the effect of changing the distance between the wire and the compass needle; (5) qualitatively studying the strength of the effect on the compass needle (held at fixed distance from the wire) when additional bulbs are inserted in the circuit either in series or in parallel with an initial single bulb; (6) studying the effect of introducing an additional battery in series with the first; (7) forming, from synthesis of the observations, the right-hand rule mnemonic for the direction of magnetic field around the current-carrying wire".

What is this "right-hand rule" just mentioned? To find out, conduct a search on the list of hits from the previous search. To do this, simply change the Search Category to "Search Hits" and have the engine look for "hand rule" (after all, we don't want to leave out the left-handed versions). A quick look through these hits provides plenty of useful information, again including graphics. Don't forget to reset the search options before your next search!

# Planning for the Activity

## Time Requirements

If the students have to make the apparatus, allow about twenty minutes to set up the lab. It should then take about 20 - 30 minutes to complete the lab activity.

Depending on available equipment, this may be done either as a demonstration or in groups.

## Materials Needed

### For the class:

- DC power supply or automobile-type battery capable of producing high current (5 amps or more)

### For each group:

- heavy-gauge (approx. #12) insulated copper wire vertically arranged to penetrate a horizontal surface such as a piece of cardboard
- magnetic compass, 1 or more
- tangent galvanometer

## Advance Preparation and Setup

Try this activity in advance to be sure the field produced by the current deflects the compass to indicate a field concentric on the wire. Parallel currents in a bundle of wire (formed by looping one long wire) may be used to enhance the effect.

Adjust the current at each station in advance, and emphasize to students that they should not change the amount of current. Show students how to safely change the direction of the current.

# Teaching Notes

Present the **WDYT** question to the students. The student responses may be explored in class discussion and noted on the board, or individually recorded by students and a discussion at the end of the lesson.

If there are not enough stations for each group, have each group come to the power supply **(FYTD #5)**, while the other groups work on mapping the magnetic field **(FYTD #2-4)** of the bar magnet.

Again, a warning regarding the power supply is in order. The settings must not be changed. Warn them of the danger of any electric source, and that electricity is not something to take for granted. If there are any students with serious heart problems, they should probably not handle the equipment.

Students sometimes will refer to the poles of a magnet, in the same sense as the poles of an electrical charge. Whereas, the electric charges of positive and negative can be isolated, the north and south poles can never be isolated. As you break a bar magnet in half, you end up with two N-S magnets, not a magnet with a north pole at both ends and a magnet with a south pole at both ends.

# Activity Overview

In this activity, the students will be observing the effects of a current on a compass. This will lay the foundation for developing an understanding of the relationship between electricity and magnetism.

## Student Objectives

### Students will:

- Describe electric motor effect and generator effect in terms of energy transformations.
- Use a magnetic compass to map a magnetic field.
- Describe the magnetic field near a long, straight current-carrying wire.

ANSWERS FOR THE TEACHER ONLY

## What Do You Think?

Answers will vary. Some answers may include: no electricity; no TVs; video games; hair dryers; etc. Look for thoughtful answers where the students may refer to life before electricity was commonplace, and some may look at it as though they were camping or living at a cabin without power.

Without motors, some students may reflect that they would have to return to the steam engine, or even earlier, where horsepower, and muscle power, provided all the advantages that are enjoyed by people today. Some may refer to the advantages found in the kitchen (blenders, mix masters, coffee grinders, etc.).

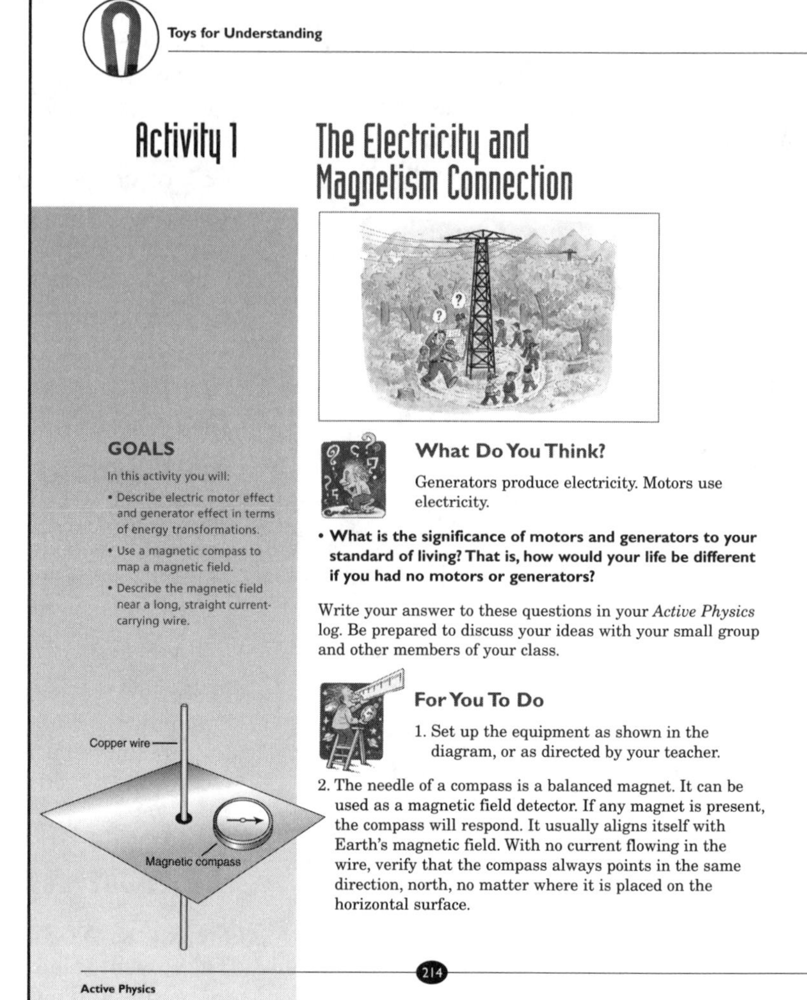

Toys for Understanding

Activity 1 The Electricity and Magnetism Connection

GOALS

In this activity you will:

- Describe electric motor effect and generator effect in terms of energy transformations.
- Use a magnetic compass to map a magnetic field.
- Describe the magnetic field near a long, straight current-carrying wire.

What Do You Think?

Generators produce electricity. Motors use electricity.

- **What is the significance of motors and generators to your standard of living? That is, how would your life be different if you had no motors or generators?**

Write your answer to these questions in your *Active Physics* log. Be prepared to discuss your ideas with your small group and other members of your class.

For You To Do

1. Set up the equipment as shown in the diagram, or as directed by your teacher.
2. The needle of a compass is a balanced magnet. It can be used as a magnetic field detector. If any magnet is present, the compass will respond. It usually aligns itself with Earth's magnetic field. With no current flowing in the wire, verify that the compass always points in the same direction, north, no matter where it is placed on the horizontal surface.



Active Physics 214

ANSWERS

## For You To Do

1. Student activity.

2. a) Student data. The compass should point in the same direction regardless of where on the horizontal surface it is placed. (Note: If some students don't see these results, it may be that they are close to a large power supply in the walls or floors of your school. This can be a teaching moment, where you can have the students try to locate the supplies.)

a) Sketch the compass direction at different places on the horizontal surface in your log.

3. Bring another type of magnet, such as a bar magnet, into the area near the compass needle.

a) Describe your observations in your log.

b) What happens to the dependable north-pointing property of a compass when the compass is placed in a region where magnetic effects, in addition to Earth's magnetic field, exist?

4. You will now make a map of the magnetic field of the bar magnet. Place the magnet on another piece of paper and trace its position. Place the compass at one location and note the direction it points. Remove the compass.

a) Put a small arrow at the location from which you removed the compass to signify the way in which it pointed.

b) Place the compass at a second location about at the tip of the first arrow. Remove the compass and place another small arrow in this location to signify the way in which the compass pointed.

c) Repeat the process at an additional 20 locations to get a map of the magnetic field of a bar magnet. Tape the piece of paper of the map in your log.

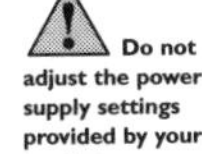

**Do not adjust the power supply settings provided by your teacher.**

5. Return the compass to the horizontal surface surrounding the wire. Observe the orientation of the compass. Send a current through the wire. The direction of the flow of electrons which make up the current in the wire is from the negative terminal of the power supply to the positive terminal. Move the compass to different locations on the horizontal surface, observing the direction in which the compass points at each location. Make observations on all sides of the wire, and at different distances from the wire.

a) Record how the compass was oriented when the bar magnet was removed.

b) Describe any pattern that you observe about how the compass behaves when it is near the current-carrying wire. Use a sketch and words to describe your observations in your log.

c) From your observations, what effect does the electric current appear to have on the wire?





Answers

## For You To Do

***(continued)***

3. a) The students should note that the compass needle will point either toward the magnet or away depending on whether the north or south pole of the magnet is pointed toward the compass.

b) At that particular spot, the magnetic field of the magnet will be larger than the magnetic field of Earth. The students can see this measurable effect by moving the magnet farther way, until it no longer has the effect of "pulling" the compass pointer.

4. a) Student data.

b) Student data.

c) The students should note a pattern like the diagram below:

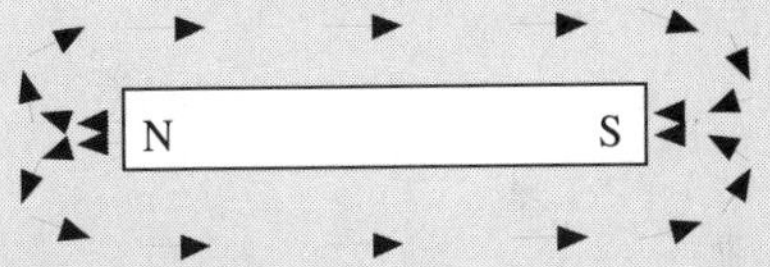


5. a) Student data.

b) Student data. Students should have a sketch showing the compass pointing in a circular path around the current-carrying wire.

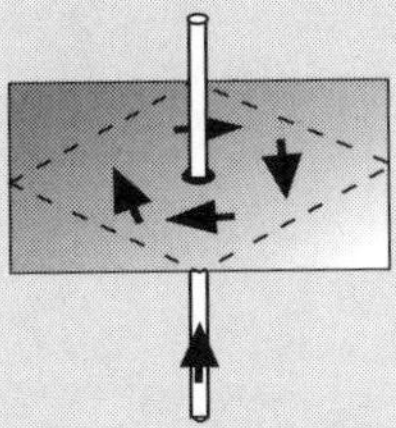

c) The electric current appears to create a magnetic field. The compass indicates the direction of the magnetic field.

ANSWERS

## For You To Do
***(continued)***

6. a) The students will note that the direction of the compass will be opposite to the previous question. Diagram will be opposite.

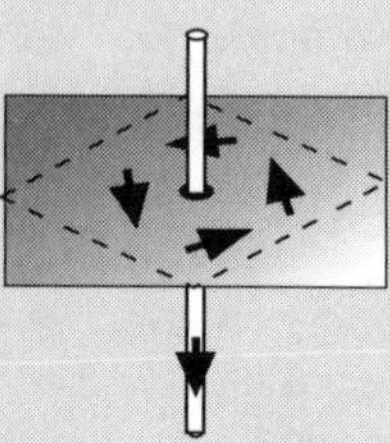

b) Students' answers will vary. Most will state that if the current is going up, the magnetic field will be directed clockwise. If the current is going down, the magnetic field is directed counter-clockwise. If they use the left-hand rule, they should say that the thumb points in the direction of the current and the fingers point in the direction of the magnetic field.

Toys for Understanding

6. Reverse the direction of the current in the wire by exchanging the contacts of the power supply. Repeat your observations.
   a) Describe the results.
   b) Make up a rule for remembering the relationship between the direction of the current in a wire and the direction of the magnetism near the wire (i.e., when the current is up, the magnetic field . . . ). Anyone told your rule should be able to use it with success. Write your rule in your log. Include a sketch. (Hint: One of the rules that physicists use makes use of your thumb and fingers.)

### Reflecting on the Activity and the Challenge

This activity has provided you with knowledge about a critical link between electricity and magnetism, which is deeply involved in your challenge to make a working electric motor or generator. The response of the compass needle to a nearby electric current showed that an electric current itself has a magnetic effect which can cause a magnet, in this case a compass needle, to experience force. You have a way to go to understand and be able to be "in control" of electric motors and generators, but you've started along the path to being in control.

### Physics To Go

1. If 100 compasses were available to be placed on the horizontal surface to surround the current-carrying wire in this activity, describe the pattern of directions in which the 100 compasses would point in each of the following situations:
   a) no current is flowing in the wire
   b) a weak current is flowing in the wire
   c) a strong current is flowing in the wire
2. If a vertical wire carrying a strong current penetrated the floor of a room, and if you were using a compass to "navigate" in the room by always walking in the direction indicated by the north-seeking pole of the compass needle, describe the "walk" you would take.

216

Active Physics

ANSWERS

## Physics To Go

1. a) All the compasses will be pointing toward the north.

   b) The students would notice that the compasses that are closest to the wire, will be pointing in a circular path surrounding the wire, in agreement with the left-hand rule. As the compasses get further and further away, they would then point toward the north.

   c) The students should notice that all of the compasses should be pointing in a circular path.

2. The path you would take would be in a circular path around the current-carrying wire. If the direction of the current is up, then the path would be clockwise. If the current is down, then the path would be counter-clockwise.

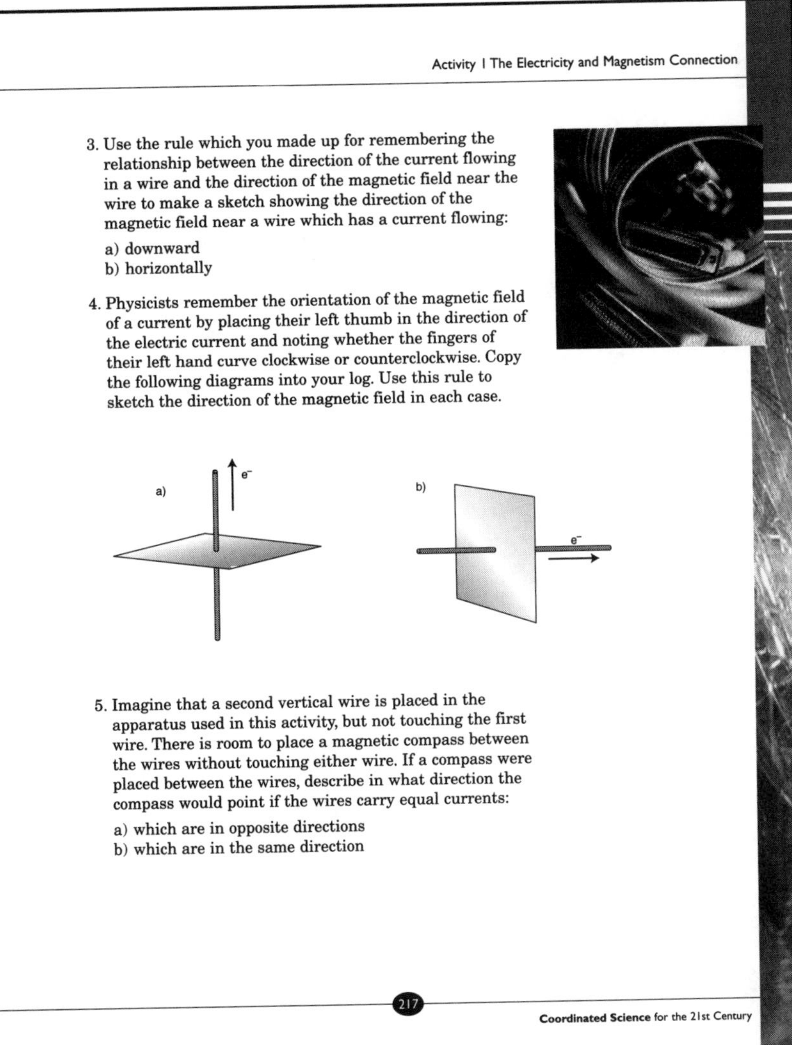



3. Use the rule which you made up for remembering the relationship between the direction of the current flowing in a wire and the direction of the magnetic field near the wire to make a sketch showing the direction of the magnetic field near a wire which has a current flowing:
   a) downward
   b) horizontally

4. Physicists remember the orientation of the magnetic field of a current by placing their left thumb in the direction of the electric current and noting whether the fingers of their left hand curve clockwise or counterclockwise. Copy the following diagrams into your log. Use this rule to sketch the direction of the magnetic field in each case.



5. Imagine that a second vertical wire is placed in the apparatus used in this activity, but not touching the first wire. There is room to place a magnetic compass between the wires without touching either wire. If a compass were placed between the wires, describe in what direction the compass would point if the wires carry equal currents:
   a) which are in opposite directions
   b) which are in the same direction



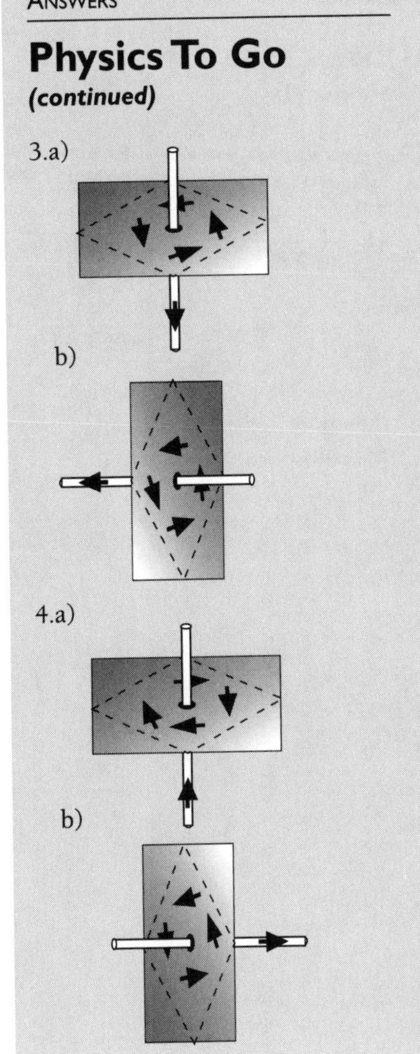

ANSWERS

## Physics To Go

***(continued)***

3.a)

b)

4.a)

b)

5.a) In this diagram, the compass will appear to go back and forth between the north and south poles.

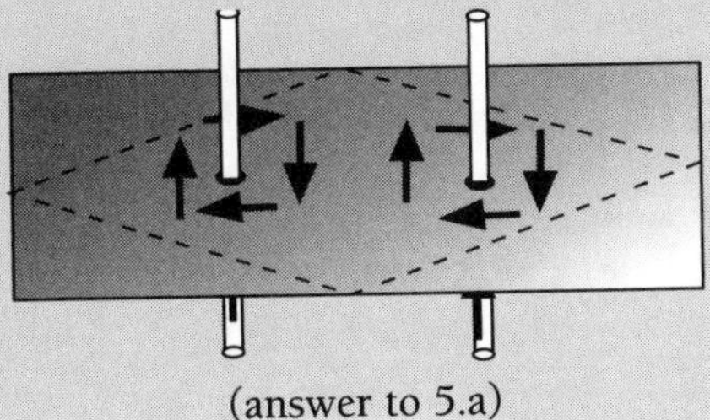

(answer to 5.a)

b) In this diagram, the compass will point in one direction (top of page).

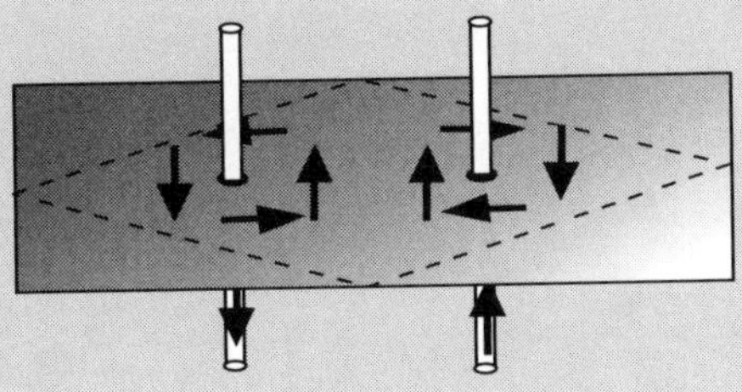

ANSWERS

## Physics To Go
***(continued)***

6. The compass will point to the left-hand side of the page. If the students don't see this, make a large loop of wire, and have them use the first left-hand rule to see the direction of the compass. They should note that wherever you put the compass inside the loop, it will point in the same direction. The compass will point in the same direction outside of the loop also. This will help them understand the next left-hand rule in the following activity.

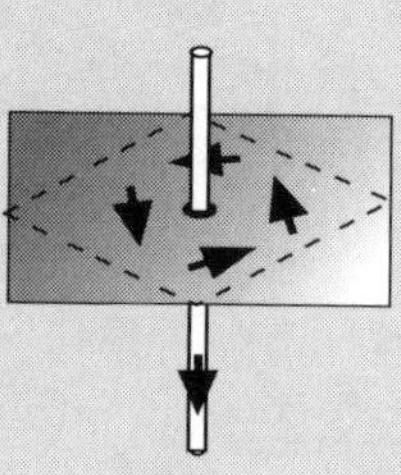

6. A hollow, transparent plastic tube is placed on a horizontal surface as shown in the diagram. A wire carrying a current is wound once around the tube to form a circular loop in the wire. In what direction would a compass placed inside the tube point? (Plastic does not affect a compass; only the current in the wire loop will affect the compass.)

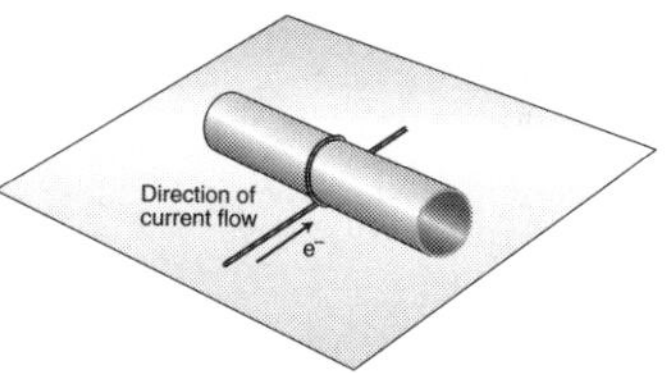

### Stretching Exercises

Use a compass to search for magnetic effects and magnetic "stuff." As you know, a compass needle usually aligns in a north-south direction (or nearly so, depending on where you live). If a compass needle does not align north-south, a magnetic effect in addition to that of the Earth is the cause, and the needle is responding to both the Earth's magnetism and some other source of magnetism. Use a compass as a probe for magnetic effects. Try to find magnetic effects in a variety of places and near a variety of things where you suspect magnetism may be present. Try inside a car, bus, or subway. The structural steel in some buildings is magnetized and may cause a compass to give a "wrong" reading. Try near the speaker of a radio, stereo, or TV. Try near electric motors, both operating and not operating.

Do not bring a known strong magnet close to a compass, because the magnet may change the magnetic alignment of the compass needle, destroying the effectiveness of the compass.

Make a list of the magnetic objects and effects that you find in your search.

ANSWERS

## Stretching Exercises

Students will enjoy this activity, because they will find that there are many different sources of magnetism, even within their own home. Most electrical appliances will cause the compass to move, but may not show an accurate reading due to AC current.

NOTES

# ACTIVITY 2
## Electromagnets

# Background Information

The physics phenomenon involved in **Activity 2** is the magnetic field of a solenoid.

A wire wound into the shape of a solenoid as shown in **For You To Do, Activity 2,** and carrying a direct current behaves similar to a bar, magnet with the ends having north and south poles. Considering short segments of the wire of the solenoid as straight pieces of wire, the direction of the magnetic field of each segment can be found using the left-hand rule introduced in the teacher **Background Information** for **Activity 1**. The combined effect of the magnetic fields of segments summed over the entire solenoid results in a magnetic field for the solenoid as shown below:

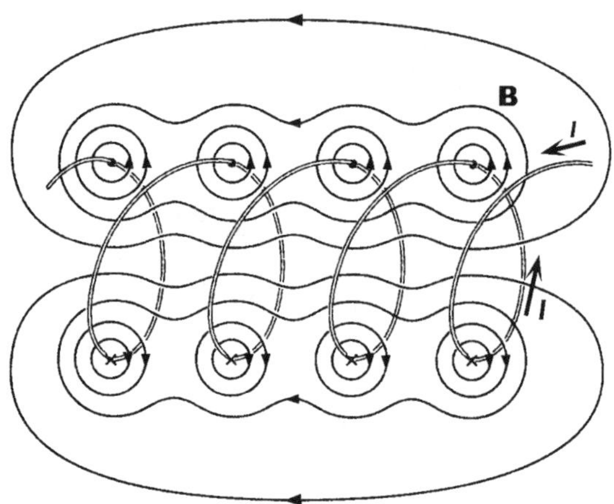


**Left-hand rule for the polarity of a solenoid.** A convenient rule for determining the magnetic polarity of a solenoid in relation to the direction of the current in the solenoid is to use the left hand to grasp the solenoid with the fingers pointing in the direction in which the current circles the windings. The extended left thumb points to the north pole of the solenoid.

**The strength of the magnetic field near the center of a solenoid.** The strength near the center (inside) a solenoid varies directly with the number of turns of wire, directly with the current flowing in the wire, and inversely with the length of the solenoid. The equation is:

$$B = \mu_0 NI/L$$

where $B$ is the magnetic field strength in Tesla, $\mu_0$ is the permeability of air or vacuum (if the core is air or vacuum) in newtons/amp$^2$, $N$ is the number of windings, $I$ is the current in amps, and $L$ is the length of the solenoid in meters.

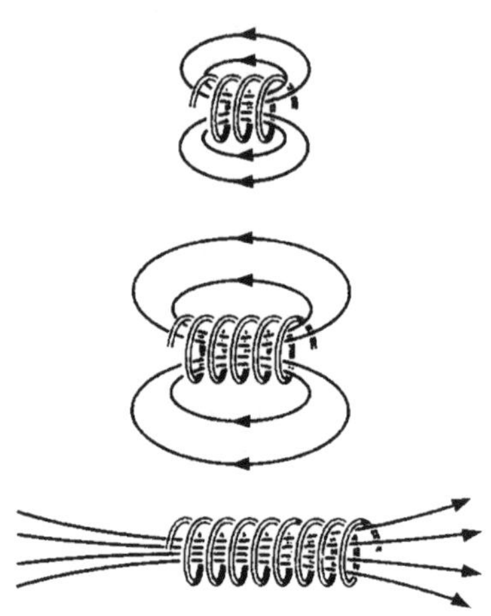

**The effect of core material on the strength of a solenoid.** If a ferromagnetic material such as iron (viz., a material having high tendency to become magnetized) is used as the core of a solenoid the magnetism of the core greatly adds to the strength of the magnetism. In such a case the constant $\mu_0$ in the above equation is replaced by the magnetic permeability, $\mu$, of the core material which commonly increases the magnetic effect by a factor of thousands.

Iron can be treated so that when used as a core material for a solenoid its magnetism "turns on and off" in concert with the electric current in the solenoid.

Notice in the diagram of the solenoid's field that inside the core of the solenoid the field lines are close together and point toward the right; outside the solenoid the field lines are farther apart and point toward the left. If a current-carrying wire were used as a probe to sample the magnetic field strength it would show that the field is stronger inside the solenoid than outside. This allows development of an alternate model for expressing the strength of a magnetic field in terms of the number of field lines which penetrate a unit of area oriented perpendicular to the direction of the field; the greater the "density" of lines penetrating a unit of area, the greater the magnetic field strength.

In the teacher **Background Information** for **Activity 1** the strength of a magnetic field was defined in terms of the force on a current-carrying wire oriented perpendicular to the direction of the field:

**1 Tesla = 1 newton/ampere-meter**

The magnetic field line model provides an alternate definition of field strength:

**1 Tesla = 1 Weber per square meter**

**= 1 Wb/m$^2$**

where the Weber (Wb) represents the number of magnetic field lines (and is called the "magnetic flux") and m$^2$ is the area penetrated by the flux [in more advanced treatments, the area is expressed as a vector; this will not be done here].

Symbolically, the alternate definition is expressed as:

$B = \phi/A$

where $B$ is the magnetic field strength in Tesla, $\phi$ is the magnetic flux (number of field lines) in Weber, and $A$ is the area in m$^2$. Therefore, the strength of a magnetic field may be expressed in two equivalent ways:

**1 Tesla = 1 newton/ampere-meter**
**= 1 Weber/square meter**

(Better to visualize magnetic fields penetrating small areas, physicists sometime prefer to express magnetic field strength in yet another unit, the "gauss" 1 Weber/m$^2$ = 10,000 gauss.)

## *Active*-ating the Physics InfoMall

A search for "electromagnet" will not be disappointing for this activity. Try it for yourself.

In **For You To Do, Step 3,** students are asked to make a prediction. The importance of the prediction should not be overlooked; indeed, predictions force students to examine their understanding of a phenomena and actively engage thought. If you were to search the InfoMall to find more about the importance of predictions in learning, you would find that you need to limit your search. Sadly, not much information exists on the InfoMall regarding predictions about electromagnets. A search for "prediction*" AND "inertia". Try it. The prediction itself is important. Too often when we observe something new, we think "that's what I thought would happen" and learn nothing. By consciously recognizing our preconceptions, we have a chance to change the misconceptions.

The **Stretching Exercise** asks about magnetic levitation. A search for "magnetic levitation" produces a wonderful hit from "Large-scale applications of superconductivity," *Physics Today*, vol. 30, issue 7, 1977.

# Planning for the Activity

## Time Requirements

The class should be conducted in 40 - 50 minutes. If there is time, you may have the students do several different solenoids, to get enough data to see the relationship to the number of windings, and the relative strength of the electromagnet. Allow another class period to do the **Inquiry Investigation.**

## Materials Needed

Prepared solenoids with taps for varying the number of turns and a variety of core materials are available from science suppliers and may be used; however, greater impact may result from students winding their own solenoids. For the "homemade" version shown, you will need:

### For each group:

- #24 magnet wire (60 ft)
- drinking straw (10)
- nail, 16 d
- paper clips (10)
- hand generator (or power supply with a variable current)
- sandpaper for removing insulation from the wire ends (6)
- magnetic compass

## Advance Preparation and Setup

Demonstrate the magnetic field of a bar magnet and a solenoid, noting similarities. Apparatus for demonstrating the fields on an overhead projector are available from several science supply houses, and employ iron filings, or transparent compasses to visualize the fields.

# Teaching Notes

Begin the lesson pursuing the students' understanding of electromagnetism. As an extension of **What Do You Think?**, demonstrate the fields of a bar magnet and solenoid (see **Materials Needed**). For the solenoid, it would be helpful to use a diagram to show that the field can be inferred by considering short wire segments of the solenoid as straight conductors, each having a surrounding circular field which contributes to the total field of the solenoid (see **Background Information**). Be sure to include in the demonstration the obvious, but important, characteristic that the electromagnet can be turned on and off, while the bar magnet's magnetism can't.

Ask students which factors may be changed to affect the strength of the electromagnet or the bar magnet, and use this to establish the purpose of the **For You To Do** activity.

Here you may wish to establish, as a matter of definition, that the north pole of an electromagnet (or any magnet) is that pole which repels the north-seeking pole of a compass needle [this, of course, dictates that Earth's north magnetic pole is a south pole! – be prepared to deal with this if you bring this up, because it is likely that a student will come up with it].

Also, you may wish to share the second left-hand rule for remembering which is the north pole of a solenoid. This will be useful for students when in the next activity they wind a coil to make a galvanometer, and when later they wind armatures for motor/generators.

SAFETY PRECAUTION: Remind students they are working with electricity, and caution must be taken. Anytime there is electricity running through a resistor wire, there can be a buildup of heat if the current is on for extended periods of time. Using a generator will help because students will not want to crank it for long periods of time. However, if you are using a power supply, make sure that the students turn it off after use, and not leave it on for long periods of time, or while they are writing up the lab activity.

NOTES

NOTES

# Activity 2 Electromagnets

**GOALS**

In this activity you will:

- Describe and explain the magnetic field of a current-carrying solenoid.
- Compare the field of a solenoid to the field of a bar magnet.
- Identify the variables of an electromagnet and explain the effects of each variable.

### What Do You Think?

Large electromagnets are used to pick up cars in junkyards.

- **How does an electromagnet work?**
- **How could it be made stronger?**

Write your answer to these questions in your *Active Physics* log. Be prepared to discuss your ideas with your small group and other members of your class.

### For You To Do

1. Wind 50 turns of wire on a drinking straw to form a solenoid as shown in the diagram on the next page. Use sandpaper to carefully clean the insulation from a short section of the wire ends to allow electrical connection of the solenoid to the generator.

ANSWERS

## For You To Do

1. Student data.

## Activity Overview

Students will be building their own electromagnet, using simple materials. They will discover the variables which give an electromagnet the force necessary to pick up large objects. There may be a need for some discussion about permanent and temporary magnets.

## Student Objectives

### Students will:

- Describe and explain the magnetic field of a current-carrying solenoid.
- Compare the field of a solenoid to the field of a bar magnet.
- Identify the variables of an electromagnet and explain the effects of each variable.

ANSWERS FOR THE TEACHER ONLY

## What Do You Think?

Electromagnets are the result of a changing electrical current in a wire surrounding a soft iron core. The greater the current flowing through the wire, or the greater the number of coils around the core, the greater the magnetic force will be.

ANSWERS

## For You To Do

***(continued)***

2. a) Student data. The compass needle will orient either toward or away from the solenoid.

   b) If the compass needle points toward the solenoid, that will be the "south" pole (north-seeking pole). If the compass needle points away from the solenoid, that will be the "north" pole.

3. a) Change the wires to the different terminals of the generator.

   b) Students may also predict that reversing the direction of the generator will have the same effect.

4. a) Student data.

5. a) Students should notice that the solenoid causes the compass needle to point in the same direction as before without the nail inside the straw. There may be a visible increase in the strength of the electromagnet now.

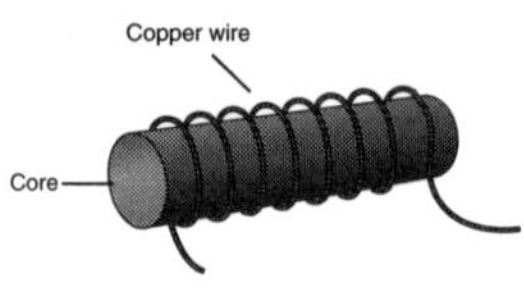


2. Carefully connect the wires from the generator to the wire ends of the solenoid. Bring one end of the solenoid near the magnetic compass and crank the generator to send a current through the solenoid. Observe any effect on the compass needle. Try several orientations of the solenoid to produce effects on the compass needle.

   a) Record your observations in your log.

   b) How can you tell the "polarity" of an electromagnet; that is, how can you tell which end of an electromagnet behaves as a north-seeking pole?

3. Predict what you can do to change the polarity of an electromagnet.

   a) Write your answer in your log.

   b) Test your prediction.

4. Use the solenoid wound on the drinking straw as an electromagnet to pick up paper clips.

   a) Record your observations in your log.

5. Carefully, slip a nail into the drinking straw to serve as a new core. Again, test the effect on a compass needle.

   a) Record your observations in your log.

6. Use the solenoid wound on the nail to pick up paper clips.

   a) Record your observations in your log.

   b) What evidence did you find that the choice of core material for an electromagnet makes a difference?

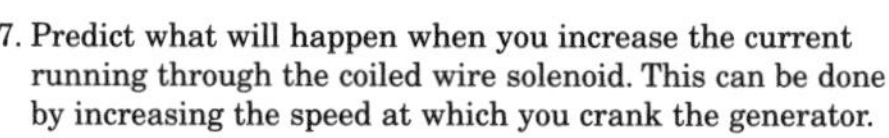

7. Predict what will happen when you increase the current running through the coiled wire solenoid. This can be done by increasing the speed at which you crank the generator.

   a) Write your answer in your log.

   b) Test your prediction by measuring how many paper clips can be picked up.

8. Predict what will happen when you increase the number of turns of wire forming the solenoid.

   a) Write your answer in your log.

   b) Test your prediction by measuring how many paper clips can be picked up.

6. a) Student data.

   b) There is an increase in the magnetic field, due to the iron core. The evidence is that the number of paper clips will be larger than before.

7. a) Student data.

   b) As the students increase the speed of the generator, there will be an increase in the number of paper clips picked up. If you have a variable current power source, you may be able to show the effect is linear by graphing the number of paper clips picked up versus the current in the power source. (Depending on your power source, the effect may be hard to see. Try this before the students attempt it.)

8. a) Student data.

   b) Again, there should be a noticeable increase in the number of paper clips picked up as you increase the number of turns. Continue to use the nail as the core, and increase the number of turns by at least 20 turns. Again, the results may be graphed.

#### Reflecting on the Activity and the Challenge

An electromagnet, often constructed in the shape of a solenoid, and having an iron core, is the basic moving part of many electric motors. In this activity you learned how the amount of current and the number of turns of wire affect the strength of an electromagnet. You will be able to apply this knowledge to affect the speed and strength with which an electric motor of your own design rotates.

#### Physics To Go

1. Explain the differences between permanent magnets and electromagnets.
2. The diagram below shows an electromagnet with a compass at each end. Copy the diagram and indicate the direction in which the compass needles will be pointing when a current is generated.

3. Which of the following will pick up more paper clips when an electric current is sent through the wire:
   a) A coil of wire with 20 turns, or a coil of wire with 50 turns?
   b) Wire wound around a cardboard core, or wire wound around a steel core?
4. Explain conditions necessary for two electromagnets to attract or repel one another, as do permanent magnets when they are brought near one another.
5. Explain what you think would happen if, when making an electromagnet, half of the turns of wire on the core were made in one direction, and half in the opposite direction.



ANSWERS

## Physics To Go

1. Temporary magnets are ferromagnetic material, which have been treated in order to turn on and off with the current. An electromagnet is also considered a temporary magnet, but as will be shown in this activity, is much stronger when there is a soft iron core. Permanent magnets are ferromagnetic material (iron, nickel, cobalt, and gadolinium) which retain their magnetic properties even when the current is removed.
2. The right side of the picture will be the north pole (compass needle will be pointing away), and the south pole (the compass needle will be pointing toward the solenoid) will be on the left side of the picture.
3. a) 50 turns.

   b) Steel core.
4. In order for the electromagnets to repel each other, the wire would have to be wound opposite to each other:

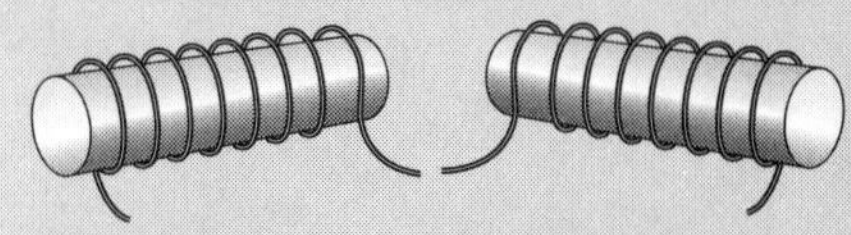

5. The currents would be in opposite directions. This would create magnetic fields in opposite directions. The strength of the electromagnet would be close to zero.

Toys for Understanding

### Stretching Exercises

1. Find out how both permanent magnets and electromagnets are used. Do some library research to learn how electromagnets are used to lift steel in junkyards, make buzzers, or serve as part of electrical switching devices called "relays." For other possibilities, find out how magnetism is used in microphones and speakers within sound systems, or how "super-strong" permanent magnets made possible the small, high-quality, headset speakers for today's portable radio, tape and CD players. Prepare a brief report on your findings.

2. Do some research to find out about "magnetic levitation." "Maglev" involves using super-conducting electromagnets to levitate, or suspend objects such as subway trains in air, thereby reducing friction and the "bumpiness" of the ride.

   a) What possibilities do "maglev" trains, cars, or other transportation devices have for the future?
   b) What advantages would such devices have?
   c) What problems need to be solved? Prepare a brief report on your research.

3. Identify as many variables as you can that you think will affect the behavior of an electromagnet, and design an experiment to test the effect of each variable. Identify each variable, and describe what you would do to test its effects. After your teacher approves your procedures, do the experiments. Report your findings.

222

ANSWERS

## Stretching Exercises

1. Student report.

2. a) Students answers will vary. Some students may report that it will reduce the amount of pollution in the atmosphere. Remind the students that the electricity to run the trains must come from some source (usually coal-fired, gas-fired or nuclear power). It might be possible to reduce the number of car and train accidents, the trains may run on time, without worrying about trains having to stop for other trains.

   b) Other advantages might be that cars would all move at the same speed, reducing the need for police; reduced need for fossil fuels; reduced noise pollution in cities; etc.

   c) The technology has not moved very quickly, and there are some problems with getting the trains to run consistently. Also, there is a great quantity of energy needed to run the trains.

# ACTIVITY 3
## Detect and Induce Currents

# Background Information

The physics phenomenon involved in **Activity 3** is electromagnetic induction.

As discussed for generators in the student text for **Chapter 4, Activity 4,** which it is suggested you read before proceeding here, an electromagnetic force, or voltage, is produced in a conductor when relative motion between the conductor and a magnetic field happens in a way that causes the conductor to "cut" magnetic flux (or, in other words, cut magnetic field lines) as shown below.

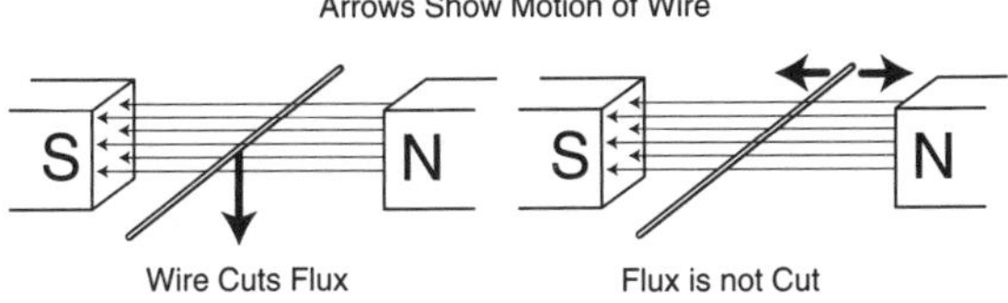


In the above example, thrusting the wire downward causes the field lines in the gap between the magnet poles to be cut by the conductor. This causes electrons in the wire to surge to the far end of the wire, giving that end of the wire an excess negative charge. The near end of the wire then has a deficiency of electrons and is positively charged. The voltage developed across the two ends of the wire is:

$$V = -\Delta\phi/\Delta t$$

where $V$ is the voltage induced, or caused to arise, between the ends of the wire, and $\Delta\phi/\Delta t$ is the number of field lines cut by the wire per unit of time. The negative sign indicates that the current which would flow if the wire were to be part of a complete circuit would produce a magnetic field, and a force, which would oppose the original thrusting action which caused the induced current.

The induced voltage, and current if the wire is part of a circuit, happens only while lines are being cut by the conductor, and the amount of induced voltage at any instant depends directly on the rate at which the magnetic field lines are being changed (cut) at that instant. Also, the direction of the induced voltage depends on the direction of the cutting action. If in the above diagram the wire had been thrust upward, the electrons would have moved in the opposite direction. Finally, if a bundle of individual wires were used above instead of a single wire, the effect of induction would be multiplied by the number of wires.

There are many ways to stage situations where a conductor cuts magnetic flux to "induce" or "generate" electricity. One way is to rotate a loop, or coil, of wire in a magnetic field, as in a generator. Another is used by students in this activity: plunging a bar magnet in and out of a solenoid. In both cases, the principle involved is the one described above; only the geometric configuration of the conductors and fields are different.

## *Active*-ating the Physics InfoMall

Our search for "electromagnet" in **Activity 2** also produced hits useful for this activity, including information on galvanometers. However, you should perform additional searches for "galvanometer," too.

This activity also provides a great chance to search the calendar cart for information on Hans Christian Oersted. Of course, you should also search the entire InfoMall for information; what were the circumstances of his discovery?

For the **Stretching Exercise**, you can do a search of the InfoMall, or you can recall that the Textbook Trove's *Household Physics* was useful for much of this book.

# Planning for the Activity

## Time Requirements

Allow at least one class period (40 – 50 minutes). If time allows, students may wish to change the number of variables in order to see different results (number of windings on the solenoid, on the galvanometer, etc.).

## Materials Needed

### For each group:

- #24 magnet wire
- alligator clip leads
- bulb base for miniature screwbase
- hand generator (or power supply with a variable current)
- miniature bulb, light
- sandpaper, fine square (6)
- tape, masking (3/4" × 60 yds)
- tube, cardboard (1 × 4)
- magnet, bar large

# Teaching Notes

Have students complete **What Do You Think?** In discussing **What Do You Think?**, mention that science, contrary to common notion, does not always proceed in deliberate ways. Sometimes accidental (serendipitous) discoveries provide breakthroughs, and Oersted's discovery involves one such occasion in the history of science.

SAFETY PRECAUTION: Regarding heating of the solenoid and the galvanometer, as in the previous activity.

# Activity 3 Detect and Induce Currents

### What Do You Think?

In 1820, the Danish physicist Hans Christian Oersted placed a long, straight, horizontal wire on top of a magnetic compass. Both the compass and the wire were resting on a horizontal surface, and both the length of the wire and the compass needle were oriented north-south. Next, Oersted sent a current through the wire, and happened upon one of the greatest discoveries in physics.

- **What do you think Oersted saw?**

**Write** your answer to this question in your *Active Physics* log. **Be** prepared to discuss your ideas with your small **group** and other members of your class.

**GOALS**

In this activity you will:

- Explain how a simple galvanometer works.
- Induce current using a magnet and coil.
- Describe alternating current.
- Recognize the relativity of motion.

### For You To Do

1. Wrap 10 turns of wire to form a coil that surrounds a magnetic compass. Wrap the wire on a diameter corresponding to the north-south markings of the compass scale, as shown in the diagram. Hold the turns of wire in place with tape, or use the method recommended by your teacher. Use sandpaper to carefully remove the insulation from a short section of the wire ends to allow electrical connection.

2. In **Step 1**, you constructed a galvanometer, a device to detect and measure small currents. Carefully connect a hand generator, a light bulb, and the galvanometer, as shown on the next page (in a series circuit). Rest the galvanometer so that the compass is horizontal, with the needle balanced, pointing north, and free to rotate. Also, turn the galvanometer, if necessary, so that the compass needle is aligned parallel to the turns of wire which pass over the top of the compass.



ANSWERS

## For You To Do

1. Student activity.
2. Student activity.

# Activity Overview

The students will build a galvanometer, test to see that it works, then show that moving a magnet through a solenoid, will produce a current. Any moving current will produce a magnetic field, and the magnetic field produced by the generator will cause the compass needle to move (depending on the direction of the current). You may want to remind the students of the left-hand rules to determine in which direction the field is, and in which direction the current is moving.

## Student Objectives

### Students will:

- Explain how a simple galvanometer works.
- Induce current using a magnet and coil.
- Describe alternating current.
- Recognize the relativity of motion.

ANSWERS FOR THE TEACHER ONLY

## What Do You Think?

Students' answers will vary. Oersted saw that when there was a current flowing through the wire, the compass needle oriented itself in a particular direction.

Chapter 4

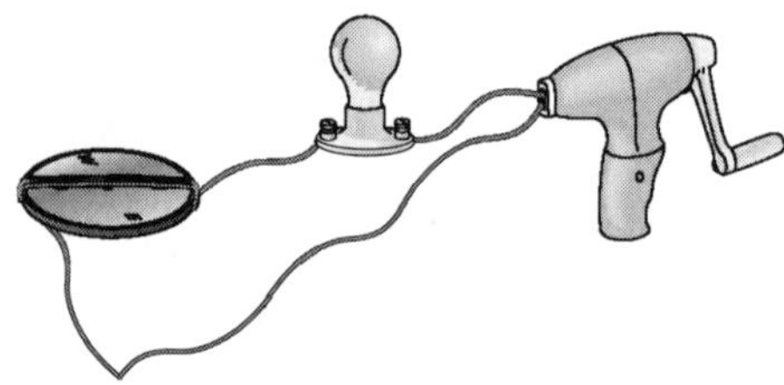

3. Crank the generator to establish a current in the circuit. Think of the compass needle as a meter such as the one in the speedometer of a car. The amount it moves corresponds to the amount of current. The glow of the light bulb verifies that current is flowing.
   a) Does the compass-needle galvanometer also indicate that current is flowing? How? In your log, use words and a sketch to indicate your answer.

4. The amount of current flowing in the circuit can be varied by changing the speed at which the generator is cranked, and the amount of current is indicated by the brightness of the light bulb. Vary the speed at which you crank the generator, and observe the galvanometer.
   a) How does the galvanometer indicate changes in the amount of current? Use words and sketches to indicate your answer.

5. Change the direction in which you crank the generator.
   a) What evidence does the galvanometer provide that changing the direction in which the generator is cranked has the effect of changing the direction of current flow in the circuit? Use words and sketches to give your answer.

6. Carefully connect each wire end of a galvanometer to a wire end of a solenoid wound on a hollow core of non-magnetic



Answers

## For You To Do *(continued)*

3. a) Due to the magnetic field created by the moving current, the needle of the compass moves. The more it moves, the greater the current.

4. a) The galvanometer indicates changes in the current by deflecting more away from equilibrium, when there is no current flowing.

5. a) The compass needle should be deflected in the other direction.

6. Student activity.

Activity 3 Detect and Induce Currents

material, such as a cardboard tube. Orient the galvanometer so that it is ready to detect current flow.

7. Hold a bar magnet in one hand and the solenoid steady in the other hand. Rapidly plunge one end of the bar magnet into the hollow core of the solenoid, and then stop the motion of the magnet, bringing the end of the magnet to rest inside the solenoid. Another person should hold the galvanometer in a steady position so that it will not be disturbed if the solenoid is moved. Observe the galvanometer during the sequence. You may need to practice this a few times.
   - a) Write your observations in your log.
8. Remove the magnet from the solenoid with a quick motion, and observe the galvanometer during the action.
   - a) Record your observations.
   - b) A current is produced! How does the direction of the current caused, or induced, when the end of the magnet is entering the solenoid, compare to the direction of the current when the magnet is leaving the solenoid?
   - c) How can you detect the direction of the current in each case?

ANSWERS

## For You To Do *(continued)*

7. a) Students should notice a deflection of the needle.

8. a) Again, the students should notice a deflection of the needle, only in the opposite direction.

   b) When going in, the current is one direction and when coming out, the current is going in the other direction.

   c) The direction of the current is observed by the deflection of the compass needle, first in one direction and then in the other.

Chapter 4

ANSWERS

## For You To Do

***(continued)***

9\. a) Students will see that the current will still be induced, but in the opposite direction of the first case.

b) As you increase the speed of the movement, the current will be increased, and the compass needle will show a greater deflection.

c) The magnet must be moving to induce a current.

d) There will be no difference, as it is the relative movement that causes the induction, not the movement of one or the other.

e) This will in effect create a greater relative movement, and therefore have the effect of increasing the induced current.

9\. Modify and repeat **Steps 7** and **8** to answer the following questions:

a) What, if anything, about the created or induced current changes if the opposite end of the bar magnet is plunged in and out of the solenoid?

b) How does the induced current change if the speed at which the magnet is moved in and out of the solenoid is changed?

c) What is the amount of induced current when the magnet is not moving (stopped)?

d) What is the effect on the induced current of holding the magnet stationary and moving the solenoid back and forth over either end of the magnet?

e) What is the effect of moving both the magnet and the solenoid?

### Reflecting on the Activity and the Challenge

In this activity you discovered that you can produce electricity. A current is created or induced when a magnet is moved in and out of a solenoid. The current flows back and forth, changing direction with each reversal of the motion of the magnet. Such a current is called an alternating current, and you may recognize that name as the kind of current that flows in household circuits. It is frequently referred to by its abbreviated form, "AC." It is the type of current that is used to run electric motors in home appliances. Part of your **Chapter Challenge** is to explain to the children how a motor operates in terms of basic principles of physics or to show how electricity can be produced from an external energy source. This activity will help you with that part of the challenge.

### Physics To Go

1\. An electric motor takes electricity and converts it into movement. The movement can be a fan, a washing machine, or a CD player. The galvanometer may be thought of as a crude electric motor. Discuss that statement, using forms of energy as part of your discussion.

ANSWERS

## Physics To Go

1\. The electric current creates a magnetic field, which induces a current in a coil of wires. This coil of wire (galvanometer) creates a magnetic field, which causes the needle to move (like poles repel, unlike poles attract). In terms of energy, the magnetic field causes a force in the wire, which moves the electrons (electrical kinetic energy). These moving electrons cause a force to act on the magnet (compass needle), thereby converting electrical energy to the kinetic energy of the needle.

2. Explain how the galvanometer works to detect the amount and direction of an electric current.
3. How could the galvanometer be made more sensitive, so that it could detect very weak currents?
4. An electric generator takes motion and turns it into electricity. The electricity can then be used for many purposes. The solenoid and the bar magnet, as used in this activity, could be thought of as a crude electric generator. Explain the truth of that statement, referring to specific forms of energy in your explanation.
5. If the activity were to be repeated so that you would be able to see only the galvanometer and not the solenoid, the magnet, and the person moving the equipment, would you be able to tell from only the response of the galvanometer what was being moved, the magnet or the solenoid or both? Explain your answer.
6. Part of the **Chapter Challenge** is to explain how the motor and generator toy works.
   a) Write a paragraph explaining how a motor works.
   b) Write a paragraph explaining how a generator works.
7. In generating electricity in this activity, you moved the magnet or the coil. How can you use each of the following resources to move the magnet?
   a) wind
   b) water
   c) steam

### Stretching Exercise

Find out about the 120-V (volt) AC used in home circuits. If household current alternates, at what rate does it surge back and forth? Write down any information about AC that you can find and bring it to class.

ANSWERS

## Stretching Exercise

AC (alternating current) is named as such due to the "back and forth" motion of the electrons in the wire. Most household circuits operate on 60 Hz (60 cycles per second). European homes operate on 50 Hz, but with a 240-V circuit. If you try to use an appliance (i.e., hair dryer, shaver, radio, etc.) designed for Europe in North America, you can cause serious damage to the appliance, and possibly to yourselves or the circuits in your home.

ANSWERS

## Physics To Go
*(continued)*

2. The galvanometer needle is deflected one way with the current moving in one direction. When the current is moving in the opposite direction the galvanometer will be deflected the other way.
3. The greater number of coils around the galvanometer will cause a greater deflection with the same current, so increasing the number of coils, will give a greater deflection for a smaller current.
4. As you move your hand in and out of the solenoid (kinetic energy) electrical energy is produced in the wire of the solenoid.
5. It is only the relative movement of the magnet and the solenoid that creates the electric current. You would not be able to tell which was being moved.
6. a) Answers will vary. Look for the changing electrical current to produce a force on the magnet inside the coils of wire (solenoid) which cause a rotor or axle to turn.
   b) Answers will vary. As you move a magnet through a coil of wire, an electrical current is produced.
7. a) Wind is the kinetic energy required to move the blades of the generator. This will cause the magnet to move through the coils of wire. This, then, generates the electrical current.
   b) The gravitational potential energy or the kinetic energy of moving water will turn the axle of the (turbine) generator, which moves a magnet through the coils of wires. Again, this produces an electrical current.
   c) Using steam drives a turbine which creates the electrical current as the magnet is turned through the coils of the solenoid.

## Activity 3 A: Demonstration

# Twin Coil Swings

### FOR YOU TO DO

The following is a demonstration to illustrate the change from mechanical energy to electrical energy. This will be the demonstration: (You may ask some students to design this and report this to the class as an alternative project. Alternately, you may have this set up as a curiosity as the students enter the classroom. Then ask for volunteers to come up and see what they can do to make one side move, by touching only the other side.)

Demonstrate that the twin coil swings, moving one to cause movement in the other due to induced current. Ask students to identify which is acting as a generator, and which as a motor. Switch to move the opposite coil – reversible?

Ask for, and try, changes in the system that will change the swings from moving the "same way" to "opposite ways". Try moving one of the magnets instead of a swing to start up the system; can you "pump" the swing if you hit the proper rhythm (the natural period of the swings)?

Insert a galvanometer to detect the induced current. What is the "E & M connection"? Expect that students will regard this system as a curiosity and want to know what makes it work. Do not offer a full explanation, but suggest that soon students will have a basis for understanding its principles of operation. If possible, leave the apparatus set up in the classroom so that students can return to it to develop explanations of it as their knowledge increases. For now, use the device to establish the meanings of the generator effect (mechanical to electrical energy) and the motor effect (electrical energy to mechanical energy).

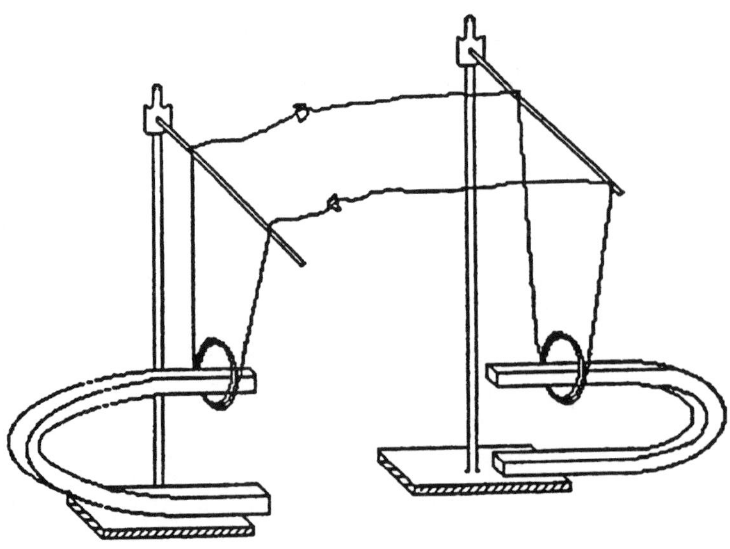

For use with *Toys For Understanding*, Chapter 4, Activity 3: Detect and Induce Currents

NOTES

# ACTIVITY 4
## AC & DC Currents

# Background Information

The principles of electromagnetic induction introduced in **Activity 3** are applied to AC and DC generators in **Activity 4**. It is suggested that you read the **Background Information** for **Activity 3** if you have not already done so.

It is also suggested that you read the detailed treatment of AC and DC generators presented in the **For You To Read** section of **Activity 4** before conducting the activity with students.

**The hertz (Hz) as a unit of frequency.** The hertz is introduced in the **Stretching Exercise** of **Activity 4** as the unit frequency; here, it is applied to AC electricity. A common misconception about this unit of measurement is shown when a frequency of, for example, 60 Hz is expressed as 60 cycles per second, or 60 vibrations per second, or 60 [fill in the noun] per second. While it is true that terms such as "cycles", "vibrations" or other descriptive nouns may enhance communication, it is essential to recognize that, by definition:

**1 hertz = (second)-1 = 1/second**

Therefore, the mathematically appropriate way to express frequency in equations is, for example, 60 Hz = 60/s; descriptive words such as "cycles" or "vibrations" are not included because they are not included in the formal definition of the hertz as a unit of measurement. Carrying such descriptive terms in the numerator of expressions of frequency within calculations lead to trouble with dimensional analysis of units because the terms do not cancel; within calculations involving equations it is best to express Hz as reciprocal seconds.

Electric power plants in the United States are required to maintain 60 Hz as the precise frequency of AC voltage distributed on the power grid. The reason for maintaining a dependable frequency is that many devices such as clocks and motors are designed to operate in synchronization with the frequency, or a multiple thereof. It should also be recognized that one complete AC cycle of 1/60 second duration contains two pulses of current, one in each direction. Therefore, the "pulse frequency" of 60 Hz electricity is 120 Hz; sometimes it can be heard being emitted from electrical devices as a "hum" corresponding to a pitch of 120 Hz.

## *Active*-ating the Physics InfoMall

We already did a search earlier for "DC generator" so now we should add to this a search for "AC generator." Again, the results are pleasing. Combine this search with the DC search, and much of this activity is covered. You can also find graphs of the generator outputs in these searches.

**For You To Do Step 4** begins working with graphs. Graphs are a tool that can be difficult for some students. Search for "difficult*" AND "graph*" in the Same Paragraph. The first hit is "Student difficulties in connecting graphs and physics: Example from kinematics," from the *American Journal of Physics*, vol. 55, issue 6, 1987. The second hit is "Student difficulties with graphical representations of negative values of velocity," from *The Physics Teacher*, vol. 27, issue 4, 1989. While these and many of the other hits are from kinematics, the findings related to graphs can still be enlightening. You should read these.

# Planning for the Activity

## Time Requirements

Allow about one 40 – 50 minute class period for this activity.

## Materials Needed

### For the class (demonstration):

- 1 3/4" buret clamp (use to support copper tube)
- 5' copper tube for cow magnet
- AC/DC demonstration generator
- magnet, cow

### For each group:

- galvanometer, 0-500 ma.

## Advance Preparation and Setup

It is assumed that limitations on equipment will require that this activity be performed as a teacher demonstration. Large, low voltage, demonstration AC and DC generators are available from science suppliers which have exposed parts, and many are convertible, serving four-ways (AC, DC generators and motors). Such a device, or the equivalent, is needed for this activity.

Be familiar with the progression of **For You To Do.** The students may or may not be able to fully grasp the concept of a sinusoidal curve. They can better appreciate it when they have seen the dimming of the bulb, and the changing of the galvanometer.

## Teaching Notes

Review with the students the result of passing a wire through a permanent magnet. They need to understand that only when the wire is moving perpendicular to the magnetic field lines will a current be produced. Ask them questions such as: What if the wire is moving at an angle? Therefore, only the perpendicular vector portion of the wire will produce the current. Have the students come up with the fact that the magnitude of the current will be less than when passed through the magnetic field lines at 90°. This will help them understand that the current is not simply on and off, but variable.

It may be difficult for all the class to see the galvanometer and the changes, so it is important to have the setup of the apparatus so that all will be able to see. If a physics software simulation program is available, you might allow the students some time to use the software during another class.

SAFETY PRECAUTION: Safety issues are the same as they are with any electronic equipment. Do not let the students operate the equipment without proper supervision.

The students may have to be reminded of the Law of Conservation of Energy. Energy cannot be created nor destroyed, only converted from one form to another. Some students may believe that the only energy that has to be added to the system of the generators of power plants is just enough to overcome the effects of friction of the mechanical parts. If this were the case, you would be creating energy from nothing which is impossible. The energy input must be greater than the energy output in order to account for the "loss" of energy due to friction.

# Activity Overview

This activity expands on the information presented in **Activity 3**.In the previous activities, we saw how the movement of a wire in a magnetic field produced a current in that wire. In this activity, the students will explore step by step, how an AC generator works. It involves the rotation of a coil of wire in a magnetic field to produce a continuous current that can be used for an electrical device.

# Student Objectives

## Students will:

- Describe the induced voltage and current when a coil is rotated in a magnetic field.
- Compare AC and DC generators in terms of commutators and outputs.
- Sketch sinusoidal output wave forms.

ANSWERS FOR THE TEACHER ONLY

# What Do You Think?

Many different answers will come from the students. Look for thoughtful responses. Some answers might include: heat, geothermal, mechanical, nuclear, light, wind, water, gas-fired hydroelectric, etc.

Toys for Understanding

## Activity 4 AC & DC Currents

**GOALS**

In this activity you will:

- Describe the induced voltage and current when a coil is rotated in a magnetic field.
- Compare AC and DC generators in terms of commutators and outputs.
- Sketch sinusoidal output wave forms.

**What Do You Think?**

In the last activity, you used human energy to produce motion to generate electricity.

- **What other kinds of energy can generate electricity?**

Write your answer to this question in your *Active Physics* log. Be prepared to discuss your ideas with your small group and other members of your class.

**For You To Do**

**AC Generator**

1. Your teacher will explain and demonstrate a hand-operated, alternating current (AC) generator. During the demonstration, make the observations necessary to gain the information needed to answer these questions:

a) When the AC generator is used to light a bulb, describe the brightness of the bulb when the generator is cranked slowly, and then rapidly. Write your observations in your log.

b) When the AC generator is connected to a galvanometer, describe the action of the galvanometer needle when the generator is cranked slowly, and then rapidly.

Active Physics 228

ANSWERS

# For You To Do

1. a) The brightness of the bulb appears to increase as the speed is increased.

   b) When the generator is cranked slowly, there is only a small deflection of the galvanometer needle. As the speed increases the needle's deflection increases. The needle also goes back and forth or positive and negative as the generator is cranked.

2. It is easier to understand the creation of a current if you think of a set of invisible threads to signify the magnetic field of the permanent magnets. The very thin threads fill the space and connect the north pole of one magnet with the south pole of the other magnet. If the wire of the generator is imagined to be a very thin, sharp knife, the question you must ask is whether the knife (the wire) can "cut" the threads (the magnetic field lines). If the wire moves in such a way that it can cut the field lines, then a current is generated. If the wire moves in such a way that it does not cut the field lines, then no current is generated.

a) Look at the diagrams of the magnetic fields shown. In which case, I, II, or III will a current be generated?

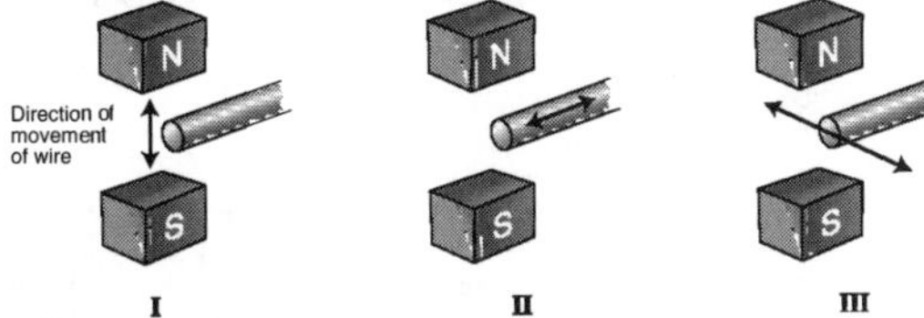


3. The following diagram shows the position of the rotating coil of an AC generator at instants separated by one-fourth of a rotation of the coil. Build a small model of the rectangular coil so that you can move the model to help you understand the drawings. The coil model can be constructed by carefully bending a coat hanger into the shape of the rectangular coil. Rest the coil between two pieces of paper—label the left paper N for the north pole of a magnet; label the right paper S for the south pole of a magnet.

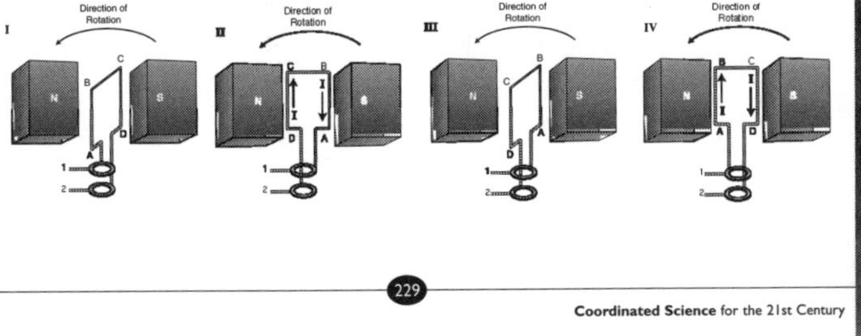




Answers

## For You To Do *(continued)*

2. a) Case III will generate a current.

3. Student activity.

4. For the purpose of analyzing the rotating coil figure, the four sides of the rectangular coil of the AC generator will be referred to as sides AB, BC, CD, and DA. Side DA is "broken" to allow extension of the coil to the rings. The "brushes," labeled 1 and 2, make sliding contact with the rings to provide a path for the induced current to travel to an external circuit (not shown) connected to the brushes. The magnetic field has a left-to-right direction (from the north pole to the south pole) in the space between the magnets in the rotating coil figure. It is assumed that the coil has a constant speed of rotation.

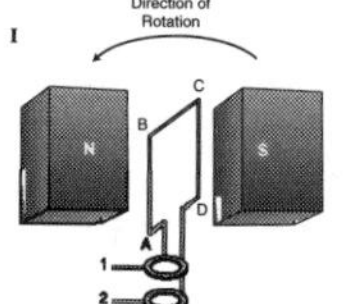


a) When the generator coil is in position I shown in the rotating coil, is a current being generated? A current is produced if the wire cuts the magnetic field lines. Record your answer and the reason for your answer in your log.

b) Use a graph similar to the one shown below. Plot a point at the origin of the graph, indicating the amount of induced current is zero at the instant corresponding to the beginning of one rotation of the coil.

c) One-fourth turn later, at the instant when the rotating coil is in position II, is a current being generated? Record your answer and the reason for your answer in your log.

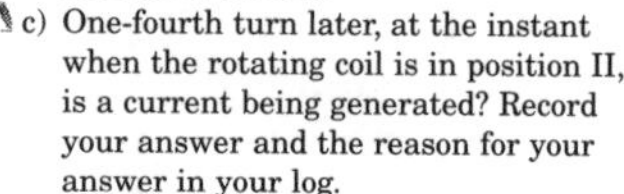

d) On your graph, plot a point directly above the $\frac{1}{4}$-turn mark at a height equal to the top of the vertical axis to represent maximum current flow in one direction.

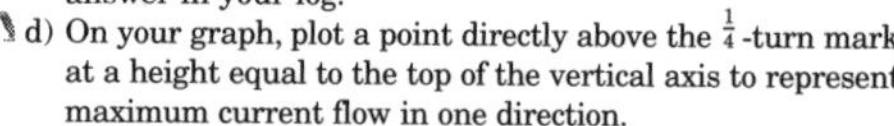

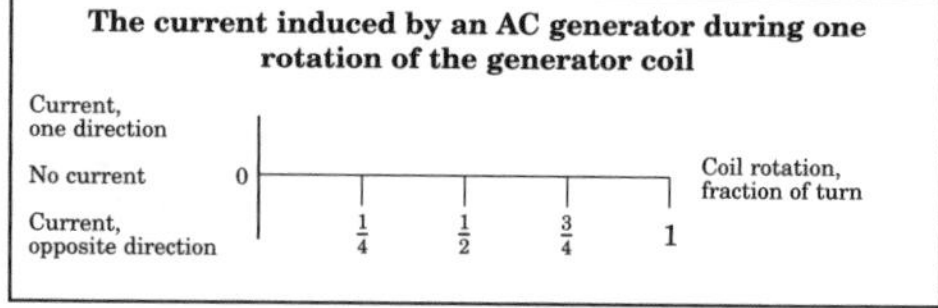




ANSWERS

## For You To Do *(continued)*

4. a) There is no current as the wire is moving parallel to the magnetic field.

b) Student activity.

c) Current is generated and should be at a maximum.

d) Student activity.

e) One-half turn into the rotation of the coil, at the instant shown in the rotating coil position III, the current again is zero because all sides of the coil are moving parallel to the magnetic field. Plot a point at the $\frac{1}{2}$ mark on the horizontal axis to show that no current is being induced at that instant.

f) At the instant at which $\frac{3}{4}$ of the rotation of the coil has been completed, shown by the rotating coil in position IV, the induced current again is maximum because coil sides AB and CD again are moving across the magnetic field at maximum rate. However, this is not exactly the same situation as shown in the rotating coil position II; it is a different situation in one important way: the direction of the induced current has reversed. Follow the directions of the arrows which represent the direction of the current flow in the coil to notice that, at this instant, the current would flow to an external circuit out of brush 2 and would return through brush 1. On your graph, plot a point below the $\frac{3}{4}$-turn mark at a distance as far below the horizontal axis as the bottom end of the vertical axis. This point will represent maximum current in the opposite, or "alternate," direction of the current shown earlier at $\frac{3}{4}$-turn.

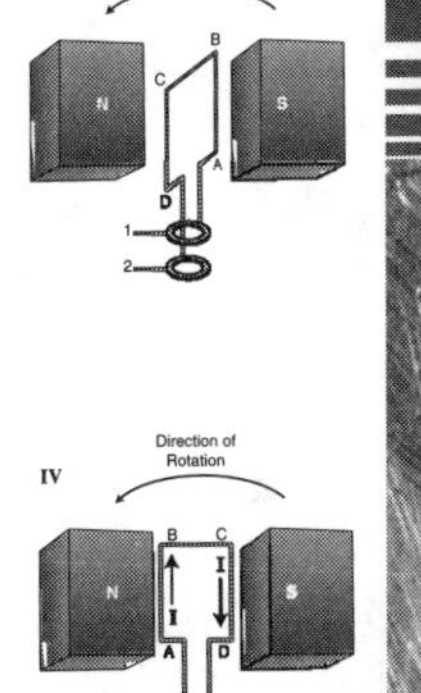


g) The rotating coil in position I is used again to show the instant at which one full rotation of the generator coil has been completed. Again, all sides of the coil are moving parallel to the magnetic field, and no current is being induced. Plot a point on the horizontal axis at the 1-turn mark to show that the current at this instant is zero.

5. You have plotted only 5 points to represent the current induced during one complete cycle of an AC generator.

a) Where would the points that would represent the amount of induced current at each instant during one complete rotation of the generator coil be plotted?

b) What is the overall shape of the graph? Should the graph be smooth, or have sharp edges? Sketch it to connect the points plotted on your graph.

c) What would the graph look like for additional rotations of the generator coil, if the same speed and resistance in the external circuit were maintained?



Answers

## For You To Do *(continued)*

e) Student activity.

f) Students will note that the current should be flowing in the opposite direction, but at a maximum.

g) Student activity.

5. a) The points should be plotted along a line that would eventually be a sinusoidal curve.

b) Students may have to be shown that the points do in fact represent a smooth curve rather than a straight line of best fit.

c) The graph will have a consistent up and down wave-like motion.

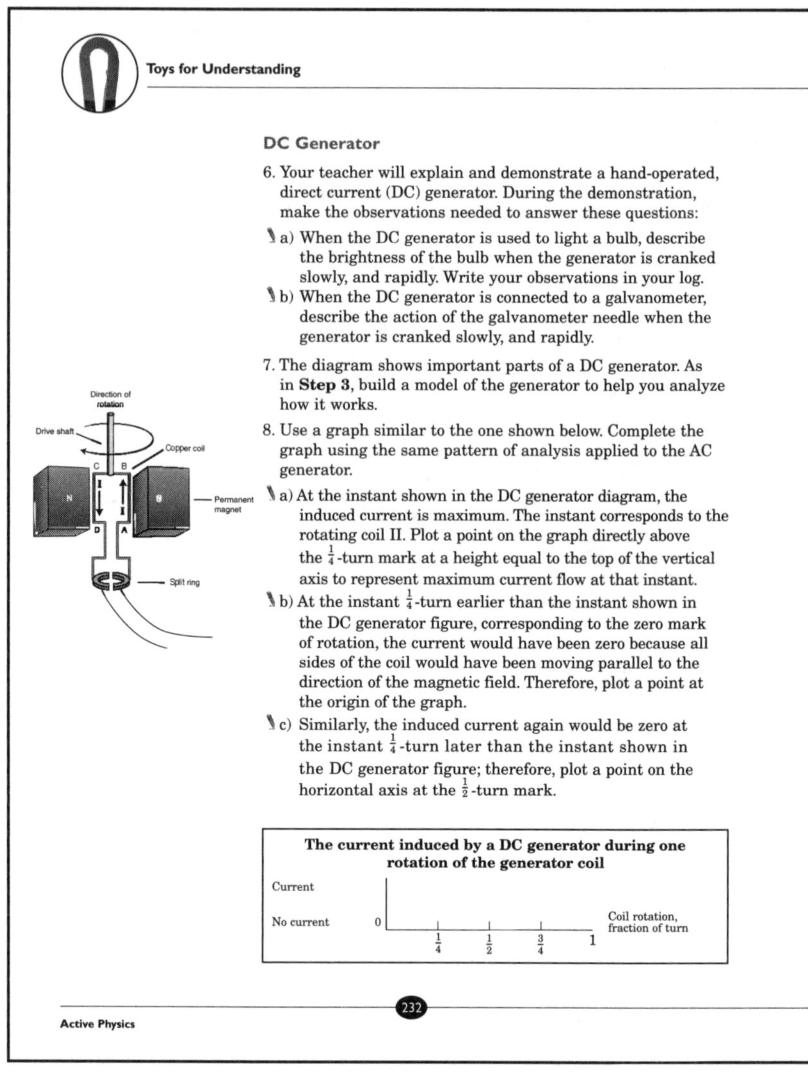



### DC Generator

6. Your teacher will explain and demonstrate a hand-operated, direct current (DC) generator. During the demonstration, make the observations needed to answer these questions:
   a) When the DC generator is used to light a bulb, describe the brightness of the bulb when the generator is cranked slowly, and rapidly. Write your observations in your log.
   b) When the DC generator is connected to a galvanometer, describe the action of the galvanometer needle when the generator is cranked slowly, and rapidly.

7. The diagram shows important parts of a DC generator. As in **Step 3**, build a model of the generator to help you analyze how it works.



8. Use a graph similar to the one shown below. Complete the graph using the same pattern of analysis applied to the AC generator.
   a) At the instant shown in the DC generator diagram, the induced current is maximum. The instant corresponds to the rotating coil II. Plot a point on the graph directly above the $\frac{1}{4}$-turn mark at a height equal to the top of the vertical axis to represent maximum current flow at that instant.
   b) At the instant $\frac{1}{4}$-turn earlier than the instant shown in the DC generator figure, corresponding to the zero mark of rotation, the current would have been zero because all sides of the coil would have been moving parallel to the direction of the magnetic field. Therefore, plot a point at the origin of the graph.
   c) Similarly, the induced current again would be zero at the instant $\frac{1}{4}$-turn later than the instant shown in the DC generator figure; therefore, plot a point on the horizontal axis at the $\frac{1}{2}$-turn mark.

**The current induced by a DC generator during one rotation of the generator coil**





ANSWERS

## For You To Do *(continued)*

6. a) Again, the students should note that the light dims and brightens regularly, with the brightness increasing as the cranking increases.

   b) The galvanometer needle will go from the equilibrium position to one side or the other. It will not go back and forth across the equilibrium as it did for the AC generator. Increasing the speed of the cranking will increase the amount of the deflection.

7. Student activity.

8. a) Student activity.

   b) Student activity.

   c) Student activity.

9. Notice the arrangement used to transfer current from the generator to the external circuit for the DC generator. It is different from the arrangement used for the AC generator. The DC generator has a "split-ring commutator" for transferring the current to the external circuit. Notice that if the coil shown in the DC generator figure were rotated $\frac{1}{4}$-turn in either direction, the "brush" ends that extend from the coil to make rubbing contact with each half of the split ring would reverse, or switch, the connection to the external circuit. Further, notice that the connection to the external circuit would be reversed at the same instant that the induced current in the coil reverses due to the change in direction in which the sides of the coil move through the magnetic field. The outcome is that while the current induced in the coil alternates, or changes direction each $\frac{1}{2}$-rotation, the current delivered to the external circuit always flows in the same direction. Current that always flows in one direction is called direct current, or DC.

   a) Plot a point on the graph at a point directly above the $\frac{3}{4}$-turn mark at the same height as the point plotted earlier for the $\frac{1}{4}$-turn mark.

   b) As done for the AC generator, find out how to connect the points plotted on this graph to represent the amount of current delivered always in the same direction to the external circuit during the entire cycle.

### Reflecting on the Activity and the Challenge

It is time to begin preparing for the **Chapter Challenge**. Now that you know how a generator works, you should begin to think about toys that might generate electricity. You should also think about how you could assemble "junk" into a toy generator, or do some research on homemade generators and motors.



ANSWERS

## For You To Do *(continued)*

9. a) Student activity.

   b) Again, the students should notice that the points should give a smooth curved line. However, the shape of the graph will differ in that the points will only be on the positive side of equilibrium in the shape of a letter "m". The pattern will appear to be a series of bumps rather than a sinusoidal curve.

Chapter 4

ANSWERS

## Physics To Go

1. a) An electric generator is used to convert mechanical energy (cranking the generator) to electrical energy.

   b) An electric motor is used to convert electrical energy to mechanical energy.

2. Direct current travels in one direction only, whereas alternating current appears to travel back and forth.

AC:

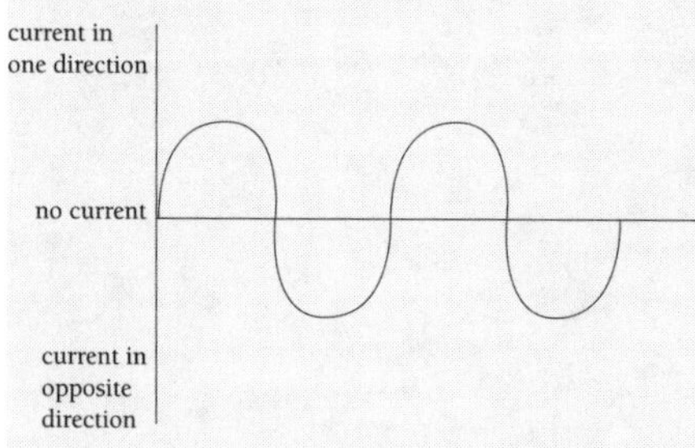


DC:

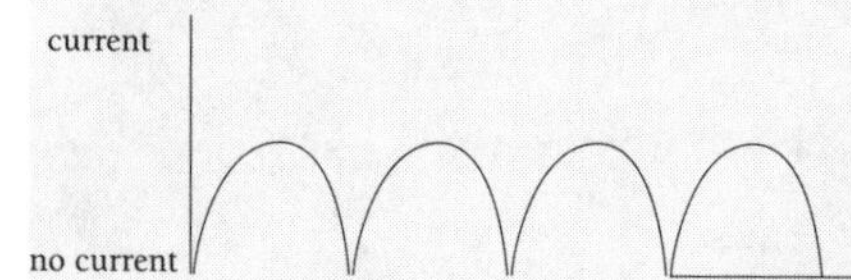


3. A current is produced in a wire only when the wire cuts through the magnetic field lines at a 90° angle. The wire must be moving in order for a current to be produced.

Toys for Understanding

### Physics To Go

1. What is the purpose of:

   a) An electric generator?

   b) An electric motor?

2. How does a direct current differ from an alternating current? Use graphs to illustrate your answer.

3. In an electric generator, a wire is placed in a magnetic field. Under what conditions is a current generated?

### Stretching Exercises

1. What is the meaning of "hertz," abbreviated "Hz," often seen as a unit of measurement associated with electricity or stereo sound components such as amplifiers and speakers?

2. What does it mean to say that household electricity has a frequency of 60 Hz?

3. Have you ever heard 60 Hz AC being emitted from a fluorescent light or a transformer?

4. Look at a catalog or visit a store where sound equipment is sold, and check out the "frequency response" of speakers—what does it mean?

5. Heinrich Hertz was a 19th-century German physicist. Find out about the unit of measurement named after him, and write a brief report on what you find.

234

ANSWERS

## Stretching Exercises

1. Hertz (Hz) is the unit of measurement for frequency.

2. In electrical appliances, it is referring to the AC circuit, which in North America is on a 60 Hz frequency. It means that the circuit is changing directions 60 times per second.

3. The sound might be referred to as a low hum. It can be heard by high voltage power lines, as well as in-ground transformers.

4. In speakers the frequency is referring to the frequency of sound. 20 Hz is generally accepted as the lowest sound heard, and depending on the individual, the highest frequency is about 12,000 to 16,000 Hz. In sound Hz is referring to the number of vibrations per second. The greater the number of vibrations, the higher the frequency.

5. Student report.

## Activity 4 A: Demonstration

# Falling Magnet

### FOR YOU TO DO

Equipment (teacher demonstration):

- rigid copper pipe, 3/4-inch diameter, 6-foot length (approx.), sold at plumbing supply stores in 10 foot lengths.
- cow magnet (or a cylindrical magnet having a diameter less than the inside diameter of the copper pipe).
- clamps and supports to hold the copper pipe in a vertical orientation with some space between the ends of the pipe and the classroom floor and ceiling.
- stepladder.

As an extension of **What Do You Think?**, ascend the stepladder and announce that you are going to drop the magnet through the copper pipe. Ask students if a current should be induced in the pipe, reminding them of the effect of relative motion between a magnet and a conductor. Also ask what effect, if any, there should be on the free fall of the magnet, neglecting any small amount of friction if the magnet hits the wall of the pipe during its fall. Then drop the magnet through the pipe. Students will observe that the fall of the magnet is profoundly retarded.

Ask students why the fall of the magnet was retarded. As a hint, suggest that they think back to using the hand generator under "load" and "no load" conditions — does an induced current "fight back" against the action that causes the induced current [Lenz's Law]? How does the energy output of a generator compare to the energy input? If the magnet was inducing a current in the copper pipe during its fall, how does that explain the retarded fall of the magnet?

# AC Generator

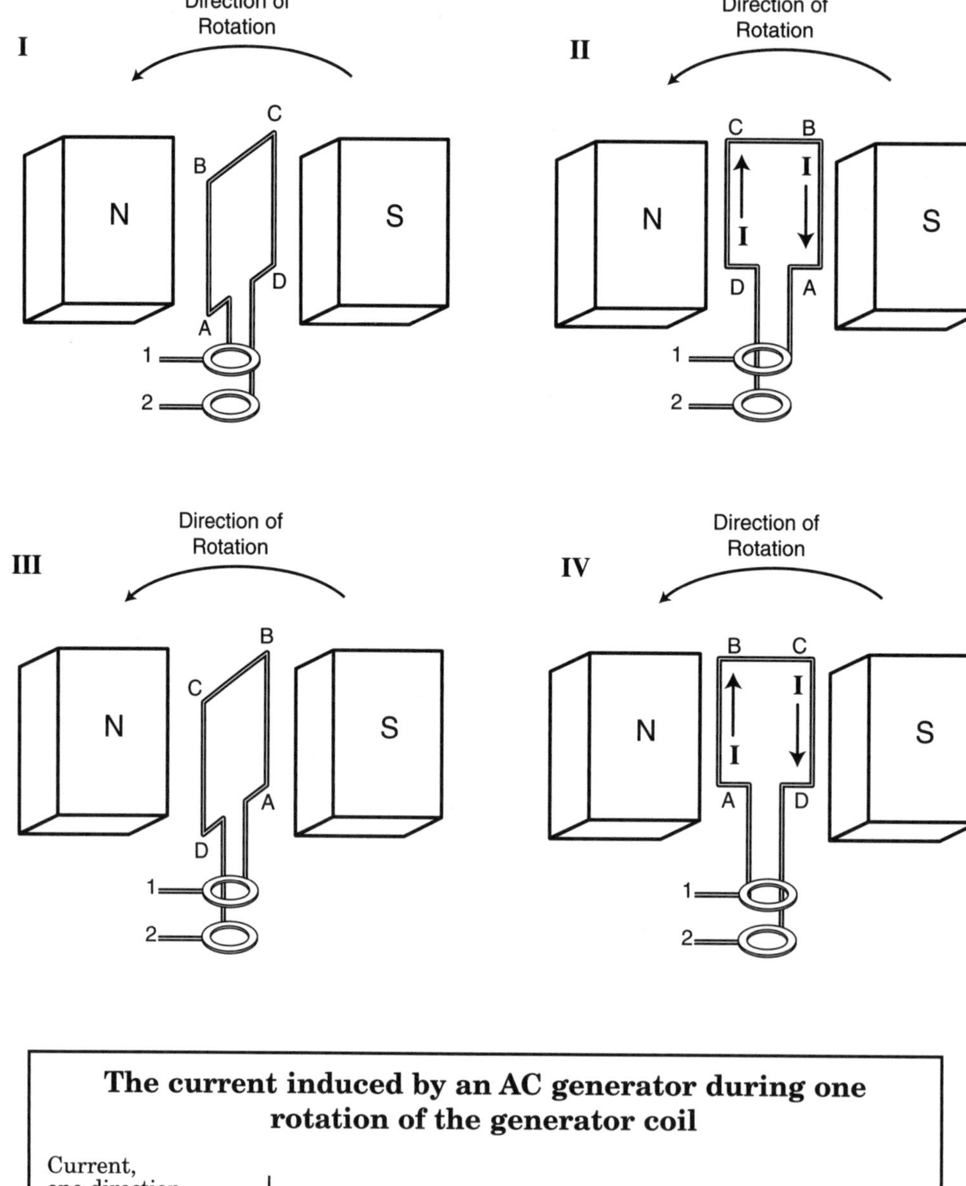


**The current induced by an AC generator during one rotation of the generator coil**



For use with *Toys For Understanding*, Chapter 4, Activity 4: AC & DC Currents

# DC Generator

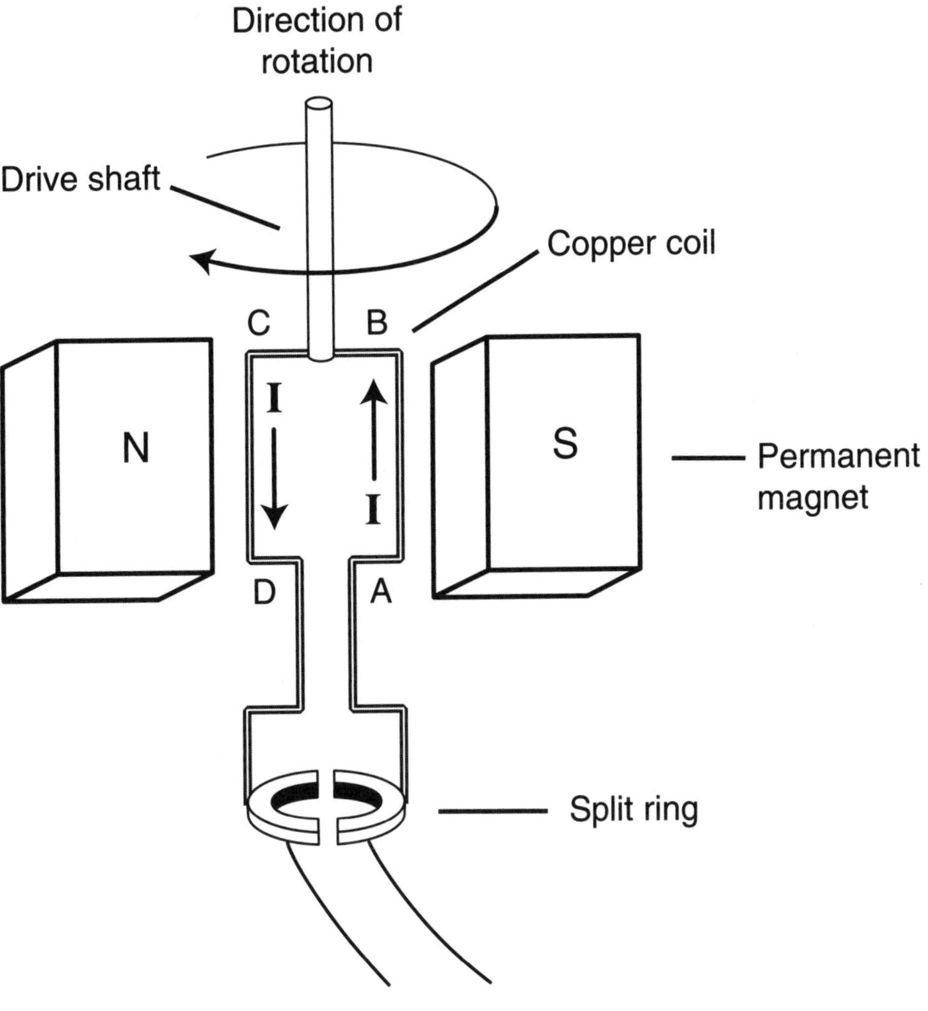


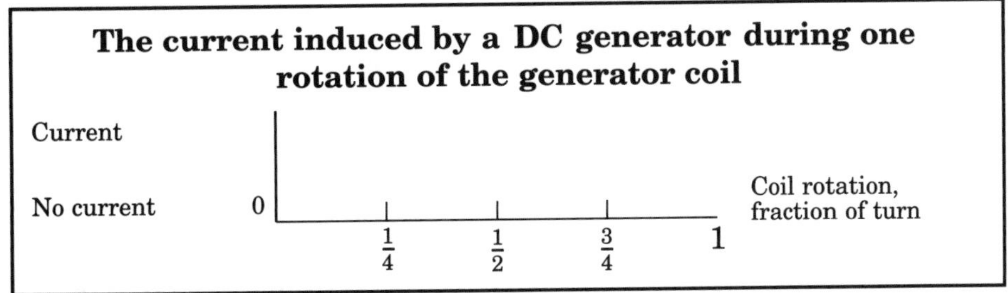


For use with *Toys For Understanding*, Chapter 4, Activity 4: AC & DC Currents

# ACTIVITY 5
## Building an Electric Motor

# Background Information

Principles of the electric motor are introduced in **Activities Five** and **Six**. The **Background Information** presented here will serve for both activities and will be limited to the DC motor.

A DC motor can be thought of as a DC generator "running backwards," and vice versa. Indeed, Gramme's discovery described in the student text for **Activity 5** shows that this is true in fact as well as principle.

How an electric motor converts electrical energy into mechanical energy is understood by considering what happens when an electric current travels in a region in which a magnetic field exists.

The diagram below shows a permanent "horseshoe" magnet which has a strong magnetic field in the region between the poles of the magnet. Magnet field lines extend from the north pole to the south pole of the magnet, as shown by the arrowheads on the field lines.

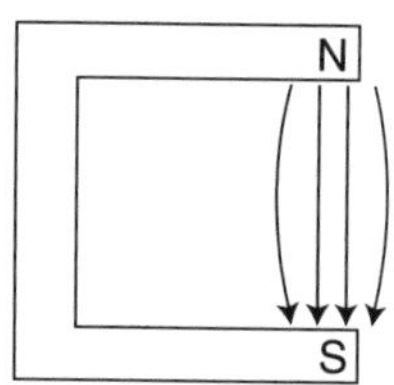


The smallest concentric circle in the diagram below represents a segment of wire viewed from one end. The "x" inside the smallest circle indicates that an electric current is flowing in the wire and has a direction away from the reader, into the page. The outer circles, with arrowheads, represent a magnetic field, caused by the current, surrounding the wire in the counterclockwise direction.

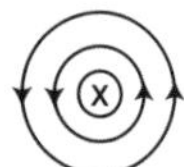

Below is shown a model which explains what happens when a current-carrying wire which has its associated circular magnetic field is placed in the magnetic field created by the permanent magnet.

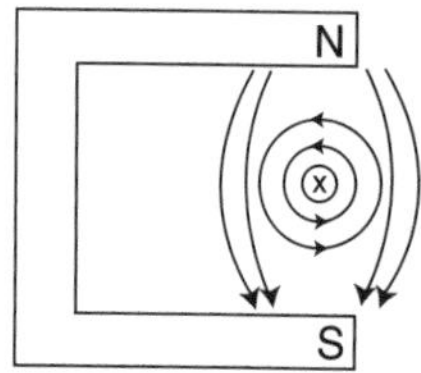


Directly to the left of the wire the magnetic field caused by the current has the same direction (downward) as the field established by the permanent magnet. Directly to the right of the wire the opposite is true; the field lines are in opposite directions. Lenz's Law suggests that forces arise to prevent changes in magnetic field lines, and, in this case, motion of the wire to the right would tend to preserve uniformity of the magnetic field; an electromagnetic force, *F*, would move the wire to the right in an attempt to move the high density field to the left of the wire to "fill in" the low density field to the right of the wire in the same way that "nature abhors a vacuum."

The amount of force on the wire in the simple case when the current passes perpendicularly through the field created by the permanent magnetic is:

$$F = BIL$$

where *F* is the force on the wire in Newtons, *B* is the strength of the magnetic field of the permanent magnet in Tesla, *I* is the current in the wire in amperes, and *L* is the length in meters along which the wire is embedded in the field.

A convenient "left-hand rule" for predicting the direction of the force in a current-carrying wire travelling through a magnetic field in a direction which is perpendicular to the direction of the magnetic field lines is shown in the diagram below:

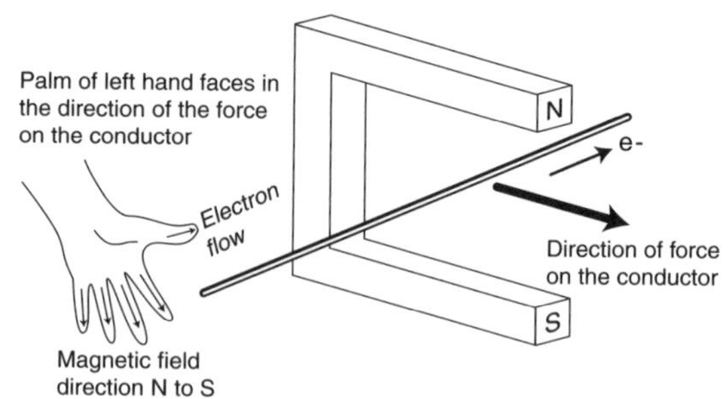


Left-hand motor rule

If the wire did indeed move to the right in the above position, it soon would exit the magnetic field, and the force would disappear. Therefore, electric motors are designed to employ a coil which rotates in a magnetic field. The above described motor effect can be applied in a "thought experiment" in which a current is fed into the coil of a DC generator to cause the coil to rotate. It is suggested that you satisfy yourself that motors and generators are basically the same in construction, the difference being that electrical energy is the output for the generator and is the input for the motor.

# Planning for the Activity

## Time Requirements

Allow at least one 40-minute period. Students may want to explore different arrangements and configurations in order to speed up and slow down the motor, etc.

## Materials Needed

### For each group:

- #24 magnet wire (60 ft)
- alligator clip leads (2)
- battery, D-cell
- cup, Styrofoam® (6)
- hand generator (or power supply with a variable current)
- magnet, ceramic ring, refrigerator (2)
- rubber band #64
- safety pins, large (2)
- sandpaper, fine square (6)
- tape, masking (3/4" × 60 yds)
- test tube, plastic (2 × 10 cm)

## Advance Preparation and Setup

Make a demonstration motor for the students to see that it is possible to build it, and show that it does work. Try to build it exactly the same way as it is in the book, and anticipate the different problems that the students might encounter.

# Teaching Notes

Take the students back to the diagram of the DC generator in **Activity 4**, and use the diagram to develop an explanation for what happens when a DC current is fed to the armature through the split ring commutator/brushes. Essentially, the motor is running backwards. There may be a need to describe to the students what is happening with regard to the creation of a magnetic field around the wire, which interacts with the magnet of the generator. This creates the force by which the rotor is turned, thus creating the motor effect.

An alert student might observe that the design for the Basic Motor does not include split rings to reverse the current in the coil each 1/2 turn. That is true, and the motor, in principle, should not work. However, the motor does work, but only because lack of symmetry of the coil causes the coil to jump during each rotation, breaking and reestablishing contact to allow the coil to keep rotating. Indeed, students who do a very careful job of shaping the coil may find that their motor will not function, and the solution is to bend the coil a bit to cause some nonsymmetry.

If a student does not bring up the lack of provision for reversing the input current each 1/2 turn, you should bring it up for discussion at the end of the activity — it will be important to deal with it, because students will need to provide a split ring system for the motor/generator which they will design and build for the next activity.

SAFETY PRECAUTION: If students will be operating the hand generator, it is important to remind them of the danger of electrical shock, and the safety procedures required.

NOTES

# Activity 5 Building an Electric Motor

**GOALS**

In this activity you will:

- Construct, operate, and explain a DC motor.
- Appreciate accidental discovery in physics.
- Measure and express the efficiency of an energy transfer.

### What Do You Think?

You plug a mixer into the wall and turn a switch and the mixer spins and spins—a motor is operating.

- **How do you think the electricity makes the motor turn?**

Write your answer to this question in your *Active Physics* log. Be prepared to discuss your ideas with your small group and other members of your class.

### For You To Do

1. Study the diagram on the following page closely. Carefully assemble the materials, as shown in the diagram, to build a basic electric motor. Follow any additional directions provided by your teacher.

ANSWERS

## For You To Do

1. Student activity.

# Activity Overview

The students will be building a simple electric motor. The more powerful the magnet, the more dramatic are the results.

## Student Objectives

### Students will:

- Construct, operate, and explain a DC motor.
- Appreciate accidental discovery in physics.
- Measure and express the efficiency of an energy transfer.

ANSWERS FOR THE TEACHER ONLY

## What Do You Think?

Students' answers will vary, but some will note that it has to do with the opposite of the generator effect. Some may even say something similar to what Faraday said, that if electricity can be made from moving a wire through a magnetic field, then why can't movement be created by a moving current in a magnetic field?

ANSWERS

## For You To Do
***(continued)***

2. a) Student data: changing polarity of the battery, more than one magnet, another magnet positioned at different locations around the wire.

3. a) Student observation. The orientation of the magnet needs to be opposite to the orientation of the other magnet in order for there to be magnetic field lines between the two magnets.

4. a) Again, the magnets must create magnetic field lines in order for there to be a motor effect.

5. a) Students' answers will vary. Changing the speed will involve an increase in the force that moves the wire. The force is influenced by the current, the magnetic field strength, and the length of the wire in the magnetic field.

6. a) Student data. They will probably discover that the faster you crank the generator the faster the motor will turn. Some other observations might be that at certain times the motor does not move. At this point, the motor is said to be in a dead zone. This is when the wire is in between the north and south poles so that the armature has an equal force acting on both sides, opposing motion.

7. a) Student answers will vary. Examples could include: washer, dryer, blender, grinder, food processor, can opener, electric razor, electric drill, electric fan.

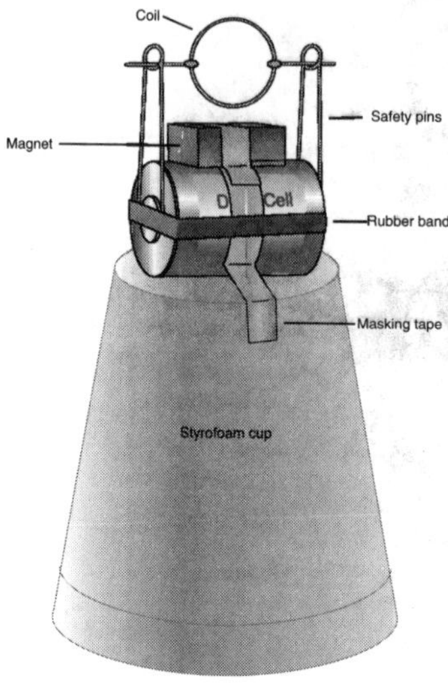


2. When your motor is operating successfully, find as many ways as you can to make the motor change its direction of rotation.
   a) Describe each way you tried and identify the ways that were successful.

3. Hold another magnet with your fingers and bring it near the coil from above, facing the original magnet, as the motor is operating.
   a) Describe what happens. Does the orientation of the second magnet make a difference?

4. Replace the single magnet with a pair of attracting magnets on top of the battery.
   a) What is the effect?

5. Think of other ways to change the speed of the motor. With the approval of your teacher, try out your methods.
   a) Describe ways to change the speed of the motor.

6. Use a hand generator as the energy source instead of the battery. You can disconnect the battery without removing it from the structure by placing an insulating material, such as a piece of cardboard, between the safety pin and the battery to open the circuit at either end of the battery. Then clip the wires from the generator to the safety pins to deliver current from the generator to the motor.
   a) Discuss what you find out.

7. Your motor turns! Chemical energy in the battery was converted to electrical energy in the circuit. The electrical energy was then converted to mechanical energy in the motor.
   a) List at least three appliances or devices where the motor spins.

236 Active Physics

8. The spin of the motor occurs because the current-carrying wire has a force applied to it. You know if something moves, a force must be applied. As you observed, when the battery connection was broken, the motor stopped turning. You know from a previous activity that a current-carrying wire creates a magnetic field. Pause for a bit to remind yourself of the behavior of magnets. Take a bar magnet and place its north pole near a compass. The compass is a tiny bar magnet that can easily turn.
   a) Draw a sketch to show the orientation of the compass.

N S

9. Shift the compass to the south pole of the bar magnet.
   a) Draw a sketch to show the orientation of the compass.

10. The north pole of the bar magnet repelled the north pole of the compass. The south pole of the bar magnet attracted the north pole of the compass. This attraction and repulsion is the result of a force on the compass. You can now investigate the force between the poles of a magnet and the magnetic field of a current-carrying wire.

8. a)–9. a)

N S

ANSWERS

## For You To Do
*(continued)*

11. a) The compasses would attract one another.

 b) The compasses would repel one another.

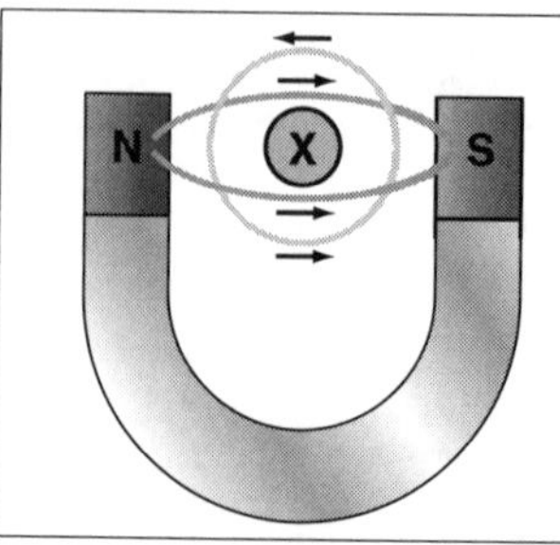


The magnetic field lines are drawn for a horseshoe magnet. The direction of the magnetic field lines is identical to a direction that a compass would point. The compass would point away from the north pole and toward the south pole. The magnetic field of the current-carrying wire is circular, as you investigated in an earlier activity. Compare the direction of this magnetic field to the direction of the magnetic field of the horseshoe magnet.

11. Think of the magnetic field lines above as small compasses.
 a) Write down whether the compasses above the wire attract one another or repel one another.
 b) Write down whether the compasses below the wire attract one another or repel one another.

12. This attraction/repulsion causes the wire to jump. There is a force on the wire. This force on the current-carrying wire is the basis for the electric motor that you built in this activity. The use of the loop of the wire allows the wire to rotate instead of jumping in the way a single wire would.

13. It is the moving electrons in the wire that create the current. In some TV sets, there is an electron beam that shoots the electrons from the back of the TV to the front. There are horseshoe magnets of a sort in the television. The moving electrons experience a force. The electrons' path is affected by the magnetic field. By varying the strength of the magnetic field, the electron beam can hit all parts of the screen and you receive a TV image.

### FOR YOU TO READ

The history of science is filled with discoveries that have led to leaps of progress in knowledge and applications. This is certainly true of physics and, in particular, electricity and magnetism. These discoveries "favor" the prepared mind. Oersted's discovery in 1820 of the magnetic field surrounding a current-carrying wire already has been mentioned. Similarly, Michael Faraday discovered electromagnetic induction in 1831. Faraday was seeking a way to induce electricity using currents and magnets; he noticed that a brief induced current happened in one circuit when a nearby circuit was switched on and off. (How would that cause induction? Can you explain it?) Both Oersted and Faraday are credited for taking advantage of the events that happened before their eyes, and pursuing them.

Michael Faraday

About one-half century after Faraday's discovery of electromagnetic induction, which immediately led to development of the generator, another event occurred. In 1873, a Belgian engineer, Zénobe Gramme, was setting up DC generators to be demonstrated at an exposition (a forerunner of a "world's fair") in Vienna, Austria. Steam engines were to be used to power the generators, and the electrical output of the generators would be demonstrated. While one DC generator was operating, Gramme connected it to another generator that was not operating. The shaft of the inactive generator began rotation—it was acting as an electric motor! Although Michael Faraday had shown as early as 1821 that rotary motion could be produced using currents and magnets, a "motor effect," nothing useful resulted from it. Gramme's discovery, however, immediately showed that electric motors could be useful. In fact, the electric motor was demonstrated at the very Vienna exposition where Gramme's discovery was made. A fake waterfall was set up to drive a DC generator using a paddle wheel arrangement, and the electrical output of the generator was fed to a "motor" (a generator running "backwards"). The motor was shown to be capable of doing useful work.

ANSWERS

## Physics To Go

1. Students' answers will vary. Some advantages: increase the field strength of the magnet in order to increase the output from the motor; maintain a constant magnetic field in order to have more control of the speed. Some disadvantages: reduction in the efficiency of the engine reduction in the energy output; need for cooling of the motor as the current running through the electromagnet will heat up.
2. Student design.
3. Student response. Many of the different kinds of materials might be children's construction toys such as Mechano™, wheels, gears, elastic bands, pulleys, string, bearings, strips of wood, plastic wheels, gears etc., popsicle sticks, toothpicks, straws, etc.
4. An electric motor operates when an electric current is moving through a magnetic field. As a moving current produces a magnetic field around the wire, it is either attracted to or repelled from the magnets in the motor, which forces the wires to move. This turns the armature. A generator works on a similar principle in that a moving wire is passed through a magnetic field and a current is induced in the wire.

### Reflecting on the Activity and the Challenge

Decision time about the **Chapter Challenge** is approaching for your group. In this activity you built a very basic, working electric motor. This is an important part of the **Chapter Challenge**. However, knowing how to build an electric motor is only part of the challenge. Your toy must be fascinating to children. You must also be able to explain how it works.

### Physics To Go

1. Some electric motors use electromagnets instead of permanent magnets to create the magnetic field in which the coil rotates. In such motors, of course, part of the electrical energy fed to the motor is used to create and maintain the magnetic field. Similarly, electromagnets instead of permanent magnets are used in some generators; part of the electrical energy produced by the generator is used to energize the magnetic field in which the generator coil is caused to turn. What advantages and disadvantages would result from using electromagnets instead of permanent magnets in either a motor or generator?
2. Design three possible toys that use a motor or a generator or both. One of these may be what you will use for your project.
3. The motor/generator you submit for the **Chapter Challenge** must be built from inexpensive, common materials. Make a list of possible materials you could use to construct an electric motor.
4. In the grading criteria for the **Chapter Challenge**, marks are assigned for clearly explaining how and why your motor/generator works in terms of basic principles of physics. Explain how an electric motor or generator operates.

NOTES

# ACTIVITY 6
## Building a Motor/ Generator Toy

# Background Information

**Background Information** for this activity was provided in the background for **Activity 5**.

### *Active*-ating the Physics InfoMall

Ideas for searches include "electric motor" and "Faraday." Most other useful information has already been found!

# Planning for the Activity

## Time Requirements

Allow at least 4 class periods, to collect, design, and build their toys. One class can be used for different groups to try the toys to see if they work and if they are fun. At least one more class will be needed to write instructions on how to build the motor/generator, and how it works in the toy.

## Materials Needed

### For each group:

- alligator clip leads (2)
- battery, D-cell
- magnet, small bar

## Advance Preparation and Setup

You will need to decide how you will acquire the items needed for groups who wish to pursue alternate designs.

# Teaching Notes

Set a time limit and communicate this to the class. You will have to set aside some time for testing the motor/generators, and for incorporating the motor in a toy. As these will be going to the HFE volunteers, the students will also need some time to prepare the instructions and explain the device to HFE inhabitants.

For assessment, you will need to decide how each group's products will be judged against the assessment criteria. Peer judging can be very effective, as long as the understanding is that they are fair. Have the students arrive at a consensus on exactly how the assessment will be done.

SAFETY PRECAUTION: Safety issues must be addressed. Working with electricity, can be dangerous (although the students are working with small voltages and amperages). There are some sharp objects involved, and students should be cautioned.

NOTES

NOTES

# Activity 6 Building a Motor/Generator Toy

**GOALS**

In this activity you will:

- Design, construct and operate a motor/generator.

### What Do You Think?

You may have heard the following expression used before: "The difference between men and boys is the cost of their toys."

- **What characteristics make an item a toy?**

Write your answer to this question in your *Active Physics* log. Be prepared to discuss your ideas with your small group and other members of your class.

### For You To Do

1. Confer within your group and between your group and your teacher about whether you will pursue, as a basis for the motor/generator kit for the **Assessment**, the motor design presented in this activity, an alternate design, or both. Whatever design(s) your group chooses to pursue, you are encouraged to be creative.

## Activity Overview

This activity is the culmination of **Chapter 4**. Students will be using their knowledge of motors and generators, to build a toy. This must be a functional toy with the criteria as set out in the **Assessment** on student page 246 and should be a fun activity for the students.

## Student Objectives

### Students will:

- Design, construct and operate a motor/generator.

ANSWERS FOR THE TEACHER ONLY

## What Do You Think?

Students' answers will vary. Many of the responses will probably include characteristics such as fun, easy to work, durable, inexpensive, not predictable, etc.

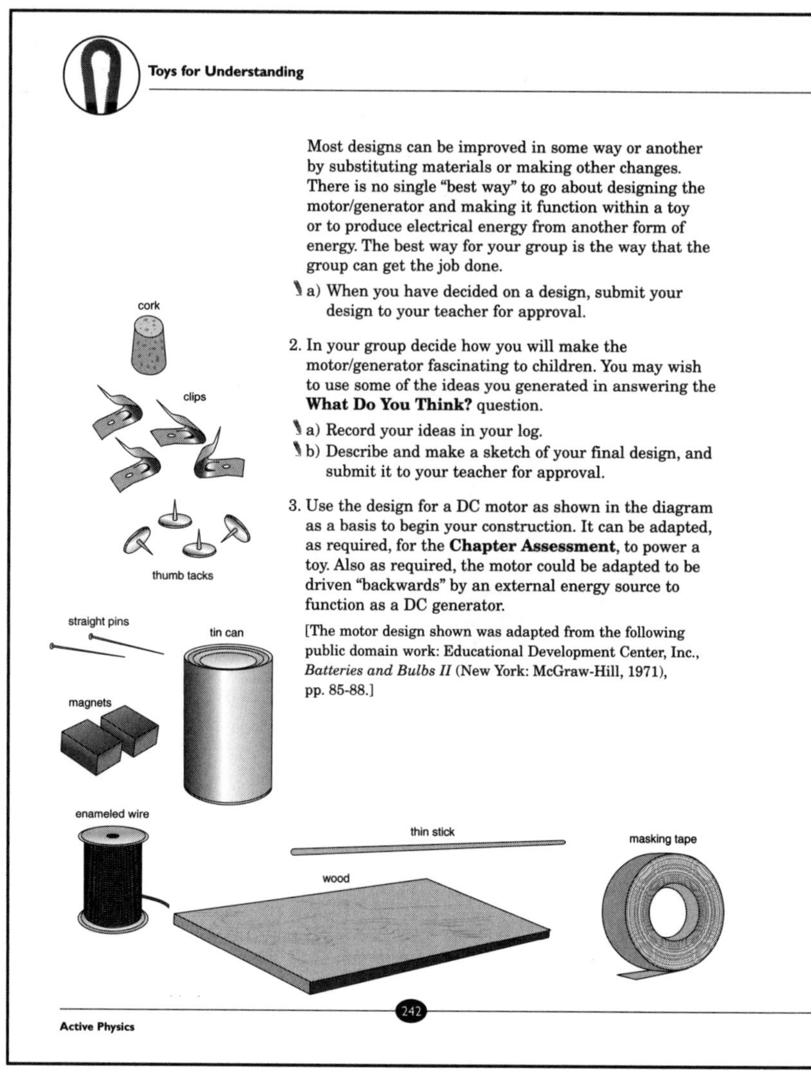


## Answers

# For You To Do

1. a) Student design of the project.

2. a) Student responses.

   b) Student design and sketch.

3. Student activity.

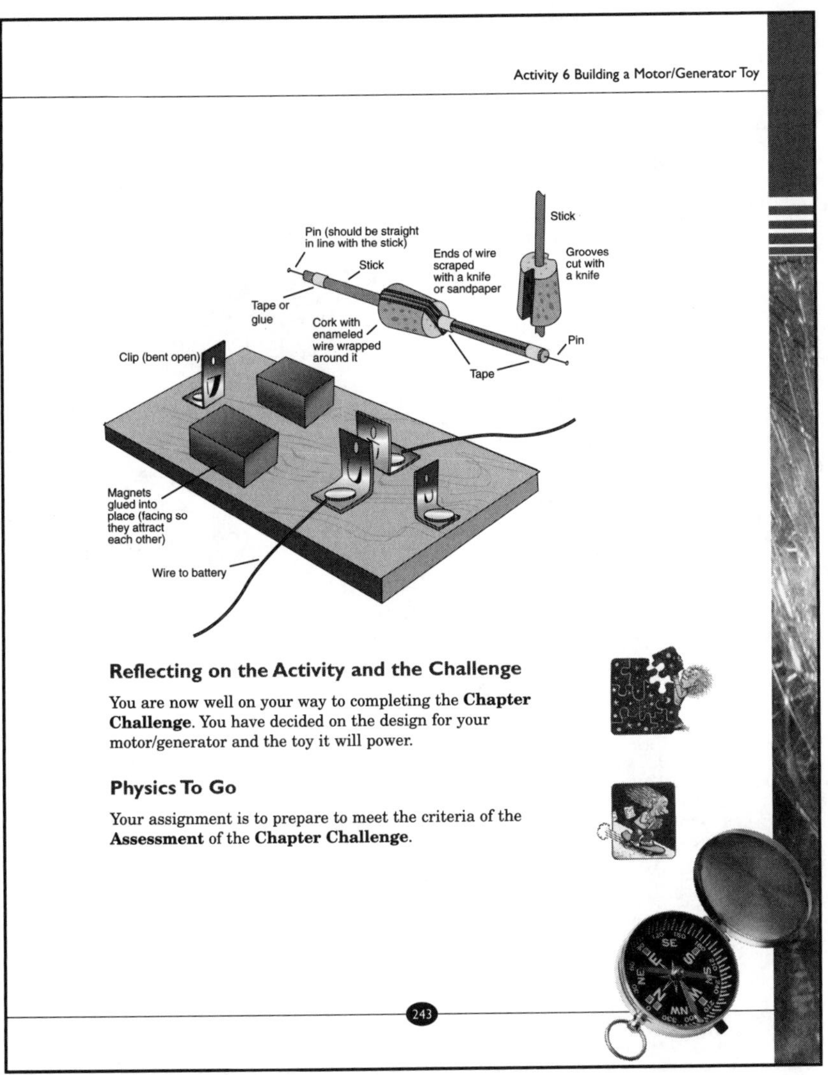


### Reflecting on the Activity and the Challenge

You are now well on your way to completing the **Chapter Challenge**. You have decided on the design for your motor/generator and the toy it will power.

### Physics To Go

Your assignment is to prepare to meet the criteria of the **Assessment** of the **Chapter Challenge**.



ANSWERS

## Physics To Go

Build the motor/generator.

# PHYSICS AT WORK

## Uriah Gilmore

### HEADED FOR THE STARS

Uriah Gilmore loved to take electric appliances apart when he was growing up. "I couldn't always get them back together," he admits, "but I was so curious I couldn't help myself. I just had to see how they worked." Fortunately, Uriah's parents supported his curiosity.

Uriah and his fellow teammates from Cleveland, Ohio's East Technical High School recently won first place at the National High School Robotics Tournament at Epcot Center in Orlando, Florida for building a robot. "We were counseled along the way by engineers from NASA," he enthusiastically explains. "We called our robot Froggy and painted it green," Uriah continues, "and we used noisemakers so it even sounded like a frog." During the final contest "Froggy" was put in a pit with two other robots and had to place balls of a certain color in a specified area. The robot who got the most balls in won the contest.

"In my sophomore year the school I was attending closed and I went to East Technical High School which was the best thing that happened to me." He entered the engineering program and became a member of the engineering team—a team that is more popular than any sports team in his school.

Uriah attended Morehouse College on a NASA scholarship. "But," he states, "it's not enough to be a good student. You also have to be involved with your school and your community." Uriah once led a march on the Cleveland, Ohio, City Hall to protest a law which threatened to fire certain teachers, including one who inspired Uriah and was responsible for the revitalization of East Technical High School.

"My ultimate goal is to travel in space and explore the galaxy," he states. A shorter term goal is to be as involved in college as he has been in high school.

Chapter 4

# Chapter 4 Assessment

Your task is to prepare a kit of materials and instructions that a toy company will manufacture. Children will use these kits to make a motor or generator, or a combination electric motor/generator. It will serve both as a toy and to illustrate how the electric motors in home appliances work or how electricity can be produced from an energy source such as wind, moving water, a falling weight, or some other external source.

Review and remind yourself of the grading criteria that you and your classmates agreed on at the beginning of the chapter. The following was a suggested set of criteria:

- **(30%) The motor/generator is made from inexpensive, common materials, and the working parts are exposed but with due consideration for safety.**
- **(40%) The instructions for the children clearly explain how to assemble and operate the motor/generator device, and explain how and why it works in terms of basic principles of physics.**
- **(30%) If used as a motor, the device will operate using a maximum of four 1.5-volt batteries (D cells), and will power a toy (such as a car, boat, crane, etc.) that will be fascinating to children.**

**OR**

- **(30%) If used as a generator, the device will demonstrate the production of electricity from an energy source such as wind, moving water, a falling weight, or some other external source and be fascinating to children.**

## Physics You Learned

Motors

Generators

Galvanometers

Magnetic field from a current

Solenoids

Electromagnets

Induced currents

AC and DC generators

# Alternative Chapter Assessment

Select the best response for each statement or question.

1. Normally, the magnetism of Earth causes a magnetic compass to point in the direction:

   a) North

   b) South

   c) East

   d) West

2. When a magnetic compass is placed near a wire carring an electric current, the direction in which the compass points is influenced by:

   a) only Earth's magnetism.

   b) only the wire's magnetism.

   c) neither Earth's magnetism nor the wire's magnetism.

   d) both Earth's magnetism and the wire's magnetism.

3. A generator transforms:

   a) electrical energy into mechanical energy.

   b) heat into mechanical energy.

   c) mechanical energy into electrical energy.

   d) mechanical energy into heat.

4. The output energy of a generator:

   a) is much greater than the input energy.

   b) is slightly greater than the input energy.

   c) is equal to the input energy.

   d) is less than the input energy.

5. The shape of the magnetic field caused by an electric current flowing in a straight wire is:

   a) along a straight line.

   b) square.

   c) circular.

   d) impossible to predict.

6. Which factor or factors listed below affects the magnetic strength of a solenoid which is carrying an electric current?

   1.The amount of the current flowing in the solenoid.

   2. The number of turns of wire of the solenoid.

   3. The length along which the turns of wire are wound on the solenoid.

   a) Factor (1) only.

   b) Factor (2) only

   c) Factor (3) only

   d) Factors (1), (2) and (3).

7. If both the current and number of turns of wire per unit of a length of a solenoid are doubled, the magnetic strength of the solenoid should:

   a) remain unchanged.

   b) decrease by a factor of two.

   c) increase by a factor of two.

   d) increase by a factor of four.

8. Which choice of core material for a solenoid will result in the strongest electromagnetic if all other factors remain equal?

   a) wood.

   b) plastic.

   c) aluminum.

   d) iron.

9. If the number of turns of wire on a solenoid is doubled along a constant length of core material, and if the amount of current flowing in the solenoid is reduced to one-half the original amount, the magnetic strength of the electromagnetic will:

   a) remain unchanged.

   b) decrease by a factor of two.

   c) increase by a factor of two.

   d) increase by a factor of four.

10. A galvanometer is meant to be used to:

    a) detect electric currents.

    b) detect magnetic fields.

    c) measure frequency.

    d) measure the efficiency of a generator.

11. The hertz is a unit of:

a) energy.

b) voltage.

c) frequency.

d) current.

12. Household electricity in the United States is:

a) DC.

b) direct.

c) inverse.

d) AC.

13. The number of complete cycles per second for household electricity in the United states is:

a) 12

b) 60

c) 120

d) 240

14. Split rings and brushes are used in a DC motor to:

a) increase speed.

b) increase efficiency.

c) increase magnetic field strength.

d) reverse current.

15. Using a conductor to cut through a magnetic field can result in:

a) increased resistance.

b) induced current.

c) power loss.

d) reduction in temperature.

16. When an electric generator is set in motion to produce electricity, the amount of energy needed to keep the generator going is:

a) only the amount needed to overcome mechanical friction in the generator.

b) greater than or equal to the energy output of the generator.

c) impossible to predict.

d) less than or equal to the energy output of the generator.

17. When an electric motor is set in motion to produce electricity, the amount of energy needed to keep the motor going is:

    a) only the amount needed to overcome mechanical friction in the motor.

    b) greater than or equal to the energy output of the motor.

    c) impossible to predict.

    d) less than or equal to the energy output of the motor.

18. Which of the conditions listed below cause electric current to flow in a coil of wire when a magnet is near the coil of wire?

    1. Only the magnet is moved.

    2. Only the coil of wire is moved.

    3. Both the magnet and the coil of wire are moved.

    a) Condition (1) only.

    b) Condition (2) only.

    c) Condition (3) only.

    d) All of the conditions.

19. If a direct current is fed into the coil of a generator which normally is used to produce DC electricity:

    a) nothing will happen.

    b) electricity will be generated as usual.

    c) the generator will function as an electric motor.

    d) it is not possible to predict what will happen.

20. An AC generator:

    a) always produces AC having a frequency of 60 Hz.

    b) produces direct current.

    c) always produces 120-V AC.

    d) produces one AC cycle during each turn of the coil.

# Alternative Chapter Assessment Answers

1. a
2. d
3. c
4. d
5. c
6. d
7. d
8. d
9. a
10. a
11. c
12. d
13. b
14. d
15. b
16. b
17. b
18. d
19. c
20. d

NOTES

# Chapter 5

# PATTERNS AND PREDICTIONS

## Chapter 5 – Patterns and Predictions

# National Science Education Standards

### Chapter Challenge

In a scenario that again involves the difference between the nature of science and pseudoscience, students are given a list of proposals presented to a funding agency. They are challenged to evaluate the scientific merit of each proposal, determining if the topic area in the proposal can be tested by experiments and the extent to which it reflects the role and importance of science in the world.

### Chapter Summary

To develop understanding of the science principles of inquiry necessary to meet this challenge, students work collaboratively on activities in which they learn to make predictions. These experiences engage students in the following content identified in the National Science Education Standards.

# Content Standards

### Unifying Concepts

- Evidence, models and explanations
- Constancy, change and measurement

### Science as Inquiry

- Identify questions and concepts that guide scientific investigations
- Use technology and mathematics to improve investigations
- Formulate and revise scientific explanations and models using logic and evidence
- Communicate and defend a scientific argument

### History and Nature of Science

- Science as a human endeavor
- Nature of scientific knowledge
- Historical perspectives

### Physical Science

- Structure and properties of matter
- Motions and forces
- Interactions of energy and matter

# Key Physics Concepts and Skills

| Activity Summaries | Physics Principles |
|---|---|
| **Activity 1: Force Fields**<br>Observing, then measuring the properties of magnets introduces the study of force fields and helps students appreciate the dilemmas that can occur when describing something that is invisible. | • Magnetic fields |
| **Activity 2: Newton's Law of Universal Gravitation**<br>In an activity in which they place a photocell on the light generated by a slide projector, students measure light intensity at various distances to uncover the inverse square law. They then apply this to Newton's Law of Universal Gravitation. | • Light intensity<br>• Newton's Law of Universal Gravitation<br>• Inverse square relationship |
| **Activity 3: Slinkies and Waves**<br>In this activity, students explore wave motion with "people waves" then with Slinkies. Students then read to learn more about wavelength, frequency, amplitude, crests, and troughs. This experience is used as a model to explain the flow of energy. | • Energy transfer<br>• Wave motion and periods<br>• Wavelengths and amplitude |
| **Activity 4: Interference of Waves**<br>Using Slinkies to model wave motion, students explore and observe the phenomena of wave interference. They expand their understanding of waves by comparing this experiment to sine waves generated on graphing calculators, circular waves in a ripple tank, laser light beams, and sound waves from identical speakers. | • Wave motion<br>• Energy transfer<br>• Wave interference |
| **Activity 5: A Moving Frame of Reference**<br>This activity introduces the concept of frames of reference by having students describe and compare observations of the same event made while standing still and while moving. They then read to learn more about how the laws of physics relate to these different descriptions of the same event. Students conduct an experiment in which they measure the speed of a car moving along a stationary board. They then measure the relative speeds when both the car and the board move. Comparing measures and observations enables students to better understand frames of reference and introduces the concept of relativity. | • Frames of reference<br>• Speed<br>• Relativity |
| **Activity 6: Speedy Light**<br>In this activity, students explore the speed of light as related to the concept of relativity by considering how to know for certain whether two clocks, large distances apart, display exactly the same time. Reading more about Einstein's theories enables them to apply these concepts to the chapter challenge. | • Simultaneous events<br>• Speed of light<br>• Relativity<br>• Frames of reference |
| **Activity 7: Special Relativity**<br>Simultaneous events and relativity set the stage for this final chapter activity in which students learn about muons. Muons, and Einstein's Theory of Special Relativity focus students on the need to consider the need for evidence from experiments to support the development of scientific theories. | • Special Relativity<br>• Muons<br>• Half-life |
| **Activity 8: The Doppler Effect**<br>In this activity, students are introduced to the Doppler Effect in an experiment in which they toss an oscillator embedded in a Nerf ball. The change in pitch as the Nerf ball moves is related to how the Doppler Effect is used to measure distance to distant stars. This is then related to the development of the Big Bang Theory. | • Doppler Effect<br>• Big Bang Theory<br>• Measuring distances in space |
| **Activity 9: Communication Through Space**<br>In this final activity, students are confronted with the extreme amount of time required for light waves to reach stars. After considering the impact of this time delay on communication with life on stars, students return to the chapter challenge and discuss what type of information is most important to send and receive. | • Speed of sound<br>• Speed of electromagnetic waves<br>• Measuring distances in space |

# GETTING STARTED WITH EQUIPMENT NEEDED TO CONDUCT THE ACTIVITIES.

## Items needed – not supplied in Material Kits

Preparing the equipment needed for each activity in this chapter is an important procedure. There are some items, however, needed for the chapter that are not supplied in the It's About Time material kit package. Many of these items may already be in your school and would be an unnecessary expense to duplicate. Please read carefully the list of items to the right which are not found in the supplied kits and locate them before beginning activities.

**Items needed – not supplied by It's About Time:**

- Chalkboard
- Projector
- Piece of chalk
- Computer
- Slinky Polarizer
- Graphing Calculator
- Michelson interferometer and laser
- Oscillator and stereo amplifier
- Chair with Wheels
- Roll of Butcher Paper

# Equipment List For Chapter 5 (Serves a Classroom of 30 Students)

| PART | ITEM | QTY | ACTIVITY |
|---|---|---|---|
| AS-1404-P3 | Amplified Speaker | 12 | 4 |
| BH-9201-C3 | Buzzer | 6 | 8 |
| BS-8200-C3 | Ball, Nerf 6" | 6 | 8 |
| BS-0790-P3 | Ball, Tennis | 6 | 5 |
| BS-1599-P3 | Battery, 9-volt | 12 | 2, 8 |
| BS-1606-C3 | Battery Clip With Solid LeadsFor 9-volt Battery | 6 | 8 |
| CA-0100-P3 | Toy Car, Battery Operated | 6 | 5 |
| CP-6081-P3 | Compass | 6 | 1 |
| GS-7706-P3 | Graph Paper Pad of 50 Sheets | 6 | 7 |
| IR-1266-P3 | Iron Filings, LB | 6 | 1 |
| LM-0112-P3 | Light Meter To Replace Galvanometer | 6 | 2 |
| MS-0354-P3 | Bar Magnet Large Pair | 6 | 1 |
| MS-1053-P3 | Metal Objects: Magnetic & Non-Mag. Pkg. | 6 | 1 |
| MS-1886-P3 | Magnets Ceramic Rings | 24 | 1 |
| MS-1425-P3 | Felt Tip Marker | 6 | 5 |
| SH-3639-P3 | Heavy Duty Slinky® | 6 | 3, 4 |
| SM-1676-P3 | Meter Stick, 100 cm, Hardwood | 12 | 2, 5 |
| SS-7722-P3 | Ball of String | 6 | 5 |
| SS-7778-P3 | Stopwatch | 6 | 5 |
| TS-6045-C3 | Tape Recorder | 6 | 8 |
| TS-6120-C3 | Roll of Adding Machine Tape | 6 | 9 |
| YS-7724-P3 | Yarn, Colored, 20 cm | 36 | 3 |
| ZZ-0027-P3 | 27 qt. Clear Plastic Storage Container | 6 | |
| | | | |
| TEACHER DEMONSTRATION ITEMS | | | |
| RH-2401-P3 | Ripple Tank Assembly | 1 | 4 |
| | | | |
| THINGS NEEDED NOT SUPPLIED | | | |
| | Chalkboard | | 2 |
| | Projector | | 2 |
| | Piece of chalk | | 2 |
| | Computer | | 3 |
| | Slinky Polarizer | | 3 |
| | Graphing Calculator | | 5 |
| | Michelson interferometer and laser | | 5 |
| | Oscillator and stereo amplifier | | 5 |
| | Chair with Wheels | | 6 |
| | Roll of Butcher Paper | | 6 |

# Organizer for Materials Available in Teacher's Edition

| Activity in Student Text | Additional Material | Alternative / Optional Activities |
|---|---|---|
| ACTIVITY 1: Force Fields p. 250 | Assessment Challenge: Inquiry Investigation, p. 478 | |
| ACTIVITY 2: Newton's Law of Universal Gravitation, p. 258 | Cartoon: The Study of Motion, p. 484<br>Cartoon: The Law of Universal Gravitation, p. 485 | Activity 2A: Galileo's Gig pgs. 482-483 |
| ACTIVITY 3: Slinkies and Waves p. 265 | Properties of Waves, p. 507 | Activity 3A: A Physicist's Guide to Taking a Bath, p. 506 |
| ACTIVITY 4: Interference of Waves, p. 274 | | |
| ACTIVITY 5: A Moving Frame of Reference, p. 280 | | |
| ACTIVITY 6: Speedy Light p. 289 | | |
| ACTIVITY 7: Special Relativity p. 296 | | |
| ACTIVITY 8: The Doppler Effect p. 302 | | Activity 8A: Doppler Effect with Water Waves, p. 551 |
| ACTIVITY 9: Communication Through Space p. 307 | | Activity 9A: Space Travel Time-Line p. 559 |

## Scenario

**Science has enriched the lives of everyone. People no longer fear the movement of the planets. Many enjoy viewing an eclipse. Science and technology have helped feed large numbers of people, and raise the standard of living of many people as well. Science and technology have also complicated lives. New problems have emerged as a result of the technologies that people have decided to use. As people learn more about the natural world through science and technology, they discover that there is more and more to know!**

248

## Challenge

Although you have grown up in a society that uses science and technology, it is difficult sometimes to distinguish between science and pseudoscience.

This challenge places you as the head of an institute that provides funding for science research. A number of groups or individuals have submitted proposals to you, all wishing funding from your institute. These include research on:

- force fields
- auras
- telekinesis
- new comets
- failure modes of complex systems
- advent of new diseases
- astrology prediction
- communicating with extraterrestrial beings
- the extinction of dinosaurs
- communication with dolphins
- prediction using biorhythms
- properties of new materials
- dowsing
- earthquake prediction
- election predictions using polling

You will choose two proposals from this list, or invent other proposals to add to the list. One of the proposals will be accepted because of its scientific merit. The other proposal will be denied because it has little or no scientific merit.

You will have to defend your selections in a position paper. You will also write letters to each of the people who submitted these studies for funding.

# Chapter and Challenge Overview

Chapter 5, *Patterns and Predictions*, engages students in the exploration of force fields, Newton's Law of Universal Gravitation, ancient and current models of the solar system, the nature of science and scientific laws compared to pseudoscience, the properties of waves, and basic ideas underpinning the Theory of Special Relativity. In each case, students are guided to explain their observations scientifically, and the value of science over pseudoscience is established.

The chapter activities provide experiences based on scientific observations and data, and thinking and processes, that students can use to successfully meet the **Chapter Challenge**. Students are reminded that science has removed fear and superstition from our daily lives. They are also reminded that science and technology have created their own special sets of problems.

In the **Chapter Challenge**, students play the role of the head of a science-research-funding institute and must determine which project from a sizable list to fund. They must also select one project to reject. Their decisions must be defended in a position paper, and they must write two letters: one to the individuals who submitted the selected project and one to the individuals who submitted the rejected project.

How will you decide which project to fund and which to deny? As you work through the chapter and think about funding, ask the following questions:

Is the area of study logical?

Is the topic area testable by experiment?

Can any observer replicate the experiment and get the same results?

Is the theory the simplest and most straightforward explanation?

Can the new theory explain known phenomena?

Can the new theory predict new phenomena?

## Criteria

Here are the standards by which your work will be evaluated:

- **The selection of proposals reflects an accurate understanding of the nature of science**
- **The selection of proposals reflects an accurate understanding of the role and importance of science in the world**
- **The selection considers all the major differences you've learned about science and pseudoscience**
- **The position paper is clearly written and accurate. Grammar and spelling are correct**
- **The letters explain your reasoning clearly, concisely, and in a businesslike fashion. Grammar and spelling are correct**

Discuss in your small groups and as a class the criteria for this performance task. For instance:

- **How much of the grade should depend on showing the scientific merit of the first idea, or the lack of scientific merit of the second idea?**
- **How much of the grade should depend on quality and clarity of the presentation?**
- **How much should depend on the letters to the hopeful researchers? How should a letter be graded?**
- **What would constitute an "A" for this project?**

Here is a sample grading rubric. You can fill out the descriptions and supply the point values.

| Criteria | Excellent max=100% | Good max=70% | Satisfactory max=50% | Poor max=25% |
|---|---|---|---|---|
| reflects an accurate understanding... points | | | | |
| role and importance of science... points | | | | |
| major differences... points | | | | |
| clearly written... points | | | | |
| letters... points | | | | |

A= points B= points C= points D= points

249

## Assessment: Challenge

Place a check mark (✓) in the appropriate box. If you would rather, you could mark it holistically, where all the statements taken together earn a mark out of 30%, for example.

You may change the criteria, by adding or deleting any of the points, determined by a class decision.

| Descriptor | Yes | No |
|---|---|---|
| **Reflects an accurate understanding of Nature of Science (25%)** | | |
| scientifically accurate and appropriate language | | |
| simple explanation of the physics involved | | |
| able to explain in simple but concise terms | | |
| evidence of research of a scientific nature | | |
| evidence of experimentation | | |
| **Role and Importance of Science in the World (25%)** | | |
| proper scientific research standards are employed | | |
| accurate record-taking and data storage | | |
| accurate graphs and tables | | |
| area of study logical | | |
| **Science and Pseudoscience: Major Differences (25%)** | | |
| standing study can be replicated | | |
| data is accurate | | |
| all data is included (i.e., data which does not work for the hypothesis is not left out) | | |
| new theory predicts new phenomena | | |
| Clarity of writing (10%) | | |
| spelling accurate | | |
| clearly written | | |
| correct grammar | | |
| **Quality of Writing (15%)** | | |
| business-like fashion | | |
| concise reasoning | | |
| clearly labeled diagrams, graphs, charts, etc. | | |

Chapter 5

For use with *Patterns and Predictions*, Chapter 5

# What is in the Physics InfoMall for Chapter 5?

This chapter of *Coordinated Science for the 21st Century* is about recognizing patterns and making predictions – among the goals of all science! That means we do not have a specific physics concept, or even a small core of concepts, to look for on the Physics InfoMall CD-ROM. Looking at the list of topics from the **Chapter Challenge**, we see that topics cover a wide range, including force fields, astronomy, astrology, forms of ESP, predictions, and communicating with animals! This will not be a simple list, with only a few books or articles, but is likely to use a large fraction of the InfoMall resources.

As you may have already discovered, the Physics InfoMall CD-ROM is a rather large data base of physics information from sources such as physics journals, textbooks, pamphlets, catalogs, history, and other physics teachers. There are thousands of journal articles, 19 textbooks, books on selected topics (such as crystal growing), thousands of homework problems (many with solutions), demonstrations, catalogs, and study guides (some are to accompany the textbooks also on the CD-ROM). This information is in text and graphic format; none of it is interactive. The volume, variety, and accessibility of the information are the real values of the InfoMall.

This may be a good time to comment on a few InfoMall characteristics. One strength of the CD-ROM is that it is dual-platform; it works on Macintosh and PCs. It operates similarly on either platform, but not identically. This bridge between *Coordinated Science for the 21st Century* and the InfoMall was produced on a MAC, with all work done on the same machine. Do not be surprised if some items appear a little different on a PC than described here – it is expected and nothing to worry about. The data base is the same for both platforms.

There are several methods for finding information. You may wish to simply browse the InfoMall, clicking on hyperlinks that seem promising. Browsing may be most useful for those who are familiar with the InfoMall; the organization for the CD-ROM is similar to a real mall – there are stores that contain related information. If you know what you are looking for, you can simply go straight to the relevant store. Since we wish to become familiar with the CD-ROM, we will concentrate on other methods for finding information, primarily the use of searches.

If you are familiar with the search engine, you may wish to skip this paragraph and the next. The search engine for the InfoMall enables the user to locate almost any word or passage in the more than 3,000 articles, 19 textbooks, etc. If you use a Macintosh (remember – the same CD-ROM works on both PC and MAC), you can search the entire data base (that is, all stores) at once. Stores should be searched one at a time on a PC. The same information can be found on either platform. There are two categories for searches: Simple and Compound. Simple searches look for a single word or phrase. Compound searches can locate up to four words or phrases that may not occur together. The Compound search also provides search options that are not available in the Simple search. Since a Compound search can always be done for a single word, a Compound search can always be used rather than a Simple search. All searches in this discussion will be done using the Compound search.

The search option can be found under the "Functions" menu. When you select "Compound Search", you will get a window that has all the options you need for just about any search you are likely to want. Among these choices are data base selection (which store or stores you want searched), category (perhaps only certain articles), and search words. You can enter any words or words you want the engine to search for. These words can be combined with the logical operators AND, OR, and NOT. That is, you can find passages that contain, for example, "force" AND "field" but NOT "corn". In addition, you can use an asterisk * as a wild character. That is, search word "fun*" will search for any word beginning with "fun", such as "fun", "funny", "funding", "fundamental", etc. Wild characters allow great freedom in searching, but as this example shows, you need to be careful. And finally, if your search is too broad, you will get "Too Many Hits" and you will have to restrict your search parameters to something that will not be found so many times on the CD-ROM. Such restriction techniques include searching in fewer stores, using fewer wild characters, or searching for less common words followed by a search of only the search hits. With a little practice, you should have very little trouble finding the information you desire.

What we will see here is a sample of what you can find on the InfoMall. This will not be a complete list of all that is useful, nor will it be an exhaustive look at every problem or concept from *Coordinated Science for the 21st Century.* You will be able to find much more than we can touch on here. For example, we will often not look at the textbooks, but you may want to check them out as alternative sources of information for you and your class.

Back to the **Chapter Challenge** . . . Some of the research topics on the list are also included in activities; others are not. For example, **Force Fields** is **Activity 1**, while auras is not a specific activity. So let's see what we can find on auras now.

The obvious search is to search all stores at once. If you have looked at the *Active-*ating the InfoMall sections before, you may have noticed we usually eliminate the Keyword Kiosk from searches, since we often know exactly what words we want to look for. In this case, there may be alternates to "auras" to search for, so let's include the Kiosk this time. If we use only the keyword "aura*" on the search, we find that this search accidentally includes items we really do not want: note that "aural" includes hearing. This means we will need to ignore some of our search results. This can be difficult if the search list is long, but this one is short enough to simply proceed. A short way down the list we find. "ESP: Teaching "scientific method" by counterexample," from the *American Journal of Physics*, vol. 43, issue 12, 1975. This may be a nice article to use for several of the items on our Challenge research list! (Coincidentally, if you find this article by going to the Articles & Abstracts Attic, choosing *American Journal of Physics*, the volume 43, and issue 12, you will notice that the article just preceding this one is "Raja Sawai Jai Singh II: An 18th century medieval astronomer," which was also on our search list, since the word "Aurangzeb" appears often in this article. Enough to got you curious just what "Aurangzeb" is?) You are urged to search for other topics from this list. The search for "aura" should be a warning for us in our searches: if the topic is not in the real of science, we may not find much on this CD-ROM. Try "dowsing" and you will find only three hits. but these may be amusing or informative, including at least one demonstration of psychic power! Topics that are good science should be here somewhere.

# ACTIVITY 1
## Force Fields

# Background Information

The discovery of magnetism is attributed to the Greeks. In fact, the name magnet is believed to be derived from Magnesia, a region in Asia Minor. The Chinese are credited with first using lodestone, a naturally-occurring magnet, for navigational purposes.

The presence of a magnet modifies the surrounding space producing a magnetic field. A magnetic field is detected by using another magnet as a test particle. If the second magnet experiences a force (actually a torque) in a region of space, we know that a magnetic field exists. Perhaps the best known magnetic field is the one produced by Earth. We see the effects of the Earth's magnetic field by observing the motion of a compass needle, which is a small magnet.

A magnetic field is represented by lines of force; the strength of the field is indicated by the number of these lines per unit area. The direction of the magnetic field at any point is indicated by a tangent to the line of force at that point. Experimentally, the direction of the magnetic field is shown by a compass. Lines of force form closed loops that continue around and through the magnet. By convention, their direction is from north pole to south pole. Magnetic lines of force do not intersect.

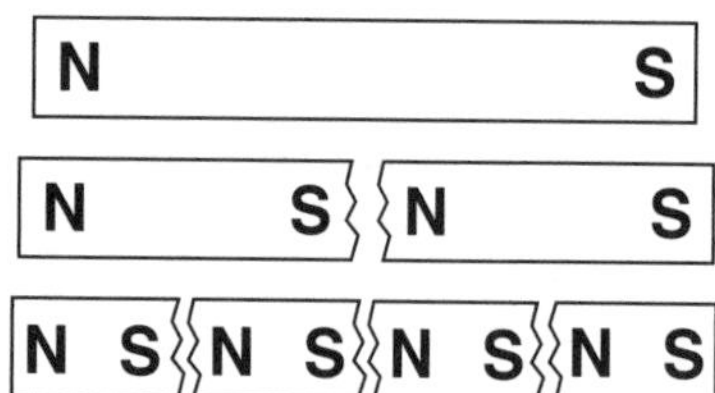


Every magnet is a dipole; that is, it has a north and south pole. Any attempt at breaking a magnet into two monopoles results in the formation of two bipolar magnets. When two magnets interact, like poles repel and unlike poles attract.

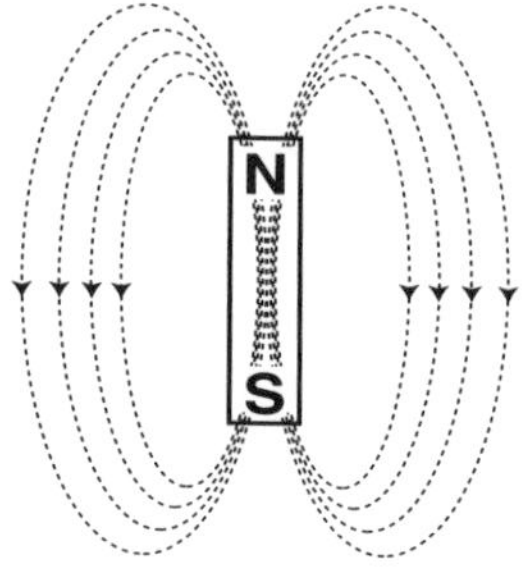


The current explanation of the Earth's magnetic field centers on the molten interior of the planet and electrical currents hypothesized as running though that molten material. Details are not well understood, however. Geologic evidence from rocks along the Mid-Atlantic rift shows that the Earth's magnetic field has flipped – reversed directions – at least 170 times in the past 17 million years. Again, the mechanism of such flips are not understood.

Is the magnetic field a real physical entity or just a useful construct? According to Einstein and Infield in *The Evolution of Physics*, "The electromagnetic field is, for the modern physicist, as real as the chair on which he sits."

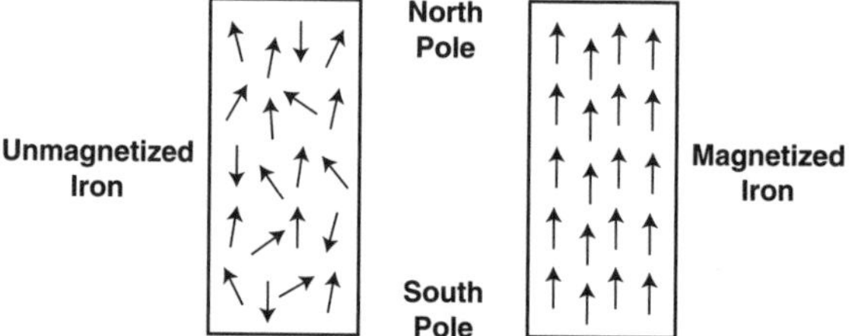


## *Active*-ating the Physics InfoMall

Some of the searches we will look at for this first activity will be specific to concepts that will only appear in this activity, but may be useful throughout this chapter. You may wish to remember some of the references we find. When you find a reference you really like, you may want to mark it with a Bookmark or Note. (Bookmarks and Notes are not covered in this *Active Physics*-InfoMall bridge, but you may want to learn how they work. The User's Guide can be found in the Utility Closet.)

The first search attempt is for "force field*". There are some hits on this list that may not be useful for anything more than noting that the term "force field" is really used in physics, not just in science fiction. But is the meaning the same?

One of two hits that looked good immediately is "Electric fields and football fields," *The Physics Teacher*, vol. 28, issue 8, 1990. According to this article, "there is remarkable similarity between athletic fields and force fields of all kinds. This similarity can be exploited when introducing students to the idea of a physical force field." Perhaps you can use some of the similarities in your classes? If you go to the Articles & Abstracts Attic to find this article, this indicates you are interested in finding information relevant to your classes. This can be dangerous – there are a large number of interesting articles in these journals, and you can lose hours just reading them. For example, there is an article on Amusement Park Physics in this same journal volume. There is no cause for worry. though; all search hits provide direct hyperlinks to the book or article. You can always browse later.

The second hit that looked great was the "Uniform Force Field vs. An Inverse Square Law Field" section of the Work and Energy chapter of *Teaching High School Physics* from the Book Basement. The title of this book alone hints at usefulness. The title of this section is also promising – we will need inverse square information in a later activity (**Activity 2**). There are also nice graphics (of graphs, no less) here. These can be copied and pasted right into your own handouts. Let's keep these in mind for later.

These two hits are not the only ones worthy of attention, but you should look for yourself. After all, we have only searched using the title of this activity; we haven't even reached **For You To Do** yet!

**For You To Do** is all about magnets. Given the single topic, you may be tempted to just go to the Demo & Lab Shop and look around. Notice that such browsing was weakly discouraged above; do it anyway. For example, with only about 2 minutes of looking (no exaggeration), we found "Three Dimensional Views of Magnetic Effects" in *Potpourri of Physics Teaching Ideas*. This is not the only great idea there, but it was an attention-grabber, since **FYTD Step 2 (c)** asks students if the magnetic field lies in only two dimensions. This is not the only topic you will see in *Potpourri of Physics Teaching Ideas* that you will find relevant, so check this out.

**FYTD Step 2 (c)** also brings to mind a very important point: the students are asked what they think, as opposed to what they saw or what an equation says. The students' preconceptions are brought to a conscious level. Making predictions can be vital to student learning, especially if the instructor is aware of these preconceptions and asks appropriate questions. Let's find some appropriate information on this topic.

If you search the InfoMall for "student difficult*" OR "student understand*" you will find several articles that deal with research into how students learn fundamental concepts in physics. A quick look at titles from this search indicates that much of the research into student difficulties has centered on mechanics. The value of such research is discussed in, for example, "A conceptual approach to teaching kinematics," *American Journal of Physics*, vol. 55, issue 5, 1987. A passage copied directly from this article states "We believe that the identification of specific student difficulties through research and the design of instructional strategies in a classroom environment have together made possible the development of curriculum that is directly responsive to student needs." Similar ideas are expressed in "Millikan Lecture 1990: What We Teach and What is Learned – Closing the Gap," *American Journal of Physics*, vol. 59, issue 4, 1991. There has not been a large amount of research into student understanding of magnetism (such research has been conducted since the InfoMall CD-ROM was produced), but the message is clear: we need to directly address known preconceptions in order to aid students in their learning. **FYTD Step 2 (c)** directly addresses one of the commonly known preconceptions some students hold.

Part of this activity is making a diagram of the magnetic field. One search that you may try is "magnetic field map". This gives only a single hit, but it is a good one! It has graphics and experiments you may want to use. The resource in Chapter 34 of *Physics for the Inquiring Mind* in the Textbook Trove. Why did we get only one hit? The search words were not very broad; simply adding an asterisk (wild character) to "map" (so it is now "map*") adds to the list. Of course, there are other choices: "field map*" or "magnetic field" for example.

**For You To Read** suggests a search you should try: "William Gilbert" AND "magnetic field" AND "earth". You may find nice graphics of the Earth's field (limited to the quality of graphics that can be compressed and fit on a single CD-ROM with this much information). Of course, you may want to search for "William Gilbert" alone to find biographical information.

The **Inquiry Investigation** has many possibilities for InfoMall searches. Want to know how an electromagnet works? Try the InfoMall. Search for "electromagnet" and you get many hits. One that should not be overlooked is near the bottom of the rather long list. This comes from an under-used part of the InfoMall – the Calendar Cart. This tells us that on "05/22/1783 William Sturgeon was born in Whittington, England. He made the first electromagnet in 1823." The Calendar Cart often has tidbits that can help add interest to some topics. One idea is to print the contents of the Calendar Cart for an entire month and mention some of the significant physics events on their anniversaries.

# Planning for the Activity

## Time Requirements

one class period

## Materials Needed

### For the class:

- chair with wheels
- roll of butcher paper

### For each group:

- compass
- iron filings, LB
- bar magnet large pair
- metal objects: magnetic and non-mag. pkg.
- magnets, ceramic rings

## Advance Preparation and Setup

Distribute small metal objects around the room for use in **Step 1 (e)**.

# Teaching Notes

You can demonstrate magnetic fields for the class. Dip a magnet into a box of paper clips. The paper clips will become temporary magnets and will line up with the field. The paper clips will be most densely packed in regions in which the field is strongest. Many modern paper clip dispensers use magnetic force to keep some clips at the top of the dispenser, in easy reach. Display one for students, if possible.

Gift shops and novelty shops also carry desktop objects that can be used to make various "magnetic sculptures." Allow students to play with such devices, if available, and explain how they work.

**Step 4** of **For You To Do** provides a very dramatic method of mapping magnetic fields. Lines of force become visible as the iron filings (tiny magnets themselves) align along field lines.

You might want to demonstrate the magnetic field of various magnets on the overhead projector. Place the magnet of the projector surface and cover it with a blank transparency sheet. Then sprinkle iron filings over the sheet.

Allow ample time for students to explore the effects of magnets on classroom objects. Direct them to stay away from computers and remove computer disks that may be positioned around the room before students begin this activity.

Caution students about using iron filings in **Step 4**. They should not breathe them in. You may also want them to wear safety goggles.

SAFETY PRECAUTIONS: Caution students about safety if they are using nails or sewing needles in **Step 5**, and be vigilant for inappropriate behavior with these objects. You may wish students to wear safety goggles.

Your students are probably familiar with magnets and their uses, although they will probably be unfamiliar with lines of force. Some students may mistakenly attribute lines of force with properties of physical entities.

As students discuss and compare the evidence and explanations for magnetic effects with the claims of psychics for "auras," you may wish to remind them of the nature of scientific evidence – how it must be repeatable by other observers.

NOTES

# Activity Overview

Students examine magnetic force fields and their effects in this activity. They read about the discovery of magnetism, the history of its scientific explanation, and the Earth as a magnet. They are challenged in the **Inquiry Investigation** to make an electromagnet.

## Student Objectives

### Students will:

- Investigate the properties of bar magnets.
- Plot magnetic fields.
- Make a temporary magnet.
- Describe the properties of magnetic fields.
- Make an electromagnet.

ANSWERS FOR THE TEACHER ONLY

## What Do You Think?

As indicated in the **Background Information**, the presence of a magnet modifies the surrounding space producing a magnetic field. A magnetic field is detected by using another magnet as a test particle. If the second magnet experiences a force (actually a torque) in a region of space, we know that a magnetic field exists.

Magnetic substances include iron, nickel, cobalt, and a few alloys which contain one or more of these metals.

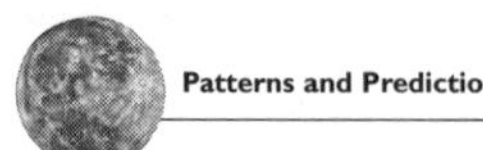

Patterns and Predictions

Activity 1 Force Fields

**GOALS**

In this activity you will:

- Investigate the properties of bar magnets.
- Plot magnetic and electric fields.
- Make a temporary magnet.
- Describe the properties of magnetic and electric fields.
- Make an electromagnet.

**What Do You Think?**

Large magnets are able to pick up cars and move them around junkyards.

- **How does a magnet work?**
- **What objects are attracted to magnets and which objects are not attracted to magnets?**

Record your ideas about these questions in your *Active Physics* log. Be prepared to discuss your responses with your small group and the class.

Active Physics 250

### For You To Do

1. Get two bar magnets from your teacher. Use them to answer the following questions:
   a) Do the magnets exert a force on one another? How do you know?
   b) Must they touch one another to exert the force?
   c) Is the force between the two magnets attractive or repulsive? Is it both? Make a drawing that shows how the forces between the magnets act.
   d) A bar magnet can be described by its ends. These ends are called magnetic poles. They are labeled N and S. Opposite poles attract each other and like poles repel. Are the poles on your magnet marked? If not, get a magnet that is marked, and use it to find out which pole on your magnet is N and which is S.
   e) Do the magnets exert a force on other objects in the classroom? Find out. Record your findings.

2. The needle of a compass is a small bar magnet. The N pole is usually painted red or shaped like an arrow. Place one of the bar magnets under a sheet of paper. Put the other magnet out of the way by moving it some distance from the paper. Move a small compass back and forth just above the surface of the paper. Be sure to move the compass back and forth in close rows so you don't miss large areas of the paper. As you move the compass, sketch the direction of the N pole of the compass at different points on the paper. The diagram you have made shows the magnetic field around the magnet under the paper. Answer the following questions in your log:
   a) The N pole of the compass points in the direction of the field. What is the direction of the magnetic field at the N pole of the bar magnet? At the S pole of the bar magnet?
   b) How does the strength of the magnetic field change with distance from the magnet? Explain your thinking.
   c) Your plot of the magnetic field is in two dimensions. Do you think that the magnetic field lies in only two dimensions? How could you test your answer?

## Answers

## For You To Do

1. a) Yes; one magnet causes a change in motion of the other magnet.
   b) No.
   c) Both; see students' drawings.
   e) Students should discover that their bar magnets can exert a force on some metal objects in the classroom.

2. a) Away from the magnet at the N pole; toward the magnet at the S pole.
   b) The strength of the magnetic force decreases with increased distance from the magnet. Students can determine this using the magnetic compass, or methods in **Step 1**.
   c) Students should speculate that magnetic force surrounds a magnet completely. Accept reasonable test ideas.

ANSWERS

## For You To Do
***(continued)***

3. a) The shape of the field depends on the shape of the magnet. The N pole of a horseshoe magnet is at the end of one arm of the U-shaped magnet; the S pole is at the end of the other arm of the magnet.

4. Students provide diagram. See below.

5. a) Students should be able to magnetize the object. They will know it is magnetized by the affect it has on other small metal objects or if it is attracted or repulsed by a small bar magnet.

**Avoid having the iron filings come into direct contact with the magnet.**

3. Repeat **Step 2** with a horseshoe magnet.
   a) Does the field depend on the shape of the magnet you use?

4. You can map a field around a magnet quickly with iron filings. In a tray place a sheet of stiff paper over the magnet. Sprinkle the filings on the paper and gently tap the paper. The filings will behave like tiny compasses. They will fall along the lines of the magnetic field and make an instant plot of the field.
   a) Sketch the magnetic field shown by the iron filings.

5. Get a metal paper clip, sewing needle, or small nail. These objects become magnets when they are stroked by a magnet. Stroke the object with a bar magnet. Stroke in one direction only.
   a) Did the object become magnetic? How do you know?

Physics Words

**electric field:** the region of electric influence defined as the force per unit charge.

**magnetic field:** the region of magnetic influence around a magnetic pole or a moving charged particle.

### FOR YOU TO READ
#### Magnetic Fields

Some forces act across space. One of these is the force between two magnets. The ancient Greeks first discovered magnetism over 2000 years ago. They found lodestone, an ore that attracted some other rocks. The name *magnet* probably comes from the region in Asia Minor where lodestone was first discovered, Magnesia. By the 10th century, the Chinese and Vikings were using magnetic compasses, which they invented, in boating.

A distinguished English physicist, William Gilbert, who was also a physician to Queen Elizabeth I, studied magnetism in detail in the late 1500s. He published a book called *De Magnete* in 1600. *De Magnete* explained how a magnet had an "invisible orb of virtue" around it. The idea of an "invisible orb of virtue" has been replaced with the theory of **magnetic fields**.

Magnets fill the region of space around themselves with a magnetic field. When another magnet enters the magnetic field, it experiences a force. Magnetic fields can be "seen" when another magnet is brought into the field and the force it experiences is observed. You did this with a magnet and compass. The direction of a magnetic field is the direction of the force on the N magnetic pole of the second magnet.

Gilbert also showed how a small ball of magnetic material could make magnetic compasses brought near them behave like compasses on Earth. His demonstrations led to the idea that the Earth behaves like a giant magnet, an idea that has been shown to be true since his time. The N pole of a magnet is attracted to the magnetic pole of the Earth in the Northern Hemisphere.

4.

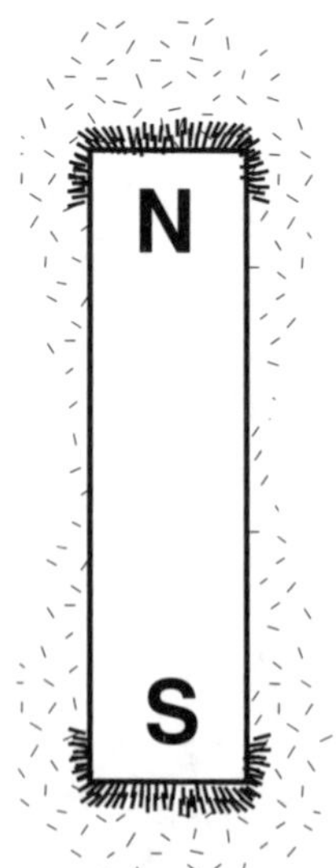

Activity 1 Force Fields

The magnetic pole of Earth in the Northern Hemisphere and the geographic North Pole are not in exactly the same place. Today they are about 1500 km from each other. Scientists have evidence that the position of the magnetic poles has changed during the history of the Earth.

All magnets have two poles. If you break a magnet in half, you'll get two magnets, each with a N pole and a S pole. Scientists are trying to break magnets into small enough pieces—atom-size pieces—to make a magnet with just one pole, but they have not been able to do so. If someone does one day, he or she may be on the short list for the Nobel Prize!

The magnetic field is invisible. While it cannot be seen, it can be measured. Anyone can use a compass to find out if a magnetic field exists. Observations of magnetic behavior are predicted by scientific theory. Magicians claim to project the powers of their minds across empty space by filling space with psychic auras. There are no known ways to detect auras. Some people say that they can sense auras. But they cannot demonstrate this "sense" to others. There is no "compass" to detect auras.

### Electric Fields

In your previous study of charges, you learned that like charges (+ + or - -) repel and unlike charges (+ - or - +) attract. You also learned that Coulomb's Law describes the force of attraction:

$$F = \frac{kQq}{R^2}$$

+Q    +q

You can think of the force on $+q$ as the interaction between the two charges $+Q$ and $+q$.

The force on a charge can also be described with the field concept. The electric field surrounding $+Q$ can be mapped in a way similar to how the magnetic field was mapped.

In mapping the magnetic field of a bar magnet, you observed what would happen to a compass placed near the bar magnet or what would happen to small iron filings (similar to tiny compasses) placed near the bar magnet.

To map the **electric fields** due to a group of charges, you can place small positive charges and observe what will happen. These small positive charges are called "test" charges since they test for the electric field.

If positive test charges were placed near the $+Q$, they would be repelled. You could draw that force of repulsion with little arrows.

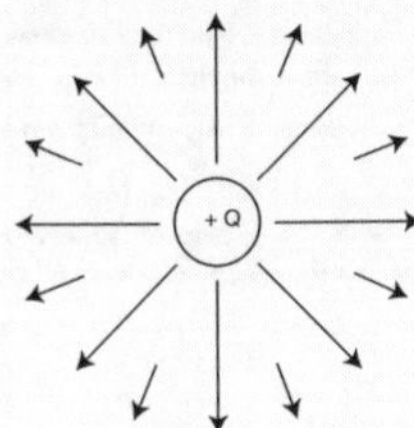


→

The lengths of the force vectors are intended to show the strength of the force. The test charges placed close to $+Q$ have a larger force than the test charges placed further from $+Q$.

The general pattern of the forces on the test charges is depicted as the electric field of the $+Q$ charge. In this case, the electric field of $+Q$ is:

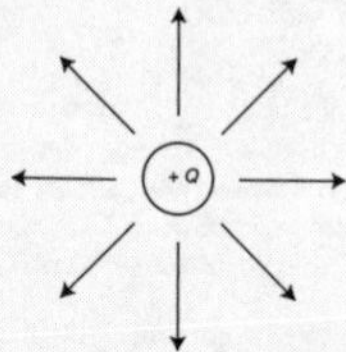


The lines tell you the following properties of the electric field or *E*– field of $+Q$:

- The direction of the electric field is the direction of the force on a + test charge.
- The electric field is strongest where the lines are close together and weaker where the lines are further apart.
- The electric field is at all points. The lines just show the field at some points.
- The electric field extends out to infinity.

The electric field lines from $+Q$ look like the points on a blowfish or like the pieces of a Koosh ball.

A similar process can be used to find the electric field of a long line of charges.

The test charge will always be pushed away from the line of charge. The force is due to the vector sum of all of the force interactions of the test charge and each charge on the line.

The electric field lines from a line of charge look like the spokes of many bike wheels lined up next to one another. In the diagram below, you don't see the lines coming at you or away from you. The lines radiate out from the line of charge in all directions. As you get further from the line of charge, the electric force gets weaker and the electric field lines spread out as the spokes of a wheel spread out.

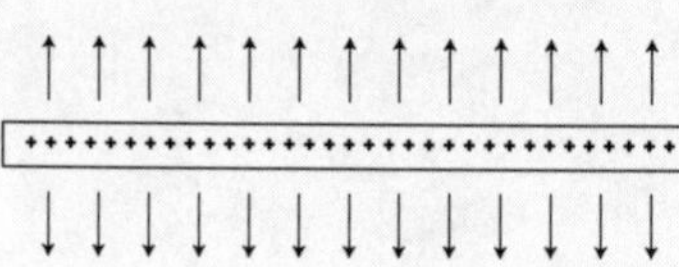

The electric field from a line of charge gets weaker as you get further from the charges, but does not weaken as much as the electric field from a spherical set of charges $+Q$.

An extremely important electric field for all sorts of electric circuits is the electric field of parallel plates. The top plate has positive charges and the bottom plate has negative charges. The force on a positive test charge will be toward the negative plate and away from the positive plate.

Since the electric field lines do not get closer or further apart within the parallel plates, you know that the force on the positive test charge is a constant force. These parallel plates are also referred to as parallel-plate capacitors.

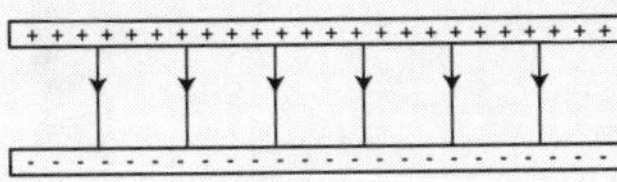

## Reflecting on the Activity and the Challenge

You know that magnetic and electric forces act across space. You were able to detect the magnetic field of a bar magnet at different distances from the magnet. The field lines are invisible, yet they can be detected by anyone using tools such as a compass or test charges. Often, people refuse to believe in things that are invisible. Can you convince someone that magnetic and electric forces and fields are real even though they are invisible? Some people believe in psychic auras. These people insist that these are also invisible and should be believed. But ordinary people with everyday tools are not able to detect auras. Psychic auras do not meet the same criteria as do magnetic and electric fields. If you funded research in psychic auras, you would expect the researchers to be able to detect auras.

ANSWERS

## Physics To Go

1. Answers will vary, but may include prior experience, experience gained in the activity, and the attraction of a magnetic compass for Earth's magnetic pole in the Northern Hemisphere.
2. The N pole of one magnet repels the N pole of another magnet; the N pole of one magnet attracts the S pole of another magnet.
3. Students provide diagrams. See below.
4. Students provide diagrams. See below.
5. Students provide diagrams. See below.



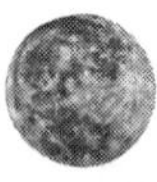

### Physics To Go

1. How do you know that magnets and electric charges act across a distance?
2. Will the N pole of one magnet attract or repel the N pole of another magnet? The S pole of another magnet?
3. Draw each of the **magnets** shown below. Then draw the magnetic field **around each magnet.**

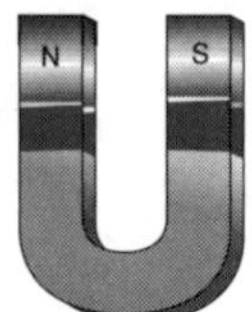

4. Copy the diagram to the left. Each circle represents a magnetic compass. Draw the compass needle for each, as it would point in its position. Use an arrow for the N end of the compass needle.
5. Copy the diagrams of electric charges below. Draw the electric field around each.

a) 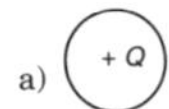

b) 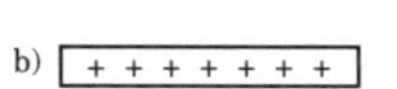

c) 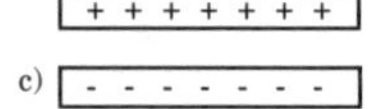


6. How does the strength of a magnetic field change with distance from a magnet?
7. A magnet is hanging from a ceiling by strings. Which way will it point?
8. The needle on a magnetic compass points to the magnetic pole of the Earth in the Northern Hemisphere. You know that like poles repel and unlike poles attract. What can you say about the magnetic pole of the Earth in the Northern Hemisphere and the N pole of the magnetic compass?



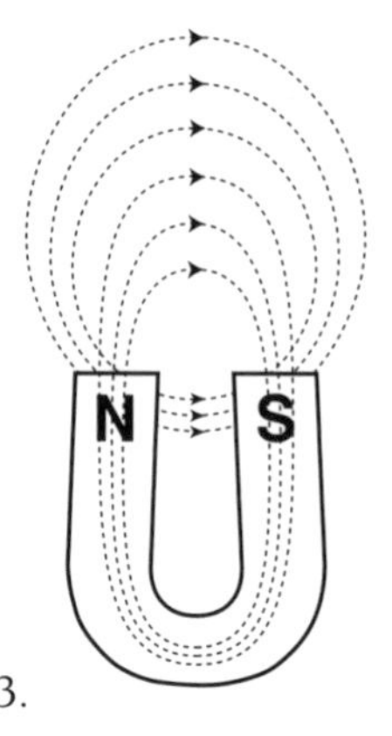


3.

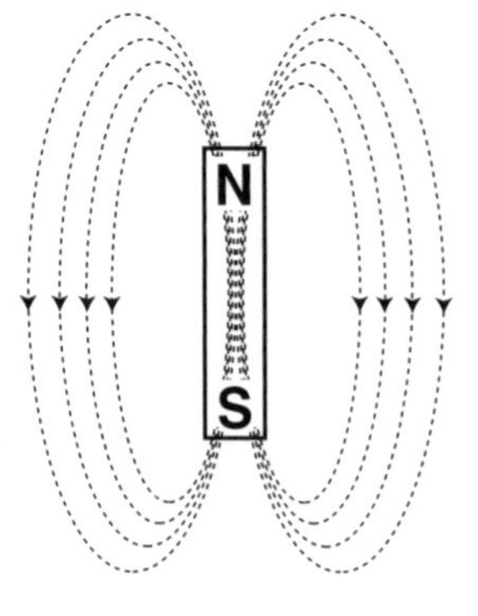


4.

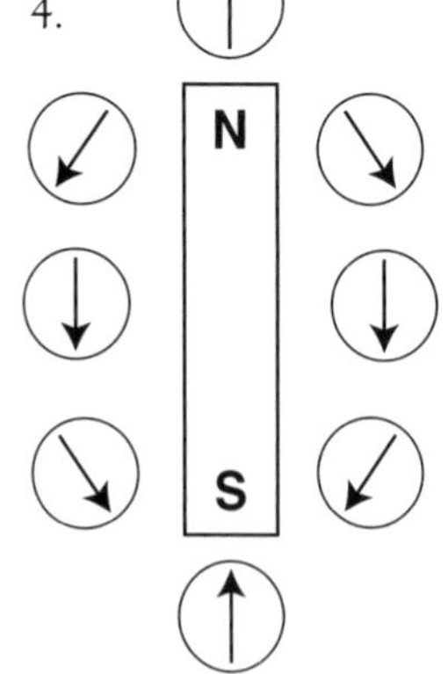


5. a)

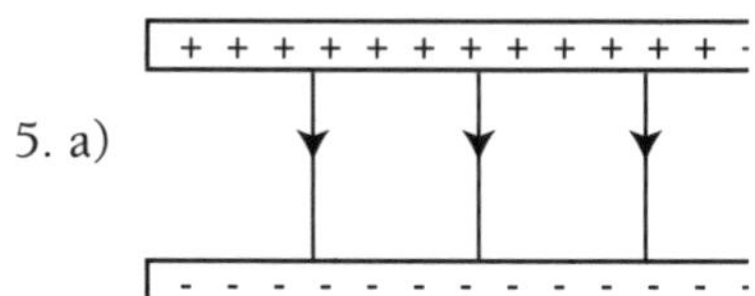

5. b)

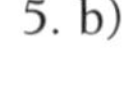

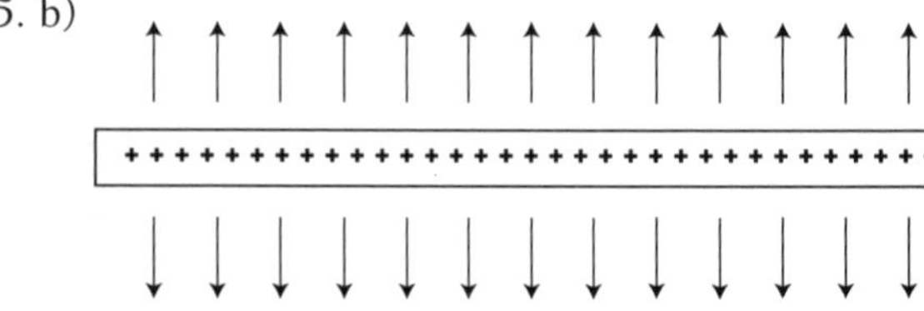

5. c)

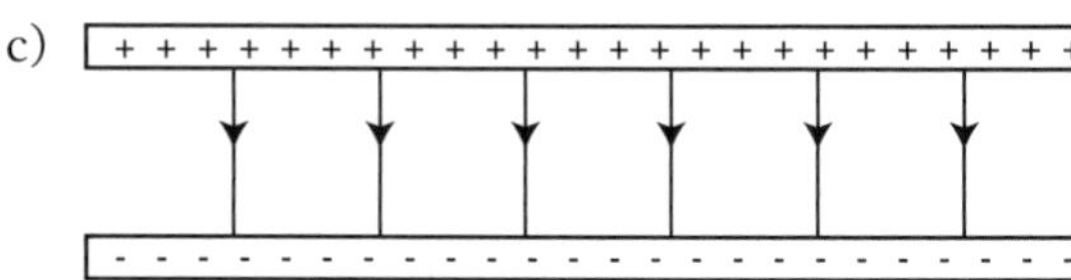

### Stretching Exercises

1. Obtain two or three household flat ("refrigerator") magnets. Use iron filings and paper or a magnetic compass to explore the magnetic field of each magnet. Draw a diagram of your findings.
2. Electricity and magnetism are different forces. However, they are related to one another. Find out how. Carefully wrap a length of wire around an iron nail. Connect each end of the wire to a battery terminal. When you attach the wires to the battery, current starts flowing through the wire. Bring a magnetic compass near the wire coils. Record your observations.

   You've made an electromagnet. Electromagnets are magnetic when current is flowing through the wire. They are the type of magnet used to pick up cars and move them around junkyards. They have many other uses, too. Research the uses of electromagnets, and report your findings to the class.



ANSWERS

## Physics To Go

***(continued)***

6. The strength of a magnetic field decreases with the distance from the magnet.
7. Their N poles will point toward the Earth's magnetic pole in the Northern Hemisphere.
8. Even though we call each pole an N pole, they cannot be the same magnetic poles because like poles repel and unlike poles attract. The pole in the Northern Hemisphere is actually equivalent to the south pole of a bar magnet.

ANSWERS

## Stretching Exercises

1. See students' diagrams. In general, the broad flat edges of the magnet are opposite poles. One of the poles is usually covered with a plastic surface, and its magnetic force cannot be measured.
2. Check students' electromagnets. Students will find that electromagnets have many varied uses. Around the home, electromagnets are found in door bells and telephones.

# Assessment Challenge: Inquiry Investigation

This rubric has some criteria, but it would be advantageous for the teacher to go over the criteria with the students, and have the students add any that may have been overlooked. Try to have the number of points be a multiple of 10, for ease of calculation of a mark out of 100. However, whatever the criteria, a mark of 100 can be given using percent. If you choose, you may want to change the marking scheme to 3 points, rather than 5.

| Descriptor | 5 | 4 | 3 | 2 | 1 |
|---|---|---|---|---|---|
| **Scientific Terminology and Accuracy** | | | | | |
| correct use of scientific terms | | | | | |
| evidence of external research and/or exploration of facts | | | | | |
| clear understanding of electromagnets | | | | | |
| compare reality to illusion using scientific terminology | | | | | |
| **Electrical Investigation** | | | | | |
| follows safety procedure | | | | | |
| evidence of background research about electricity | | | | | |
| reference to at least two scientists involved with the discovery of the properties of electricity | | | | | |
| **Magnetic Investigation** | | | | | |
| evidence of background research about magnetism | | | | | |
| historical references to magnetism | | | | | |
| reference to at least two scientists involved with the discovery of the properties of magnetism | | | | | |
| **Uses of Electromagnets** | | | | | |
| at least five different uses | | | | | |
| each of the uses are explained well | | | | | |
| research about history of electromagnets | | | | | |
| reference to at least two scientists involved with the discovery of the relationship between electricity and magnetism | | | | | |
| **Written Quality** | | | | | |
| easy to read | | | | | |
| well laid out | | | | | |
| concise, simple language | | | | | |
| uses diagrams where appropriate | | | | | |

For use with *Patterns and Predictions*, Chapter 5, Activity 1: Force Fields

NOTES

# ACTIVITY 2
## Newton's Law of Universal Gravitation

# Background Information

As with a magnet, a mass distorts the space that surrounds it. Each bit of matter produces a gravitational field. An object placed in this field will experience an attractive force. Unlike the magnetic force, gravity is strictly attractive and acts on all objects.

The gravitational force is the weakest of all the fundamental forces. Consequently, the gravitational force between ordinary objects such as baseballs and people is extremely small. Gravity becomes an important force only when at least one of the objects is extremely massive. We feel the force of Earth's gravity on us because of the Earth's gigantic mass. The force of gravity provides the "cosmic glue" that holds solar systems and galaxies together.

Sir Isaac Newton proposed that the gravitational force was universal: it acts on all masses everywhere in the universe. He showed that the magnitude of the gravitational force is directly related to the product of the two masses involved in the interaction. He also demonstrated that gravity depends on the distance separating two bodies in a unique way. Gravity obeys the "inverse square law" – the attractive force decreases inversely with the square of the distance separating two masses. When the distance between two masses doubles, the force of attraction is reduced by one-fourth, and so on. As with any force obeying the inverse square law, gravity's range is infinite.

Newton verified his law of universal gravitation by comparing the acceleration of a terrestrial object to the acceleration of the Moon. Newton knew that the ratio of the distance from the center of Earth to the Moon to the radius of Earth was roughly 60:1. Hence the ratio of the acceleration should be $1/60^2$, or 1/3600, which it is.

In 1798 Henry Cavendish confirmed Newton's Law of Universal Gravitation and measured the gravity constant, $G$. His apparatus consisted of two spheres of known mass at either end of a light rod suspended by a thin fiber from the center of the rod. He had earlier found the small force that was needed to twist the fiber. By bringing a third sphere close to one of the suspended spheres, he was able to measure the force of gravity between the spheres.

Dowsing has been practiced for centuries, to search for minerals as well as water. There is no scientific explanation for its claimed "success," and the scientific community largely remains skeptical of dowsers' claims. One idea put forth is that dowsing results from electromagnetic forces between the dowsers' hands and the material being sought.

## *Active*-ating the Physics InfoMall

This activity describes another force field, the gravitational field. **What Do You Think?** asks about zero-gravity. Yet in **Physics Talk**, we see that the gravitational force between two objects (such as the Earth and fish in orbit) is never zero. A straightforward calculation shows that there is plenty of gravity acting on orbiting fish. How, then, are orbiting fish treated as if there is no gravity? Let's see what the InfoMall says. Search for "weightlessness". (Even such a strange word works.) Of course, our *Coordinated Science for the 21st Century* book does not use the word "weightlessness," but students often think of astronauts as weightless. It is a good idea to be prepared for questions about this.

The third hit from this search ("Fun in Space," *American Journal of Physics*, vol. 28, issue 8, 1960) says "the force of gravity 200 miles above the surface of the Earth is only 10% less than it is on the Earth's surface. Even at 4000 miles the gravity is reduced only to one-quarter of its value on the Earth's surface; and at 8000 miles, to one-ninth. Since it is obviously gravity that holds a satellite in a circular orbit, and since the Earth's gravity is even strong enough out at the distance of the Moon – 240,000 miles – to hold the Moon in its orbit, the weightlessness in an Earth satellite is evidently not caused by the absence of gravity." How does one explain this? Look this one up and see what it says.

The first hit, from *Potpourri of Physics Teaching Ideas* (in the Demo & Lab Shop) has some interesting demonstrations. The second hit, "Physics of living in space: A new course," *American Journal of Physics*, vol. 49, issue 8, 1981, may also interest you or your students.

As you go through **For You To Do**, you encounter photocells and galvanometers. If you want information on the operation of these devices, or additional uses for them (or even information on how to order them), simply search using "photocell" or "galvanometer". There is plenty of information on these devices.

In **Step 8**, we see reference to the inverse square relation. As luck would have it, we did a search earlier that found something useful. That was the

"Uniform Force Field vs. An Inverse Square Law Field" section of the Work and Energy chapter of *Teaching High School Physics* from the Book Basement. This includes a graph of an inverse square relation. If we don't want to rely on luck, we do a search for "inverse square" and find several hits that mention Newton, gravity, light intensity, and even our reference from the Book Basement.

In **For You To Read**, we read a little about Henry Cavendish. A search using his last name provides many excellent articles, including "The Cavendish experiment as Cavendish knew it," *American Journal of Physics*, vol. 55, issue 3, 1987. Or, in the Calendar Cart, we find that on "10/10/1731 Henry Cavendish was born in Nice, France. In 1798, he determined the value for the gravitational constant *G*. With this number, he then calculated the mass of the Earth." There is more information than you will need, but you may still want to look at some of it.

Later in **For You To Read**, we find dowsing. We searched for that earlier, but you may want to check it out again.

**Physics Talk** introduces us to the equation often used to summarize Newton's Universal Law of Gravitation. An easy choice for information on this Law is the Textbook Trove. There are books at the conceptual, algebra, and calculus levels to suit almost any needs. Additional information can be found on the InfoMall in another place that can be easily overlooked. If you look at the InfoMall Entrance screen (the one with the Mall directory map), the right-hand side is lined with buttons. One of these is an additional link to the Calendar Cart. The lowest button takes us to the Equations Dictionary. Click on this, and you will get a menu. Choose, in order, mechanics, dynamics, then gravity, and gravitational forces. You will see a hyperlink to the Universal Law of Gravitation. This gives you the equation, plus access to examples. The search engine does the work and finds a few selected examples. Of course, you can always find your own. Plus, there is still the Problems Place in case you want worked out examples or more homework-type problems (for **Physics To Go**).

# Planning for the Activity

## Time Requirements

one class period

## Materials Needed

### For the class:

- chalkboard
- projector
- piece of chalk

### For each group:

- battery 9-volt
- light meter to replace galvanometer
- meter stick, 100 cm, hardwood

## Advance Preparation and Setup

Use the masking tape or cardboard and tape to square off the lamp of the projector so that a square of light is delivered during this activity. You may also want to premark the measured positions for the projector using masking tape on the floor of the room. Check that the photocell/galvanometer apparatus is working properly.

# Teaching Notes

You may wish to review scientific notation and its use in simple operations before student are assigned the **Physics To Go** questions.

Engage volunteers in setting up the equipment and positioning the projector. Others may work in groups to predict the results to be found with each new distance of the projector from the board. You may wish to have one student record data on the board.

Caution students moving around the room to step over the projector cord and stand clear of the projector, which might become hot during the activity.

In everyday life, gravity seems to cause different objects to fall at different rates. Autumn leaves drift slowly to Earth whereas an acorn may drop like a rock. Observations such as these lead people to believe that more massive objects fall faster than light ones. Aristotle was the most respected proponents of this school of thought. Galileo was the first to demonstrate that objects of different masses fell equal distances in the same time. Like Aristotle, your students may cling to the misconception that heavy objects fall faster than light ones. If necessary, provide experimental evidence to the contrary and discuss the effects of surface area and air resistance on various falling objects.

# Activity 2 A

# Galileo's Gig

Encourage students to discuss their ideas and enter them in their logs. If necessary, ask probing questions to lead students in seeing the value of measurable evidence.

## WHAT DO YOU THINK?

- Is the gravity of the Earth the same as the magnetism of the Earth? Explain.

Record your ideas about this question in your *Active Physics* log. Be prepared to discuss your responses with your group and the class.

## FOR YOU TO DO

1. Do you think that the acceleration of objects near the Earth's surface changes with the material the objects are made of? Design a simple experiment to justify your answer.

   a) Record your prediction in your log.

   b) Record your experimental design in your log.

   c) After your teacher has approved your experiment, carry out the investigation. Record your results in your log.

2. Do you think that the acceleration of objects near the Earth's surface depends on the mass of the object? Again design a simple experiment to justify your answer.

   a) Record your prediction in your log.

   b) Record your experimental design in your log.

   c) After your teacher has approved your experiment, carry out the investigation. Record your results in your log.

3. Set up the ticker-tape timer. Put the timer on a table and thread a strip of paper tape through the timer.

4. Attach the other end of the tape to a heavy object that doesn't have much air resistance. It's conventional here to use a rock, but various small objects work equally well, as long as the object weighs much more than the tape in the timer so that friction between the tape and the timer won't mess up your measurement. Different members of the class should choose objects with differing masses and made of different materials.

For use with *Patterns and Predictions*, Chapter 5, Activity 2: Newton's Law of Universal Gravitation

5. Turn on the timer and drop the object. Make three tapes for different objects to help you eliminate the effect of air resistance on your measurement. In this activity, we wish to obtain a numerical value for the acceleration due to the Earth's gravity, so we shall have to be careful about using numbers. Measure all distances in meters and all times in seconds. Remember that 6-dot intervals correspond to 1/10 of a second.

6. For each object, measure the distance covered by each group of 6-dot intervals on the tape. Divide these distances by the time taken (1/10 second). This will give you the speed of the object in each time interval.
   a) Record your calculations in your log.

7. Calculate the change in speed between adjacent time intervals. Again divide this change by 1/10 second (because each measured change in speed also happened in 1/10 second) to obtain the acceleration of the falling object. It may help to use a data table like the one below. To obtain results, take the average acceleration for each falling object.
   a) Record your calculations in your log.

Time interval:__________________________seconds

| Distance of 6-dot intervals in meters | Distance divided by time interval = speed | Change in speed | Change in speed divided by time = acceleration |
|---|---|---|---|
| | | | |
| | | | |
| | | | |

   b) Which result do you think is closest to the acceleration due to gravity?
   c) Do you trust one result more than another? Why?
   d) Do your measurements agree with those of others in the classroom? Why do you think they agree or disagree?
   e) You've measured the acceleration due to gravity. Do you think it's the same everywhere on the surface of the Earth? Why or why not?
   f) Does the acceleration due to gravity vary for different masses? Is the acceleration the same for falling rocks and elephants?

GALILEO SAID THAT OBJECTS MOVE IN A STRAIGHT LINE AT CONSTANT SPEED UNLESS THEY ARE ACTED ON BY AN OUTSIDE FORCE

ALL OBJECTS TAKE THE SAME TIME TO FALL THE SAME DISTANCE

ONE DAY GALILEO BET THE SCHOLARS THAT TWO DIFFERENT WEIGHTS WOULD HIT THE GROUND AT THE SAME TIME

GALILEO RELEASED THE WEIGHTS AT EXACTLY THE SAME TIME

BOOM!

THEY HIT AT THE SAME TIME!!!

IT PAYS TO UNDERSTAND MOTION!

Galileo lived in the 1600s

- Do you think that the last panel accurately depicts the values of Galileo?
- What values are assumed in the cartoon?
- Why do you think Galileo risked imprisonment in explaining his ideas?

For use with *Patterns and Predictions*, Chapter 5, Activity 2: Newton's Law of Universal Gravitation

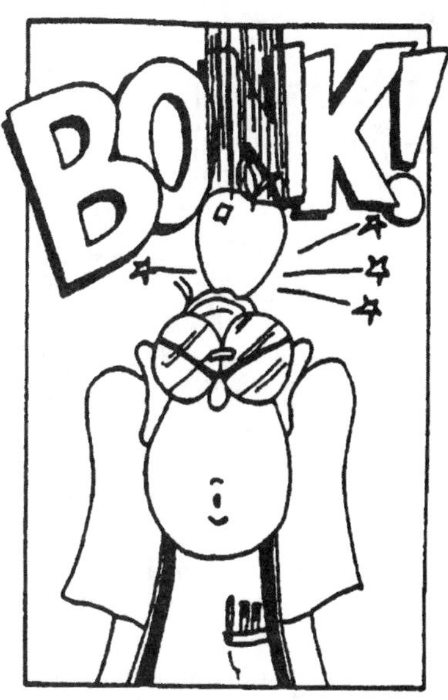

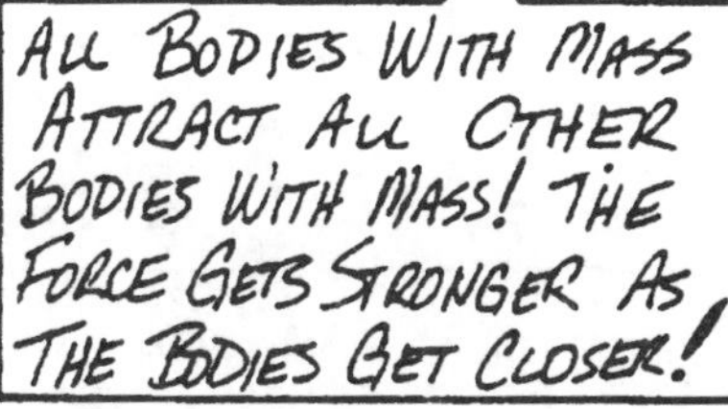

Create the last panel.

For use with *Patterns and Predictions*, Chapter 5, Activity 2: Newton's Law of Universal Gravitation

# Activity Overview

Students discover a pattern in both the spreading out of light as distance from the source increases and the reduced intensity of light as distance from the source increases, phenomena analogous to the reverse square law exhibited by the gravity force. They graph and analyze the patterns. Finally, they calculate acceleration due to gravity on Earth from given data, graph their calculations, and examine the pattern. In the **Stretching Exercise**, students determine if the magnetic force obeys the inverse square law, and in the **Inquiry Investigation** students explore the "force" claimed by dowsers that enables them to find water underground.

## Student Objectives

### Students will:

- Explore the relationship between distance of a light source and intensity of light.
- Graph and analyze the relationship between distance of a light source and intensity of light.
- Describe the inverse square pattern.
- Graph and analyze gravity data.
- State Newton's Law of Universal Gravitation.
- Express Newton's Law of Universal Gravitation as a mathematical formula.
- Describe dowsing and state why the practice is not considered scientific.

Patterns and Predictions

# Activity 2 Newton's Law of Universal Gravitation

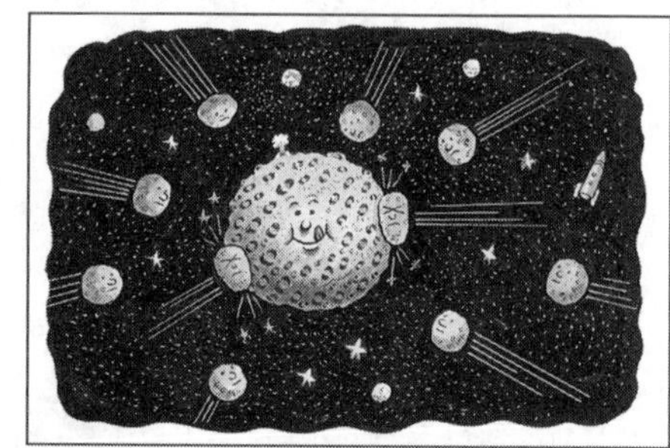

**GOALS**

In this activity you will:

- Explore the relationship between distance of a light source and intensity of light.
- Graph and analyze the relationship between distance of a light source and intensity of light.
- Describe the inverse square pattern.
- Graph and analyze gravity data.
- State Newton's Law of Universal Gravitation.
- Express Newton's Law of Universal Gravitation as a mathematical formula.
- Describe dowsing and state why the practice is not considered scientific.

**What Do You Think?**

Astronauts on many Shuttle flights study the effects of zero-gravity. Fish taken aboard the Shuttle react to "zero-gravity" by swimming in circles.

- **How would a fish's life be different without gravity?**
- **Does gravity hold a fish "down" on Earth?**

Record your ideas about these questions in your *Active Physics* log. Be prepared to discuss your responses with your small group and the class.

**For You To Do**

1. Place a projector 0.5 m from the chalkboard. Insert a blank slide. Turn on the projector.
2. Use chalk to trace around the square of light on the board.

Active Physics  258

ANSWERS FOR THE TEACHER ONLY

## What Do You Think?

Like all things on Earth, fish are subject to gravitational forces. Their existence is also influenced by the buoyant force of the water in which they live. Modern bony fish have a swim bladder which operates much like a balloon. The swim bladder is filled with oxygen, nitrogen, and carbon dioxide. It helps the fish maintain buoyancy and permits it to change depth with ease. By adding or removing gas, the fish can adjust its buoyancy. Without gravity, there is no reason for a preferred direction based on density. The fish will float in the water at any level. "Zero-gravity" refers to a free-fall situation.

3. Place the photocell in one corner of the light square. Attach it to the galvanometer as directed by your teacher. The photocell and galvanometer measure light intensity. The more light that strikes the cell, the greater the current reading on the galvanometer.
   a) Copy the table below in your log. Record the distance to the board, current in galvanometer, and length of a side of the square.

| Distance to board (m) | Distance squared | Current in galvanometers (A) | Side of square (cm) | Area of square ($cm^2$) |
|---|---|---|---|---|
| | | | | |
| | | | | |
| | | | | |

4. Move the projector to a position 1 m from the board. Adjust the projector so that the original square of light sits in one corner of the new square of light.
   a) Enter the data into the table in your log.
5. Repeat **Step 4** with the projector at distances of 1.5 m, 2 m, 2.5 m, and 3 m.
   a) Enter the data into the table in your log.
6. Graph the current in the galvanometer versus distance. Label this graph Graph 1.
   a) Is Graph 1 a straight line?
   b) What does a straight line on the graph tell you?

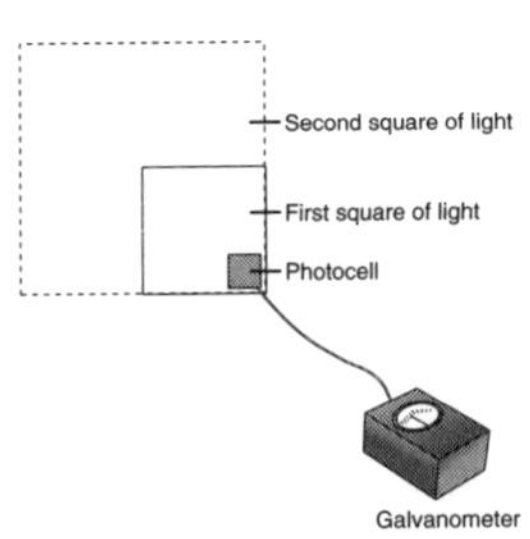




ANSWERS

## For You To Do

1.–2. Student activity.

3. a)–5. a) Students record their data in their logs.

6. a) No, it is a curved line.

   b) A straight line tells you that a relationship is directly proportional. The relationship on this graph is inversely proportional.

ANSWERS

## For You To Do
*(continued)*

7. a) Answers will vary, based on students' data, but should reflect the inverse square law.

8. a) It is a curve similar to that graphed in Step 7.

   b) Answers will vary. The pattern shows an inversely proportional relationship.

**Physics Words**

**acceleration:** the change in velocity per unit time.

**gravity:** the force of attraction between two bodies due to their masses.

**Inverse square relation:** the relationship of a force to the inverse square of the distance from the mass (for gravitational forces) or the charge (for electrostatic forces).

7. Light intensity decreases with distance as the light from the source spreads out over larger areas. The light is literally spread thin. The light intensity at any one spot increases as the area gets smaller and decreases as it gets larger. This observation is an example of a pattern called the inverse square relation. In an inverse square relation, if you double the distance the light becomes $\frac{1}{2^2}$ or $\frac{1}{4}$ as bright. If you triple the distance, the light becomes $\frac{1}{3^2}$ or $\frac{1}{9}$ as bright. If you increase the distance by 5 times, the light becomes $\frac{1}{5^2}$, or $\frac{1}{25}$ as bright. If you increase the distance by 10 times, the light becomes $\frac{1}{10^2}$, or $\frac{1}{100}$ times as bright.

   a) How closely does your data reflect an inverse square relation?

**Acceleration Due to the Earth's Gravitational Field at Different Heights**

| Height above Sea Level (km) | Acceleration due to Gravity (m/s²) |
|---|---|
| 0 | 9.81 |
| 3.1 | 9.76 |
| 11 | 9.74 |
| 160 | 9.30 |
| 400 (Shuttle orbit) | 8.65 |
| 1600 | 6.24 |
| 8000 | 1.92 |
| 16,000 | 0.79 |
| 36,000 (geosynchronous orbit for communications satellite) | 0.23 |
| 385,000 (orbit of the Moon) | 0.003 |

8. Compute the distances from the center of the Earth (6400 km below sea level). Plot these distances vs. acceleration in a graph. Draw the best possible curve through the points on the graph. Label this graph Graph 2.

   a) Does the data form a pattern?

   b) Is the pattern familiar to you? Give evidence for your conclusion.

Activity 2 Newton's Law of Universal Gravitation

## FOR YOU TO READ

### An Important Pattern

You've seen one pattern in this activity. But you've seen it in two different ways. In **Steps 1** through **8** you found that light intensity becomes less as the light source is moved further away. In **Step 7,** you've seen that **acceleration** due to **gravity** becomes less as an object moves further from the surface of the Earth. Both are examples of the **inverse square relation.** Although light is not a force, the effect of distance on its behavior in this activity is like that of the effect of distance on the force of gravity. That is, the behavior of light in this activity is analogous to the behavior of gravity. In simple terms for gravity, the inverse square relation says that the force of gravity between two objects decreases by the square of the distance between them.

### Mapping the Earth's Gravitational Field

In **Activity 1,** you mapped the magnetic field around a bar magnet using a compass as a probe. You also read about electric fields. In this activity, you used data on acceleration due to gravity to map the Earth's gravitational field. The probe is the acceleration of a falling mass. To see the pattern of Earth's gravitational field, you needed data from satellites. The gravitational field changes very slowly near the surface of the Earth. The pattern is very difficult to see using surface data.

Newton's Law of Universal Gravitation describes the gravitational attraction of objects for one another. Isaac Newton first recognized that all objects with mass attract all other objects with mass.

Experiments show that objects have mass and that the Earth attracts all objects. Newton reasoned that the Moon must have mass, and that the Earth must also attract the Moon. He calculated the acceleration of the Moon in its orbit and saw that the Earth's gravity obeyed the inverse square relation. It is a tribute to Newton's genius that he then guessed that not only the Earth but all bodies with mass attract each other.

Almost 100 years passed before Newton's idea that all bodies with mass attract all other bodies with mass was supported by experiments. To do so, the very small gravitational force that small bodies exert on one another had to be measured. Because this force is very small compared to the force of the massive Earth, the experiments were very difficult. But in 1798, Henry Cavendish, a British physicist, finally measured the gravitational force between two masses of a few kilograms each. He used the tiny twist of a quartz fiber caused by the force between two masses to detect and measure the force between them.

**Newton's Law of Universal Gravitation states:**

**All bodies with mass attract all other bodies with mass.**

**The force is proportional to the product of the two masses and gets stronger as either mass gets larger.**

**The force decreases as the square of the distances between the two bodies increases.**

→

### Physics and Dowsing: Comparing Forces

Dowsing is a way some people use to locate underground water. It is claimed to work on an apparent "attraction" between running water and a dowsing rod carried by a person. All dowsers claim to feel a force pulling the rod towards water, and many claim to feel unusual sensations when they cross running water. In the 19th century, many dowsers described the force on the rod as an electric force. No evidence supports this idea. In fact, there is no scientific theory to explain any attraction between running water and a dowsing rod.

Despite the skepticism of the scientific community about dowsing, it is widely used in the United States. Even a national scientific laboratory has used dowsers! But the United States Geological Survey has investigated dowsing and finds no experimental evidence for it. Statistics show that the success rate could be a result of random events. Even if experimental evidence supported the success of dowsing, there is no theory to predict its operation. In order to be accepted as scientific, a phenomenon must be reproducible in careful experiments. Its effects must be predictable by a theory. Also, the theory must give rise to other predictions that can be tested by experiments.

### PHYSICS TALK

#### Newton's Law of Universal Gravitation in Mathematical Form

Complex laws like Newton's Law of Universal Gravitation may look easier in mathematical form. Let $F_G$ be the force between the bodies, $d$ be the distance between them, $m_1$ and $m_2$ the masses of the bodies and $G$ be a universal constant equal to $6.67 \times 10^{-11}$ N·m²/kg².

You can express Newton's Law of Universal Gravitation as

$$F_G = \frac{G\, m_1 m_2}{d^2}$$

You can see that the equation says exactly the same thing as the words in a much smaller package.

### Reflecting on the Activity and the Challenge

In this activity you determined experimentally how light intensity varies with distance. By plotting measured data, you found that gravity follows an identical pattern. You detect gravity by measuring the acceleration of objects falling at specific locations. Patterns help you understand the world around you. Light follows the inverse square relation and so does gravity. You can detect gravity with masses. You can detect magnetic fields with compasses. But you cannot detect the "attraction" claimed by dowsers. There are no detectors for that! You will be required in the **Chapter Challenge** to differentiate between the measured gravity and its inverse square nature and the dowser's claim of measurement. This activity helped you to understand one difference between science and pseudoscience.

### Physics To Go

1. How would the light intensity of a beam from a projector 1 m from a wall change if the projector was moved 50 cm closer to the wall?
2. The gravitational force between two asteroids is 500 N. What would the force be if the distance between them doubled?
3. A satellite sitting on the launch pad is one Earth radius away from the center of the Earth ($6.4 \times 10^6$ m).
   a) How would the gravitational force between them be changed after launch when the satellite was two Earth radii ($1.28 \times 10^7$ m) from the center of the Earth?
   b) What would the gravitational force be if it was $1.92 \times 10^7$ m from the center of the Earth?
   c) What would the gravitational force be if it was $2.56 \times 10^7$ m from the center of the Earth?
4. Why does everyone trust in gravity?
5. Why doesn't everyone trust in dowsing?



ANSWERS

## Physics To Go

1. It would increase by a factor of 4.
2. It would decrease by 1/4.
3. a) It is reduced by 1/4.
   b) It is reduced by 1/9.
   c) It is reduced by 1/16.
4. We trust gravity because it can be measured and explained by scientific theory, and because it consistently holds in everyday experience.
5. Everyone does not trust dowsing because it cannot be explained by scientific theory and its success depends on the dowser, that is, it is not repeatable by anyone.

ANSWERS

## Physics To Go
***(continued)***

6. a) The water on the side of the Earth facing the Moon is closer to the Moon than is the center of the Earth.

   b) Because it is closer, the water on the side of the Earth facing the Moon is pulled by the Moon's gravity toward the Moon more than the center of the Earth.

   c) Answers will vary, but should demonstrate some understanding of the effect of the Moon's gravity on the Earth's surface water facing the Moon. Students may also suggest, correctly, that the Earth's surface water on the side facing away from the Moon is effected less by the Moon's gravity than is the center of the Earth, thus it also mounds up.

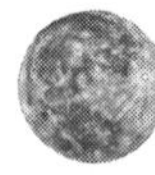



6.

a) Which is closer to the Moon, the middle of the Earth or the water on the side of Earth facing the Moon?
b) Use your answer to **a)** to propose an explanation for the uneven distribution of water on Earth's surface, as shown in the diagram.
c) Suggest an explanation for high tides on the side of the Earth facing the Moon.

### Stretching Exercises

1. To locate underground water, a dowser uses a Y-shaped stick or a coat hanger bent into a Y. The dowser holds the Y by its two equal legs with the palms up and elbows close to his or her sides. The long leg of the Y is held horizontal. The dowser walks back and forth across the area he or she is searching. When he or she crosses water, the stick jerks convulsively and twists so hard that it may break off in the dowser's hands. Dowsers claim to be unaware of putting any force on the stick. Most observers think that the motion of the stick is probably due to the unconscious action of the dowser.

   According to records of those who believe in dowsing, approximately 1 in 10 people should have the ability to dowse. Do you have dowsing ability? Try this activity to find out. Can you prove that you're a dowser to a classmate? What would constitute proof?

2. Does the inverse square relation apply to magnetic force? Work with your group to plan an experiment to find out. State your hypothesis, and describe the method to test it. If your teacher approves your experimental design, try it. Report your results to the class.

ANSWERS

## Stretching Exercises

1. Students' paragraphs should reflect the ideas that a dowser can be affected by signals unknowingly given by those around him or her or that prior knowledge could affect results. Students' opinions should reflect the notion that dowsing does not meet the requirements of science because experimental results cannot be replicated by anyone, bias cannot be eliminated from such experiments, and, most importantly, there is no theory to explain the results of dowsing.

2. Check students' experimental plans and their results. Students will discover that the inverse square relation applies to a magnetic force, as the response of a test object to the field decreases with distance from the magnet.

NOTES

# ACTIVITY 3
## Slinkies and Waves

# Background Information

A wave may be defined as a disturbance that travels through a medium. There are essentially two types of waves: mechanical and electromagnetic. Water waves, sound waves, a pulse traveling along a rope or Slinky®, and a tremor in the Earth are examples of mechanical waves. Light and radio waves are examples of electromagnetic waves. Mechanical waves require a material medium; air, water, steel springs, or fabric are examples. Electromagnetic waves can travel in a vacuum. As a mechanical wave passes through a medium, the particles of the medium move about some equilibrium position, but do not suffer any permanent displacement. After a wave has passed through a medium, the medium returns to its previous undisturbed state.

Waves may be further categorized as transverse, longitudinal, or torsional. In transverse waves, particles in the medium move perpendicular to the motion of the disturbance. A pulse on a Slinky or along a rope are examples. As longitudinal waves, such as sound, pass through a medium, particles in the medium vibrate back and forth parallel to the direction in which the wave is traveling. Torsional waves, which are not covered in this unit, are waves of twist. A twist that originates at the end of an extended garden hose may be observed traveling along the length of the hose.

Speed of a mechanical wave is determined solely by the medium through which it travels. Electromagnetic waves all travel with the same speed through a vacuum, but travel more slowly through transparent material such as glass or water.

A wave is characterized by its amplitude, wavelength, frequency, and period. The amplitude is the height (or depth) of the wave from the medium's equilibrium position. The wavelength is the distance between adjacent crests, adjacent troughs, or in general any two points on the wave that are in phase. The frequency of a wave is the number of complete waves that pass a point per second. The period of a wave is the time required for a complete wave to pass one point in space.

All waves exhibit similar behavior: they reflect, refract, diffract, and interfere. Waves may pass through each other and emerge completely unchanged.

## *Active*-ating the Physics InfoMall

For **What Do You Think?**, you can search for "Tacoma Narrows" and find several references, including "Resonance, Tacoma Narrows Bridge Failure, and Undergraduate Physics Textbooks," *American Journal of Physics*, vol. 59, issue 2, 1991; and "11/07/1940 The Tacoma Narrows Bridge collapsed on a typical breezy day in Tacoma Narrows, Washington. This event was filmed, and later used to study resonance and its effects upon construction" from the Calendar Cart.

Need some good information on waves? Of course, the textbooks will have plenty of information. "Waves" is a pretty broad topic, so you will need to make your searches reasonably specific, such as "Slinky wave*". This search found a versatile mount for Slinky wave demonstrations. If you search for "waves" you will get "Too Many Hits" unless you look in only one store per search. For example, if you search the Articles & Abstracts Attic, you find that there are MANY articles that mention waves, but not in a useful manner. Far down the list, you will find "People Demos," from *The Physics Teacher*, vol. 21, issue 3, 1983. This has a demonstration you can use with your students to make them part of the wave.

If the **Physics Talk** terms need to be expanded, you can either browse a textbook, or search for the terms you want more information on. For more examples of the equation relating velocity to frequency and wavelength, you can go to the Equations Dictionary (the lowest button on the InfoMall Entrance screen), or go to the Problems Place and find any section on waves.

For demonstrations, you can either browse or search the Demo & Lab Shop. There are many (perhaps hundreds) of ideas there. Some are quite simple, others require special equipment. You may find ideas to extend the **Stretching Exercise**.

# Planning for the Activity

## Time Requirements

one class period

## Materials Needed

### For the class:

- computer

### For each group:

- heavy duty Slinky®
- yarn, colored, 20 cm
- Slinky polarizer

## Advance Preparation and Setup

Find an open area which will accommodate groups of students working with Slinkies.

# Teaching Notes

The people wave activity works best if you give specific directions to students sitting in rows. Direct the first row of students to raise their hands in unison, then lower them in unison. As that row of students lower their hands, the next row of students raise their hands. As the second-row students lower their hands, the third-row students raise their hands, and so on across the room. Adapt your directions for the wave made by standing and sitting.

You can extend this part of the activity by having students make a longitudinal people wave. Line students up so that they are facing the back of another student with just more than two-steps distance between them. Direct the student at the end of the line to take a step toward the second student. He or she then takes a second step toward the second student as the second student takes a step toward the third student. Then the first student steps back toward his or her original position as the second student takes a second step toward the third student and the third student takes a first step toward the fourth student. On the next count, the first student steps back into his or her original position, the second student steps back toward the first student; the third student takes a second step toward the fourth student, and the fourth student steps toward the fifth student. This people wave is actually a fairly good model of a sound wave.

Students may need assistance in designing the experiment called for in **Step 4 (b)** of the **For You To Do** activity. You might guide them to see that they could generate waves of notably different heights, and use the height of the waves to distinguish between them in the activity.

Find a large enough space to accommodate students working with Slinkies. Preview the activity with the entire class before beginning, then travel from group to group to ensure that each group is performing the activity appropriately. If it is possible to videotape the motions, it will be both entertaining and educational.

Many of your students will hold the notion that the particles in the medium through which a wave is passing move with the wave. Film footage, or personal experience, with large ocean waves, supports the misconception. As students are working with the Slinkies, probe this issue with them. While they will clearly see that their notion does not apply to people waves and the waves along a Slinky, they may rely on the misconception to explain water waves or sound waves.

Encourage students to state what they have learned about waves in the activity, including their observations and measurements. Have them discuss why measurements are important in their understanding of waves, and, record their ideas in their logs.

NOTES

# Activity 3 Slinkies and Waves

**GOALS**

In this activity you will:

- Make a "people wave."
- Generate longitudinal and transverse waves on a Slinky.
- Label the parts of a wave.
- Analyze the behavior of waves on a Slinky.
- Compare longitudinal and transverse waves.
- Define wavelength, frequency, amplitude, and period of a wave.
- Measure the speed of various waves on a Slinky.

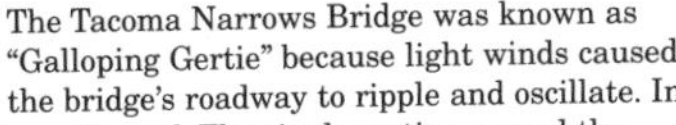

**What Do You Think?**

The Tacoma Narrows Bridge was known as "Galloping Gertie" because light winds caused the bridge's roadway to ripple and oscillate. In 1940 the bridge collapsed. The ripple motion caused the structure to break.

- **Have you ever crossed a bridge that was rippling? How secure did you feel?**
- **If you were an engineer, how would you test the strength of bridges?**

Record your ideas about these questions in your *Active Physics* log. Be prepared to discuss your responses with your small group and the class.



# Activity Overview

Students examine the behavior of waves by making "people waves" and with a Slinky. The "people wave" portion of the activity provides a kinesthetic approach to the study of waves; many students will benefit from this first activity with waves. Definitions of longitudinal and transverse waves are introduced, as is the terminology used in physics to describe waves.

## Student Objectives

### Students will:

- Make a "people wave."
- Generate longitudinal and transverse waves on a Slinky.
- Label the parts of a wave.
- Analyze the behavior of waves on a Slinky.
- Compare longitudinal and transverse waves.
- Define wavelength, frequency, amplitude, and period of a wave.
- Measure the speed of various waves on a Slinky.

ANSWERS FOR THE TEACHER ONLY

## What Do You Think?

Most students will have experienced the sensation of a bridge vibrating, especially when larger trucks are covering the expanse.

Engineers build and test models to determine the type of forces they can withstand.

ANSWERS

## For You To Do

1. a) In the first wave, arms moved up and down; in the second, bodies moved up and down.

   b) From one side of the room to the other.

   c) No.

   d) Accept all reasonable answers.

2. a) From one end of the Slinky to the other end.

   b) Back and forth, perpendicular to the direction of the pulse.

   c) Back and forth.

   d) No.

   e) See students' sketches.

   f) The speed of the pulse remained the same as it moved.

   g) It comes back in the other direction; it is reflected.

   h) The speed of the pulse does not depend on the size of the pulse. Answers will vary but should all be the same.

Patterns and Predictions

### For You To Do

1. With your classmates, make a "people wave," like those sometimes made by fans at sporting events.
   - Sit on the floor about 10 cm apart. At your teacher's direction, raise, then lower your hands. Practice until the class can make a smooth wave.
   - Next, make a wave by standing up and squatting down.

   a) Which way did you move?
   b) Which way did the wave move?
   c) Did any student move in the direction that the wave moved?
   d) What is a wave?

2. Work in groups of three. Get a Slinky® from your teacher. Two members of your group will operate the Slinky; the third will record observations. Switch roles from time to time.

   Sit on the floor about 10 m apart. Stretch the Slinky between you. Snap one end very quickly parallel to the floor. A pulse, or disturbance, travels down the Slinky.

   a) Which way does the pulse travel?
   b) Look at only one part of the Slinky. Which way did that part of the Slinky move as the pulse moved?
   c) Mark a coil on the Slinky by tying a piece of colored yarn around the coil. Send a pulse down the Slinky. Describe the motion of the coil tied with yarn.
   d) Send a pulse down the Slinky. Watch the pulse as it moves. Does the shape of the pulse change?
   e) Sketch the Slinky with a pulse moving through it.

   f) Does the speed of a pulse appear to increase, decrease, or remain the same as it moves along the Slinky?
   g) What happens to a pulse when it reaches your partner's end of the Slinky?
   h) Shake some pulses of different sizes and shapes. Does the speed of a pulse depend on the size of the pulse? Use a stopwatch to time one trip of pulses of different sizes.

3. Instead of sending single pulses down the Slinky, send a wave, a continuous train of pulses, by snapping your hand back and forth at a regular rate.

Active Physics 266

a) Sketch the wave. Use the diagram below to label the parts of the wave on your sketch.

The crests of this wave are its high points. The troughs are its low points. The wavelength of this wave is the distance between two crests or between two troughs.

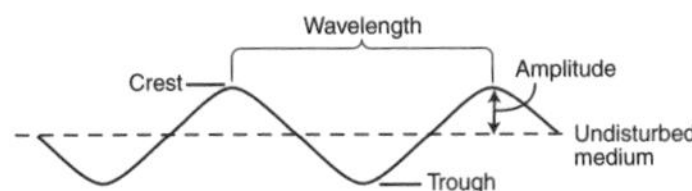


b) How does the number of wave crests passing any point compare to the number of back-and-forth motions of your hand?

4. Lift one end of the Slinky and drop it rapidly. You've sent a vertical pulse down the Slinky. If both ends of the Slinky are lifted and dropped rapidly, the vertical pulses will be sent down the Slinky and will meet in the middle. Try it.

a) What happens when the pulses meet in the middle?

b) Do the pulses pass through each other or do they hit each other and reflect? Perform an experiment to find out.

5. Gather up 7 or 8 coils at the end of the Slinky and hold them together with one hand. Hold the Slinky firmly at each end. Release the group of coils all at once.

a) Describe the pulse that moves down the Slinky.

b) Sketch the pulse.

c) In which direction do the coils move? In which direction does the pulse move?

6. You can use a Slinky "polarizer" to help you determine the direction of oscillation of the transverse wave. A Slinky "polarizer" is a large slit that the Slinky passes through.

Have one student hold the "polarizer" vertically. Have another student send a vertical pulse down the Slinky by moving her arm in a quick up and down motion.



ANSWERS

## For You To Do
***(continued)***

3. a) See students' sketches.

b) The number of wave crests passing any point is the same as the number of back-and-forth motions of your hand.

4. a) Accept reasonable answers; students should note that the amplitude of the pulse made where the pulses from each wave meets is greater than the amplitude of either wave.

b) Students' experiments should demonstrate that the waves pass through each other.

5. a) The pulse moving down the Slinky is a clump of coils.

b) See students' sketches.

c) The coils move back and forth in the direction that the wave moves (parallel to the wave). The pulse moves from one end of the Slinky to the other end.

ANSWERS

## For You To Do
*(continued)*

6. a) The pulse travels through the "polarizer" unaffected.

7. a) The pulse is not transmitted through the "polarizer."

8. a) Student predictions will vary. Encourage students to record their predictions before they actually do the experiment.

9. a) The pulse is not transmitted.

10. a) The pulse travels through the "polarizer" unaffected.

11. a) The light passes through the filter.

12. a) When the filters are 90° to one another the light is blocked.

a) Describe what happens to the pulse on the far side of the "polarizer."

7. Continue to have one student hold the "polarizer" vertically. Have another student send a horizontal pulse down the Slinky by moving his arm is a quick side-to-side motion.

   a) Describe what happens this time to the pulse on the far side of the "polarizer."

8. Predict what might happen if a diagonal pulse were sent toward the vertical "polarizer."

   a) Record your prediction.

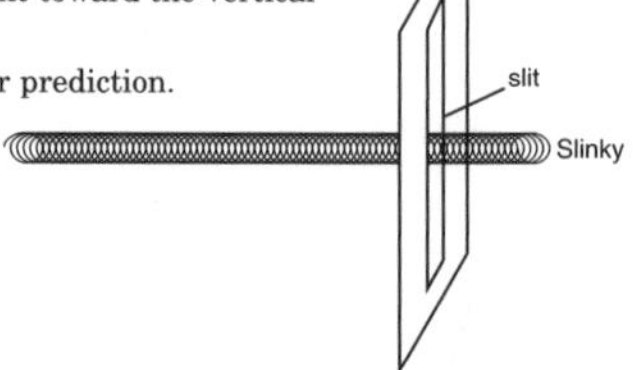


9. Send the diagonal pulse toward the vertical "polarizer."

   a) Record your observations.

10. Gather up 7 or 8 coils at the end of the Slinky and hold them together with one hand. Hold the Slinky firmly at each end. Release the group of coils all at once.

    a) Did this pulse travel through the "polarizer"?

11. You may have heard of polarizing sunglasses. Your teacher will supply you with two pieces of polarizing film that could be used for sunglasses. Hold each polarizing filter to the light and record your observation. Make a sandwich of the two pieces of polarizing filter and hold it to the light.

    a) Record your observations.

12. Rotate one polarizing filter in the sandwich by 90°. Hold it to the light.

    a) Record your observations.

13. The Slinky "polarizer" can be used as a way to understand the light polarizers. If a vertical Slinky pulse were sent through two vertical slits, you would expect that the pulse would pass through the first slit and then the second slit.

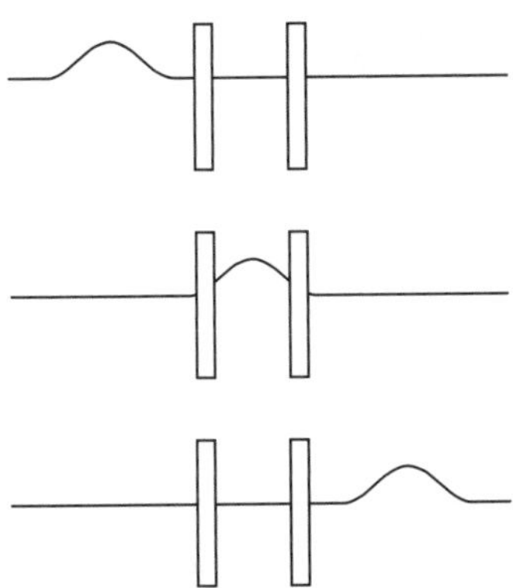

14. Draw a similar set of sketches to show what would happen if the first slit were vertical and the second slit were horizontal.

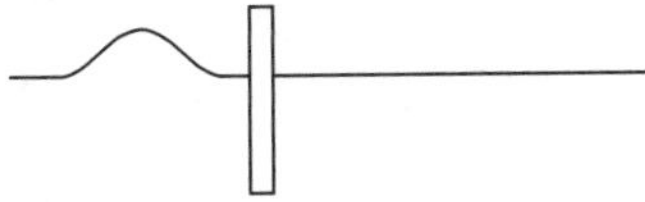

The light behaved in a similar way. When the two polarizers were parallel, the light was able to travel through both. When the two polarizers were perpendicular to each other, the "vertical" light could get through the first polarizer, but that light could not get through the second polarizer.

14.

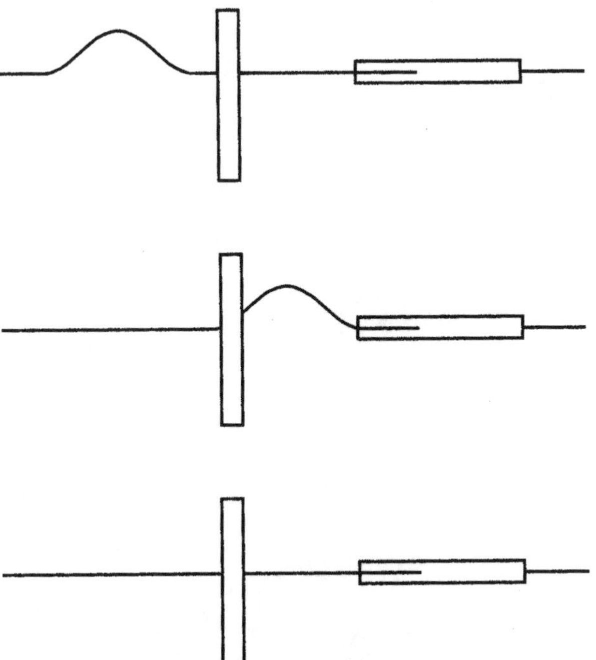

Physics Words

**transverse pulse or wave:** a pulse or wave in which the motion of the medium is perpendicular to the motion of the wave.

**longitudinal pulse or wave:** a pulse or wave in which the motion of the medium is parallel to the direction of the motion of the wave.

**wavelength:** the distance between two identical points in consecutive cycles of a wave.

**frequency:** the number of waves produced per unit time; the frequency is the reciprocal of the amount of time it takes for a single wavelength to pass a point.

**velocity:** speed in a given direction; displacement divided by the time interval; velocity is a vector quantity; it has magnitude and direction.

**amplitude:** the maximum displacement of a particle as a wave passes; the height of a wave crest; it is related to a wave's energy.

**period:** the time required to complete one cycle of a wave.

## PHYSICS TALK

### Describing Waves

You've discovered features of transverse waves and longitudinal waves. In **transverse waves,** the motion of the medium (the students or the Slinky) is perpendicular to the direction in which the wave is traveling (along the line of students or along the Slinky). In **longitudinal,** or **compressional waves,** the medium and the wave itself travel parallel to each other. Four terms are often used to describe waves. They are wavelength, frequency, amplitude, and period.

**Wavelength** is the distance between one wave and the next. It can be measured from the top part of the wave to the top part of the next wave (crest to crest) or from the bottom part of one wave to the bottom part of the next wave (trough to trough). The symbol for wavelength is the Greek letter lambda ($\lambda$). The unit for wavelength is meters.

**Frequency** is the number of waves that pass a point in one unit of time. Moving your hand back and forth first slowly, then rapidly to make waves in the Slinky increases the frequency of the waves. The symbol for frequency is $f$. The units for frequency are waves per second, or hertz (Hz).

The **velocity** of a wave can be found using wavelength and frequency. The relationship is shown in the equation

$$v = f\lambda.$$

The **amplitude** of a wave is the size of the disturbance. It is the distance from the crest to the undisturbed surface of the medium. A wave with a small amplitude has less energy than one with a large amplitude. The unit for amplitude is meters.

The **period** of a wave is the time for a complete wave to pass one point in space. The symbol for period of a wave is $T$. When you know the frequency of a wave, you can easily find the period.

$$T = \frac{1}{f}$$

FOR YOU TO READ

### Waves and Media

Waves transfer energy from place to place. Light, water waves, and sound are familiar examples of waves. Some waves need to travel through a medium. Water waves travel along the surface of water; sound travels through the air and other material. As they pass, the disturbances move by but the medium returns to its original position. The wave is the disturbance. At the point of the disturbance, particles of the medium vibrate about their equilibrium positions. After the wave has passed, the medium is left undisturbed.

Light waves, on the other hand, can travel through the vacuum of space and through some media. Radio waves, microwaves, and x-rays are examples of waves that can travel through vacuums and through some media. These waves are transverse waves.

### Polarization

Vertical disturbances and transverse waves can travel through a polarizer. Two parallel polarizers will allow these waves to continue to travel. If the two polarizers are perpendicular, then no wave is transmitted through the polarizers. You observed the polarization of light. This is evidence that light is a transverse wave. As you observed with the Slinky, compressional (longitudinal) waves are not affected by polarizing filters.

## Reflecting on the Activity and the Challenge

Energy can move from one place to another. For example, throwing a baseball moves energy from the thrower to the catcher. Energy can also move from one place to another without anything moving. In the waves you made with a Slinky, no part of the Slinky moves across the room but the energy gets from one side to the other. This is true of sound waves as well. The activity provides evidence for this "unusual" concept of energy moving without "stuff" moving. Waves in a Slinky, water waves, and sound waves travel through media. Yet, light can travel through empty space. Trust in the wave model lets physicists create a theory for light.

Think about sending thoughts as waves. If a research team proposed this, they would have to explain how they would test this idea. They would also have to show that the study was valid and could produce reliable and repeatable results.

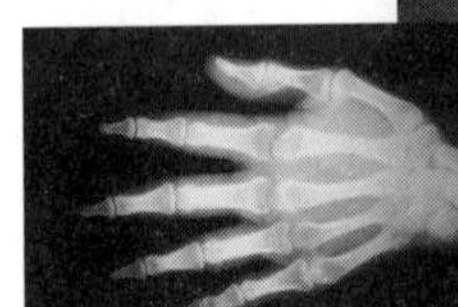

ANSWERS

## Physics To Go

1. In a transverse people wave, the people move up and down while the wave moves across the room.
2. a) The pulse moved from one end of the Slinky to the other end.
   b) No.
   c) It reversed direction and came back (reflected).
3. a) Check students' drawings.
   b) Transverse wave.
4. Check students' drawings
   a) Longitudinal wave.
   b) This type of wave must travel through a medium; it cannot travel through empty space. The disturbance and the velocity are parallel to one another.
5. a) See students' sketches. When they meet, the pulses "add together." If two crests meet, the amplitude will be greater than that of either wave's crest; if a crest and trough meet, the amplitude will be less than that of the either wave.
   b) See students' sketches. The waves look like they did before the waves met.
6. Students' phrasing will vary, but all answers should reflect an understanding that frequency is the number of waves that pass a point in one unit of time while period is the time it takes for a complete wave to pass, and that period is the inverse of frequency.
7. Frequency and wavelength; $v = f \times \lambda$
8. 7 m/s
9. 1/3 s
10. Answers will vary

Patterns and Predictions

### Physics To Go

1. Compare the direction in which people move in a people wave and the way the wave moves.
2. You sent a pulse down a Slinky.
   a) Which way did the pulse move?
   b) Did the shape of the pulse change as it moved?
   c) What happened to the pulse when it reached the end of the Slinky?

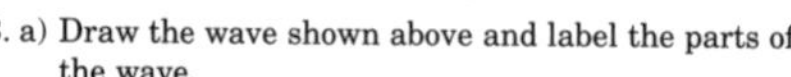

3. a) Draw the wave shown above and label the parts of the wave.
   b) What kind of wave is this?

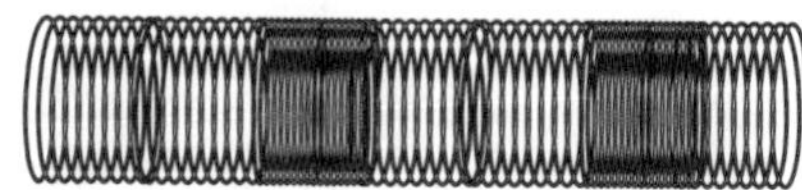

4. a) What kind of wave is this?
   b) Describe the movement of the wave and the movement of the medium.
5. Two pulses travel down a Slinky, each from opposite ends, and meet in the middle.
   a) What do the pulses look like when they meet? Make a sketch.
   b) What do the pulses look like after they pass each other? Make a sketch.
6. In your own words, compare frequency and period.
7. What determines the speed of a wave?
8. Find the velocity of a 2-m long wave with a frequency of 3.5 Hz.
9. Find the period of a wave with the frequency of 3 Hz.
10. Find the frequency of your favorite AM and FM radio stations.

Active Physics

272

### Stretching Exercises

1. Perform this activity and answer the questions. You'll need a basin for water or a ripple tank, a small ball, a ruler, and a pencil.
   - Fill the basin or ripple tank with water. Let the water come to rest so that you are looking at a smooth liquid surface as you begin the activity. If possible, position a "point source" of light above the water basin or tank. The light will help you see the shadows of the waves you produce.
   - Touch the surface of the water with your finger.

   a) Describe the wave you produced and the way it changes as it travels along the water surface.
   - Drop the ball in the water and watch closely.

   b) What happens at the point where the ball hits the water?
   - Drop the ball from different heights and observe the size of the mound of water in the center.

   c) What happens to the size of the mound as the height of the drop increases?

   d) Describe the pattern in which the waves travel.
   - Drop the ruler into the water.

   e) Does the shape or size of the wave maker affect the shape of the wave produced? How?
   - Make waves by dipping one finger into the water at a steady rate.

   f) What is the shape of the wave pattern produced?
   - Now vary the frequency of dipping.

   g) Describe what happens to the distance between the waves as the rate of dipping your finger increases.

   h) Describe what happens to the distance between the waves as the rate of dipping your finger decreases.

   i) Express the results of your observations in terms of wavelength and frequency of the waves.

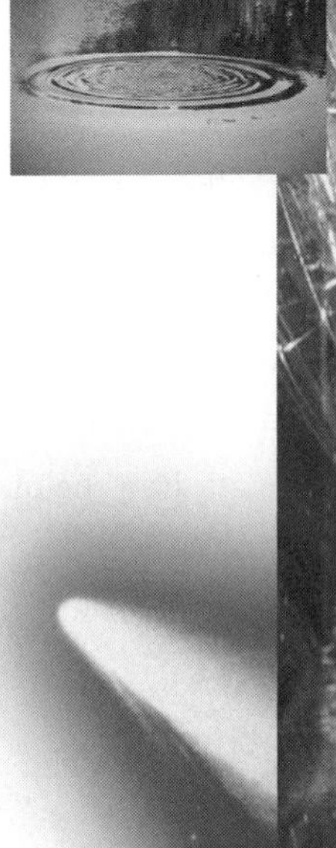

2. Find or create a computer simulation that will allow you to explore the behavior of waves in slow motion and stop action. If possible, "play" with these simulations in order to get a better sense of the behavior of waves. Demonstrate the simulation to your teacher and others in the class.



ANSWERS

## Stretching Exercises

1. a) The wave traveled out from the finger in a circle and bounced off the side of the basin, moving back toward the center of the circle.

   b) Water comes up along the sides and over the ball.

   c) The size of the mound (splash) increases as the height from which the ball is dropped increases.

   d) In circles out from the ball, bouncing back when they hit the side of the basin and moving back toward the center of the circle.

   e) The shape of the wave maker changes the shape of the wave. The ruler produces waves that travel out in straight lines from its sides.

   f) A concentric circle.

   g)-h) The distance between the waves decreases as the finger is dipped more rapidly, and it decreases as the rate of dipping the finger slows down. (Students may also report setting up a destructive interference pattern as new waves meet reflected waves.)

   i) Students' answers should reflect that wavelength increases as frequency decreases and wavelength decreases as frequency increases.

2. Allow students time to develop or use a computer simulation of waves, then describe what they have observed. You may wish to assign this investigation to teams of two students per classroom computer.

## Activity 3 A

# A Physicist's Guide to Taking a Bath

Physicists claim that physics is everywhere in the world around you. Bathing offers an ideal opportunity to observe the properties of water waves.

### FOR YOU TO DO

1. Fill the bathtub with water. If you are using a hot tub, turn off the whirlpool. You will need a few aids such as a marble or small ball, a soap dish or pan, a ruler or a pencil, a washcloth, and (optional) a rubber duck. Let the water come to rest so that you are looking at a smooth liquid surface as you begin the activity.

2. Initially, touch the surface of the water with your finger.

a) Does this produce a wave? If so, describe the wave and the way it changes as it travels along the tub.

It might happen, depending on the lighting in the area of the tub, that you will be able to see the shadows of waves cast on the bottom of the tub. If so, observe the shadows as well. They may be easier to see than the waves that cause them. A "point source" of light above the tub will make wave shadows visible.

3. Next, drop the marble or the ball in the water and watch closely. Drop the marble from different heights and observe the size of the mound of water in the center.

a) Record your observations. What happens at the point where the marble hits the water?

b) What happens to the mound as the height of the drop increases?

c) In what pattern do the waves produced by dropping the marble travel? What happens to the shape of the waves if you drop the ruler, the soap dish or the rubber duck?

d) Does the shape or size of the wave maker affect the shape of the wave produced? How?

4. Use the ruler or pencil to make straight waves that reflect against the side of the tub. First send waves directly at the side of the tub.

a) How does this wave reflect?

5. Now send the waves in at various angles.

a) Describe the reflected waves that result.

6. Produce waves by dipping your finger into the water at a steady rate.

a) What is the shape of the wave pattern produced?

7. Now vary the frequency of dipping.

a) Describe what happens to the distance between the waves as the rate of dipping your finger increases and as the rate decreases.

b) Express the results of your observations in terms of wavelength and frequency of the waves.

For use with *Patterns and Predictions*, Chapter 5, Activity 3: Slinkies and Waves

# Properties of Waves

| Wave Term | Definition |
|---|---|
| *Crest* | point of maximum positive disturbance |
| *Trough* | point of maximum negative disturbance |
| *Amplitude* | maximum disturbance from equilibrium |
| *Wavelength* | the distance from crest to crest of a wave; represented by the Greek letter lambda ($\lambda$) |
| *Frequency* | the number of waves which pass a point in a given time; measured in hertz (cycles/ sec) |
| *Period* | time for a complete wave to pass one point in space |
| *Speed* | speed of one point on the disturbance |
| *Interference* | two waves add so that their sum is either larger or smaller than the amplitude of either–a defining characteristic of waves |
| *Superposition* | waves add so that their amplitudes at any given point also add |
| *Medium* | material through which the wave travels |
| *Longitudinal waves* | disturbance is in the same direction as the wave is traveling |
| *Transverse waves* | disturbance is perpendicular to the direction in which the wave is traveling |
| *Standing wave* | a wave formed by interference of two waves that stays in one place |
| *Node* | point of no disturbance in a standing wave |
| *Antinode* | maximum disturbance in a standing wave |
| *Reflection* | bouncing back of a wave from an obstacle |
| *Refraction* | changing of a wave's travel when it passes from one medium to another |
| *Damping* | force that reduces a wave's amplitude as it travels |



For use with *Patterns and Predictions*, Chapter 5, Activity 3: Slinkies and Waves

# ACTIVITY 4
## Interference of Waves

# Background Information

Waves may pass through each other and emerge completely unchanged. While the two disturbances are occupying the same portion of the medium, however, they interfere. That is, the amplitude at each point of the medium is the algebraic sum of the amplitudes of the individual waves.

When two crests meet, both amplitudes are positive, and constructive interference occurs. That is, the resultant amplitude is greater than the amplitude of either component wave. For example, two wave crests with amplitudes of 4 cm and 3 cm produce a resultant amplitude of 7 cm. Destructive interference occurs when a crest and a trough meet. Destructive interference results in a composite wave whose amplitude is less than that of the larger component wave. For example, a crest with amplitude of 4 cm and a trough with amplitude 3 cm will result in a crest with amplitude 1 cm. When a wave meets its mirror image, the waves completely cancel each other out.

Interference patterns occur when two sets of waves pass through a medium. One-dimensional interference patterns knows as standing waves are produced when two identical wave trains moving in opposite directions interfere. Standing waves are characterized by nodes, points of no displacement, and antinodes, regions of maximum displacement. Nodes are the result of total destructive interference.

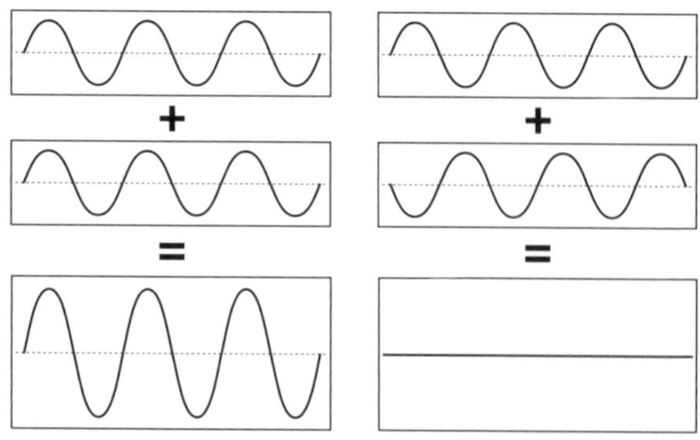

These waves add together These waves cancel each other

A two-dimensional interference pattern is produced when two point sources bob up and down in a pool of water. The resulting nodal and antinodal lines are hyperbolic in shape. A three-dimensional sound interference pattern can be produced in air by two loudspeakers being driven in phase.

The wave nature of light was verified in 1801 when Thomas Young performed his famous two-slit interference experiment. The interference of light waves is used to improve the performance of camera lenses and is the basis for holography.

## *Active*-ating the Physics InfoMall

In **Activity 3**, we found that there is plenty on the InfoMall for help with waves. Interference of waves is in the same boat – you should have no trouble finding demos, information, exercises, or just graphics with a simple search or a little browsing. Many of the ideas presented for **Activity 3** work just as well here; in fact, "interference of waves" is a topic that appears in many of the same articles and demonstrations. This includes standing waves, of course. And many of the textbooks carry full explanations and graphics, too.

One sample from the earlier search is "Two-source interference transparency," from *The Physics Teacher,* vol. 7, issue 2, 1969.

**For You To Do Step 7** mentions the Michelson Interferometer. Simply search for "interferometer" and you will not be disappointed. You get explanations with graphics with this single search. However, due to the limited space on the CD-ROM, a quality graphic of interference fringes is difficult, if not impossible, to find.

The **Stretching Exercise** suggests looking for holography, Dennis Gabor, and others. Successful searches include "hologra*", "gabor", "maiman", "leith", and "upatnieks". You can also find plenty on polarizers.

# Planning for the Activity

## Time Requirements

two class periods

## Materials Needed

### For the class:

- ripple tank assembly

### For each group:

- amplified speakers
- heavy duty Slinky®

## Advance Preparation and Setup

Preview **Steps** 4, 6, and 7 of the **For You To Do** activity. Procedures for different graphing calculators will vary slightly. Find an area for **Step 6** that will accommodate the class and allow each student to experience the interference of sound waves; a field out of doors may be best.

# Teaching Notes

The **What Do You Think?** questions are designed to stimulate thinking about the activity. Have students share their answers and accept all answers without correction.

For **Step 4** of the **For You To Do** activity, work with students and their graphing calculators to set up the graphs, enter data, and display the coordinates of various points on the waves. The latter can be done using the cursor and a setting that displays the coordinates, or by using the TRACE feature.

Make sure that each student has an opportunity to participate in each step of the activity. Check the standing waves produced on the Slinkies.

Encourage students to discuss and record in their logs the difference between magic and science, and the value of scientific theories and evidence to explain apparent "magic." Tell them that they will use these notes in the **Chapter Challenge**.

# Activity Overview

Students examine wave interference using a Slinky, a graphing calculator, two radios, and laser light. Interference is a unique property of waves. It is observed in both one- and two-dimensional media. Standing waves on Slinkies are used to introduce constructive and destructive interference. Then students generate standing waves on a graphing calculator. Nodes and antinodes are then examined in two-dimensional interference patterns produced on the surface of water and in the air with sound waves. The wave nature of light is illustrated using a Michelson interferometer. Each of these experiences provide students with a better understanding of the nature of waves.

## Student Objectives

### Students will:

- Generate waves to explore interference.
- Identify the characteristics of standing waves.
- Generate and identify parts of water waves.
- Experience interference of sound waves.

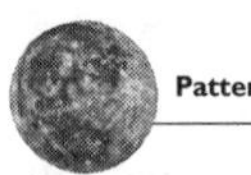

Patterns and Predictions

# Activity 4 Interference of Waves

**GOALS**

In this activity you will:

- Generate waves to explore interference.
- Identify the characteristics of standing waves.
- Generate and identify parts of water waves.
- Experience interference of sound waves.

### What Do You Think?

After a cost of millions of dollars, the Philharmonic Hall in New York City had to be rebuilt because the sound in the hall was not of high enough quality. Now named Avery Fisher Hall, it has excellent acoustics.

- **What does it mean to have "dead space" in a concert hall?**
- **What is the secret to good acoustics?**

Record your ideas about these questions in your *Active Physics* log. Be prepared to discuss your responses with your small group and the class.

### For You To Do

1. Work with two partners. Two of you will operate the Slinky and one will record the observations. Switch roles from time to time. Stretch the Slinky to about 10 m. While one end of the Slinky is held in a fixed position, send a pulse down the Slinky by quickly shaking one end.

Active Physics 274

ANSWERS FOR THE TEACHER ONLY

## What Do You Think?

"Dead space" is a result of destructive interference of waves. For longitudinal waves, such as sound waves, this occurs when a compression overlaps a rarefection.

Acoustics are the qualities of a room or auditorium that determine how well sound is heard. The acoustics of a room depend on the shape of the room, the contents of the room, and the composition of the ceiling, floor, and walls. In a large empty room, sounds will echo, producing poor quality. Rugs and furniture improve the quality of sound. In concert halls, even the density of the audience will make a difference to the acoustics. In designing an auditorium, the characteristics of the walls must produce a good balance of absorption and reflection of sound. Many "decorative" features in concert halls are frequently strategically placed in an effort to improve acoustics.

a) What happens to the pulse when it reaches the far end of the Slinky?

2. Send a series of pulses down the Slinky by continuously moving one of its ends back and forth. Do not stop. Experiment with different frequencies until parts of the Slinky do not move at all. A wave whose parts appear to stand still is called a standing wave.

3. Set up the following standing waves:
   - a wave with one stationary point in the middle
   - a wave with two stationary points
   - a wave with three stationary points
   - a wave with as many stationary points as you can set up

4. You can simulate wave motion using a graphing calculator.

   Follow the directions for your graphing calculator to define a graph and set up the window. Use the following for the Y-VARS, and select FUNCTION. Use the Y= button to enter these values:

   $Y_1 = 4 \sin x$

   $Y_2 = 4 \sin x$

   $Y_3 = Y_1 + Y_2$

   Press GRAPH to view the waves.

   a) Describe the two waves you see on the screen.

   b) Can you see that $Y_3$ is equal to $Y_1 + Y_2$?

   c) Use the vertical axis on the screen to find the amplitude of the crest of each wave. How do they compare?

   You can edit the $Y_1$ and $Y_2$ functions to show waves $Y_1$ and $Y_2$ moving from the left to the right, as follows:

   $Y_1 = 4 \sin (x - \pi/4)$

   $Y_2 = 4 \sin (x + \pi/4)$

   $Y_3 = Y_1 + Y_2$

   d) How many waves do you see on the screen? Compare the amplitude of the third wave to those of the first two waves.

   Edit again.

ANSWERS

## For You To Do

1. a) It bounces back (is reflected).

4. [Note: Students will observe the first wave plotted, (Y1). The second wave (Y2) is identical to the first, and students will not be able to see this wave graphed on their calculator. The third wave (Y3) shows constructive interference.]

a) (Y1) and (Y2) have less amplitude than (Y3).

b) Yes.

c) The amplitude of (Y3) is twice that of (Y1).

d) 3 waves. It is greater.

Chapter 5

ANSWERS

## For You To Do
***(continued)***

4\. e) They form a standing wave (amplitude 4).

f) See students' drawings. There are 7 nodes.

g) (*Y*1) and (*Y*2) are slightly out of sync. *Y*3 shows the destructive interference between the waves.

h) See students' drawings. The nodes are points which have always remained zero as the waves passed.

i) The amplitude of *Y*1 and *Y*2 is 4. The amplitude of *Y*3 is less.

j) The amplitude of *Y*1 is greater than that of *Y*2. The amplitude of *Y*3 is greater than that of either wave.

k) See students' drawings. The nodes are points which have always remained zero as the waves passed.

l) The amplitude of *Y*1 and *Y*2 is 4; that of *Y*3 is 8. *Y*3 is twice that of *Y*1 or *Y*2.

*Y*1 and *Y*2 have the same amplitude, but they do not create a standing wave because they are out of sync.

6.–7. Student activities.

$Y_1 = 4 \sin (x - \pi/2)$
$Y_2 = 4 \sin (x + \pi/2)$
$Y_3 = Y_1 + Y_2$

e) Describe the waves you see on the screen. Look for locations on the waves that always remain zero. These locations are called nodes.

f) Draw the waves you see on the screen on graph paper. Label the nodes.

Edit again.

$Y_1 = 4 \sin (x - 3\pi/4)$
$Y_2 = 4 \sin (x + 3\pi/4)$
$Y_3 = Y_1 + Y_2$

g) Describe the waves you see on the screen.

h) Draw the waves you see on the screen on graph paper. Label the nodes.

i) Compare the amplitude of each wave. How does the amplitude of the third wave compare to that of the first and the second wave?

Edit again.

$Y_1 = 4 \sin (x - \pi)$
$Y_2 = 4 \sin (x + \pi)$
$Y_3 = Y_1 + Y_2$

j) Describe the waves you see on the screen.

k) Draw the waves you see on the screen on graph paper. Locate the positions of the nodes.

l) Measure the amplitude of the first wave. What is the amplitude of the second wave? How do they compare?

5\. Use a ripple tank to explore what happens when two sources of circular water waves "add together" in the tank.

6\. As directed by your teacher, set up two speakers to explore what happens when two identical single tone sounds are broadcast.

7\. As directed by your teacher, use a double slit to explore what happens when two beams of laser light are "added."

**Physics Words**

**destructive interference:** the result of superimposing different waves so that two or more waves overlap to produce a wave with a decreased amplitude.

**constructive interference:** the result of superimposing different waves so that two or more waves overlap to produce a wave with a greater amplitude.

**standing wave:** a stationary wave formed by the superposition of two equal waves passing in opposite directions.

## FOR YOU TO READ

### Wave Interference

The wave that you sent down the Slinky was reflected and traveled back along the Slinky. The original wave and the reflected wave crossed one another. In the previous activity, you saw that waves can "add" when they pass one another. When waves "add," their amplitudes at any given point also "add." If two crests meet, both amplitudes are positive and the amplitude of the new wave is greater than that of the component waves. If a crest and trough meet, one amplitude is positive and one is negative. The amplitude of the resulting wave will be less than that of the larger component wave. If a wave meets its mirror image, both waves will be canceled out.

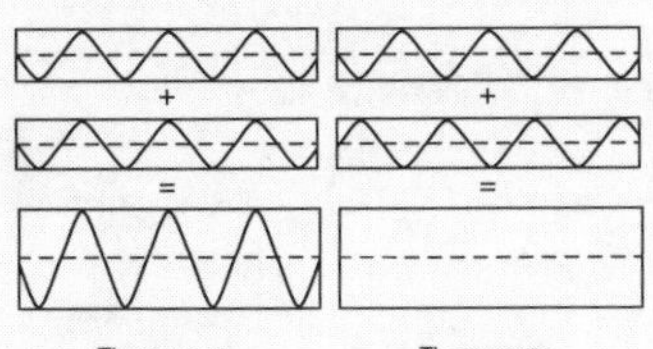


These waves add together. These waves cancel each other.

In this activity, you created a pattern called a **standing wave.** Two identical waves moving in opposite directions interfere. The two waves are constantly adding to make the standing wave. Some points of the wave pattern show lots of movement. Other points of the wave do not move at all. The points of the wave that do not move are called the nodes. The points of the wave that undergo large movements are called the antinodes.

The phenomena that you have observed in this activity is called wave interference. As waves move past one another, they add in such a way that the sum of the two waves may be zero at certain points. At other points, the sum of the waves produces a smaller amplitude than that of either wave. This is called **destructive interference.** The sum of the waves can also produce a larger amplitude. This is called **constructive interference.** The formation of nodes and antinodes is a characteristic of the behavior of all kinds of waves.

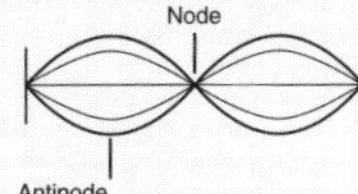




Chapter 5

ANSWERS

## Physics To Go

1. A standing wave is made by two identical waves that move in opposite directions and interfere with each other, producing nodes and antinodes.
2. Students' answers should reflect that the amplitudes of waves that meet can be added together to get constructive interference (adding, thus increasing the amplitude) or destructive interference (subtracting, thus decreasing the amplitude).
3. They must have the same amplitude, and the crest of one wave must meet with the trough of the other wave.
4. See students' drawings.
5. The distance between nodes is equal to 1/2 the wavelength of the wave. Reasons should reflect an understanding that the positions where the waves cross the line of the undisturbed medium form the nodes.
6. Students should propose that the interference cancels out bright areas caused by scattering light that might show up on the photograph.
7. You would hear the sound from the other speaker. Students' explanations will vary but should reflect their understanding that if the interference was removed, the remaining sound would be heard.

### Reflecting on the Activity and the Challenge

Imagine you were told that adding one sound to another sound in a space could cause silence. Would you believe that light plus light can create interference fringes, where dark lines are places where no light travels? You might have thought such strange effects are magic. In a Slinky, a wave traveling in one direction and a wave traveling in the opposite direction create points on the Slinky that do not move at all. That is experimental evidence for the interference of waves. Now you know that dead spaces and dark lines can be explained by good science. You can approve funding to study phenomena that appear strange as long as some measurements on which all observers can agree are used in supporting the claims.

### Physics To Go

1. What is a standing wave?
2. Describe in your own words how waves can "add."
3. What properties must two pulses have if they are to cancel each other out when they meet on a Slinky?
4. Make a standing wave using a Slinky or a graphing calculator. Draw the wave. Label its nodes and antinodes.
5. What is the distance, in wavelengths, between adjacent nodes in a standing wave pattern? Explain your thinking.
6. In photography, light can scatter off the camera lens. A thin coating is often placed on the lens so that light reflecting off the front of the thin layer and light reflecting off the lens will interfere with each other. How is this interaction helpful to the photographer?
7. Two sounds from two speakers can produce very little sound at certain locations. If you were standing at that location and one of the speakers was turned off, what would happen? How would you explain this to a friend?

8. Makers of noise reduction devices say the devices, worn as headsets, "cancel" steady noises such as the roar of airplane engines, yet still allow the wearer to hear normal sounds such as voices. How would such devices work? What principles of waves must be involved?

### Stretching Exercises

An optical hologram is a three-dimensional image stored on a flat piece of film or glass. You have probably seen holograms on credit cards, in advertising displays, and in museums or art galleries. Optical holograms work because of the interference of light. Constructive interference creates bright areas, and destructive interference, dark areas. Your eyes see the flat image from slightly different angles, and your brain combines them into a 3-D image.

Find out how holograms are made. Describe the laboratory setup for making a simple hologram. If your teacher or you have the equipment, make one!

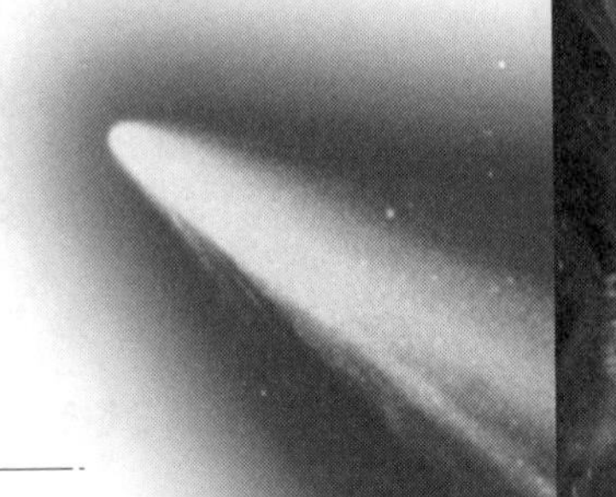



ANSWERS

## Physics To Go

***(continued)***

8. They would produce sounds that would interfere with the sounds from the "noise." Interference would occur at these wavelengths, but not at other wavelengths of sound.

ANSWERS

## Stretching Exercises

Dennis Gabor invented holography in 1947, but his holograms were of poor quality because he did not have a good source of light of just one wavelength. Theodore Maiman invented lasers in 1960. Leith and Upatnieks used laser light to produce holograms in 1961.

There are a variety of holograms, each produced by slightly different methods. Students' answers might, therefore, vary. If the equipment for making holograms is available, allow time for students to use it.

Holograms have a variety of uses, commercial and industrial. Students' lists will vary.

# ACTIVITY 5
# A Moving Frame of Reference

## Background Information

A frame of reference is a coordinate system from which we make observations and measurements. The most common and frequently used reference system is the Earth and objects attached to it. Most of the time when we say we are moving we mean we are changing our position with respect to the Earth. However, the Earth is only one of an infinite number of reference frames that we can use to make observations. We are often caught off guard when we inadvertently use another reference frame. While waiting for their train to leave the station, passengers are sometimes fooled into believing that they are moving as the train next to them pulls away. If a car next to you at a stoplight rolls backward, you may feel as if you are moving forward into traffic.

The path that a moving object takes may also be altered by your frame of reference. A ball dropped by a person in a car moving at a constant velocity falls straight down as viewed by the passenger in the car. A pedestrian watching from the street sees the ball move in a curved path. Similarly, when a passenger in an amusement park ride that moves in a circular path throws a ball straight ahead, he or she sees the ball return along a curved path. An observer in the gallery above sees the ball continue in a straight line.

We may use the previous examples to identify two types of reference frames. The car moving at a constant velocity is called an inertial frame of reference; the rotating frame of the amusement park ride is called a non-inertial frame of reference.

Galileo's principle of relativity states that the laws of physics are the same in all inertial frames. All frames are equivalent and no experiment will reveal your true state of motion or rest. This is another way of saying that there is no preferred frame of reference.

### *Active*-ating the Physics InfoMall

This activity describes what happens when two observers see the same things from different frames. The obvious search is nearly the best one; try a twist on the use of wild characters by searching for "frame* of reference". It seems unlikely that even this search engine can handle a wild character in the middle of a phrase; on the contrary, the engine runs as smoothly as ever, and we find several excellent hits for both "frame" and "frames" of reference! These include some graphics you may wish to use.

Note that when one frame of reference is moving, this can lead to relative motion, another topic to search for. If you do this, you can find items that the previous search did not uncover.

The **Stretching Exercise** is about Einstein. The searches above will produce some information, but there is plenty more to be found. If you search for "Einstein" you will quickly get "Too Many Hits" and you will need to limit your search. Try just the Book Basement and the Calendar Cart. The Cart alone provides over 30 hits!

## Planning for the Activity

### Time Requirements

two class periods

### Materials Needed

#### For the class:

- graphing calculator
- Michelson interferometer and laser
- oscillator and stereo amplifier

#### For each group:

- ball, tennis
- toy car, battery operated
- meter stick, 100 cm, hardwood
- ball of string
- felt tip marker
- stopwatch

### Advance Preparation and Setup

Arrange to use a long, wide corridor, the gym, or a smooth paved surface out of doors for this activity. Assure the stability of the moving vehicle you choose.

# Teaching Notes

Assist students in planning how to move the cart at a constant velocity, and how to accelerate the cart in a straight line. A stopwatch and various landmarks should be suggested and used. If necessary, set up the landmarks (possibly cones positioned at equal distances) astride the path of the cart.

If you have equipment for class groups of four, students can work in their groups. For best understanding, have each student take a turn on the cart. If not possible, repeat the activity so that all students have participated as either stationary observers or the moving observer.

High speeds are not necessary. Coach students to move the vehicle at a slow constant speed, and help them determine methods to accelerate the cart to a greater, yet safe, speed.

SAFETY PRECAUTIONS: Student safety on the moving vehicle is the major safety concern. A low dolly or wagon usually prove more stable than a chair or lab cart. Make sure the surface is free of debris, and demonstrate a safe way to stop the moving vehicle if its motion seems out of control.

Students may resist the notion that one frame of reference is no more valid than another. Their experience riding on the cart in the activity will begin to dispel this notion, but continue to remind them of their observations and ask probing questions to uncover their continued adherence to this misconception.

Encourage students to review the observations they made during the **For You To Do** activity and discuss the differences in observations made by stationary observers and the moving observer. Guide them to explain how these observations can be resolved so as to be sensible. Have them record their ideas in their logs for use in the **Chapter Challenge**.

# Activity Overview

Students make observations of motion from three different reference points, a stationary reference, while moving at a constant velocity, and while accelerating in a straight line, and discover that no one frame of reference is "better" than another.

## Student Objectives

### Students will:

- Observe the motion of a ball from a stationary position and while moving at a constant velocity.
- Observe the motion of a ball from a stationary position and while accelerating in a straight line.
- Measure motion in a moving frame of reference.
- Make predictions about motion in moving frames of reference and test your predictions.
- Define relativity.
- Define frame of reference and inertial frame of reference.
- Reconcile observations from different frames of reference.

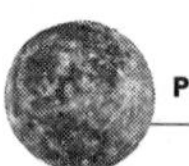

Patterns and Predictions

## Activity 5 A Moving Frame of Reference

**GOALS**

In this activity you will:

- Observe the motion of a ball from a stationary position and while moving at a constant velocity.
- Observe the motion of a ball from a stationary position and while accelerating in a straight line.
- Measure motion in a moving frame of reference.
- Make predictions about motion in moving frames of reference and test your predictions.
- Define relativity.
- Define frame of reference and inertial frame of reference.
- Reconcile observations from different frames of reference.

**What Do You Think?**

As you sit in class reading this line, you are traveling at a constant speed as the Earth rotates on its axis. Your speed depends on where you are. If you are at the Equator, your speed is 1670 km/h (1040 mph). At 42° latitude, your speed is 1300 km/h (800 mph).

- **Do you feel the rotational motion of the Earth? Why or why not?**
- **What evidence do you have that you are moving?**

Record your ideas about these questions in your *Active Physics* log. Be prepared to discuss your responses with your small group and the class.

Active Physics 

ANSWERS FOR THE TEACHER ONLY

## What Do You Think?

The rotational motion of the Earth cannot be felt. Our frame of reference appears to be stationary and we perceive the movement of objects around us.

The movement of the stars and Moon in the sky gives evidence that our frame of reference may be moving if one assumes the stars are stationary.

### For You To Do

1. When you view a sculpture you probably move around to see the work from different sides. In this activity, you'll look at motion from two vantage points—while standing still and while moving. Get an object with wheels that is large enough to hold one of your classmates seated as it rolls down the hall. You might use a dolly, lab cart, a wagon, or a chair with wheels.

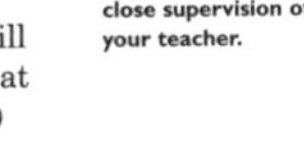

This activity should be done under close supervision of your teacher.

2. Choose a student to serve as the observer in the moving system. Have the observer sit on the cart and practice pushing the cart down the hall at constant speed. (This will take a little planning. Find a way to make the cart travel at constant speed. Also, find a way of controlling that speed!)

3. Once you can move the cart at a constant speed, give the moving observer a ball. While the cart is moving at constant speed, have the moving observer throw the ball straight up, then catch it.

a) How does the person on the cart see the ball move? Sketch its path as he or she sees it.

b) How does a person on the ground see the ball move? Again, sketch the path of the ball as he or she sees it.



## Answers

# For You To Do

1.–2. Student activity.

3. a) The person in the cart sees the ball move in straight lines up and down.

b) A person on the ground sees the ball move in a curved path.

ANSWERS

## For You To Do
*(continued)*

4.a) The moving person sees the ball move in a curved path. See students' sketches.

b) The person on the ground sees the ball move in straight lines up and down. See students' sketches.

4. With the observer on the cart traveling at constant speed, let a student standing on the ground throw the ball straight up and catch it.
   a) How does the moving observer see the ball move? Sketch its path.
   b) How does a person on the ground see the ball move? Sketch its path.
5. Work in groups for the following steps of this activity, as directed by your teacher. Get a wind-up or battery-powered car, two large pieces of poster board or butcher paper, a meter stick, a marker, string, and a stopwatch from your teacher.

   Use a marker to lay out a distance scale on the poster board. Be sure to make it large enough so that a student walking beside the poster board can read it easily.

   Next, lay out an identical distance scale along the side of the classroom or in a hall.

   Attach a string to the poster board and practice moving it at a constant speed.
6. Place the toy car on the poster board and let it move along the strip. Measure the speed (distance/time) of the car along the poster board as the board remains at rest. Try the measurement several times to make sure that the motion of the car is repeatable.
   a) Record the speed.
7. Move the poster board at constant speed while the car travels on the board. Focus on the car, not on the moving platform. Measure the speed of the car relative to the poster board when the board is moving.
   a) Record the speed.
   b) Compare the speed of the car when its platform is not moving and when its platform is moving.
   c) Do your observations and measurements agree with your expectations?

8. Work with your group to make two simultaneous measurements. Measure the speed of the board relative to the fixed scale (the scale on the floor) and the speed of the car relative to the fixed scale. The second measurement can be tricky. Practice a few times. It may help to stand back from the poster board.
   a) Record the measurements.

9. Next, measure the speed of the poster board and the car relative to the fixed scale while moving the board at different speeds. Make and complete a table like the one below.

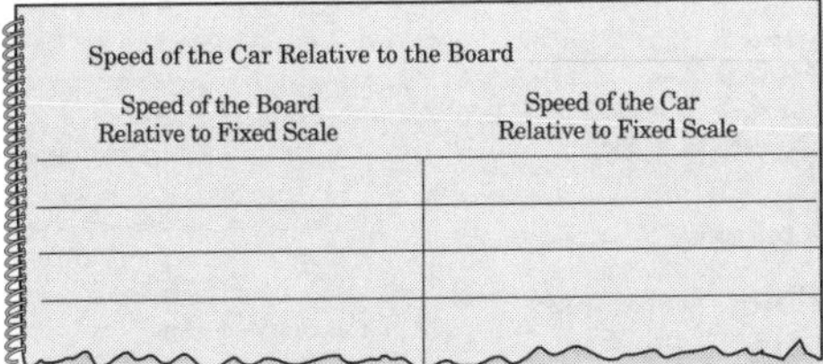

| Speed of the Car Relative to the Board | |
|---|---|
| Speed of the Board Relative to Fixed Scale | Speed of the Car Relative to Fixed Scale |
| | |
| | |
| | |
| | |

   a) Work with your group to state a relationship between the speed of the car relative to the fixed scale, its speed relative to the board, and the speed of the board relative to the fixed scale. Describe the relationship in your log. Also explain your thinking.
   b) What do you think will happen if the car moves in the direction opposite to the direction the board is moving? Record your idea. Now try it. Do the results agree with your predictions?
   c) Plan an experiment in which the car is moving along the poster board, the poster board is moving, and the car remains at the same location. Try it. Record the results.



ANSWERS

## For You To Do
### *(continued)*

9. b) Students should predict that the motion of the car relative to the fixed scale will be less than in **Step 9. a)** when the car and platform are moving in opposite directions. Results should agree with predictions.

   c) For this experiment, the speed of the car and the platform must be the same but in opposite directions. Results will vary.

ANSWERS

## For You To Do
*(continued)*

9. d) Students should predict that the car moving perpendicular to the direction of the platform's motion will travel with the platform in relation to the fixed scale. The results should agree with predictions.

   e) No.

d) What will happen if the car travels perpendicular to the direction in which the board is moving? Record your ideas. Now try it. Do the results agree with your predictions?

e) When the car travels perpendicular to the motion of the frame of reference, does the motion of the board affect your measurement of the car's speed?

### FOR YOU TO READ
#### Frames of Reference

A frame of reference is a coordinate system from which observations and measurements are made. Your usual frame of reference is the surface of the Earth and structures fixed to it.

Have you experienced two frames of reference at once? Many large public spaces have banks of escalators to transport people from one floor to another. If two side-by-side escalators are moving in the same direction and at the same speed, and you and a friend step onto these escalators at the same time, you will seem to be standing still in relation to your friend. From the frame of reference of your friend, you are not moving. From the frame of reference of a person standing at the base of the escalator, you are both moving.

As you saw in the activity, there are other frames of reference. For a person moving at a constant velocity, the vehicle is the local frame of reference. In a moving train or plane or car, the local frame of reference is the train or plane or car. When you are moving at a constant velocity, the local frame of reference is easier to observe than the frame of reference fixed to the Earth. If you drop an object in front of you while moving at a constant velocity in an airplane, it will fall to the floor in front of you. If the plane is traveling at 300 m/s, how do you explain the motion of the dropped object? Because you and the object are moving at a constant velocity, the object and you act as if you and it were standing still!

Would you be surprised to know that one frame of reference is not "better" than another? No matter what your frame of reference, if you are moving at a constant velocity, the laws of physics apply.

**Relativity** is the study of the way in which observations from moving frames of reference affect your perceptions of the world.

Relativity has some surprising consequences. For example, you cannot tell if your frame of reference is moving or standing still compared

Physics Words

**relativity:** the study of the way in which observations from moving frames of reference affect your perceptions of the world.

to another frame of reference, as long as both are moving at constant speed in a straight line. Newton's First Law of Motion states that an object at rest will stay at rest, and an object in motion will stay in motion unless acted on by a net outside force. Newton's First Law holds in each frame of reference. Such a frame of reference is called an **inertial frame of reference.**

If you are in a frame of reference traveling at a constant velocity from which you cannot see any other frame of reference, there is no way to determine if you are moving or at rest. If you try any experiment, you will not be able to determine the velocity of your frame of reference. This is the first postulate in Einstein's Theory of Relativity. Think of it this way: Any observer in an inertial frame of reference thinks that he or she is standing still!

### Reflecting on the Activity and the Challenge

Different observers make different observations. As you sit on a train and drop a ball, you see it fall straight down—its path is a straight line. Someone outside the train observing the same ball sees the ball follow a curved path, a parabola, as it moves down and horizontally at the same time. However, a logical relation exists between different observations. If you know what one observer measures, you can determine what the other observer measures. This relation works for any two observers. It is repeatable and measurable. Pseudoscience requires special observers with special skills. No relation or pattern exists between them. Different explanations can be accepted for the same phenomenon and it's still science. Your **Chapter Challenge** is to distinguish between different explanations that are science and different explanations that have no basis and are pseudoscience.

**Physics Words**
inertial frame of reference: unaccelerated point of view in which Newton's Laws hold true.

### Physics To Go

1. A person walking forward on the train says that he is moving at 2 miles per hour. A person on the platform says that the man in the train is moving at 72 miles per hour.
   a) Which person is correct?
   b) How could you get the two men to agree?

ANSWERS

## Physics To Go

1. a) Both are correct.

   b) Point out that they are measuring from two different frames of reference. To the man in the train, the train is his frame of reference, and within that frame of reference he measures his movement. To the man on the platform, the train is moving at 70 mi/h relative to his position and the man is moving in the same direction at 2 mi/h; added together, the man on the train is moving at 72 mi/h.

2. 90 mi/h
3. Answers should reflect subtracting the persons speed toward the end of the plane from the speed of the plane forward.
4. a) 2200 km/h

   b) 1400 km/h

   c) 3000 km/h
5. Before he is over the target.
6. a) We see the Sun in the sky from the frame of reference of the Earth, and we observe the Sun move across the sky in an east to west direction.

   b) The Earth is rotating on its axis, so Earth is the moving frame of reference relative to the Sun. Because of the Earth's rotation, a person is moved from a position in which he or she could not see the Sun to a position in which the Sun is visible in the east. As the Earth continues to move, the position of the observer moves so that he or she is "below" the Sun – the Sun is overhead, and eventually taken out of view of the Sun, as it "sets" in the west.

2. If you throw a baseball at 50 miles per hour north from a train moving at 40 miles per hour north, how fast would the ball be moving as measured by a person on the ground?
3. You walk toward the rear of an airplane in flight. Describe in your own words how you would find your speed relative to the ground. Explain your thinking.
4. A jet fighter plane fires a missile forward at 1000 km/h relative to the plane.
   a) If the plane is moving at 1200 km/h relative to the ground, what is the velocity of the missile relative to the ground?
   b) What is the velocity of the missile relative to a plane moving in the same direction at 800 km/h?
   c) What is the velocity of the missile relative to a target moving at 800 km/h toward the missile?

5. A pilot is making an emergency air drop to a disaster site. When should he drop the emergency pack: before he is over the target, when he is over the target, or after he has passed the target?
6. Each day you see the Sun rise in the east, travel across the sky, and set in the west.
   a) Explain this observation in terms of your frame of reference.
   b) Compare the observation to the actual motions of the Sun and Earth.

7. How would you explain relativity to a friend who is not in this course. Outline what you would say. Then try it. Record whether or not you were successful.
8. Explain this event based on frame of reference. You are seated in a parked car in a parking lot. The car next to you begins to back out of its space. For a moment you think your car is rolling forward.

## FOR YOU TO READ

### A Social Frame of Reference

Physics is sometimes a metaphor for life. Just as physicists speak of judging things from a frame of reference, a frame of reference is also used in viewing social issues. For example, a Black American*, one of the authors of this chapter, shared the following story about choosing a career, because his frame of reference conflicted with that of his father.

"I was born in Mississippi in 1942, the place where my parents had spent their entire lives. My father lived most of his life during a period of "separate-but-equal," or legal segregation. He believed that the United States would always remain segregated. So when I was choosing a career, all of his advice was from that frame of reference.

"On the other hand, my frame of reference was changing. To me, the United States could not stay segregated and remain a world power. The time was 1962, about 10 years after the Supreme Court had made its landmark Brown vs. Topeka School Board decision. I reasoned that the opportunities for black people would be greatly expanded.

"Both my parents had encouraged me to get as much education as possible. My mother always said that "the only way to guarantee survival is through good education." I had a master's degree and was teaching in a segregated college. I thought I would need a Ph.D. to stay in my profession, and decided to quit my job and go back to school. That decision brought on an encounter with my father that I shall never forget.

"My father did not say good-bye on the day I left home for graduate school. Our frames of reference had moved very far apart. The possibility of becoming a professor at a white college or university, particularly in the South, was not very high. My father could not understand why I needed a Ph.D. After all, I could have a good life in our segregated system without quitting my prestigious job to return to graduate school.

"As was usual for him, my father eventually supported my decision. At his death in 1989, however, he still had not fully accepted my frame of reference."

*The author uses the term Black American instead of African-American because the use of that term also shows social changes in frames of reference.



7. Students' ideas and outlines will vary, but should include the ideas of relativity and include some of the examples provided in the activity.
8. As you sit in the parked car, your frame of reference includes objects in the car and outside of the car, including a car parked next to you. When that car begins to move, you momentarily do not know which object is moving, your car or the other car.

ANSWERS

## Stretching Exercises

1. Student reports will vary. Einstein was born in 1897 in Ulm, Germany, and grew up in Munich. Students might report the story that Einstein was not a good student; some scholars dispute this story. Einstein made his first big discovery when he was working as a patent-office clerk in Bern, Switzerland. As a result, he received an honorary doctorate, and moved into university teaching and continued his important work in relativity. Einstein, along with many other leading scientists, left Germany with the rise of Hitler. He took a position at Princeton University. As an adult, he was a pacifist, yet he is considered as instrumental in the development of the atomic bomb because, along with other scientists of the time, he signed a letter to then president Franklin D. Roosevelt encouraging America's development of the bomb and because his famous relativity equation, $E = mc^2$ showed the potential for a bomb.

2. Students' stories will vary. Accept all reasonable applications of the idea of differing frames of reference.

### Stretching Exercise

1. The famous scientist Albert Einstein is noted for his Theory of Relativity. Research Einstein's life. What kind of a student was he? What was his career path? When did he make his breakthrough discoveries? What were his political beliefs as an adult? What role did he play in American political history? Report your findings to the class.

2. A Social Frame of Reference tells the story of one man's encounter with different ideas about society, or social frames of reference. Write a short story that illustrates what happens when two people operate from different frames of reference. Your story can be based on your own experience, or it can be fiction.

NOTES

# ACTIVITY 6
## Speedy Light

# Background Information

Based on everyday experience, the Galilean velocity addition makes perfect sense. However, at velocities approaching the speed of light, the Galilean velocity transformations break down and a new approach to adding velocities is needed. For example, if we envision a rocket traveling at 90% of the speed of light with relation to the Earth suddenly turning on its "headlights," an observer on Earth might expect to see the light move at $c + 0.9c$. However, this expectation does not agree with experiment. Einstein provided a way to add in a way that is consistent with both experiment and everyday experience. When relativistic velocity transformations are applied to velocities much less than the speed of light, they yield results consistent with the Galilean transformations.

The Special Theory of Relativity is based on two fundamental principles. First, the speed of light is the same for all observers in inertial frames of reference. Second, all laws of nature are obeyed in all inertial systems. Thus, relativity provides a way of understanding some very special situations; it does not replace Newton's view of the universe.

Light is not transmitted instantaneously. It has a finite speed. The light we see from the Sun takes 8 minutes to reach Earth. When we see the Sun, we see the Sun as it was 8 minutes earlier. When we see the nighttime sky, we are viewing light that was released from the stars years, hundreds of years, or thousands of years ago. There may be young stars whose light we cannot yet see. As a result of this reasoning, "now" does not truly exist. The past, present, and future are due to some degree to the separation of space as much as time.

If "now" does not truly exist, the notion of simultaneity comes under question. Einstein used the example of a high-speed train hit by a lightning bolt at each end of the train to explain this idea. To an observer on the ground, the bolts may appear to hit the train at the same moment. To an observer in the train, however, the bolt at the front of the train will be seen before the bolt at the end of the train as the train moves forward and the bolt at the front of the train has less distance to travel to the moving observer than does the bolt at the back of the train. Additionally, the faster the train is moving, the greater the measured interval between the arrival of the two bolts to the moving observer.

# Planning for the Activity

## Time Requirements

one class period

## Materials Needed

- no materials needed

## Advance Preparation and Setup

None

# Teaching Notes

Encourage spirited discussion of the questions posed in the **For You To Do** activity. Be nonjudgemental during the discussion, but ask probing questions if students seem to be losing their focus on the problem.

Have students work in groups of 3 or 4.

The idea of spontaneous events will be hard to shake for most students. The activity and discussion surrounding it should help raise questions that challenge the prior conceptions of simultaneity.

Encourage students to discuss the ideas in this section, sharing ideas and debating conclusions. Students should ultimately conclude that repeatable observations and data that can be agreed upon are a requirement of science. Have them record their ideas and conclusions in their logs.

Activity 6 Speedy Light

# Activity 6 Speedy Light

**GOALS**

In this activity you will:

- Perform a thought experiment about simultaneous events.
- State Einstein's two Postulates of Special Relativity.

### What Do You Think?

It takes light 8 **minutes** to travel from the Sun to **the Earth.** If **the Sun** suddenly went dark, no one would know **for** 8 **minutes**.

- **If an event happened on Mars and on Earth at the "same time," what would that mean?**
- **How would a person on Mars report the event to a person on Earth?**

Record your ideas about these questions in your *Active Physics* log. Be prepared to discuss your responses with your small group and the class.

# Activity Overview

Students perform a gedanken experiment concerning simultaneous events.

# Student Objectives

## Students will:

- Perform a thought experiment about simultaneous events.
- State Einstein's two Postulates of Special Relativity.

ANSWERS FOR THE TEACHER ONLY

# What Do You Think?

See the discussion in the **Background Information** regarding the concept of "now" and the simultaneity of events. At the "same time" means that some observer between Earth and Mars sees the events at the "same time." It does not mean that all observers see the event at the same time. The person on Mars sees the Mars event first.

ANSWERS

## For You To Do

1. a) Students' answers will vary. Some might suggest that he determine the time taken to travel from one clock to another and set the clocks accordingly. Others might say that he cannot set the clocks to chime at the same time.

   b) Students' answers will vary. Some will suggest that he could hear the clocks if he was equidistant from each of them, and there were no barriers between he and the clocks to slow the motion of the sound waves.

   c)-d) Students' answers will vary. With increasing distances, the idea of setting all clocks to chime at the same instant as outlined in a), above, is increasingly unlikely. Similarly, the likelihood of hearing the clocks simultaneously at such distances becomes remote.

2.–3. Students' answers will vary, but should reflect the idea that because it takes a finite time to travel to each "clock" to set it, the setting of clocks at such great distances approaches impossibility. Also, the ability to hear each clock at the same instant is diminished by the distance between the observer, the clocks and the finite speed of electromagnetic radiation.

Patterns and Predictions

### For You To Do

Work with your group to solve these problems. Record your thinking and conclusions in your log.

1. An old, slow-moving man has a large house with a grandfather clock in each room. He has no wristwatch and he cannot carry the clocks from one room to another. He wants to set each clock at 12 noon. He finds that by the time he sets the second clock, it is no longer noon because it takes time to get from the first clock in one room to the second clock in another.

   a) How can he make sure that all the clocks in the house chime the same hour at the same instant?

   b) Would he hear all the chimes at the same instant?

   Imagine that his house is huge—100 km × 100 km × 100 km.

   c) How can he set all the clocks to chime the same hour at the same instant?

   d) Would he hear all the chimes at the same instant?

2. Your group is put in charge of a solar system time experiment. You send clocks to Mercury, Venus, Mars, and Jupiter. These clocks "chime" by sending out radio waves. It takes many hours for the "chime" to travel between planets.

   a) How can you set all the clocks to "chime" at the same hour at the same instant?

   b) Would you "hear" all the clocks at the same instant?

3. Your group gets another mission. You are to send clocks to distant stars. These clocks "chime" by sending out pulses of light. It takes hundreds of years for the light to travel between the different stars.

   a) How can you set all the clocks to "chime" at the same hour at the same instant?

   b) Would you "hear" all the clocks at the same instant?

Active Physics 290

## FOR YOU TO READ

### The Theory of Special Relativity

The speed of light is $3.0 \times 10^8$ m/s (meters per second) in a vacuum, or 186,000 miles per second. The speed of light is represented by the symbol *c*. So, $c = 3.0 \times 10^8$ m/s. If light could travel around Earth's equator, it would make over 7 trips each second! The very great speed of light makes it difficult to measure changes in the speed of light caused by motion of frames of reference that are familiar to you on Earth.

In the early part of the 20th century, Einstein showed that light does not obey the laws of speed addition that we have seen in objects on Earth's surface. His theory predicted that light traveled at the same speed in all frames of reference, no matter how fast the frames were moving relative to one another.

To understand exactly how startling this result was, let's use an example. Recall measuring the speed of objects in a moving frame of reference in the previous activity. The following sketch shows a woman standing on a moving cart and throwing a ball forward.

A man watches from the roadside. The speed of the cart relative to the road is 20 m/s.

The speed of the ball relative to the cart is 10 m/s. How fast is the ball traveling according to the man by the side of the road?
(20 m/s + 10 m/s = 30 m/s.)

In the second sketch, the ball is replaced by a flashlight.

Once again the cart travels at speed 20 m/s relative to the road. The light travels at speed *c* relative to the cart. How fast does the light travel according to the man at the side of the road? (Take time to discuss your thinking with your group!)

Imagine the cart traveling at 185,000 mi/s relative to the road. The light travels at 186,000 mi/s. How fast does the light travel according to the man at the side of the road?

As a young clerk in the Swiss patent office, Albert Einstein postulated that the speed of light in a vacuum is the same for all observers. Einstein recognized that light and other forms of electromagnetic radiation (including x-rays, microwaves, and ultraviolet waves) could not be made to agree with the laws of relative motion seen on Earth. Einstein modified the ideas of relativity to agree with the theory of electromagnetic radiation. When he did, he →

uncovered consequences that have changed the outlook of not only physics but the world.

The basic ideas of Einstein's **Theory of Special Relativity** are stated in two postulates:

- **The laws of physics are the same in all inertial frames of reference. (Remember that inertial frames of reference are those in which Newton's First Law of Motion holds. This automatically eliminates frames of reference that are accelerating.)**
- **The speed of light is a constant in all inertial frames of reference.**

The first postulate adds electromagnetism to the frames of reference discussed. Its implications become clear when you begin to ask questions. Is the classroom moving or standing still? How do you know? Remember that an observer in an inertial frame of reference is sure that he or she is standing still. An observer in an airplane would be convinced that he or she is standing still and that your classroom is moving. The meaning of the first postulate is that there is no experiment you can do that will tell you who is really moving.

The second postulate, however, produces results that seem to defy common sense. You can add speeds of objects in inertial frames of reference. But you cannot add the speed of light to the motion of an inertial frame of reference.

### What Are Simultaneous Events?

Like the old man in the **For You To Do** activity, light travels at a finite speed. Although it travels very rapidly, it takes time for light to get from one place to another. Just as the old man had a problem setting his clocks at the same time, physicists have a problem saying when two events happen at the same time.

The speed of light in a vacuum is always the same. Physicists say that two events are simultaneous if a light signal from each event reaches an observer standing halfway between them at the same instant. You can demonstrate this idea in your classroom. An observer standing midway between two books would see them fall at the same instant if their falls were simultaneous. It is a little more difficult to imagine an observer midway between classrooms in two different time zones avidly watching for falling books, but—in principle—the experiment is possible.

An experiment that could be done but would be very difficult to carry out can be replaced by what is called a gedanken, or thought, experiment. Physicists use gedanken experiments to clarify principles. If the principle is called into question, experimenters can always try to conduct the actual experiment, although it may be very difficult to do so. In the activity, you performed a gedanken experiment.

### Reflecting on the Activity and the Challenge

Einstein's second postulate is that any observer moving at any speed would measure the speed of light to be $3.0 \times 10^8$ m/s. This postulate and his first postulate leads to the idea that simultaneity depends on the observer. You cannot say whether two events in different places occurred at the same time unless you know the position of the observer. For one observer, event A and event B happen at the same time while for a second observer, event A happens before event B. Why should you trust such a strange theory? Why should you trust in new ideas about space and time? You can trust them because they are supported by experimental results.

Can you be in two places at the same time? Should you fund a research project to test this out? If the proposal produces measurements and observations that can be used as evidence, you could fund it. If the proposal requires observations that only certain people are "qualified" to make, or data that cannot be agreed on, you should not fund it.

### Physics To Go

1. How long does it take a pulse of light to:
   a) cross your classroom?
   b) travel across your state?
2. Calculate the number of round trips between New York City and Los Angeles a beam of light can make in one second. (New York and Los Angeles are 5000 km or 3000 miles apart.)
3. The fastest airplanes travel at Mach 3 (3 times the speed of sound). If the speed of sound is 340 m/s, what fraction of the speed of light is Mach 3?
4. The Earth is about 150 million kilometers from the Sun. Use 365 days as the length of 1 year, and think of the Earth's orbit as a circle. Find the speed of the Earth in its orbit. What fraction of the speed of light is the Earth's orbital speed?



ANSWERS

## Physics To Go

1. a)-b) Answers will vary, but can be reached by dividing the distance across the classroom or the state by the speed of light.
2. One trip (one way) takes $1.66 \times 10^{-2}$ s, so light could make 60 trips in 1 second.
3. 1/30,000
4. 1/10,000

ANSWERS

## Physics To Go
***(continued)***

5. a)-b) Both astronauts would report that the other traveled past at the speed of light.

   c) Both are correct.

6. a) 2 mph

   b) 72 mph

   c) 42 mph

   d) 112 mph

   e) Each observer is correct because each is measuring motion based on his or her own inertial frame of reference.

5. Try this gedanken experiment to clarify the consequences of Einstein's postulates.

   Armin and Jasmin are astronauts. They have traveled far into space and, from our frame of reference, they now pass each other at 90% of the speed of light. Their ships are going in straight lines, in opposite directions, at constant speeds. The astronauts each see their own ship as standing still. (Remember, observers in inertial frames of reference think that they are standing still.)

   a) How does Armin describe the motion of the two ships?
   b) How does Jasmin describe the motion of the two ships?
   c) Is Armin's or Jasmin's description of the motion correct? What is a correct description? (Did you think that your frame of reference is the correct one? Is your frame of reference "better" than Armin's or Jasmin's?)

6. A train is traveling at 70 mph in a straight line. A man walks down the aisle of the train in the direction that the train is traveling at a speed of 2 mph relative to the floor of the train. What is the man's speed as measured by:

   a) the passengers on the train?
   b) a man standing beside the track?
   c) a passenger in a car on a road parallel to the track traveling in the same direction as the train at a speed of 30 mph?
   d) a passenger in a pickup truck on the parallel road traveling in the opposite direction from the train at a speed of 40 mph?

   Each of the above measurements has produced a different result. Who is telling the truth? Explain your answer.

### Stretching Exercises

The sound of two radios will reach you at a time depending on your position relative to the radios. The sound will seem simultaneous only when you are at the midpoint between the radios.

Place two radios on opposite sides of the room or, if possible, out of doors and far apart and away from traffic. Tune them to the same station. Then move around listening to the two radios until you find a position where you hear the sounds from both radios. Answer the following questions:

a) Is there only one place in the area where the radios are playing the same words at the same time?
b) Describe that place, or those places, in terms of their location(s) compared to the locations of the radios.
c) How would this experiment be different if light from flashlights were used instead of sound from radios?

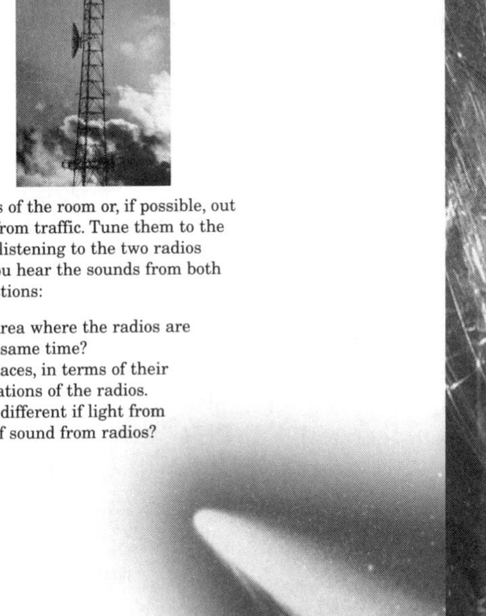



ANSWERS

## Stretching Exercises

a) Students should report that they can hear both radios simultaneously in only one spot.

b) Answers should reflect their observations.

c) Because light travels so fast and the distances involved in this activity are so small, students should suggest that they would see the flashes of light simultaneously from any position.

# ACTIVITY 7
## Special Relativity

# Background Information

Special relativity offered physicists of the 21st century possibly their greatest challenge, a new way of looking at space, time, and mass. Although Einstein's two Postulates of Special Relativity may seem quite innocuous, making the speed of light a constant means that quantities that were assumed to be absolutes in Newtonian mechanics – length, time, and mass – are relative.

Observers in two inertial frames see objects as contracted along the direction of relative motion. This effect was experimentally verified by the Mt. Washington muon experiment. The muons were able to reach the ground during their short lifetimes because to them the journey down the mountain was less than measured by an observer in the Earth's frame of reference.

Einstein predicted and experiments verify that the time interval between two events will be different in different frames of reference. Time dilation, or "time enlargement," means that time intervals are no longer measured by moving clocks. In other words, a moving clock ticks more slowly than a clock at rest. Because the laws of physics are the same in all reference frames, observers in both frames will observe the clock, in what appears to be the moving frame, running slowly.

Time dilation was first verified in the famous muon experiment outlined in this activity. Further confirmation has come from other experiments. For example, in 1977 it was demonstrated that muons in a CERN particle accelerator had extended, or dilated, lifetimes, and clocks flown around the globe on jet airplanes have been shown to run very slightly slower than those that remained on the ground.

The mass of an object at rest with respect to an observer is called the object's rest mass. According to Einstein, the mass of an object depends on the object's velocity. The mass of an object in motion, its relativistic mass, is greater than its rest mass. The reality of this consequence is constantly being affirmed in particle accelerators around the world.

Special Relativity also predicts that mass and energy are equivalent. That is, matter may be transformed into energy and energy into matter. This relationship is embodied in the famous equation $E = mc^2$. The equivalence of mass and energy is demonstrated in the operation of nuclear power plants and in the awesome release of energy in an atomic explosion.

## *Active*-ating the Physics InfoMall

Special Relativity! You can find as much information on Special Relativity as you could want on the InfoMall. Whether you want information on muons, twin paradoxes, simultaneity, or half lives, you should have no trouble (unless your search is too broad and you have to sift through too many hits). Or you may want to browse. For example, go to the Textbook Trove and select *Tipler's Elementary Modern Physics*, a relatively new textbook (published in 1992). Chapter 1 is Relativity, with graphics! Don't stop there; *Modern College Physics* also has a section on this topic. So do other texts. Just look until you find one that you like.

As we hope you can see, there is an incredible amount of information to be found using the Physics InfoMall. The problem is usually not IF– you can find the information you want; the problem is usually deciding which source you want to use.

# Planning for the Activity

## Time Requirements

One class period

## Materials Needed

### For each group:

- graph paper pad of 50 sheets

## Advance Preparation and Setup

None

# Teaching Notes

The **What Do You Think?** questions are designed to stimulate thinking about the activity. Have students share their answers and accept all answers without correction.

Circulate as students work on their graphs in **Step 2** of the **For You To Do** activity to ensure that they are plotting the data correctly. Suggest that they set up the graph using a small interval scale for number of muons; 25-muon intervals works well.

Have students do this pencil-and-paper activity in groups of 3 or 4.

Understandably, students may find themselves confused by the idea of time dilation, and may harbor the misconception that time as experienced on Earth is somehow "right" and that other "times" are different and "not right." This misconception may surface during class or group discussions: guide students holding this notion to see that all frames of reference, thus all "times" are equally valid.

Discuss the twin paradox with students and relate it to time dilation experienced by muons. Explore physicists thinking in accepting such a strange story: its prediction is supported by evidence, there is an uncomplicated theory that explains it, and the theory could be disproven by experimental evidence. Have students record their ideas in their logs.

# Activity Overview

Students further explore special relativity using the half-life of muons, a subatomic particle, to establish a clock. They use the clock to predict the time it should take for muons to travel a specific distance, but because muons travel at close to the speed of light, their predictions do not match experimental evidence.

## Student Objectives

### Students will:

- Plot a muon clock based on muon half-life.
- Use your muon clock and the speed of muons to predict an event.
- Identify the ways that special relativity meets the criteria of good science.

Patterns and Predictions

# Activity 7 Special Relativity

GOALS

In this activity you will:

- Plot a muon clock based on muon half-life.
- Use your muon clock and the speed of muons to predict an event.
- Identify the ways that special relativity meets the criteria of good science.

**What Do You Think?**

Einstein's Theory of Special Relativity predicts that time goes more slowly for objects moving close to the speed of light than for you. If you could travel close to the speed of light, you would age more slowly than if you remained on Earth. This prediction doesn't fit our "common sense."

- Does this prediction make sense to you? Explain your thinking.
- What do you mean by "common sense"?

Record your ideas about these questions in your *Active Physics* log. Be prepared to discuss your responses with your small group and the class.

Active Physics 296

ANSWERS FOR THE TEACHER ONLY

## What Do You Think?

You may wish to introduce Einstein's quote for the definition of common sense:

"Common sense is the collection of prejudices acquired by age eighteen."
– *Albert Einstein*

## For You To Do

1. A muon is a small particle similar to an electron. Muons pour down on you all the time at a constant rate. If 500 muons arrive at a muon detector in one second, then 500 muons will arrive during the next second.

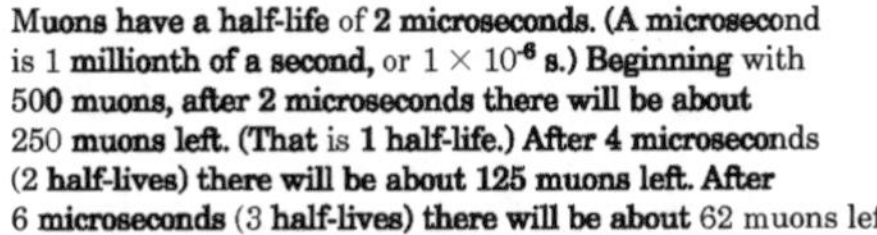

Muons have a half-life of 2 microseconds. (A microsecond is 1 millionth of a second, or $1 \times 10^{-6}$ s.) Beginning with 500 muons, after 2 microseconds there will be about 250 muons left. (That is 1 half-life.) After 4 microseconds (2 half-lives) there will be about 125 muons left. After 6 microseconds (3 half-lives) there will be about 62 muons left.

a) How many muons would be left after 4 half-lives?

2. The half-life of muons provides you with a muon clock. Plot a graph of *the number of muons* versus *time*. Use 500 muons as the size of the sample. This graph will become your clock.

a) If 125 muons remain, how much time has elapsed?

b) If 31 muons remain, how much time has elapsed?

c) If 300 muons remain, how much time has elapsed?

d) If 400 muons remain, how much time has elapsed?

3. Measurements show that 500 muons fall on the top of Mt. Washington, altitude 2000 m. Muons travel at 99% the speed of light or $0.99 \times 3.0 \times 10^{8}$ m/s.

a) Calculate the time in microseconds it would take muons to travel from the top of Mt. Washington to its base.

b) Use your calculation and the muon clock graph to find how many muons should reach the bottom of Mt. Washington.

4. Experiments show that the actual number of muons that reach the base of Mt. Washington is 400.

a) According to your muon clock graph, how much time has elapsed if 400 muons reach the base of Mt. Washington?

b) By what factor do the times you found differ?

c) Suggest an explanation for this difference.

ANSWERS

## For You To Do

1. a) 31 muons

2. a) 4 microseconds

b) 8 microseconds

c) 1.8 microseconds

d) 1 microsecond

3. a) 6.7 microseconds

b) 50 muons

4. a) 1 microsecond

b) 6.7

c)-d) Answers will vary.

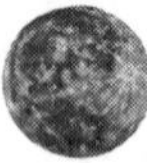



**Physics Words**

**muon:** a particle in the group of elementary particles called leptons (not affected by the nuclear force).

Albert Einstein had an answer. The muon's time is different than your time because muons travel at about the speed of light. He found that the time for the muon's trip (at their speed) should be 0.8 microseconds. That is the time that the muon's radioactive clock predicts.

d) As strange as that explanation may sound, it accurately predicts what happens. Work with your group to come up with another plausible explanation.

## FOR YOU TO READ

### Special Relativity

Physicists of this century have had a difficult decision to make. They could accept common sense (all clocks and everyone's time is the same), but this common sense cannot explain the data from the **muon** experiment. They could accept Einstein's Theory of Special Relativity (all clocks and everyone's time is dependent on the speed of the observer), which gives accurate predictions of experiments, but seems strange. Which would you choose, and why?

The muon experiment shows that time is different for objects moving near the speed of light. You calculated that muons would take 7 microseconds to travel from the top of Mt. Washington to its base. Experiments show that the muons travel that distance in only 0.8 microseconds. Because of their speed, time for muons goes more slowly than time for you!

Time is not the only physical quantity that takes on a new meaning under Einstein's theory. The length of an object moving near the speed of light shrinks in the direction of its motion. If you could measure a meter stick moving at 99% of the speed of light, it would be shorter than one meter!

Perhaps the most surprising results of Einstein's theory are that space and time are connected and that energy and mass are equivalent. The relationship between energy and mass is shown in the famous equation $E = mc^2$. Put in simple words, increasing the mass of an object increases its energy. And, increasing the energy of an object increases its mass. This idea has been supported by the results of many laboratory experiments, and in nuclear reactions. It explains how the Sun and stars shine and how nuclear power plants and nuclear bombs are possible.

Meter stick traveling at near the speed of light, as seen from Earth

Meter stick on Earth's surface

**Physics and Pseudoscience**

Physics, like all branches of science, is a game played by rather strict rules. There are certain criteria that a theory must meet if it is to be accepted as good science. First, the predictions of a scientific theory must agree with all valid observations of the world. The word *valid* is key. A valid observation can be repeated by other observers using a variety of experimental techniques. The observation is not biased, and is not the result of a statistical mistake.

Second, a new theory must account for the consequences of old, well-established theories. A replacement for the Theory of Special Relativity must reproduce the results of special relativity that have already been solidly established by experiments.

Third, a new theory must advance the understanding of the world around us. It must tie separate observations together and predict new phenomena to be observed. Without making detailed, testable predictions, a theory has little value in science.

Finally, a scientific theory must be as simple and as general as possible. A theory that explains only one or two observations made under very limited conditions has little value in science. Such a theory is not generally taken very seriously.

The Theory of Special Relativity meets all the criteria of good science. When the relative speeds of objects and observers are very small compared to the speed of light, time dilation, space contraction, and mass changes disappear. You are left with the well-established predictions of Newton's Laws of Motion. On the other hand, all the observations predicted by the Theory of Special Relativity have been seen repeatedly in many laboratories.

By contrast, psychic researchers do not have a theory for psychic phenomena. The psychic phenomena themselves cannot be reliably reproduced. Psychic researchers are unable to make predictions of new observations. Thus, physicists do not consider psychic phenomena as a part of science.

## Reflecting on the Activity and the Challenge

One of the strangest predictions of special relativity is that time is different for different observers. Physicists tell a story about twins saying good-bye as one sets off on a journey to another star system. When she returns, her brother (who stayed on Earth) had aged 30 years but she had aged only 2 years. Commonsense physicists trust this far-fetched idea because there is experimental evidence that supports it. The muon experiment supports the idea. There is no better explanation for the events in the muon experiment than the Theory of Special Relativity. The theory is simple but it seems to go against common sense. But common sense is not the final test of a theory. Experimental evidence is the final test.

One of the criteria for funding research is whether the experiment can prove the theory false. If muons had the same lifetime when at rest and when moving at high speed, the Theory of Special Relativity would be shown to be wrong. Many theories of pseudoscience cannot be proven false. According to pseudoscientific theories, any experimental evidence is okay. There is no way to disprove the theory. Any evidence that doesn't fit causes the "pseudoscientists" to adjust the theory a bit or explain it a bit differently so that the evidence "fits" the theory.

The proposal you will fund should be both supportable and able to be disproved. The experimental evidence will then settle the matter—either supporting the theory or showing it to be wrong.

## Physics To Go

1. Use the half-life of muons to plot a graph of the number of muons vs. time for a sample of 1000 muons.
   a) If 1000 muons remain, how much time has elapsed?
   b) If 250 muons remain, how much time has elapsed?
   c) How many muons are left after 6 half-lives?
   d) How many muons are left after 8 half-lives?
2. If the speed of light were 20 mph . . .

   You don't experience time dilation or length contraction in everyday life. Those effects occur only when objects travel at speeds near the speed of light relative to people observing them. Imagine that the speed of light is about 20 mph. That means that observers moving near 20 mph would see the effects of time dilation and space contraction for objects traveling near 20 mph. Nothing could travel faster than 20 mph. As objects approach this speed, they would become increasingly harder to accelerate.

   Write a description of an ordinary day in this imaginary world. Include things you typically do in a school day. Use your imagination and have fun with the relativistic effects.



ANSWERS

# Physics To Go

1. a) No time has elapsed.
   b) 4 microseconds.
   c) 18 muons.
   d) 4 muons.
2. Students' descriptions will vary. They should demonstrate an understanding of the effects of time dilation and length contraction.

# ACTIVITY 8
## The Doppler Effect

# Background Information

There are numerous examples of the Doppler Effect. Anyone who has watched an auto race has heard the distinctive drop in pitch as a car goes by the microphone.

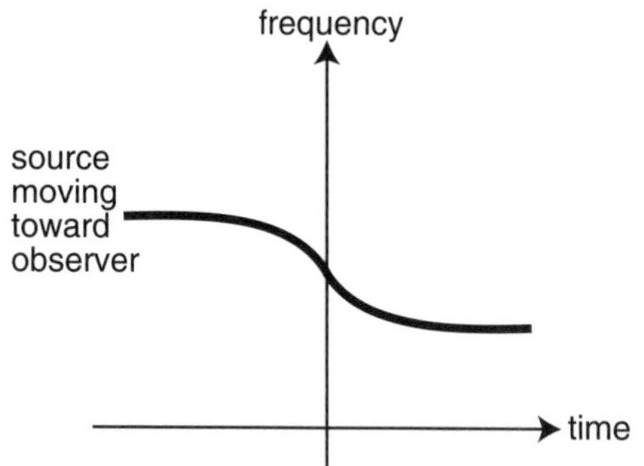


The same effect can be heard while riding on a train and passing a crossing signal – the pitch of the ding-ding-ding drops distinctly as the train passes the crossing. Remember that when the pitch we hear decreases, that means the frequency of the sound wave also decreased. Notice that in the race-car example, the sound source is moving, whereas in the train example, the observer is moving. The physics is perfectly symmetrical.

It is easy to derive an equation for the amount of the shift. Suppose a sound source produces circular waves, as shown in the drawing, with wavelength $\lambda$. The source begins to move to the right. For waves to the right of the source, the source is moving in the same direction as the wavefronts, so the wavelength in this direction is decreased. Let $S$ be the speed of sound. Since $f = S/\lambda$, decreasing the wavelength increases the frequency. To find out how much the frequency increases, we simply calculate the observed wavelength, which is reduced by the distance the source travels during one cycle of the wave. One cycle of the wave takes a wave period $T$, which is the inverse of the frequency $f$. In this time, the source has moved a distance $V \times T$, where $V$ is the source speed.

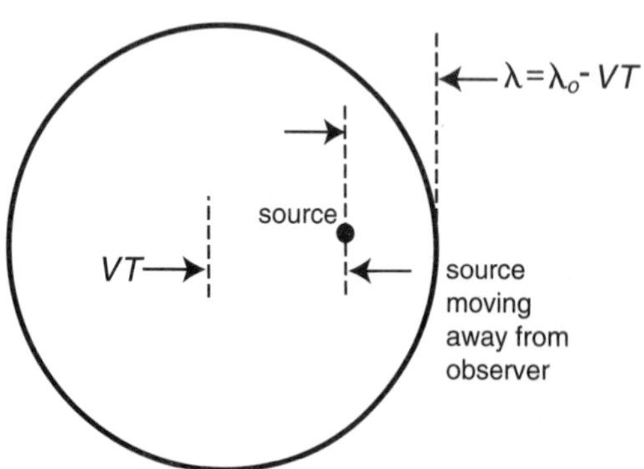


$\lambda$ is simply the "at-rest" wavelength $\lambda_0$ minus $V \times T$:

$$\lambda = \lambda_0 - VT$$

The observed frequency is $f$. Converting to frequency, using $\lambda = S/f$ and $T = 1/f$, gives

$$\frac{S}{f} = \frac{S}{f_0} - \frac{V}{fz_0} = \frac{I}{fz_0}[S - V]$$

Solving for $f$, the observed frequency,

$$f = f_o \left[\frac{S}{(S - V)}\right]$$

Notice that the observed frequency is expressed as the original frequency times a dimensionless factor. This factor goes to infinity as $V$ goes to $S$, which expresses the fact that all the wavefronts pile up right in front of the source when the source moves with the speed of the waves.

Doppler-shift measurements on the light from distant galaxies led to one of the most important discoveries in the history of astronomy – the recession of the galaxies. Astronomers identify the spectral lines of certain elements, such as hydrogen, in the light from galaxies. These lines have the same spacing as laboratory measurements on Earth, but all the lines are shifted to longer wavelengths, towards the red end of the spectrum. A longer wavelength means a lower frequency, so these galaxies are receding from Earth. When the distance to these galaxies is measured independently, a pattern emerges: the more distant the galaxy, the faster the galaxy is moving away. This "recession of the galaxies" is an important part of the evidence for the Big Bang theory. Incidentally, Einstein developed an equation in his Theory of General Relativity (his theory of gravity) that predicted the expansion of the universe. But since he believed the universe to be static, he introduced an extra term into his equation that made his result come out the way he wanted. Unfortunately, he missed a stunning prediction.

## *Active*-ating the Physics InfoMall

The Doppler Effect is something you can find easily on the InfoMall. If you want to perform a classroom demonstration of the Doppler Effect, consider the "Simple laboratory demonstration of the Doppler shift of laser light," in the *American Journal of*

*Physics*, vol. 53, (1985). This was found using a search with keyword "Doppler."

A somewhat dated, but still reasonable, reference you may want to check is "The Doppler and echo Doppler effect," *American Journal of Physics*, vol. 12, (1944). This helps explain, for example, Doppler Ultrasound. It also has good graphics that you may want to use. (Don't forget – the contents of the InfoMall can be copied into a word processor for your use.)

If you want an equation, go to the InfoMall Entrance and click the icon in the lower right corner. This will take you to the Equation Dictionary. Choose "Waves and Sound," which will provide additional options. Next choose "Waves" then "Behavior of Waves." The Doppler Effect will be listed. Clicking this will show you the equations, along with definitions for the variables used. There is also a hyperlink to examples. If you choose this, the InfoMall will do the search for you and find a few examples of how to use the equation.

**Step 4** of **For You To Do** uses graphs. You may wish to search for "student difficult*" AND "graph*". This produces several wonderful hits, including "Student difficulties with graphical representations of negative values of velocity," in *The Physics Teacher*, vol. 27, issue 4 on the InfoMall. Also found in that search is "Student difficulties in connecting graphs and physics: Example from kinematics," in the *American Journal of Physics*, vol. 55, issue 6. Don't let these titles make you think that graphs are a bad thing! It is good to be aware that students do not always understand graphs, a valuable tool in the study of physics.

# Planning for the Activity

## Time Requirements

two class periods

## Materials Needed

### For each group:

- buzzer
- battery, 9-volt
- battery clip with solid leads for 9-volt battery
- ball, Nerf™, 6"
- tape recorder

## Advance Preparation and Setup

You will have to assemble the oscillator. Drill a hole in one end of the plastic box for the cord. Mount the battery-holder and buzzer on a small piece of wood. Connect the buzzer wires to the battery clip. If the sound is too loud, place tape over the buzzer. You may also want to tape the battery in place to make sure it stays in the battery holder. Cut a slot in the Nerf ball, stuff the circuit inside, and cover the opening with tape.

# Teaching Notes

Have the students share any experiences they have had with the Doppler Effect (attending auto races, listening on a moving train to railroad crossing bells). Ask the students to listen closely to the buzzer tone as they toss the Nerf ball back and forth. You can try making a harness of sturdy net material and whirling the Nerf ball around in a circle. Be sure the ball is securely mounted in the harness. If the students make a tape of the horn of a moving car, suggest that they put the waveform of the sound on an oscilloscope screen and measure the period. Incidentally, ask the students to listen on the tape for the Doppler shift of the engine noise of cars passing by.

Enjoy tossing the Nerf ball.

SAFETY PRECAUTION: If you swing the Nerf ball in a circle, be sure it is securely mounted in the netting. Keep students back at a safe distance. If the students make the measurement of the car horn sounds, give them plenty of instruction on safety. The students doing the recording must stay well out of the road. The car must stay within the speed limit.

It can be difficult for students to make the connection between the Doppler Effect in sound and in light. Many students have very little sense of sound waves or wavefronts. Visualizing light as a wave is even more difficult. You may be able to help them visualize the Doppler Effect by suggesting that, as the source moves towards the observer, the wavefronts pile up in front of the source, so the frequency increases.

# Activity Overview

To understand how astronomers measure the distance to distant galaxies, students explore the Doppler Effect. They observe a tone produced by a small oscillator tossed from student to student in the lab. Then they observe the shift in the sound frequency as a car moves by with the horn blowing. Through reading, they learn how astronomers observe the Doppler shift in spectral lines to obtain the "red shift" of distant galaxies and measure the incredible distances in the universe.

## Student Objectives

### Students will:

- Describe red shift.
- Sketch a graph.
- Observe changes in pitch.
- Calculate with a formula.

ANSWERS FOR THE TEACHER ONLY

## What Do You Think?

For simplicity, think about just one frequency in all the sound of the roar of the motor. First picture the car at rest. One crest after another moves toward the observer, with successive crests separated by the wavelength of the sound. Now imagine the car moving at high speed towards the observer. One crest heads toward the observer, but the car itself is also moving toward the observer, so when the second crest is emitted, the car has reduced the distance to the previous crest.

The result is a reduction in the wavelength and a corresponding increase in the frequency and, when perceived by the ear and brain, in the pitch of the sound. If a car ever breaks the sound barrier, all the crests will pile up and generate a sonic boom.

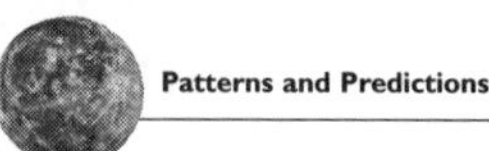


Activity 8 The Doppler Effect

GOALS

In this activity you will:

- Describe red shift.
- Sketch a graph.
- Observe changes in pitch.
- Calculate with a formula.

What Do You Think?

You have probably heard the sound of a fast-moving car passing by you.

- **Why is there a change in tone as the car moves by?**

Record your ideas about this question in your *Active Physics* log. Be prepared to discuss your responses with your small group and with your class.

For You To Do

1. Listen to a small battery-powered oscillator. It makes a steady tone with just one frequency. The oscillator is fastened inside a Nerf™ ball for protection.

Active Physics

ANSWERS

## For You To Do

1. Student activity.

Activity 8 The Doppler Effect

2. Stand about 3 m away from your partner. Toss the oscillator back and forth between you. Listen to the pitch as the oscillator moves. As you listen, observe how the pitch changes as the oscillator moves.
   a) How is the oscillator moving when the pitch is the highest?
   b) How is the oscillator moving when the pitch is the lowest?
3. Stop the oscillator so you can listen to its "at rest" pitch.
   a) With the oscillator moving, record how the pitch has changed compared to the "at rest" pitch. How has the pitch changed when the oscillator is moving towards you?
   b) How has the pitch changed when the oscillator is moving away from you?
4. Look at the graph axes shown. The axes show pitch vs. velocity. When the velocity is positive, the oscillator is moving away from you. When the velocity is negative, the oscillator is moving towards you.

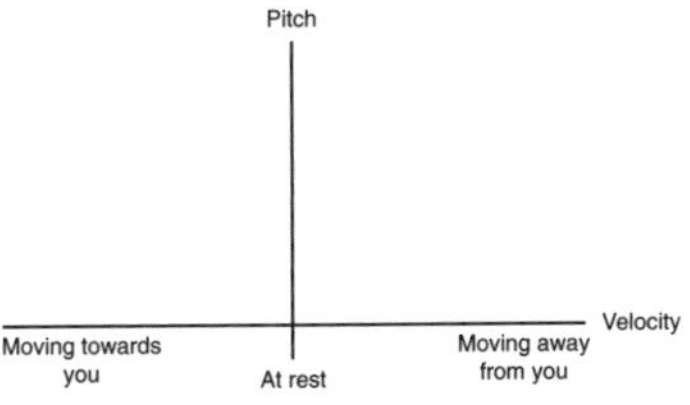


   a) On a similar set of axes in your log, sketch a graph of your pitch observations. Explain your graph to the other members of your group.
5. You can do an outdoor Doppler lab using the horn of a moving car as the wave source. Tape-record the horn when the car is at rest next to the tape recorder. Then, with the driver of the car maintaining an agreed-upon speed, tape the sound of the horn as the car passes. Have the driver blow the horn continuously, both as the car approaches and as it moves away. Be very careful to stay away from the path of the car.



## Answers

## For You To Do

*(continued)*

2. a) Toward you.
   b) Away from you.
3. a) The pitch increases.
   b) The pitch decreases.
4. a) See graph below.
5. Have students record what they hear. (The pitch of the horn drops as the car goes by.)

4. a)

pitch

velocity

moving towards you

at rest

moving away from you

Chapter 5

ANSWERS

## For You To Do

***(continued)***

6. a) When $v$ is positive, $(s - v) < s$,

so $\left(\frac{s}{s-v}\right) > 1$

and the frequency increases when the car moves toward you.

6. You can determine the observed frequency by matching the recorded tone to the output of an oscillator and loudspeaker. Use this formula:

$$f = f_0\left(\frac{s}{(s-v)}\right)$$

$f_0$ = frequency when car is at rest

$v$ = speed of the car

$s$ = speed of sound = 340 m/s

a) When the car is moving toward you, $v$ is positive. When the car is moving away from you, $v$ is negative. Use the equation to calculate the speed of the car from the data you collected.

### FOR YOU TO READ

#### Measuring Distances Using the Doppler Effect

Astronomers measure distances to stars in two different ways. One way is with parallax, but this method works only for the nearest stars. For all other stars astronomers apply the Doppler effect. They use the Doppler shift of spectral lines. The next-nearest galaxy is Andromeda, more than a million light-years away.

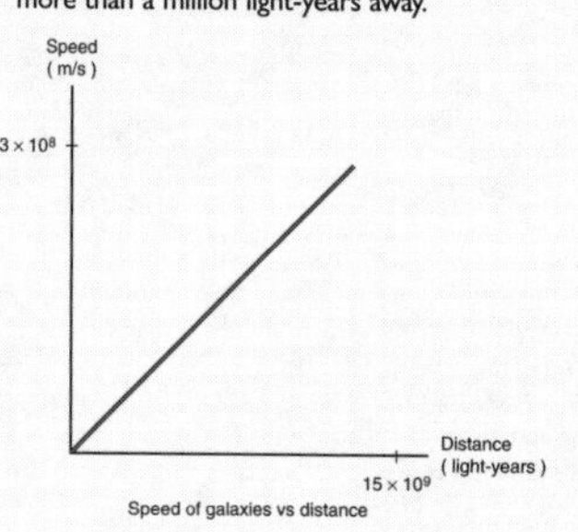


Speed of galaxies vs distance

Astronomers have observed galaxies at far greater distances, up to about 12 billion light-years away. These incredible distances are measured by observation of the absorption lines of light. These lines are consistently Doppler-shifted towards the red end of the spectrum, and the result is called the "red shift."

All the lines are shifted toward longer wavelengths. Since this is a shift towards lower frequencies, the galaxies are moving away from Earth. By measuring the size of the shift, astronomers find the speed of distant galaxies. Different galaxies move away at different speeds, but with a clear pattern. The farther away the galaxy, the faster it is moving away, as shown in the graph.

Astronomers explain this result with the Big Bang theory, which says that the universe began in an explosion about 15 billion years ago. After the explosion, the matter in the galaxy continued to move apart, even after the galaxies formed.

### Reflecting on the Activity and the Challenge

You have learned that the pitch of a sound changes if the source of the sound is moving toward you or away from you. This is called the Doppler effect for sound. You also learned that there is a Doppler effect for light where the frequency or color of the light would change if the source of the light were moving. Measurements of sound frequency can be used to determine the speed of the source of sound. Measurements of light frequencies from distant galaxies can be used to determine the speed of the galaxies. The speed of galaxies moving away from Earth has been shown to relate to the distance of the galaxies from Earth. A measurement of light frequency and the Doppler effect can be used to measure distances. Measuring speeds through the use of changes in frequency of sound or light is good science. Some people may say that hearing a person's voice indicates to them whether that person is kind or gentle. If this were to have a scientific basis, you would have to conduct experiments. You should be able to contrast the experimental evidence you have for the Doppler effect and the lack of evidence you have for finding out what a person is like by their voice as you decide the kinds of proposals you may fund.

### Physics To Go

1. a) If a sound source is moving towards an observer, what happens to the pitch the observer hears?
   b) If a sound source is moving towards an observer, what happens to the sound frequency the observer measures?
2. a) If a sound source is moving away from an observer, what happens to the pitch the observer hears?
   b) If a sound source is moving away from an observer, what happens to the sound frequency the observer measures?
3. a) If you watch an auto race on television, what do you hear as the cars go by the camera and microphone?
   b) Sketch a graph of the pitch you hear vs. time. Make the horizontal axis of your graph the time, and the vertical axis the pitch. (Hint: Don't put any numbers on your axes. Label the time when the car is going right by you.)

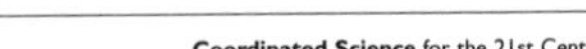



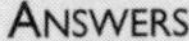

ANSWERS

## Physics To Go

1. a) The pitch increases.
   b) The frequency increases.
2. a) The pitch decreases.
   b) The frequency decreases.
3. a) The pitch of the motor noise decreases as the car moves by.
   b)-c) See graph in **Backround Information** and below.

3. b)

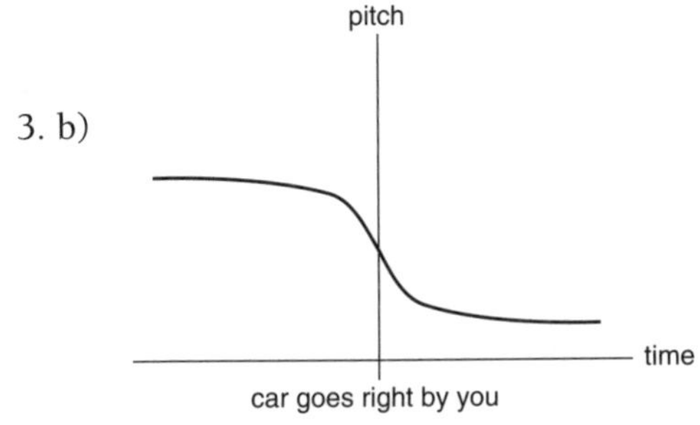


3. c)

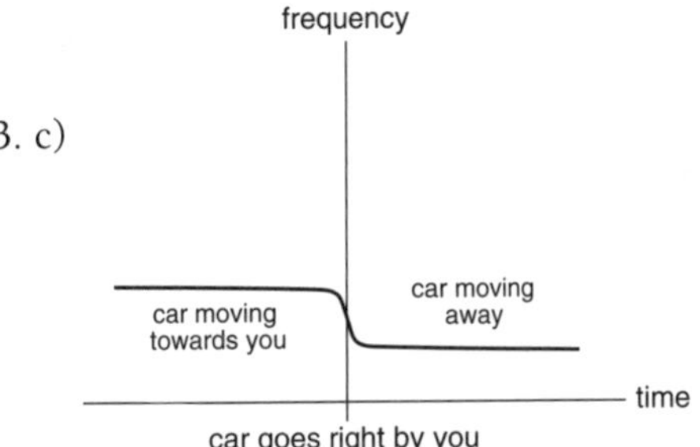

## Answers

# Physics To Go
***(continued)***

4. a) The shift in pitch and frequency would be greater.
   b) The shift in pitch and frequency would be less. See graphs below.
5. a) Red light, because the product of wavelength and frequency is a constant.
   b) Lower
   c) Lower
   d) Lower
   e) Away from us.

4. b)

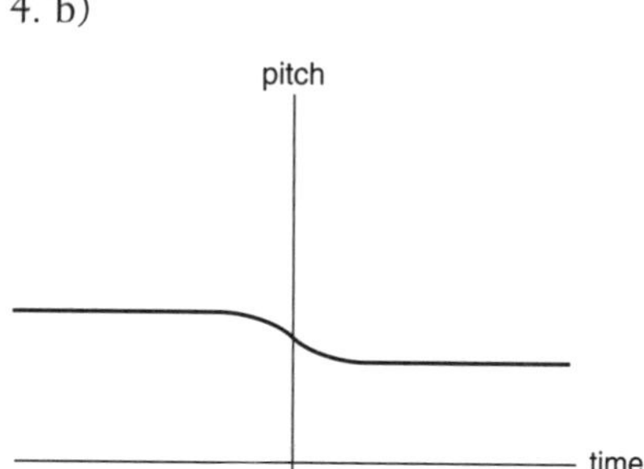


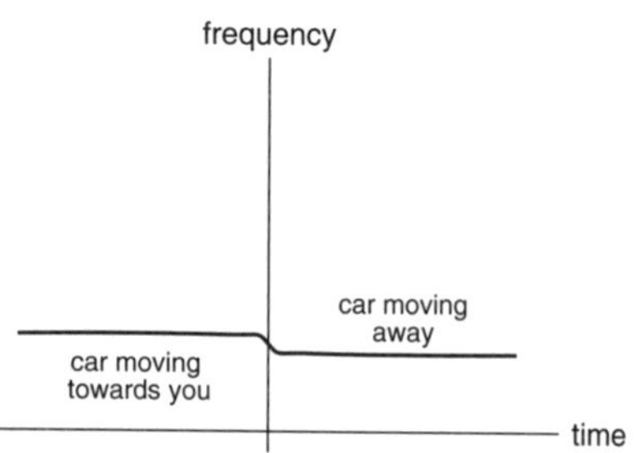

c) Sketch a graph of the frequency you observe vs. time. As in **Part (b)**, label the time when the car is going right by you. (Hint: Don't put any numbers on your axes.)

4. a) In **Question 3** above, what would happen to your graphs if the speed of the racing car doubled? Make a sketch to show the change.
   b) What would happen to your graphs if the speed of the racing car was cut in half? Make a sketch to show the change.

5. a) Red light has a longer wavelength than blue light. Which light has the lower frequency? You will need the equation:

   wave speed = wavelength × frequency

   Show how you found your answer.
   b) When the oscillator moved away from you, was the pitch you heard lower or higher?
   c) When the oscillator moved away from you, was the frequency you heard lower or higher?
   d) If light from a distant galaxy is shifted towards the red, is it shifted to a lower or a higher frequency?
   e) If the light is shifted towards the red, is the galaxy moving away from Earth or towards Earth?

### Stretching Exercise

Watch a broadcast of an auto race. Listen closely to the cars as they zoom past the microphone. Use the Doppler effect to explain your observations.

## Answers

# Stretching Exercise

The pitch drops as the cars go by, so the frequency of their sound drops too, as expected.

## Activity 8 A

# Doppler Effect with Water Waves

### FOR YOU TO DO

Set up ripple tanks. Have students make ripples by dipping their finger in the water at about 1 Hz. Have them move their finger smoothly along the water source as they dip. Ask what happens to the wavelength they observe. (If they move their finger faster than the wave speed, their finger will leave a wake, like a motorboat.)



For use with *Patterns and Predictions*, Chapter 5, Activity 8: The Doppler Effect

# ACTIVITY 9
# Communication Through Space

## Background Information

Our solar system contains some 100 billion stars. Many of these stars have systems of planets, and some of those planets probably have the conditions that can support life. Perhaps life has evolved in our own galaxy. Will we be able to make contact? Oddly enough, our chances of success are limited by the lifetime of our own civilization. Imagine that we search for life by sending out radio signals. Imagine we begin by searching the nearby stars and then work our way out progressively from the Earth. If we search in a sphere of radius $r$, the volume of that sphere is $\frac{4}{3}\pi r^3$, which increases more rapidly the further out we go. Consequently, as we search further, we encounter far more stars that might harbor life, so the odds strongly favor searching as far away from the Earth as possible. We could look out at least 30,000 light-years just within our own galaxy. What would happen if we sent a signal and another civilization replied? The reply might arrive 50,000 years after we sent the signal. And suppose we sent a signal to Andromeda, the nearest galaxy, at a distance of 2.2 million light-years. Would there still be a technological civilization on Earth in 4.4 million years to receive a reply? We get little help with this question from our own history, since our civilization has been technological for fewer than 100 years. How long will our civilization last? The answer to this question may determine whether or not we make contact with extraterrestrial life.

Here are some important distances:

| | |
|---|---|
| distance across the solar system: | 11 light-hours |
| distance to the nearest star: | 4.2 light-years |
| distance across the galaxy: | 100,000 light-years |
| distance to the nearest galaxy: | 2.2 million light-years |
| distance to the edge of the universe: | 15 billion light-years |

And here are some important times:

| | |
|---|---|
| the Big Bang: | 15 billion years ago |
| the formation of the Earth: | 4.5 billion years ago |

Originally, the entire universe was confined to a single point. The Big Bang was the explosion that created the universe in a blast of radiation, which was so hot that at first, matter could not exist. As the radiation expanded and cooled, subatomic particles acquired a stable existence and, later, atoms as well. Matter collected into galaxies, which are still rushing apart today and producing the characteristic red shifts – one important piece of evidence supporting the Big Bang Theory. The other piece of evidence is the microwave background, residual microwave radiation from the beginning of the universe, when the temperature was so high that all the energy in the universe was radiation.

### *Active*-ating the Physics InfoMall

This is another self-explanatory activity. There are a few things you might want to search for, though. For example, **Physics To Go Step 5** asks about spaceships. Search for "space travel" to find a few nice bits from the InfoMall. Of course, you might think of other great searches, and you are encouraged to try them all!

## Planning for the Activity

### Time Requirements

one class period

### Materials Needed

**For each group:**

- roll of adding machine tape

### Advance Preparation and Setup

Cut the adding machine tape into 1 m lengths.

## Teaching Notes

Probably the greatest challenge for the students in this activity is thinking in terms of light travel-time. You can help the students by discussing the progression of travel-times the students calculate in **Step 2**. Also, when the students make the time-line of Earth history, you can point out the age of the Earth is 4.5 billion years. You may have to suggest a logarithmic graph, with each scale marking indicating a multiple of ten in the time.

Activity 9 Communication Through Space

# Activity 9 Communication Through Space

**GOALS**

In this activity you will:

- Calculate time delays in radio communications.
- Express distances in light travel-time.
- Solve distance-rate-time problems with the speed of light.

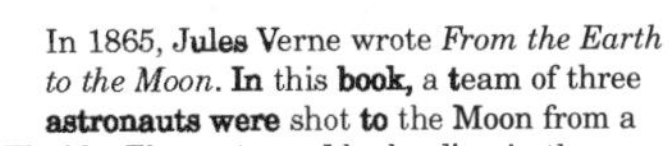

**What Do You Think?**

In 1865, Jules Verne wrote *From the Earth to the Moon*. In this book, a team of three astronauts were shot to the Moon from a cannon in Florida. They returned by landing in the ocean. Verne correctly anticipated many of the details of the Apollo missions.

- **How well do you think *Star Trek* predicts the future?**

Record your ideas about this question in your *Active Physics* log. Be prepared to discuss your responses with your small group and with your class.



# Activity Overview

To assess the difficulty in communicating with any extraterrestrial life, students examine a table of astronomical distances. They make a time-line of Earth history to record the past changes in the Earth over the last 50 million years. The students then consider what future changes will occur on Earth during the long times needed to communicate with a distant civilization.

## Student Objectives

### Students will:

- Calculate time delays in radio communications.
- Express distances in light travel-time.
- Solve distance-rate-time problems with the speed of light.

ANSWERS FOR THE TEACHER ONLY

## What Do You Think?

When a starship goes in warp drive, *Star Trek* violates the laws of physics. The theory of relativity stipulates that matter cannot be accelerated to the speed of light (or beyond). In addition, "beaming up" requires the complete conversion of matter into energy, possible only in matter-antimatter annihilation. Also, the genetic information from the DNA would be destroyed by the "transporter."

So it seems unlikely that *Star Trek* predicts the future. There are some aspects of *Star Trek*, like multicultural teams working together, that make sense. Computer technologies are also plausible.

ANSWERS

## For You To Do

1. a) This idea is impractical. Each time the sound reflects from the walls of the tube, some of the sound is absorbed. No sound could be heard at the other end.

   b) 5000 km $\times 10^3$ / 340 m/s = $1.5 \times 10^4$ s = a little more than four hours.

   c) The time calculated in **Part (b)** is much shorter than the time required to send light to the next galaxy.

2. a) to the Sun: 500 s = 8.3 min

   to Jupiter: $2.7 \times 10^3$ s = 45 min

   to Pluto: $2 \times 10^4$ s = 5.5 hours

   to the nearest star: $1.3 \times 10^8$ s = 4.2 light-years

   to the center of our galaxy: $7.4 \times 10^{11}$ s = 24,000 light-years

   to the Andromeda galaxy: $7.1 \times 10^{13}$ s = 2.2 million light-years

   to the edge of the observable universe: $0.5 \times 10^{18}$ s = $16 \times 10^9$ years

   b) Double the above times plus whatever time is required for the extraterrestrial to create the reply.

### For You To Do

1. Alexander Graham Bell's grandson suggested a simple way to talk to Europe long-distance. He recommended placing a long air tube across the bottom of the Atlantic Ocean. He believed that if someone spoke into one end of the tube, someone else at the other end would hear what was said.

   a) Do you think this is practical? Give reasons for your answer.

   b) If the sound could be heard in Europe, how long would it take to send a message? (Hint: The distance to Europe is about 5000 km, and the speed of sound is about 340 m/s.)

   c) Compare this time with the time to communicate with extraterrestrials in the next galaxy using light. The nearest galaxy is Andromeda, which is about two million light-years away. (It takes light about two million years to get from Earth to Andromeda.)

2. The highest speed ever observed is the speed of light, $3.0 \times 10^8$ m/s. In addition, a basic idea of Einstein's Theory of Relativity is that no material body can move faster than light. Radio waves also travel at the speed of light. If Einstein is correct, there are serious limitations on communication with extraterrestrials. Look at the table of distances below. These are distances from the Earth.

| | |
|---|---|
| to the Sun: | $1.5 \times 10^{11}$ m |
| to Jupiter: | $8 \times 10^{11}$ m |
| to Pluto: | $6 \times 10^{12}$ m |
| to the nearest star: | $4 \times 10^{16}$ m |
| to the center of our galaxy: | $2.2 \times 10^{20}$ m |
| to the Andromeda galaxy: | $2.1 \times 10^{22}$ m |
| to the edge of the observable universe: | $1.5 \times 10^{26}$ m |

   a) How long would it take to send a message using radio waves to each place?

   b) How long would it take to send this message and get an answer back?

3. A real-life problem occurred when the Voyager spacecraft was passing the outer planets. NASA sent instructions to the spacecraft but had to wait a long time to find out what happened. The ship had to receive the instructions, take data, and send the data back home.
   - a) If the spacecraft was at Jupiter, how long would it take for the message to travel back-and-forth?
   - b) If this spacecraft was at Pluto, how long would it take for the message to travel back-and-forth?

4. Make a time-line of Earth history. For the scale of your time-line, make six evenly spaced marks.
   - a) Label the time-line like the one shown.

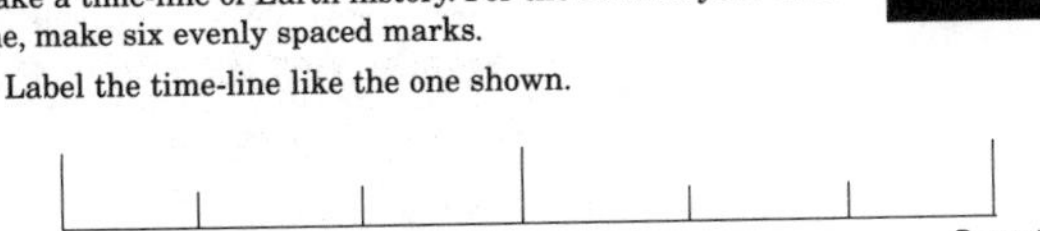


   - b) On your time-line, label interesting events in Earth's history that occurred during these times. Possibilities include the end of the last Ice Age (10,000 years ago), the evolution of the modern horse (50 million years ago), the evolution of humans (3 million years ago), the Iron Age (1000 BC), the Stone Age (8000 BC), the Middle Ages in Europe (13th century), the beginning of civilization (3000 BC), and the spread of mammals over the Earth (50 million years ago).(Dates given are approximations.)

5. Many scientists believe that intelligent life would most likely be thousands or millions of light-years away.
   - a) How would this affect two-way communication?
   - b) If you asked a question, how long would it be before a response came back? Would you be able to receive the response?
   - c) What questions would you ask? (Note: Think about the distances involved.)
   - d) What kind of answers might you expect?
   - e) What changes have occurred on Earth over this time period?
   - f) What changes would you expect on Earth before the answer came?
   - g) Is two-way communication possible over such distances? Is it practical? Is it likely?



ANSWERS

## For You To Do

***(continued)***

3. a) The message would take 90 minutes to travel back-and-forth (assuming the spacecraft replies immediately).

   b) It would take 11 hours.

4. a)-b) Notice how the most recent times are impossible to represent. One million years ago is only about one millimeter to the left of the present mark.

5. a) It would take so long that it is hard to imagine how it would ever happen.

   b) Thousands or millions of years; no.

   c)-d) Answers will vary.

   e)-f) Perhaps catastrophic changes.

   g) It seems unlikely.

ANSWERS

## Physics To Go

1. a) It takes 200 m / 340 m/s = 0.59 s for the sound to travel from one gong to another. It would take 1.08 s, plus the reaction time of the other student, about 0.5 s, for the sound to return to you.

   b) We will have to wait until our message reaches other terrestrial life and continue waiting until their reply reaches us.

2. a) Yes.

   b) No. At that distance, it could be in the Andromeda galaxy.

3. a) 1.3 s

   b) 8.3 minutes

   c) 5.5 hours

   d) 4.3 years

   e) 100,000 years

   f) More than a million years

   g) 15 billion years

### Reflecting on the Activity and the Challenge

This activity helped you to understand how much time it would take for a light signal to travel from Earth to other places in the solar system, the galaxy, or the edge of the universe. There are many reasons why contact with another life form on another planet would be valuable. You may wish to consider the merits of scientific proposals that seek to communicate with other life forms. You will want to consider whether the proposals take into account the difficulties of sustained communication. You will have to consider the type of communication expected and whether the proposal understands that a response to the simplest question to a life form near the nearest star would take at least six years. If a proposal states that it can communicate faster than the speed of light, they would have to explain how this would be possible since no technique is now known that permits this.

### Physics To Go

1. a) The speed of sound is about 340 m/s in air. You and another student take gongs outside about 200 m apart. You hit the gong. After hearing the sound of your gong, the other student hits the other gong. How long is it before you hear the sound of the other gong?
   b) How is this experiment similar to the problem of communicating with extraterrestrial life?

2. a) If extraterrestrial life is probably 1000 light-years away, would it be within this galaxy?
   b) If extraterrestrial life is likely probably several million light-years away, would that be within this galaxy? Could it be in the Andromeda galaxy? (Note: This galaxy has over 100 billion stars.)

3. a) The Moon is $3.8 \times 10^8$ m from the Earth. How long does it take a radio wave to travel from the Moon to the Earth?
   b) The Sun is $1.5 \times 10^{11}$ m from the Earth. How long does it take a light wave to travel from the Sun to the Earth?
   c) Pluto is about $6 \times 10^{12}$ m from the Sun. How long does it take a light wave to travel from the Sun to Pluto?
   d) The nearest star is 4.3 light-years away from Earth. How

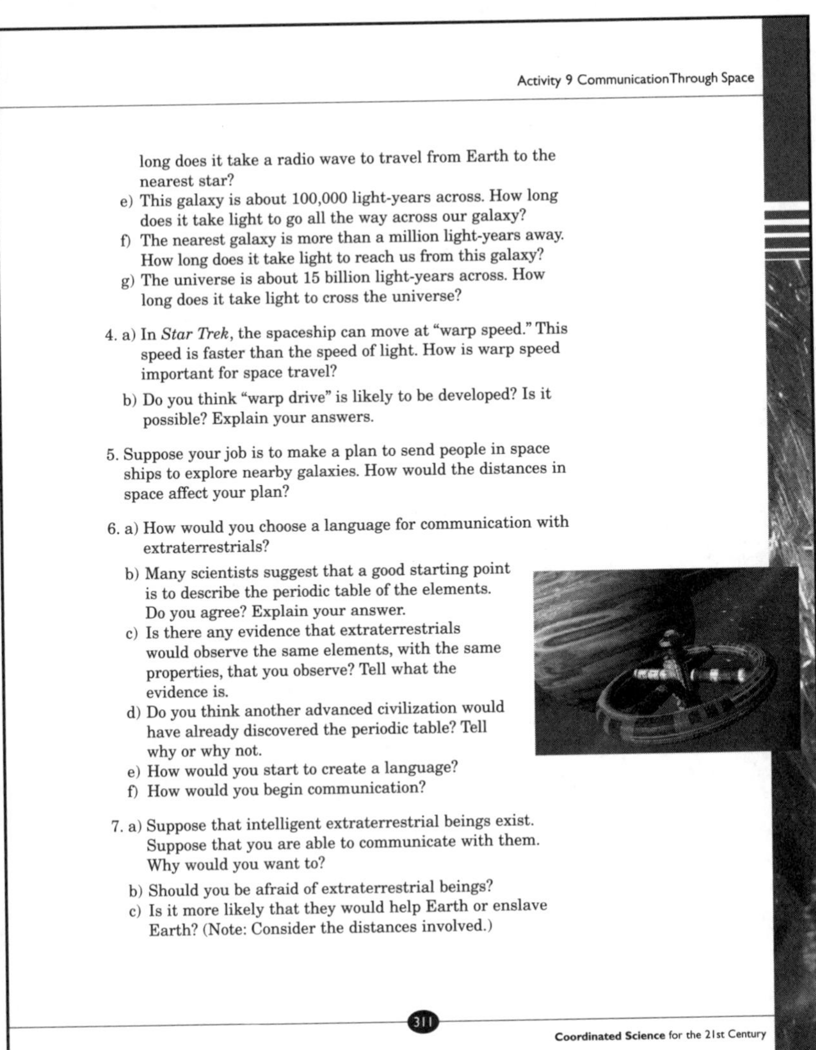



long does it take a radio wave to travel from Earth to the nearest star?

e) This galaxy is about 100,000 light-years across. How long does it take light to go all the way across our galaxy?

f) The nearest galaxy is more than a million light-years away. How long does it take light to reach us from this galaxy?

g) The universe is about 15 billion light-years across. How long does it take light to cross the universe?

4. a) In *Star Trek*, the spaceship can move at "warp speed." This speed is faster than the speed of light. How is warp speed important for space travel?

b) Do you think "warp drive" is likely to be developed? Is it possible? Explain your answers.

5. Suppose your job is to make a plan to send people in space ships to explore nearby galaxies. How would the distances in space affect your plan?

6. a) How would you choose a language for communication with extraterrestrials?

b) Many scientists suggest that a good starting point is to describe the periodic table of the elements. Do you agree? Explain your answer.

c) Is there any evidence that extraterrestrials would observe the same elements, with the same properties, that you observe? Tell what the evidence is.

d) Do you think another advanced civilization would have already discovered the periodic table? Tell why or why not.

e) How would you start to create a language?

f) How would you begin communication?

7. a) Suppose that intelligent extraterrestrial beings exist. Suppose that you are able to communicate with them. Why would you want to?

b) Should you be afraid of extraterrestrial beings?

c) Is it more likely that they would help Earth or enslave Earth? (Note: Consider the distances involved.)



ANSWERS

## Physics To Go

***(continued)***

4. a) The speed of light is a serious limitation on space travel to the site of another civilization. Even traveling at the speed of light, the nearest star would require 4.3 years just to get there.

b) According to the Theory of Special Relativity, it is not possible.

5. It would be necessary to plan to sustain life for almost 4.4 million years (assuming the voyagers wanted their descendents to return).

6. a) Base it on something we and they have in common, like an understanding of the periodic table or the spectral lines of hydrogen.

b) Yes. An advanced civilization would have to understand the periodic table to have the technology to communicate with us.

c) Yes. We see the spectral lines of many elements in the light of distant stars and galaxies.

d) See answer to **Part (b)**.

e) Begin describing the elements of the periodic table.

f) Send a binary message with radio waves.

7. a) Because we are curious.

b) Not afraid, but perhaps careful.

c) The distances are so large that they might not feel threatened.

8. a) We know that civilizations had already existed for thousands of years, but we consider their technology to be primitive compared to ours.

b) It would not be up-to-date, because much will have happened in the 2000 years since the travelers left Earth.

c) Even more change would have occurred. It might be worthwhile to those who were alive at the time of the contact with the extraterrestrials.

ANSWERS

## Physics To Go
***(continued)***

9. a)-c) Answers will vary.

   d) Ideally, yes.

10. a)-b) Answers will vary.

    c) *Star Trek* presents scientific investigation accurately. However, relativity prohibits travel faster than the speed of light. Also, "beaming" persons or objects from one place to another violates the conservation of matter (during the beaming process).

8. a) What is known of the Earth of 2000 years ago?
   b) It takes 2000 years for a spaceship to travel to a star. When the travelers arrive at the star, would their information about the Earth be up-to-date? Explain why or why not.
   c) If the trip to another star took 10,000 years, would such a trip be worthwhile? Explain why or why not.

9. A record was sent into space in an effort to communicate with extraterrestials.
   a) If you were on the team designing the record, what music would you include?
   b) What photographs would you include?
   c) What drawings would you include?
   d) Have you fairly represented the majority of the world with your choices?

10. a) Make a list of movies, books, and TV shows that involve trips to other parts of the galaxy or extraterrestrials visiting the Earth.
    b) Very briefly describe the plot of the story.
    c) How accurately is science represented?

### Stretching Exercises

1. Read the Carl Sagan book, *Contact*, or watch the movie. What features of the book and movie have you considered in this chapter? What features have been ignored?

2. Look up the messages that were placed on the Pioneer and Voyager spacecraft. Make a report to the class on how this plaque communicated information about humans.